Rotating Machinery: Technology and Applications

Rotating Machinery: Technology and Applications

Editor: George Rivera

NY RESEARCH PRESS
New York

Published by NY Research Press
118-35 Queens Blvd., Suite 400,
Forest Hills, NY 11375, USA
www.nyresearchpress.com

Rotating Machinery: Technology and Applications
Edited by George Rivera

International Standard Book Number: 978-1-63238-851-3 (Hardback)

Cataloging-in-Publication Data

Rotating machinery : technology and applications / edited by George Rivera.
p. cm.
Includes bibliographical references and index.
ISBN 978-1-63238-851-3
1. Rotors. 2. Rotational motion. 3. Machinery--Vibration. 4. Rotors--Dynamics. I. Rivera, George.
TJ1058 .R68 2022
621.82--dc23

Contents

Preface

Rotating machinery constitutes a major and integral component of a broad range of mechanical systems in a variety of areas such as industrial plants, vehicles and household appliances. They are generally made up of two main parts, the stator and the rotor. The stationary part of the machinery is known as the stator while the non-stationary part is called the rotor. A few examples of rotating machines are steam turbines, gas turbines, gearboxes and aircraft engines. This book, with its detailed analyses and data, will prove immensely beneficial to professionals and students involved in this area at various levels. The topics included in it on rotating machinery are of utmost significance and bound to provide incredible insights to readers. Coherent flow of topics, student-friendly language and extensive use of examples make it an invaluable source of knowledge.

The researches compiled throughout the book are authentic and of high quality, combining several disciplines and from very diverse regions from around the world. Drawing on the contributions of many researchers from diverse countries, the book's objective is to provide the readers with the latest achievements in the area of research. This book will surely be a source of knowledge to all interested and researching the field.

In the end, I would like to express my deep sense of gratitude to all the authors for meeting the set deadlines in completing and submitting their research chapters. I would also like to thank the publisher for the support offered to us throughout the course of the book. Finally, I extend my sincere thanks to my family for being a constant source of inspiration and encouragement.

Editor

1

Evaluation of Wind Energy Potential as a Power Generation Source in Chad

Djamal Hissein Didane, Nurhayati Rosly, Mohd Fadhli Zulkafli, and Syariful Syafiq Shamsudin

Department of Aeronautical Engineering, Faculty of Mechanical and Manufacturing Engineering, Universiti Tun Hussein Onn Malaysia (UTHM), Parit Raja, 86400 Batu Pahat, Johor, Malaysia

Correspondence should be addressed to Djamal Hissein Didane; dhdfils@gmail.com

Academic Editor: Li Ren

Long-term wind speed data for thirteen meteorological stations, measured over a five-year period, were statistically analyzed using the two-parameter Weibull distribution function. The purpose of this study is to reveal for the first time the wind power potentials in Chad and to provide a comprehensive wind map of the country. The results show that the values of the shape and scale parameters varied over a wide range. Analysis of the seasonal variations showed that higher wind speed values occur when the weather condition is generally dry and they drop considerably when the weather condition is wet. It was also observed that the wind speed increases as one moves from the southern zone to the Saharan zone. Although the wind power at each site varies significantly, however, the potentials of most of the sites were encouraging. Nevertheless, according to the PNNL classification system, they are favorable for small-scale applications only. A few stations in the middle of Sudanian and Sahel regions are found to be not feasible for wind energy generation due to their poor mean wind speed. The prevailing wind direction for both Saharan and Sahel regions is dominated by northeastern wind, while it diverged to different directions in the Sudanian zone.

1. Introduction

The fast growing population in developing countries and their lack of access to electricity supply particularly in rural or remote areas make some of these nations face the challenge of generating more energy sources and establishing a new form of energy supply structure in an effort to meet current and future increasing electricity demands. Permanent electricity supply is considered as one of the major factors responsible for sustainable economic and social development of a nation [1]. In Chad, for example, although the country is an oil producer, access to energy is extremely limited and incessant power outage is still an everyday burden that undermines its development possibilities and needs a rapid action with a view to coping with the minimum requirement of the twenty-first century's lifestyle.

As such, the majority of people in Chad tend to rely on traditional biomass for cooking and heating purposes particularly in rural areas as it is the case for many African nations [2]. A recent projection from the World Energy Outlook (WEO) [3] reported that more than 80% of the population in sub-Saharan Africa and 50% of the population in developing Asia depend on traditional biomass for cooking and heating. This situation has become more critical due to high fuel price in the country (considered the highest in the region) and the ever-increasing rate of the desert encroachment towards the Sahel region which forces the government to set some policies that prohibit deforestation and use of charcoal for cooking. Additionally, the government is also creating a green belt from the extreme east to the west of the country in order to halt the desertification process and to protect the region from economic and environmental impacts of land degradation.

Currently, a great deal of extensive research on wind energy is taking place almost all over the world due to the exceptional benefits that wind energy could offer. Many regional countries like Morocco, Egypt, Tunisia, Algeria, and so forth are already harvesting wind energy by establishing wind farms in the open Sahara desert and coastal areas [4, 5]. Chad was not an exception to embrace such sustainable renewable energy source as an alternative energy since the preliminary study in the region shows that the wind conditions are favorable for wind energy adoption [5].

Chad experiences wind and sunlight throughout the year. Thus, it is interesting to explore the abundant untapped potentials of the wind and the possibilities of using the Wind Energy Conversion System (WECS) as a source of electricity in the effort to fulfill the energy requirements in the country and to reduce this exclusive dependency on biomass and fossil fuels. However, although the region is found to be situated in the midst of enormous wind power potentials, to date, there is no extensive study undertaken to assess the prospects of wind energy as an alternative source in the country, except in [6] which covered only a single city (Ndjamena). Thus, this article could be the first of its kind to investigate the potentiality of wind power in Chad and it provides an initial assessment which together contributes towards the establishment of a comprehensive wind map/pattern for the country which may hopefully become the main wind energy reference for future development of wind power projects in the region.

2. Geographical Nature of Chad

Chad is a landlocked country with a 1,284,000 km^2 area, situated in Central Northern Africa. It is the 5th largest in the African continent and 21st in the world in terms of land mass. It is bordered by Libya form the north, Sudan from the east, the Central African Republic from the south, Niger from the west, and Cameroon and Nigeria from the southwest. It lies at 7–23° north of the equator and its larger part is Sahara desert. The vast majority of the area is flat with some plateaus around the country.

Unlike many countries in the world, the climate in Chad is highly varying throughout the country due to nonuniform geographical regions. These regions could be categorized into three in terms of geographical aspects as in Figure 1. The southern part is fertile agricultural land and is covered by tropical savanna, with a climate with longer raining season of about six months of precipitation. The raining season starts as early as May and ends around October. Meanwhile, the winter starts from November until February, and the summer takes place between February and May [7].

The region in the middle has an arid subtropical climate and it forms a belt of about 500 kilometers wide in the center of the country between the Sahara and Sudanian savanna. The raining season in this region begins a bit late with almost three months of rainfall duration. The first water drops are seen in June and then raining ends in September, while winter is between October and February. The Sahara desert occupies the northern part of the country with less population though it is the largest region as shown in Figure 1. Most of the people here are farmers and/or ranchers. Rain is less likely to be seen in this locality and these people only depend on underground water for the living. Thus, only a hotter summer or colder and dry winter exists in this region [7].

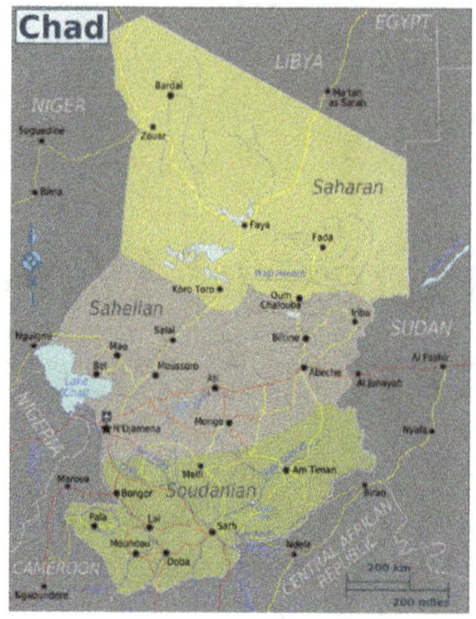

Figure 1: Geographical map of Chad [8].

3. Wind Data Collection

The wind data for this study are collected from the National Meteorological General Administration (Direction Générale de la Météorologie Nationale) at the Hassan Djamous International Airport in N'Djamena. The geographical coordinates of the thirteen major meteorological stations across the country are furnished in Table 1. The data extensively cover a wide range of five-year period and recorded in a monthly averaged form, at the standard 10 m height above ground level. These thirteen locations represent different geographical and climatological conditions, namely, Faya-Largeau, which is found in the Saharan zone; Abéché, Ati, Bokoro, Mao, Mongo, and Ndjamena, which are located in the Sahelian zone; and Am Timan, Bousso, Doba, Pala, Moundou, and Sarh, which are situated at the Sudanian zone as shown in Figure 1.

Furthermore, although most of the data involved in this study are recent, the last observation made by some stations was in late 1978. This might be attributed to the fact that some of these stations are no longer recording such data in a consistent manner. However, due to the need to uncover the potential of these regions, they are being considered in this study by selecting five consecutive years, as it was learned that old or new data have a minor effect on the wind data assessment process as long as the study period is long enough [9, 10]. More so, [11, 12] stated that observation of less than 30 years may inherently cause variations in the long-term average, but for the sole purpose of resource assessment such period may be sufficient. However, covering longer periods would provide confidence to the interested investors that the wind potentials will be available in the coming years besides giving a more reliable evaluation [9, 10].

4. Analysis Procedure

It is commonly agreed that assessment of wind potential of a site using solely the conventional meteorological wind speed data is not sufficient. This is because depending on average wind speed alone to estimate the potential of a site might be

Table 1: Physical features of the meteorological stations.

Name of station	Coordinates: Latitude (N) Deg.	Latitude (N) Min.	Longitude (E) Deg.	Longitude (E) Min.	Elevation (m)
Abéché	14	40	20	50	545
Am Timan	11	02	20	16	433
Ati	13	14	18	18	332
Bokoro	12	23	17	04	300
Bousso	10	29	16	43	335
Doba	08	42	16	50	387
Faya-Largeau	17	55	19	06	235
Mao	14	08	15	18	327
Mongo	12	10	18	40	431
Moundou	08	37	16	04	429
N'Djaména	12	08	15	02	295
Pala	09	22	14	55	467
Sarh	09	09	18	22	365

misleading [13]. This is due to the fact that the wind speed frequency distribution and persistence at a site are not constant and they vary significantly with time. As such, merely taking the average wind speed for a specific period of time does not take such consideration into account. Thus, statistical methods are used to reveal the wind characteristics of a potential site. Such statistical assessment of individual terrains or as a whole will collectively assist in establishing a clear wind map or wind pattern for a region while providing more detailed characteristics of the wind at the site such as wind distribution, stability, and direction, as the wind is known for its inherently intermittent behavior due to seasonal and temporal variations. The FirstLook software is commonly used to determine specific locations with higher wind power potentials, while the RIAM-COMPACT numerical model is responsible for suggesting the appropriate location to install a WECS during micrositing analysis of terrains.

4.1. Weibull Distribution Function. Many statistical methods are available to closely estimate the wind power potentials and wind speed characteristics of a given terrain. Weibull distribution is among the widely accepted approaches to statistically assess the wind behavior at a particular site clearly [14]. This technique consists of several methods to evaluate the shape and scale parameters of Weibull, depending on the form of data available and the level of sophistication required [15]. In this method, the variations of wind speed are characterized by probability density function (PDF) (see (1)) and cumulative distribution function (CDF) (see (5)) [16]. The PDF indicates the fraction of time or the probability of which the wind speed prevails at a certain direction and the CDF shows the probability of which the wind speed is equal to or lower than the average speed [17].

Thus, in the current study, the meteorological data are statistically analyzed using the effective two-parameter Weibull distribution function. This approach is particularly useful in studying wind speed characteristics and wind energy density. It is typically adopted due to its simplicity, flexibility, and ability to show good agreement with the observed data [18–26]. In addition, the majority of the known commercial software packages for Annual Energy Production (AEP) estimation are based on the two-parameter Weibull distribution function [27].

The two parameters are the dimensionless shape function, k (see (2)), and the scale function, c (see (3)), in meters per second unit. They, respectively, indicate the stability and strength of the wind at the site of interest [28].

$$f(v) = \left(\frac{k}{c}\right)\left(\frac{v}{c}\right)^{k-1} \exp\left[-\left(\frac{v}{c}\right)^{k}\right], \quad (k > 0,\ v > 0,\ c > 1). \tag{1}$$

$f(v)$ is the probability of observing wind speed v. c is the Weibull scale parameter and k is the dimensionless Weibull shape parameter and they are given by [15] as follows:

$$k = 0.83\overline{V}^{0.5} \tag{2}$$

$$c = \frac{\overline{V}}{\Gamma(1 + 1/k)}. \tag{3}$$

Γ is the gamma function and $\overline{V}$ is the average wind speed and can be expressed as

$$\overline{V} = \frac{1}{n}\left[\sum_{i=1}^{n} v_i\right]. \tag{4}$$

The cumulative distribution, given as in the equation below, is the integral of the probability density function:

$$F(v) = 1 - e^{-(v/c)^k}. \tag{5}$$

4.2. Wind Speed Carrying Maximum Energy. $V_{\max.E}$ represents the wind speed that carries the maximum wind energy and it is determined using the shape and scale parameters of the Weibull distribution. It is highly significant as it can expose the maximum energy possible at the given site, and it is expressed as in (6) [29]. The wind speed that carries maximum energy is different from the highest wind speed and the average wind speed because it takes into account the scale and shape parameters of the Weibull distribution which are responsible for indicating the persistence, stability, and strength of the wind at the site under investigation as was mentioned earlier.

Thus, such estimation is quite important for the sake of revealing the wind speed that may carry the highest energy which could probably assist in knowing the maximum amount that can be expected from a particular site. Moreover, it is also used in selecting a suitable wind turbine or an appropriate rated wind speed [30]. As wind turbines perform efficiently at their rated wind speed, [31] suggests that the rated wind speed should be close to the wind speed carrying maximum energy:

$$V_{\max.E} = c\left(1 + \frac{2}{k}\right)^{1/k}. \quad (6)$$

4.3. Wind Power Density. Generating electricity from kinetic energy via wind that is flowing through a blade swept area, A, is proportional to the cube of the velocity and it can be described as follows in (7). A wind turbine starts to operate at the starting speed and produces useful energy at the cut-in speed. The corresponding power increases relative to the increase of the wind speed until the cut-out speed of the wind turbine. The maximum power is generated when the wind turbine is operated at the rated speed and power becomes constant when it operates between the rated speed and cut-out speed. At higher wind speeds above the cut-out speed, the turbine stops to avoid damage due to the higher wind velocity.

$$P(v) = \frac{1}{2}A\rho v^3. \quad (7)$$

The wind power density based on Weibull distribution analysis is calculated using the following equation [32]:

$$\frac{P}{A} = \int_0^{\infty} \frac{1}{2}\rho v^3 f(v)\,dv = \frac{1}{2}\rho c^3 \Gamma\left(\frac{k+3}{k}\right), \quad (8)$$

where ρ is the air density (1.225 kg/m^3) at sea level with a mean temperature of 15°C and 1 atmospheric pressure.

4.4. Wind Energy Density. After knowing the wind power density of a given site, one of the important wind characteristics could be estimated which is the wind energy density. It is the product of wind power density for a certain duration or period of time which can reveal the amount of energy density that could be expected at the site under study. The wind energy density for the desired time can easily be determined using the following equation [32]:

$$\frac{E}{A} = \frac{1}{2}\rho c^3 \Gamma\left(\frac{k+3}{k}\right)T, \quad (9)$$

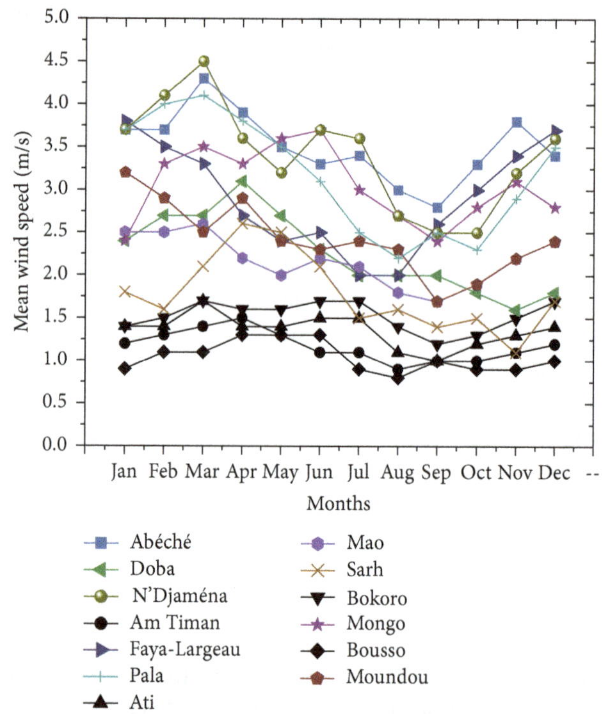

FIGURE 2: Monthly variation of wind speeds for all the selected stations.

where T is the desired time; for example, for wind energy density for a time period of one month or a year, the time, T, can be taken as 720 h or 8640 h, respectively.

5. Results and Discussion

5.1. Wind Speed Variation. The potential of wind energy in Chad was investigated at thirteen stations for a period of five years. Knowledge of monthly wind speed variations is essential to draw a clear picture of the seasonal wind speed behavior at the potential site. These fluctuations are significant in designing and selecting an appropriate WCES, energy storage, and load scheduling. These variations were determined using (4) and presented in Figure 2.

The trends of monthly and yearly mean wind speed for all the stations under investigation exhibit almost a similar pattern. The monthly mean wind speed varies in the range of 2 to 4 m/s for the Saharan region and in the range of 1 to 4 m/s for the Sahel and tropical Savanna regime. It was also observed that the wind speed tends to increase significantly when the weather condition is generally dry (November to May) and decreases considerably when the weather conditions are wet (June to October). It was learned that the highest demand for electricity is also at the dry period [6]. However, a few stations such as Bousso, Bokoro, and Am Timan maintained almost similar monthly mean wind speed throughout the year.

In general, it was established that stronger wind speed is available as one moves from south to north, towards the Sahara as depicted in Figure 3. In other words, some stations at the Savanna region have demonstrated lower wind speed

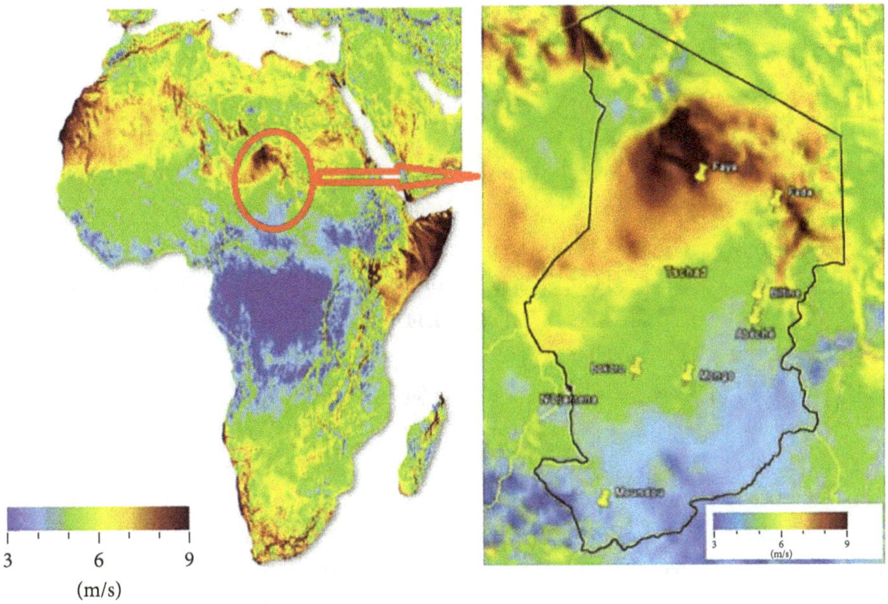

FIGURE 3: Wind speed pattern in Chad.

compared to the Sahel and Saharan regions with the exception of Ati and Bokoro. This could partly be attributed to the fact that the southern region is tropical with heavy forest and the land becomes clearer as one moves towards the north. However, it was specifically found that the stations at the middle of the Sahel and Sudanian zones showed the lowest mean wind speed with an annual average wind speed of less than 2 m/s as it is evident in Figures 2 and 3. This suggests that this area is less likely to be a potential location for wind energy harvesting.

Moreover, the seasonal wind speed variations among the three regions become more noticeable particularly during winter, though this variation is marginal between the Sahel and Saharan regions. The month March showed the highest mean wind speed with the value of 4.5 m/s, and August showed the lowest mean wind speed value of 0.8 m/s.

The results presented also showed that, within five years under study, the highest corresponding annual average wind speed at 10 m height from the ground is approximately 3.5 m/s, and it was found in the eastern part of the country at Abéché specifically, while the minimum is found at Bousso with a wind speed of 1.0 m/s. Thus, according to the PNNL (Pacific Northwest National Laboratory) classification system, wind energy potentials in Chad could only be used for small-scale applications due to the current wind conversion technologies and cost factor [32]. As such, this could probably solve the incessant power outage if a WECS is installed at every household and would also be quite useful for the ranchers who depend only on wells and use conventional methods to pump water.

However, it is worth noting that although the presented station from the Saharan region (Faya-Largeau) showed a bit lower wind speed compared to some stations in the Sahel region, data for wind speed at 30 m height proves otherwise. Thus, the Sahara zone could be considered more promising as depicted in Figure 3.

5.2. Wind Power and Energy Density. Although wind speed characteristics are essential in drawing a clear picture of the wind potentials of a site, wind power density is believed to be a better indicator than wind speed. In this study, the wind power and energy density are evaluated using the Weibull equations (8) and (9), respectively, and tabulated in Table 3. The results illustrated in Figure 4 show that the highest value of wind power density was found at Abéché (69 W/m^2) followed by Ndjamena (65 W/m^2) and Pala (55 W/m^2), whereas the lowest wind power density was observed at Bousso, Am Timan, and Ati with the value of 7, 8, and 10 W/m^2, respectively. On the other hand, the wind energy density ranged between 59 kWh/m^2/year and 603 kWh/m^2/year. The city of Abéché again showed the maximum value of energy with more than 600 kWh/m^2/year and Bousso presented the minimum. Based on the presented data, it could be deduced that the maximum power density in Chad is around 70 W/m^2 and it is found in the eastern part of the country.

As wind speeds during the dry season are higher as compared to other periods of the year [6], interestingly, the corresponding power and energy density are also higher in this season. This clearly indicates that a WECS would produce more energy during this period for all the stations at the three distinctive geographical regions. It is also interesting to note that although some stations showed a higher value of monthly mean wind speed, the yearly mean was low due to significant seasonal variations of the site, which lead to low yearly power output. This gives an insight into using a WECS as an alternative for energy generation at such particular period of the year for that particular site.

TABLE 2: Monthly Weibull parameters (k, c).

Station	Parameter	Month											
		Jan	Feb	Mar	Apr	May	Jun	Jul	Aug	Sep	Oct	Nov	Dec
Abéché	k	1.59	1.59	1.72	1.65	1.56	1.50	1.54	1.43	1.39	1.52	1.61	1.52
	c	4.10	4.10	4.85	4.40	3.94	3.63	3.82	3.28	3.09	3.70	4.22	3.77
Am Timan	k	0.91	0.94	0.99	1.02	0.96	0.87	0.86	0.79	0.82	0.83	0.85	0.89
	c	1.17	1.27	1.44	1.56	1.32	1.05	1.00	0.81	0.88	0.91	0.98	1.10
Ati	k	0.97	0.99	1.08	0.99	0.97	1.01	1.02	0.86	0.82	0.89	0.96	0.97
	c	1.36	1.44	1.75	1.44	1.34	1.49	1.53	1.00	0.88	1.10	1.32	1.36
Bokoro	k	0.99	1.00	1.08	1.03	1.06	1.07	1.08	0.96	0.89	0.93	1.01	1.08
	c	1.41	1.46	1.75	1.58	1.68	1.73	1.75	1.34	1.10	1.22	1.49	1.75
Bousso	k	0.80	0.88	0.86	0.93	0.93	0.94	0.77	0.74	0.81	0.80	0.79	0.84
	c	0.81	1.05	1.00	1.22	1.24	1.29	0.74	0.67	0.86	0.83	0.81	0.95
Doba	k	1.28	1.37	1.36	1.46	1.35	1.24	1.17	1.16	1.16	1.12	1.04	1.10
	c	2.59	2.97	2.93	3.44	2.92	2.45	2.09	2.07	2.07	1.90	1.61	1.82
Faya-Largeau	k	1.62	1.55	1.50	1.37	1.27	1.31	1.16	1.16	1.33	1.42	1.53	1.60
	c	4.29	3.87	3.61	3.00	2.54	2.71	2.09	2.11	2.83	3.25	3.77	4.15
Mao	k	1.31	1.31	1.34	1.22	1.18	1.22	1.21	1.11	1.08	1.15	1.24	1.28
	c	2.71	2.71	2.83	2.31	2.14	2.31	2.26	1.85	1.73	2.02	2.38	2.59
Mongo	k	1.28	1.50	1.55	1.52	1.57	1.58	1.44	1.36	1.28	1.38	1.45	1.37
	c	2.61	3.61	3.87	3.70	4.01	4.08	3.35	2.97	2.59	3.06	3.40	3.02
Moundou	k	1.48	1.42	1.31	1.40	1.28	1.25	1.27	1.26	1.08	1.15	1.23	1.27
	c	3.52	3.23	2.76	3.16	2.61	2.45	2.54	2.52	1.77	2.04	2.35	2.54
N'Djaména	k	1.60	1.69	1.75	1.58	1.47	1.60	1.57	1.36	1.31	1.31	1.49	1.56
	c	4.17	4.64	5.03	4.03	3.49	4.13	3.99	2.93	2.71	2.71	3.59	3.96
Pala	k	1.59	1.66	1.68	1.63	1.54	1.45	1.31	1.22	1.30	1.27	1.41	1.54
	c	4.08	4.50	4.59	4.29	3.84	3.40	2.73	2.33	2.66	2.52	3.21	3.86
Sarh	k	1.10	1.06	1.19	1.35	1.31	1.20	1.00	1.03	0.99	1.00	0.88	1.06
	c	1.82	1.68	2.21	2.88	2.71	2.26	1.46	1.58	1.44	1.46	1.08	1.70

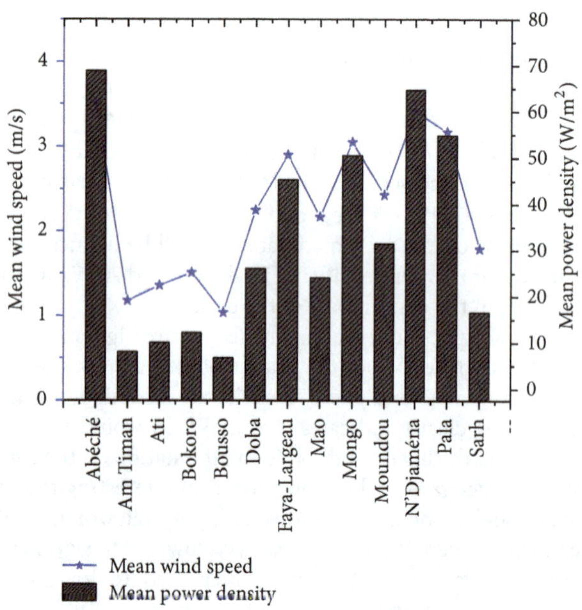

FIGURE 4: Comparison of yearly mean wind speeds and mean wind power density.

5.3. Weibull Distribution. The monthly mean scale parameter c (m/s) and shape parameter k (dimensionless) of the Weibull distribution for all the stations under study are determined using (2) and (3) and furnished in Table 2. It was observed that the shape parameter has smaller temporal variation compared to the scale parameter. The monthly values of the parameter k range between 1 and 1.6 for almost all of the stations, except for a few stations in the Savanna and Sahel zones which exhibit less than one. This indicates that such stations have poor stability and poor persistence compared to the others. The scale parameter c, which is slightly greater than average wind speed, ranges from 2 to 5 m/s for nearly all of the stations in the Sahara and Sahel zones and from 1 to 4 for most of the stations in the Sudanian region. The highest and lowest values of the dimensionless shape parameter k were observed to be 1.80 and 0.67, respectively, and they were found at Ndjamena and Bousso. This is perhaps due to the more open and flat areas in the Sahel and Saharan regions. Thus, these higher values of the scale parameter denote that these sites are windy.

The frequency and cumulative distributions of monthly average wind speed for all the stations are presented, respectively, in Figures 5 and 6. It is obvious that all curves have

Table 3: Yearly Weibull parameters (k, c), average wind speed, wind speed carrying maximum energy, wind power density, and wind energy density (at 10 m height).

Station	Parameters					
	k	c	$\bar{v}$	$V_{\max,E}$	P/A (W/m^2)	E/A (kWh/m^2)
Abéché	1.56	3.91	3.52	6.65	68.80	602.72
Am Timan	0.90	1.13	1.18	4.11	8.04	70.39
Ati	1.00	1.34	1.36	4.26	10.06	88.09
Bokoro	1.02	1.52	1.51	4.42	12.21	106.97
Bousso	0.85	0.96	1.04	4.02	6.73	58.95
Doba	1.24	2.40	2.25	5.20	26.03	228.0
Faya-Largeau	1.41	3.19	2.90	5.95	45.29	396.72
Mao	1.22	2.32	2.17	5.11	24.07	210.85
Mongo	1.45	3.36	3.05	6.11	50.54	442.77
Moundou	1.29	2.62	2.43	5.41	31.41	275.17
N'Djaména	1.53	3.79	3.42	6.54	64.63	566.18
Pala	1.48	3.51	3.17	6.26	54.80	480.01
Sarh	1.11	1.86	1.79	4.70	16.58	145.24

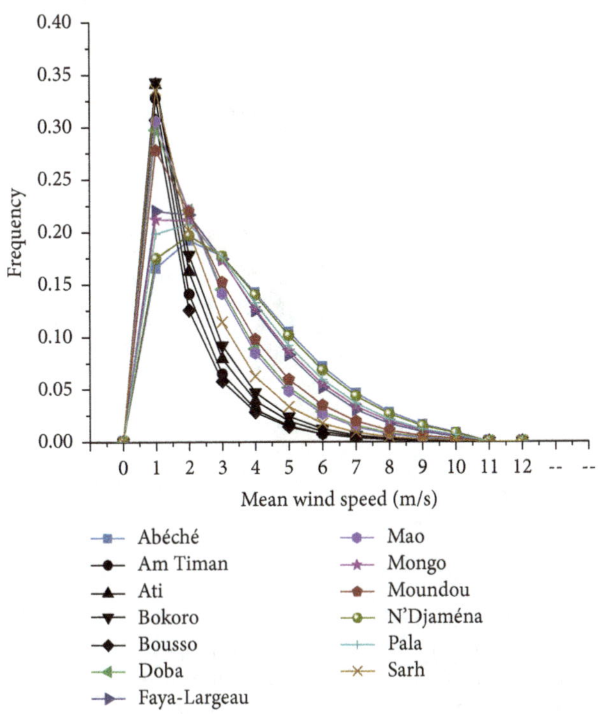

Figure 5: Frequency distributions of monthly mean wind speed for all stations.

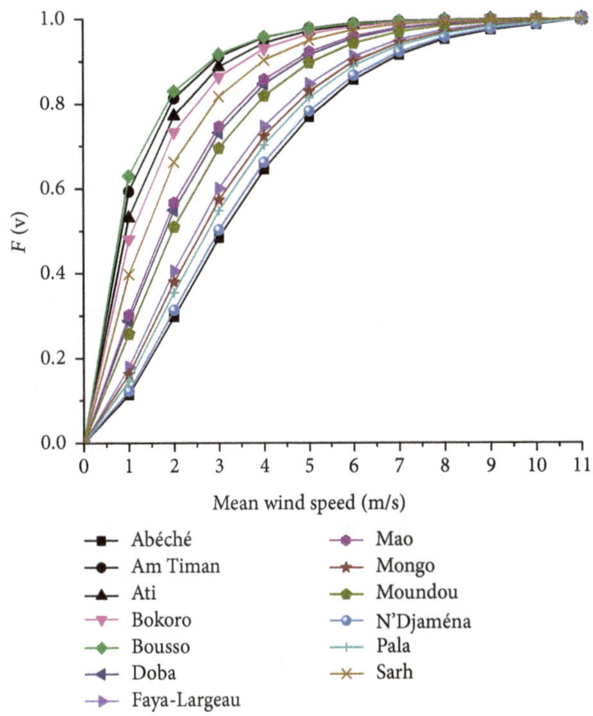

Figure 6: Cumulative distributions of monthly mean wind speed for all stations.

a similar tendency of wind speeds on both frequency and cumulative density. However, the frequency peaks in the range of 2 to 3 m/s for the two regions in the upper part of the country and between 1 and 2 m/s for the lower part of the country. This indicates that most of the wind energy lies in these ranges. This information is very practical and can be used to determine the amount of power which can be generated in a given speed band and could also help in selecting a suitable wind turbine for each particular site.

5.4. Polar Diagrams. Evaluation of wind direction helps to expose the impact of the geographical features on the wind and to obtain the prevailing direction and magnitude of the most frequent wind. In Figures 7–12, METAR data of the year 2014 and polar diagrams for six potential sites in Chad are presented. These locations are from the three distinctive geographical zones. It was observed that the prevailing wind direction varied from zone to zone and within the same zone as well. As such, the prevailing wind direction for the stations

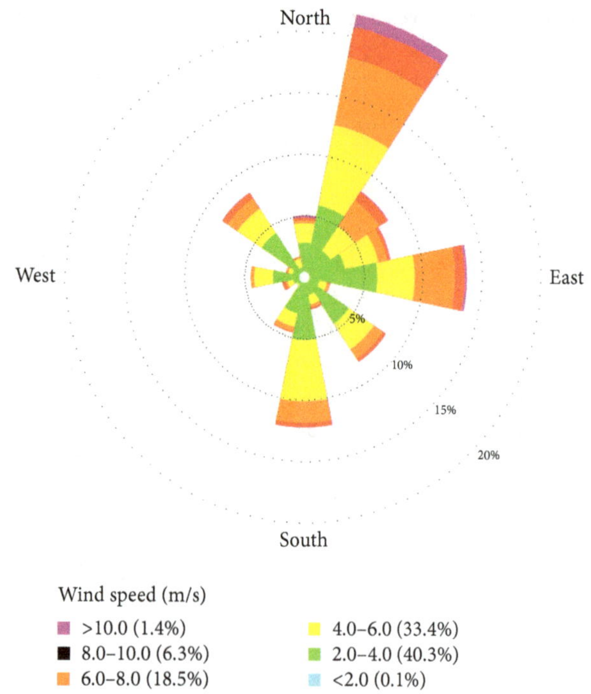

FIGURE 7: Polar diagram: wind direction for Abéché.

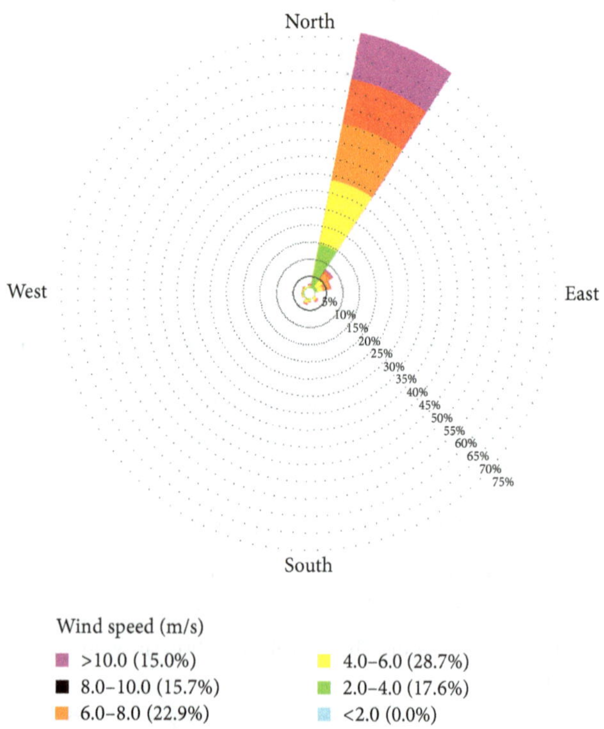

FIGURE 8: Polar diagram: wind direction for Faya-Largeau.

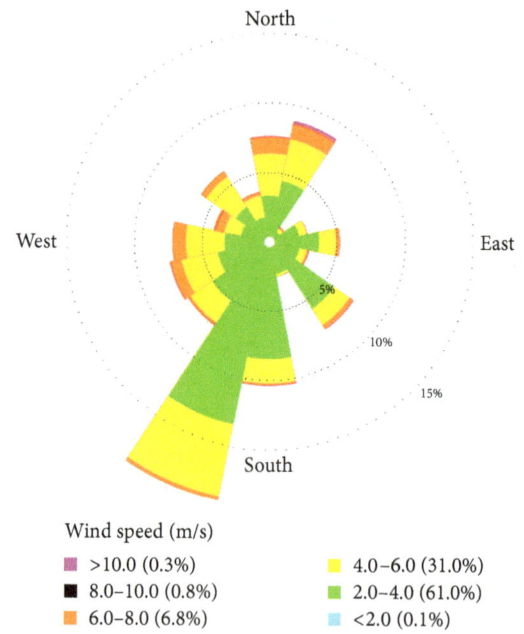

FIGURE 9: Polar diagram: wind direction for Moundou.

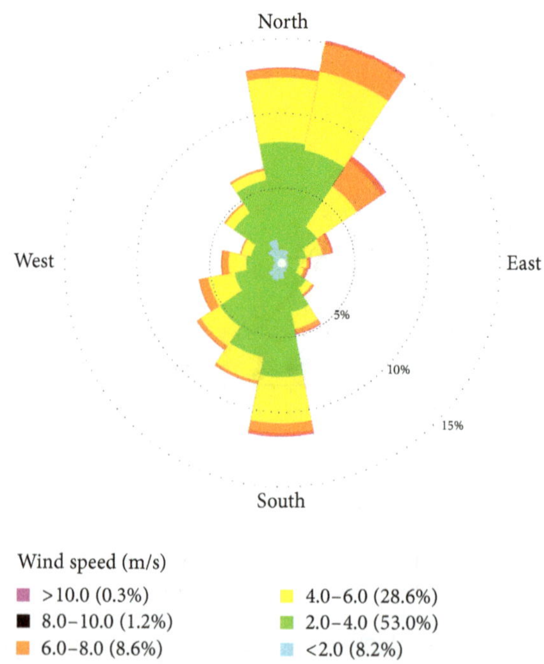

FIGURE 10: Polar diagram: wind direction for Ndjamena.

in the Saharan and Sahel zones was northeast throughout the year, while in the Sudanian zone the direction prevailed at various directions. In Moundou, for example, the most frequent wind direction was dominated by the southwestern wind. Meanwhile, in Pala and Sarh, the wind direction prevailed, respectively, in north and south directions throughout 2014.

6. Conclusion and Future Direction

In this study, the monthly and yearly wind speed distribution and wind power density for thirteen meteorological stations in Chad were evaluated. The novel two-parameter Weibull distribution function was employed to analyze the five-year period data for each site. While the data used in this study are being published for the first time, they are collected for the purpose of studying the wind energy potential and to have a comprehensive wind database or a wind map in Chad.

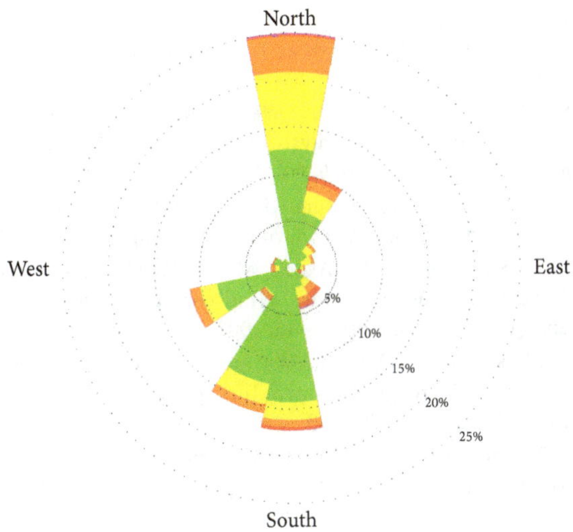

Wind speed (m/s)

- >10.0 (0.3%)
- 8.0–10.0 (2.3%)
- 6.0–8.0 (11.8%)
- 4.0–6.0 (21.9%)
- 2.0–4.0 (63.5%)
- <2.0 (0.2%)

FIGURE 11: Polar diagram: wind direction for Pala.

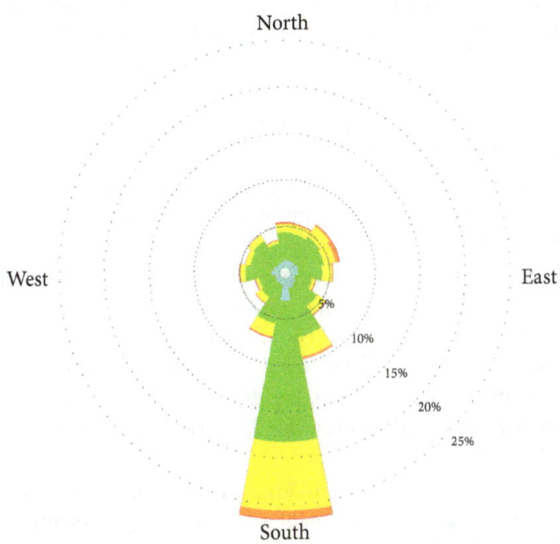

Wind speed (m/s)

- >10.0 (0.3%)
- 8.0–10.0 (0.4%)
- 6.0–8.0 (2.6%)
- 4.0–6.0 (22.2%)
- 2.0–4.0 (59.5%)
- <2.0 (15.1%)

FIGURE 12: Polar diagram: wind direction for Sarh.

The seasonal variations of the mean wind speed data show that higher wind speeds are available when the weather condition is generally dry, that is, November to May, and lower wind speed is found when the weather condition is wet, that is, June to October. This trend is applicable for all the thirteen sites under investigation. Interestingly, it was learned that higher electrical energy demand in the country is also in this period (dry season).

Moreover, the data presented also revealed that the monthly mean wind speed varies in a wide range in each geographical zone. In the Sahel zone, for example, it ranged between 1 and 4 m/s, though most of the stations in this region have shown reasonably strong wind speed of more than 2 m/s. The same behavior was also noticed at the Sudanian zone.

It was observed that the sites in the middle of Sahel and Sudanian zones tend to show lower mean wind speed compared to the others. However, in general, higher wind potentials are witnessed as one moves towards the Saharan region. It is also worth noting that the peak monthly mean wind speed was 4.5 m/s and it was found in Ndjamena in the month of March. Meanwhile, the lowest monthly average wind speed was less than a unit and was found in Bousso.

Based on yearly averaged data, the most recommended site for wind energy generation in Chad would be Abéché, followed by Ndjamena, Pala, Mongo, and Faya-Largeau, which are the stations that possess mean wind speed of more than 3 m/s at 10 m height above ground level. These sites exhibited wind power density in the range of 45–69 W/m^2 and corresponding wind energy density values ranged between 400 and 600 kWh/m^2/year. However, although the wind energy potential in Chad is promising for most of the sites under study for small-scale applications, it is concluded that few stations in the middle of Sudanian and Sahel regions (Am Timan, Bousso, Bokoro, and Ati) are not feasible for wind energy generation due to their weak mean wind speed of around 1 m/s.

In terms of wind direction, the prevailing wind direction for both Saharan and Sahel region was dominated by northeastern wind. However, the wind direction in every station at the Sudanian region prevailed at different directions. The most probable wind speed direction was southwest in Moundou, dominated by northern wind in Pala and southern wind in Sarh.

It is recommended in the future to develop a wind map at the other parts of the Saharan region and some locations close to Lake Chad. The data should be recorded at various heights above ground level and it will then be used for developing wind atlas for Chad which will encourage the development of wind energy projects in the country.

Conflicts of Interest

The authors declare that there are no conflicts of interest regarding the publication of this paper.

Acknowledgments

The authors would like thank Universiti Tun Hussein Onn Malaysia for providing financial support of this research under Grant U416. The authors also wish to express their sincere gratitude to the National Meteorological General Administration (ASECNA) in Chad for the provision of the meteorological data.

References

[1] Y. S. Mohammed, M. W. Mustafa, and N. Bashir, "Status of renewable energy consumption and developmental challenges in Sub-Sahara Africa," *Renewable and Sustainable Energy Reviews*, vol. 27, pp. 453–463, 2013.

[2] K. S. P. Kumar and S. Gaddada, "Statistical scrutiny of Weibull parameters for wind energy potential appraisal in the area of northern Ethiopia," *Renewables: Wind, Water, and Solar*, vol. 2, no. 1, p. 1, 2015.

[3] International Energy Agency (IEA), World Energy Outlook, 2016, http://www.worldenergyoutlook.org/resources/energy-development/energyaccessdatabase.

[4] A. S. Ahmed, "Wind energy as a potential generation source at Ras Benas, Egypt," *Renewable and Sustainable Energy Reviews*, vol. 14, no. 8, pp. 2167–2173, 2010.

[5] A. D. Mukasa, E. Mutambatsere, Y. Arvanitis, and T. Triki, "Development of wind energy in Africa," Report, African Development Bank, 2013.

[6] D. H. Didane, A. A. Wahab, S. S. Shamsudin, N. Rosly, M. F. Zulkafli, and S. Mohd, "Assessment of wind energy potential in the capital city of Chad, N'Djamena," *AIP Conference Proceedings*, vol. 1831, no. 1, p. 020049.

[7] C. McSweeney, M. New, and G. Lizcano, UNDP Climate Change Country Profiles, Chad, 2010, http://www.geog.ox.ac.uk/research/climate/projects/undp-cp/UNDP_reports/Chad/Chad.hires.report.pdf.

[8] Wikivoyage, 2016 https://en.wikivoyage.org/wiki/Chad.

[9] S. Rehman, "Wind energy resources assessment for Yanbo, Saudi Arabia," *Energy Conversion and Management*, vol. 45, no. 13-14, pp. 2019–2032, 2004.

[10] R. Kose, M. A. Ozgur, O. Erbas, and A. Tugcu, "The analysis of wind data and wind energy potential in Kutahya, Turkey," *Renewable and Sustainable Energy Reviews*, vol. 8, no. 3, pp. 277–288, 2004.

[11] I. Y. Lun and J. C. Lam, "A study of Weibull parameters using long-term wind observations," *Renewable Energy*, vol. 20, no. 2, pp. 145–153, 2000.

[12] R. M. R. Kainkwa, "Wind speed pattern and the available wind power at Basotu, Tanzania," *Renewable Energy*, vol. 21, no. 2, pp. 289–295, 2000.

[13] M. Siti, M. Norizah, and M. Syafrudin, "The evaluation of wind energy potential in peninsular Malaysia," *International Journal*, vol. 2, no. 4, 2011.

[14] S. Mathew and G. S. Philip, Eds., *Advances in Wind Energy and Conversion Technology*, vol. 20, Springer, Berlin, Germany, 2011.

[15] C. G. Justus, W. R. Hargraves, A. Mikhail, and D. Graber, "Methods for estimating wind speed frequency distributions," *Journal of Applied Meteorology*, vol. 17, no. 3, pp. 350–353, 1978.

[16] D. Wood, "Small wind turbines," in *Advances in Wind Energy Conversion Technology*, pp. 195–211, Springer, Berlin, Germany, 2011.

[17] M. R. Islam, R. Saidur, and N. A. Rahim, "Assessment of wind energy potentiality at Kudat and Labuan, Malaysia using Weibull distribution function," *Energy*, vol. 36, no. 2, pp. 985–992, 2011.

[18] A. A. Shata and R. Hanitsch, "Evaluation of wind energy potential and electricity generation on the coast of Mediterranean Sea in Egypt," *Renewable Energy*, vol. 31, no. 8, pp. 1183–1202, 2006.

[19] P. Ramírez and J. A. Carta, "Influence of the data sampling interval in the estimation of the parameters of the Weibull wind speed probability density distribution: a case study," *Energy Conversion and Management*, vol. 46, no. 15, pp. 2419–2438, 2005.

[20] K. Ulgen and A. Hepbasli, "Determination of Weibull parameters for wind energy analysis of Izmir, Turkey," *International Journal of Energy Research*, vol. 26, no. 6, pp. 495–506, 2002.

[21] A. S. S. Dorvlo, "Estimating wind speed distribution," *Energy Conversion and Management*, vol. 43, no. 17, pp. 2311–2318, 2002.

[22] M. Y. Sulaiman, A. M. Akaak, M. A. Wahab, A. Zakaria, Z. A. Sulaiman, and J. Suradi, "Wind characteristics of Oman," *Energy*, vol. 27, no. 1, pp. 35–46, 2002.

[23] V. M. Karsli and C. Geçit, "An investigation on wind power potential of Nurdağı-Gaziantep, Turkey," *Renewable Energy*, vol. 28, no. 5, pp. 823–830, 2003.

[24] R. Tchinda and E. Kaptouom, "Wind energy in Adamaoua and North Cameroon provinces," *Energy Conversion and Management*, vol. 44, no. 6, pp. 845–857, 2003.

[25] A. N. Celik, "A statistical analysis of wind power density based on the Weibull and Rayleigh models at the southern region of Turkey," *Renewable Energy*, vol. 29, no. 4, pp. 593–604, 2004.

[26] E. K. Akpinar and S. Akpinar, "An assessment on seasonal analysis of wind energy characteristics and wind turbine characteristics," *Energy Conversion and Management*, vol. 46, no. 11, pp. 1848–1867, 2005.

[27] O. A. Jaramillo and M. A. Borja, "Bimodal versus Weibull wind speed distributions: an analysis of wind energy potential in La Venta, Mexico," *Wind Engineering*, vol. 28, no. 2, pp. 225–234, 2004.

[28] D. Weisser, "A wind energy analysis of Grenada: an estimation using the 'Weibull' density function," *Renewable Energy*, vol. 28, no. 11, pp. 1803–1812, 2003.

[29] M. Jamil, S. Parsa, and M. Majidi, "Wind power statistics and an evaluation of wind energy density," *Renewable Energy*, vol. 6, no. 5, pp. 623–628, 1995.

[30] S. O. Oyedepo, M. S. Adaramola, and S. S. Paul, "Analysis of wind speed data and wind energy potential in three selected locations in South-East Nigeria," *International Journal of Energy and Environmental Engineering*, vol. 3, no. 1, p. 7, 2012.

[31] S. Mathew and K. P. Pandey, "Analysis of wind regimes for energy estimation," *Renewable Energy*, vol. 25, no. 3, pp. 381–399, 2002.

[32] A. Keyhani, M. Ghasemi-Varnamkhasti, M. Khanali, and R. Abbaszadeh, "An assessment of wind energy potential as a power generation source in the capital of Iran, Tehran," *Energy*, vol. 35, no. 1, pp. 188–201, 2010.

2

Transition Process from Diffuser Stall to Stage Stall in a Centrifugal Compressor with a Vaned Diffuser

Nobumichi Fujisawa and Yutaka Ohta

Department of Applied Mechanics and Aerospace Engineering, Waseda University, 3-4-1 Okubo, Shinjuku-ku, Tokyo 169-8555, Japan

Correspondence should be addressed to Yutaka Ohta; yutaka@waseda.jp

Academic Editor: Gerard Bois

The transition process from a diffuser rotating stall to a stage stall in a centrifugal compressor with a vaned diffuser was investigated by experimental and numerical analyses. From the velocity measurements, it was found that the rotating stall existed on the shroud side of the diffuser passage in the off-design flow condition. The numerical results revealed the typical vortical structure of the diffuser stall. The diffuser stall cell was caused by the systematic vortical structure which consisted of the tornado-type vortex, the longitudinal vortex at the shroud/suction surface corner (i.e., leading edge vortex (LEV)), and the vortex in the throat area of the diffuser passages. Furthermore, the stage stall, which rotated within both the impeller and diffuser passages, occurred instead of the diffuser stall as the mass flow rate was decreased. According to the velocity measurements at the diffuser inlet, the diffuser stall which rotated on the shroud side was shifted to the hub side. Then, the diffuser stall moved into the impeller passages and formed the stage stall. Therefore, the stage stall was caused by the development of the diffuser stall, which transferred from the shroud side to the hub side in the vaneless space and expanded to the impeller passages.

1. Introduction

Centrifugal compressors equipped with vaned diffusers are widely used in industry because of their high pressure-rise characteristics. However, they are likely to result in unsteady phenomena, such as surges and rotating stalls, as compared with vaneless diffusers. The unstable operating range is enlarged by surges and rotating stalls that have the potential to cause serious accidents. Turbochargers are generally used for a wide operating range, including off-design conditions. However, centrifugal compressors are not subjected to similar operating conditions. Therefore, understanding the flow physics behind rotating stalls is important for improving high pressure compressor performance and for safe operation in off-design flow condition.

Many researchers have focused on rotating stalls in axial compressors [1–4]. In addition, several articles have also been devoted to the study of rotating stalls in centrifugal compressors [5–8]. In recent years, there has been a renewed interest in diffuser rotating stalls that occur in centrifugal compressors with vaned diffusers. For example, Spakovszky reported that four-cell backward travelling modal waves occur in the vaneless space prior to the surge [9]. Everitt et al. reported the generation mechanism of spike stall inception by numerical analysis [10]. The short wavelength stall inception in centrifugal compressors with vaned diffusers was caused by the separation on the shroud side of the diffuser leading edge and radial flow reversal throughout the vaneless space. As Bousquet et al. accurately pointed out, the growth of the amplitude of a modal wave rotating in the vaneless space led to occurrence of the diffuser rotating stall; this was evident from the numerical analysis [11]. The authors also have reported that the unsteady behavior of the diffuser stall was influenced by the unsteady vortices at the leading edge of the diffuser vane during the off-design operation [12, 13].

Furthermore, several researchers have reported that the rotating stall changed its scale and behavior in accordance with the decrement of mass flow rate in turbomachinery. In axial machines, Outa et al. showed that the flow blockages, which were generated and developed separately in the respective cascades of the rotor and stator, get coupled with each other and propagate as a stage stall cell with

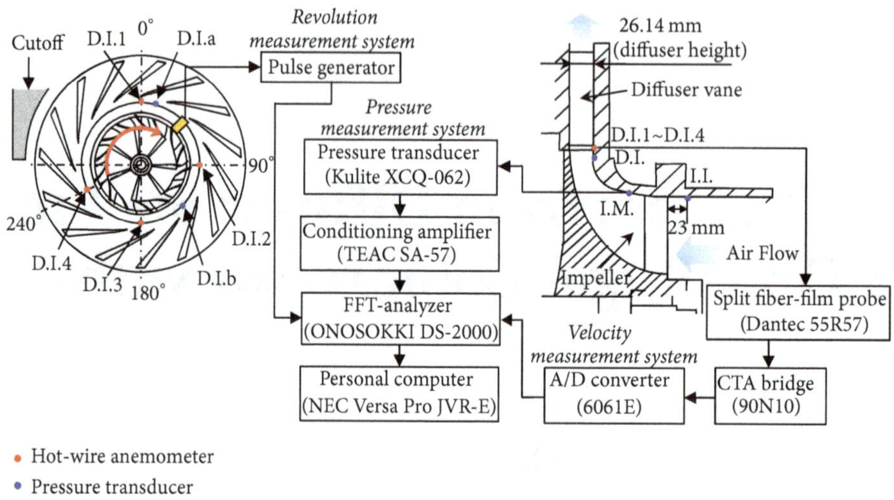

Figure 1: Experimental apparatus and measuring system.

decreasing mass flow rate [14]. Yoshida et al. investigated the rotating stalls that occur in the impeller and diffuser passages [15]. The behavior of the rotating stall was found to be dependent on the impeller-diffuser radial gap. Mizuki and Oosawa experimentally reported unsteady flow patterns in a centrifugal compressor with a vaneless diffuser during rotating stall and surge [16]. The part-span stall at the leading edge of the impeller changed to a full-span stall, as the flow rate decreased. Moreover, the strong influence of the full-span stall at the impeller inlet appeared during the surge. Zheng and Liu also examined the flow physics of rotating stalls and surge through their experiments [17]. The behavior of rotating stalls and surge was found to be dependent on the rotor speed and mass flow rate. However, very few researchers have reported on the transition process of the rotating stall via a detailed investigation of the internal flow field.

This study investigated the transition from the diffuser rotating stall to the stage stall in the vaned diffuser of a centrifugal compressor during the off-design operation. First, we focused on the generation mechanism of the diffuser stall by the both experimental and numerical analysis. The detached eddy simulation (DES) analysis was carried out to investigate the vortical structure of the diffuser stall. Secondly, we also investigated the transition characteristics of rotating stalls by the experiments.

2. Experimental Apparatus and Procedure

2.1. Experimental Apparatus. The compressor being tested had a centrifugal design based on that of a turbocharger used for marine diesel engines. The dimensions of the compressor are listed in Table 1 and the geometric configuration of the compressor and the main measurement system is illustrated in Figure 1. During the experiments, the rotational speed, N, of the compressor was set to $6000\ \mathrm{min}^{-1}$. The test impeller was an open-shroud type with seven main and splitter blades, and its inlet and outlet diameters were 248 and 328 mm, respectively. The shroud clearance of the impeller was approximately 1.0% of the impeller inlet blade width, $B_1 = 105.6$ mm.

Table 1: Dimensions of tested compressor.

Tested centrifugal compressor		
Rotational speed	N	$6000\ \mathrm{min}^{-1}$
Mass flow rate	G	1.64 kg/s
Pressure ratio	p_{t5}/p_{t1}	1.1
Impeller		
Number of blades (main + splitter)	Z	14 (7 + 7)
Inlet diameter	D_1	248 mm
Outlet diameter	D_2	328 mm
Exit blade width	B_2	26.14 mm
Diffuser		
Blade shape		Wedge
Number of vanes	V	15
Leading edge Diameter	D_3	360 mm
Trailing edge Diameter	D_4	559 mm
Diffuser width	B_4	26.14 mm

The vaned diffuser was employed in the experiments. The vaned diffuser had 15 wedge-type diffuser vanes located between two parallel diffuser walls of height, B_4, of 26.14 mm. The distance between the impeller exit and diffuser leading edge, L_{id}, was set to 16 mm. The original wedge-type diffuser vane (ODV) was employed in the experiments. The vanes were attached to the shroud surface of the channel diffuser.

The volute was designed to give the circumferential uniform pressure at the design operation. Near to the stall point ($\phi = 0.14$), the flow is decelerated in the circumferential direction and the pressure rises around the volute. However, the fluctuation of circumferential pressure in vaneless space at the off-design point is ±50 Pa at most, from the pressure measurements. Therefore, the unsteady behavior of diffuser stall was not highly affected by the volute design.

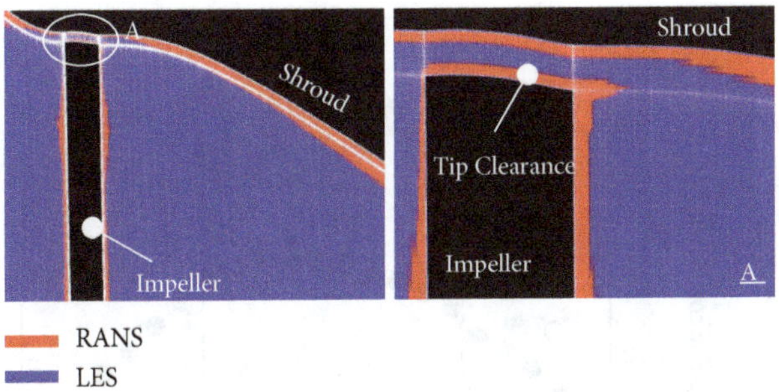

(a) Impeller passage

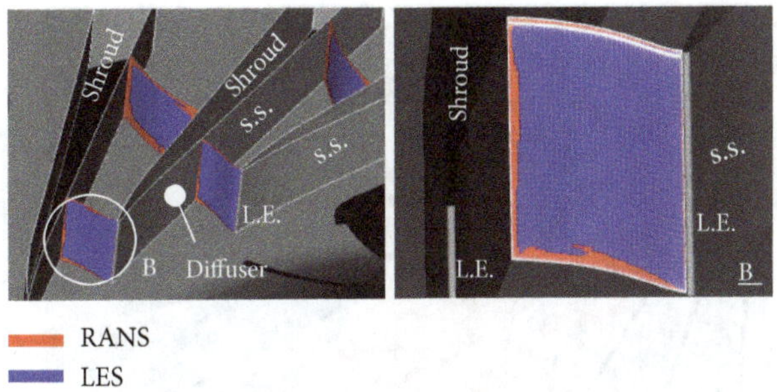

(b) Diffuser passage

FIGURE 2: RANS/LES region within impeller and diffuser passages.

2.2. Measuring Methods. The operation point of the compressor was set using a butterfly valve installed at the outlet duct. The mass flow rates were calculated using an orifice flow meter and a thermocouple installed at the outlet duct end. The pressure-rise characteristics were measured using a differential pressure transducer (Yamatake JTD920A) installed at the exit of the compressor scroll. The error of differential pressure probe was less than ±0.1%. The measuring bandwidth is from 2.0 kPa to 14 MPa. The total pressure rise at the compressor exit was calculated by the pressure rise at the exit of the compressor scroll and the meridional velocity which was calculated by mass flow rate.

The pressure and velocity measuring systems for investigating unsteady phenomena are also illustrated in Figure 1. The pressure transducers (Kulite, XCQ-062-25A) with an error within ±0.1% were mounted flush with the shroud surface of the compressor. The diameter of pressure probe head was 1.7 mm. The pressure was measured at the following points along the impeller passage: impeller inlet (I.I.), impeller midsection (I.M.), and diffuser inlet (D.I.a and D.I.b). The diameter of above three points is $0.757D_2$, $0.758D_2$, and $1.024D_2$, respectively. Furthermore, the spanwise distribution of the impeller-discharge flow was measured in the diffuser circumferential direction (D.I.1–D.I.4), using by a split-film probe (DANTEC, 55R57) with a constant-temperature anemometer (DANTEC, Streamline 90N10). The length of split-film probe was 8 mm and the film length was 1.25 mm. The measurement location $D = 339$ mm was located between the impeller exit and the diffuser leading edge. The four measurement points are placed at 0 deg, 90 deg, 180 deg, and 240 deg. The impeller-discharge flow measurement was traversed from the hub to the shroud side every 5% of the diffuser width (20 points). The unsteady pressure and velocity traces were measured at a sampling frequency of 25.6 kHz. The temporal resolution is about 20 samples per blade passing period.

3. Computational Procedure

3.1. Governing Equations. The present simulations were conducted using the in-house computational fluid dynamics (CFD) code, which had been validated for various turbomachinery flows. The code solved the governing equation of continuity, a three-dimensional compressible Navier-Stokes equation, an energy equation, and an equation for the state of an ideal gas. The convective flux was evaluated by the flux difference splitting (FDS) [18], which is extended to the third order using the monotonic upwind scheme for conservative laws (MUSCL) interpolation [19]. The viscous flux was determined as a second-order central difference using Gauss' theorem. The matrix free Gauss-Seidel (MFGS) implicit algorithm was employed for the time integration [20]. The DES approach, which is a hybrid scheme involving large eddy simulation (LES) and RANS [21], was employed in the turbulent modelling. The DES was based on the shear stress transport (SST) k-ω turbulence model [22]. This model incorporates a dependency on the local turbulence length scale. The model constant C_{DES} was set to the value recommended by Strelets et al. Both Coriolis and centrifugal forces were considered to be inertial force terms in the relative coordinate system. The RANS and LES regions of both the impeller and diffuser passages are illustrated in Figure 2. This DES simulation code could transform the RANS-mode in regions near the wall into the LES-mode inside the fluid volume.

3.2. Computational Domains. The computational domains employed in the numerical simulations are illustrated in Figure 3. The grid system included all the flow passages (14 impeller and 15 diffuser passages). The computational domain was divided into three regions: the moving impeller, stationary diffuser, and the moving tip clearance regions. The volute region was not included in this simulation. For the whole domain including the tip clearance region, the mesh was generated with the multiblock structured grid using AutoGrid5 ver9.1 (NUMECA International). In the full annulus analysis, the impeller (including the clearance region) and diffuser system had 26.2 and 24.8 million cells, respectively. The computational grid incorporated seven main and splitter blades and 15 diffuser vanes and had 51.0 million cells. The cell width at the walls was 0.1 μm, which corresponded to a $y+$ parameter that was approximately equal to one along all solid surfaces. From the grid dependency test in partial annulus analysis, the grid density was sufficient in order to capture the unsteady behavior of vortices near the diffuser vane.

3.3. Boundary Conditions. At the inflow boundary, the total pressure and temperature were fixed. At the outflow boundary, a throttle resistance with a load coefficient was assumed

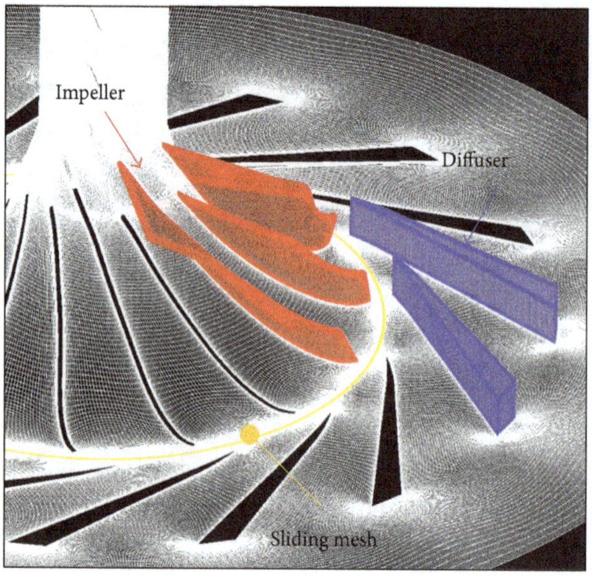

FIGURE 3: Overview of computational grid.

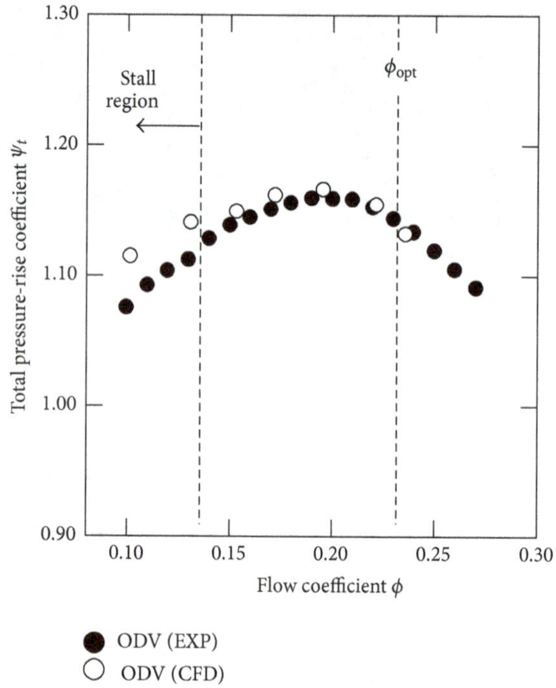

FIGURE 4: Compressor performance.

to exist between the boundary and the external atmosphere. The mass flow rate was not fixed at the outlet boundary because we assumed the throttle resistance with a load coefficient in order to simulate the velocity and pressure fluctuation caused by stall. We already observed the mass flow fluctuation at the exit of the compressor in both experiments and numerical simulations. Therefore, the static pressure at the outlet boundary was related to the mass flow rate through the boundary. Across the sliding boundary separating the moving impeller and stationary diffuser frames, the most recent data on one side was interpolated to obtain the data for the opposite side by using a sliding mesh for unsteady simulation. Nonslip and adiabatic conditions were adopted for the wall conditions.

4. Results and Discussion

4.1. Stall Characteristics of Tested Compressor. The experimental and numerical results of compressor performance are exhibited in Figure 4. The numerical total pressure-rise characteristics were obtained from DES unsteady analysis. The flow and total pressure-rise coefficients are defined as follows:

$$\phi = \frac{60Q}{\pi^2 D_2^2 B_2 N},$$
$$\Psi_t = \frac{60^2 \Delta p_t}{\rho \pi^2 D_2^2 N^2/2}. \quad (1)$$

The ODV experimental results were denoted by the black circles in the figure. The simulations were conducted at seven operating points from the design point (ϕ_{opt} = 0.24) to the deep stall point (ϕ = 0.10). The time-averaged results of the unsteady DES analysis were in good agreement with the measured results obtained using the ODV, with the exception of results at stall point (ϕ = 0.14 and 0.10). One reason for this is that the additional loss from the volute was not concluded in this numerical simulation. Particularly at deep stall point (ϕ = 0.10), the compressor performance data obtained numerically, using CFD, was higher in value than that obtained experimentally. The resulting question about the performance remarkably mismatch at ϕ = 0.10 is considered later in detail.

To investigate the unsteady characteristics of rotating stall, the impeller-discharge flow was measured with a split-film anemometer at ϕ = 0.14 and 0.10. The spatial distribution of the FFT (Fast Fourier Transformation) results in the spanwise direction is shown in Figure 5. The vertical axis represents the intensity of radial velocity fluctuation. The measurement of flow coefficient, ϕ = 0.14, shows that the large intensity of velocity fluctuations occurred at 25 Hz near the shroud side. We have already reported that these unsteady phenomena at 25 Hz, which were equivalent to 25% of the impeller rotational speed, were caused by the rotating stall within the diffuser passages [12]. Therefore, this disturbance, as illustrated in Figure 5(a), was considered to result from the diffuser stall fluctuations. When the mass flow decreased (ϕ = 0.10), the diffuser stall fluctuation was not found and a large intensity velocity fluctuation occurred at 22 Hz in Figure 5(b). From a previous study [23], we found that the large intensity fluctuations, called "stage stall," rotated within the impeller and diffuser passages. In comparison with the diffuser stall, the stage stall cell had a much larger intensity of velocity fluctuation over the entire span, especially near the hub side. In addition, from the pressure experiments, the number of diffuser and stage stall cell was found to be one. Thus, the rotating stall behavior of the tested compressor varied with the decreasing mass flow rate.

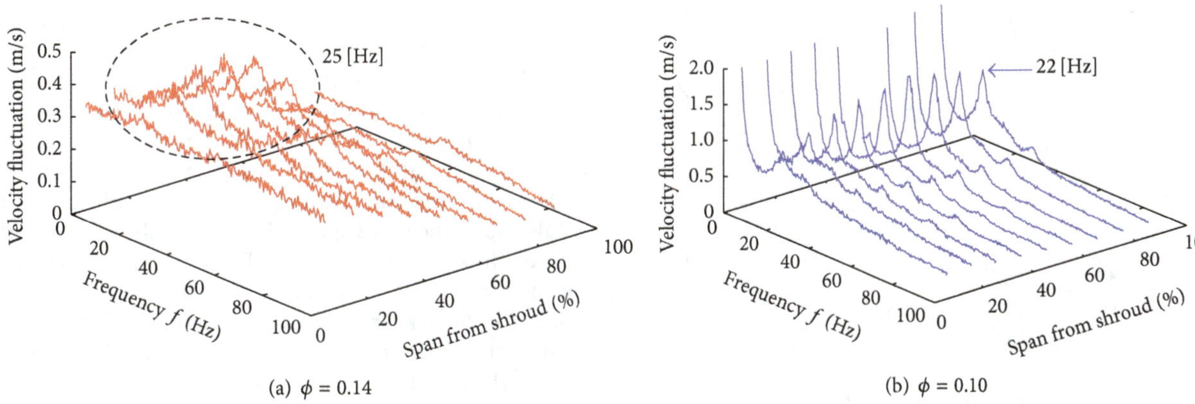

FIGURE 5: Distribution of velocity fluctuation in spanwise direction (ODV).

4.2. Structure of Diffuser Stall Cell. First, the rotational structure of the diffuser stall cell at flow coefficient $\phi = 0.14$ was investigated using numerical analysis. To investigate the unsteady characteristics of rotating stall, the casing wall static pressure fluctuation was measures at two positions (D.I.a and D.I.b). The experimental and numerical casing pressure traces are presented in Figure 6. The red lines represent the low-pass filtered traces and the low-pass filter frequencies were set to 30 Hz. From the both experimental and numerical traces, the disturbance that was propagated in vaneless space can be observed. The rotational speed of this disturbance was approximately 25% of impeller speed. The unsteady phenomena at 25 Hz, which were equivalent to 25% of impeller rotational speed, were caused by the diffuser rotating stall. In addition, the influence of impeller stall fluctuation could be excluded because the scale of disturbances caused by the impeller stall were much smaller than that of the diffuser stall at diffuser inlet from Figure 5. Therefore, the numerical results for diffuser stall were in good agreement with the results obtained by the experiments.

Furthermore, the instantaneous casing pressure fluctuations in the vaneless space at nondimensional times t^* = 0.0–31.5 are illustrated in Figure 7. The vertical axis is each diffuser vane, which was labelled with numbers from one to fifteen. The low casing static pressure region, which is represented by a black dashed circle, propagated to the next diffuser passage as time passed. The rotational speed of this region was approximately 25% of the impeller rotational speed. Thus, it may be concluded that this low static pressure region was caused by the diffuser rotating stall cell. Furthermore, the radial velocity fluctuation measured at the diffuser inlet is illustrated in Figure 8. The red and blue lines represent the diffuser passages numbers 1-2 and 7-8, respectively. The red dashed line, as illustrated in Figure 8, indicates the average of radial velocities calculated for all diffuser passages. As shown in this figure, the radial velocity of the diffuser passage numbers 7-8 was lower than the average radial velocity, because the diffuser rotating stall cell was approaching the diffuser passage numbers 7-8 from the nondimensional time, t^*_a = 0.0, as illustrated in Figure 7. At the nondimensional time, t^*_c = 4.5, the radial velocity of

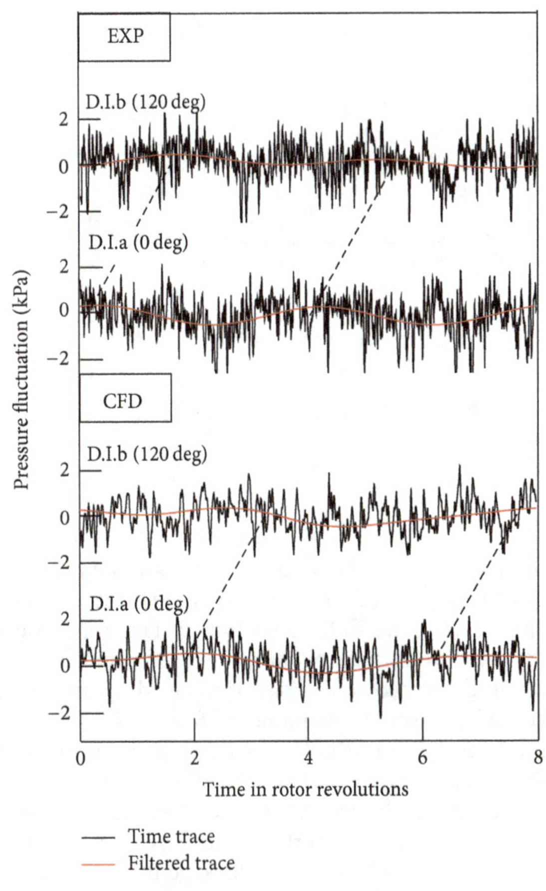

FIGURE 6: Casing wall pressure traces measured at diffuser inlet circumferential direction (D.I.a and D.I.b).

the diffuser passage numbers 7-8 was significantly decreased. Therefore, these results suggested that the flow within the diffuser passage was stalled by the blockage effect of the diffuser stall cell formation.

Then, the instantaneous vortical structure of the low static pressure region at the nondimensional time, t^*_c = 4.5, is illustrated in Figure 9. The vortex structure within several stalled diffuser passages numbers 7–11, as illustrated

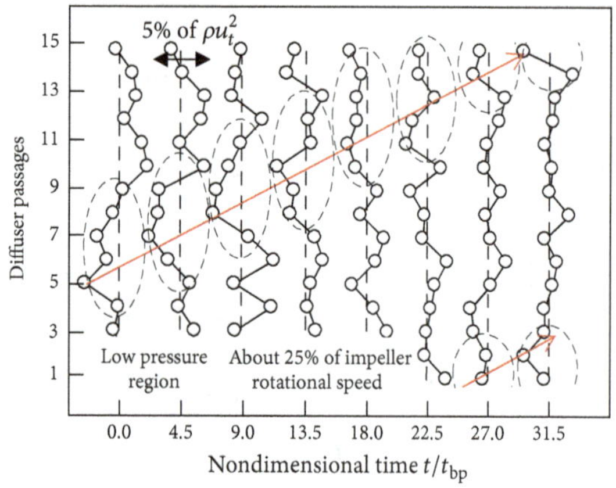

FIGURE 7: Casing static pressure fluctuations at diffuser inlet from nondimensional time $t^* = 0.0$ to $t^* = 31.5$ (CFD).

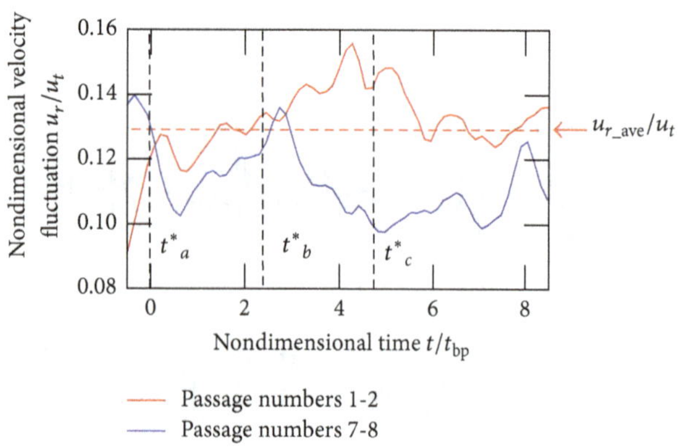

FIGURE 8: Radial velocity fluctuation measured at diffuser inlet (CFD).

in Figure 7, is visualized by Q-definition [24] and colored by nondimensional helicity. In the flow fields at the nondimensional time, $t^*_c = 4.5$, the tornado-type vortex connecting the shroud wall and diffuser suction surface was observed on the suction surface of diffuser vanes numbers 7 and 9. Longitudinal vortex was also developed on the suction surface of diffuser vanes numbers 8 and 10. This vortex is called the LEV. The LEV was developed and it formed a substantial blockage within the diffuser passages at off-design points. Furthermore, the other vortical structure was found near the pressure surface of diffuser vanes numbers 8 and 10. This vortex formed a blockage in the throat area, which induced a strong reversal flow near the shroud side. Therefore, it was determined that the developed vortical structure was formed within the stall cell.

To understand the unsteady behavior of the diffuser stall cell, the instantaneous vortical structure within diffuser passages numbers 7-8 at nondimensional times t^*_a, t^*_b, and t^*_c was visualized. The vortex core colored by nondimensional helicity within diffuser passages numbers 7-8 is illustrated in Figure 10. First, the tornado-type vortex on the suction surface of the diffuser vane was generated at the nondimensional time $t^*_a = 0.0$. The tornado-type separation was caused by the positive incidence angle to the diffuser vane near the shroud side, which had resulted from the stall cell approaching diffuser passages numbers 7-8. Then, the tornado-type vortex was developed and the LEV was observed at the shroud/suction surface corner at the nondimensional time $t^*_b = 2.1$. The developed LEV was extended to the next diffuser vane (number 8) because of the high adverse pressure gradient within the diffuser passage. Additionally, the LEV leg was observed on the suction surface of the diffuser vane number 7. The LEV was caused by the secondary flow effect on the suction surface of the diffuser vane. The flow near the suction surface rolled up to the shroud side because of the vertical pressure gradient and formed the longitudinal vortex at the shroud/suction surface corner. At the nondimensional time, $t^*_c = 4.5$, the LEV interacted with the next diffuser leading edge and formed the throat area blockage, which induced the strong reversal flow near the shroud side. Accordingly, the separation vortex was considered as having been generated on the suction surface of diffuser vane number 8 with the passage of time, because the incidence angle to diffuser vane number 8 was highly positive. The newly formed tornado-type vortex and LEV were also observed near the leading edge of diffuser vane number 7.

For an improved understanding of the generation mechanism of the vortices near the diffuser vane, the time-averaged distribution and isolines of the static pressure on the diffuser suction surface and casing wall are shown in Figure 11. The vectors on the suction surface and casing wall represent the pressure gradient vectors. Furthermore, the limiting streamlines on the diffuser suction surface are illustrated in Figure 12. Near the diffuser leading edge, the strength of vertical pressure gradient on the suction surface gradually increased from the hub to midspan because the tornado-type vortex resulted in the low-pressure region near the shroud side. From the limiting streamlines of the suction surface, the focus point of the tornado-type vortex leg was to be observed at the diffuser leading edge. In addition to this, the suction surface streamlines rolled up from the hub side to the shroud side because of the high vertical pressure gradient near the diffuser leading edge. As illustrated in Figure 11(b), the direction of circumferential pressure gradient vector on the shroud wall was strongly adverse near the suction surface of the diffuser leading edge, which was the origin of the tornado-type separation vortex. The pressure gradient vectors in the throat area were directed to the impeller exit and leading edge of the succeeding diffuser vane. This is because of the static pressure recovery within the diffuser passage.

To summarize, the structure of the diffuser stall is illustrated in Figure 13. Initially, the tornado-type vortex connecting the shroud wall and diffuser suction surface was generated on the suction surface of the diffuser vane. The LEV developed with the growth of the tornado-type vortex and it approached the succeeding diffuser vane because of the high adverse pressure gradient near the shroud side. Then, the developed LEV interacted with the succeeding diffuser leading edge and formed the throat area blockage with the passage of time. The blockage in the throat area induced the radial reversal flow of high magnitude near the shroud side, and the next separation vortex was formed because of the back flow. Finally, the throat area blockage

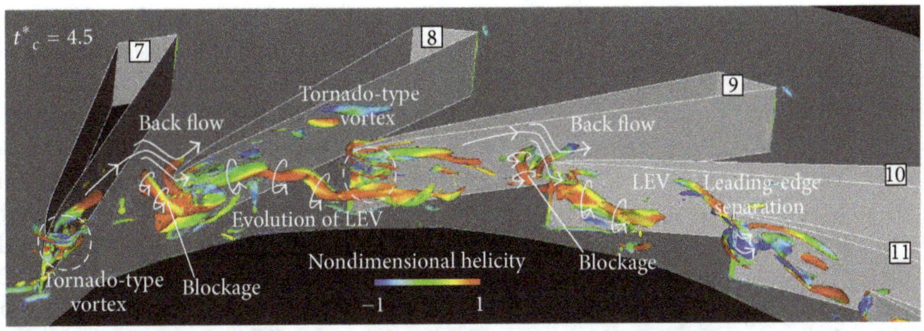

FIGURE 9: Instantaneous vortical structure of diffuser stall within diffuser passages number 7 to number 11.

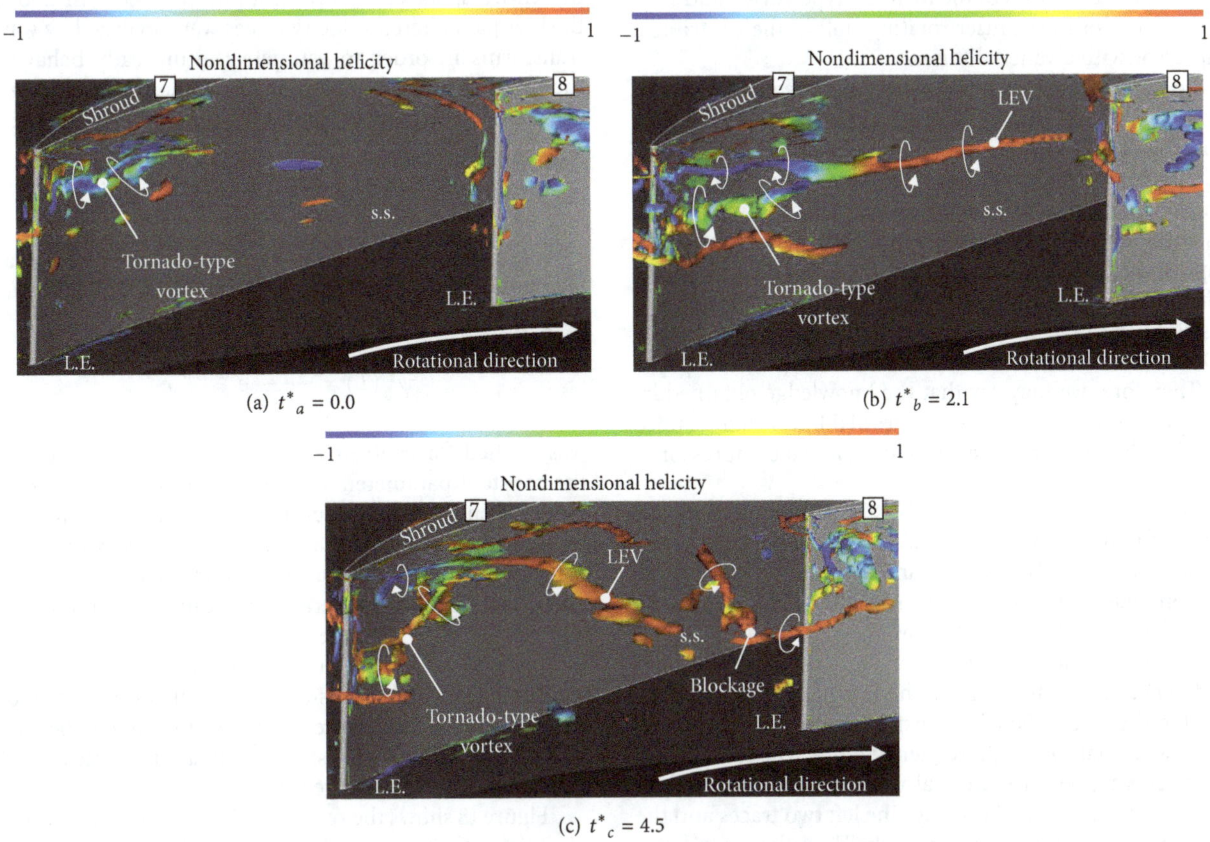

(a) $t^*_a = 0.0$

(b) $t^*_b = 2.1$

(c) $t^*_c = 4.5$

FIGURE 10: Instantaneous vortical structure within diffuser passage numbers 7-8 at nondimensional time t^*_a, t^*_b, and t^*_c.

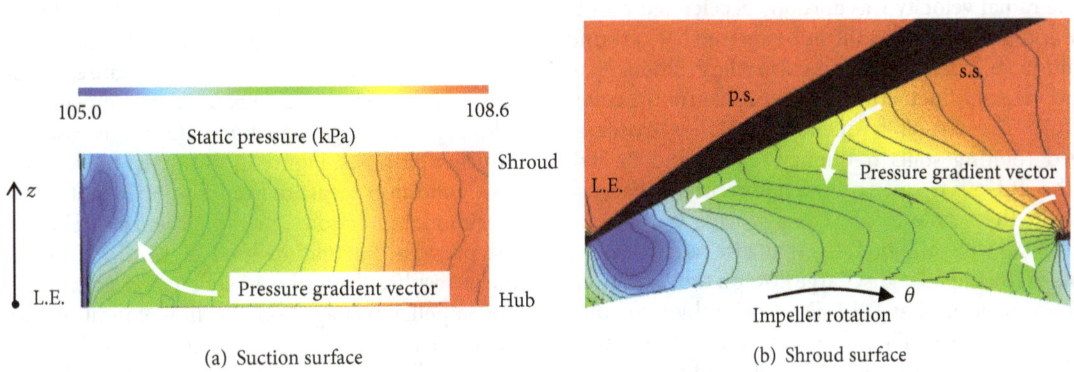

(a) Suction surface

(b) Shroud surface

FIGURE 11: Distribution of static pressure on diffuser suction surface and casing wall.

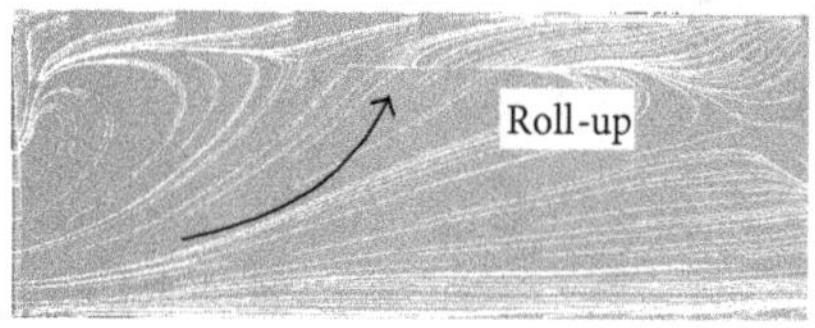

FIGURE 12: Limiting streamlines on diffuser suction surface.

resulted in the tornado-type vortex and LEV. The diffuser stall cell, which was formed by the systematic vortical structure and which consisted of the tornado-type vortex, LEV, and throat blockage, propagated to the succeeding diffuser vane. Therefore, the evolution of the tornado-type vortex and LEV was the cause of the diffuser rotating stall in the centrifugal compressor with a vaned diffuser.

The tested compressor is the low-pressure ratio centrifugal compressor; however the results of vortical structure of diffuser stall cell can be generalized not only low-pressure fluid machinery but also high pressure compressors. For example, Everitt et al. [10] reported the similar unsteady vortices at the diffuser leading edge in a high speed centrifugal compressor, as with the case of our paper. Furthermore, we confirmed that the LEV had the same form at high speed and the size of the LEV was increased at low mass flow late by CFD analysis, as we have mentioned in our previous paper [12]. Therefore, we may say that the knowledge obtained by this paper about the vortical structure of diffuser stall was also applied to the high pressure ration centrifugal compressor.

4.3. Transition to the Stage Stall. Next, the generation mechanism of the stage stall cell was investigated by experimental analysis at $\phi = 0.10$. The velocity and pressure measurements were conducted to reveal the detailed flow fields, when the stage stall occurred. Figure 14 shows the measurement results of radial velocity and casing wall pressure traces (30 Hz low-pass filter) at the diffuser inlet. The low-pass filter was conducted by the inverse Fourier transformation to the measured spectra which cut the high frequency range. The top left and right figures represent the radial velocity traces on the hub side and shroud side, respectively. The left two traces and the right one are measured simultaneously. These figures indicate that the radial reversed flow basically occurred on the shroud side and that the intensity of pressure fluctuation was small in the vaneless space (blue background color: condition (a)). However, the radial velocity was abruptly accelerated on the shroud side and decelerated on the hub side (red background color: condition (b)). Then, there was also a large intensity of pressure and velocity fluctuations (green background color: condition (c)). The cut-off frequency (30 Hz) is closer to the observed rotating stalls frequency (25 Hz and 22 Hz). However, the cut-off frequency was appropriate, because we focused on the whole intermittent process (conditions (a)–(c)) in this study. In addition, the spanwise radial velocity distributions of the impeller-discharge flow on the conditions (a) and (b) are measured at D.I.1–D.I.4 using by two split-film probes, as shown in Figure 15. One probe was set on the hub side in order to capture the stall conditions (a) to (c). Another velocity probe was set to measure the velocity distribution on each condition. On the flow condition (a), the flow on the shroud side was reversed at all measuring points. Furthermore, the reversed flow at D.I.1–D.I.4 moved from the shroud side to the hub side on condition (b). Thus, it was assumed that the flow condition in vanless space at $\phi = 0.10$ was divided into three behaviors and switched unsteadily.

The power spectra of the velocity fluctuations on each condition in the vaneless space are shown in Figure 16. The velocity fluctuations (measured at 25 Hz) caused by the diffuser stall were found near the shroud side on condition (a). Then, the diffuser rotating fluctuations transferred to the hub side on condition (b). Finally, the large intensity of the velocity fluctuation caused by the stage stall was found over the entire span, especially near the hub side. In addition, the low-pass filtered velocity traces were analyzed by wavelet transforms in order to estimate the unsteady behavior of the rotating stalls appearing in the compressor. The wavelet transform of a velocity signal, $u(t)$, is defined as follows:

$$W(a,b) = \frac{1}{\sqrt{a}} \int \psi^{*}(T)\, u(t)\, dt, \tag{2}$$

where t is the time, $\psi^{*}(T)$ is a mother wavelet, a is a scaling factor, b is a time transformation factor, and $T = (t-b)/a$. In this work, the Gabor mother wavelet of

$$\psi^{*}(T) = \frac{1}{\sqrt{2\pi a}} e^{-T^2/2\sigma^2} e^{-iT} \tag{3}$$

was applied for capturing rotating stall disturbances. Here, the constant parameter, σ, is defined as $\sigma = 32$ in this work. Figure 17 shows the contour of the wavelet transform $W(a,b)$ for the low-pass filtered radial velocity traces on the hub side. The vertical axis indicated the frequency, f. When the radial velocity was decelerated, the diffuser stall fluctuation was appeared. Then, the unsteady disturbances at 25 Hz disappeared, and the stage stall fluctuation was found on condition (c). Finally, the stage stall fluctuation was not found on condition (a). Therefore, it was considered that the diffuser stall which rotated on the shroud side suddenly moved to hub side and evolved into stage stall.

Figure 18 shows the results of simultaneous measurement of radial velocity in vaneless space and casing wall pressure traces (30 Hz low-ass filter) within impeller passages. The top figure is the radial velocity trace on the hub side. The bottom three figures represent the casing pressure traces in the meridional direction, I.I., I.M., and D.I.a. From the results at I.I. and I.M., the scale of pressure fluctuation in condition (b) was larger than condition (a), because the diffuser stall was considered to be expanded to impeller passages. In addition, the stage stall fluctuation was also found in the impeller passages. Figure 19 shows the power spectra of each flow condition measured on the impeller shroud wall in the meridional direction. The red, green, and blue lines indicate the pressure fluctuation level of I.I., I.M., and D.I.a. From these results, the diffuser stall cell was not found within the impeller passages on the flow condition (a). However, the pressure fluctuation of the diffuser stall occurred within the impeller passages on condition (b). Thus, the diffuser stall was considered to expand into the impeller passages

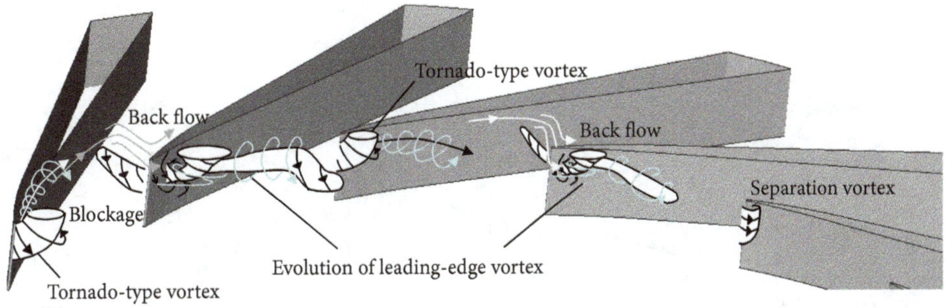

FIGURE 13: Illustration of rotating mechanism of diffuser stall cell.

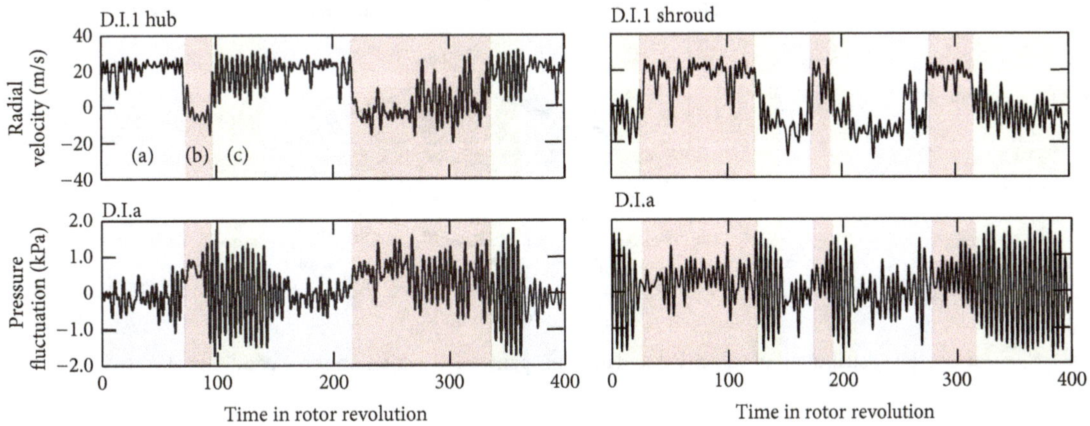

FIGURE 14: Radial velocity and pressure fluctuation traces at diffuser inlet ($\phi = 0.10$, 30 Hz low-pass filter).

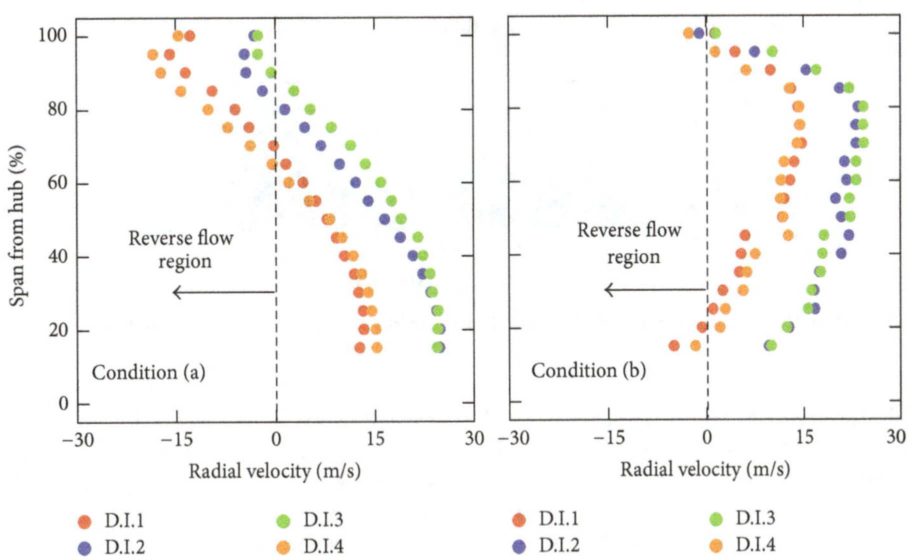

FIGURE 15: Spanwise radial velocity distributions of the impeller-discharge flow on conditions (a) and (b).

before the stage stall occurred. Finally, the stage stall cell was formed by the expansion of the diffuser stall on condition (c). Furthermore, the pressure fluctuations measured at 55 Hz were found from the power spectra of I.I. on conditions (a) and (b). Based on the previous report [12, 13], this 55 Hz pressure fluctuation was caused by the impeller rotating stall.

In the axial compressor, it was reported that the stall cells, which consisted of multiple vortices, merged into a single vortex cell. The single vortex cell blocked the flow in the rotor and stator passages as the mass flow rate decreased [14]. The suggested stall cell structure at the early stage in the axial compressors was similar to the diffuser stall cell observed

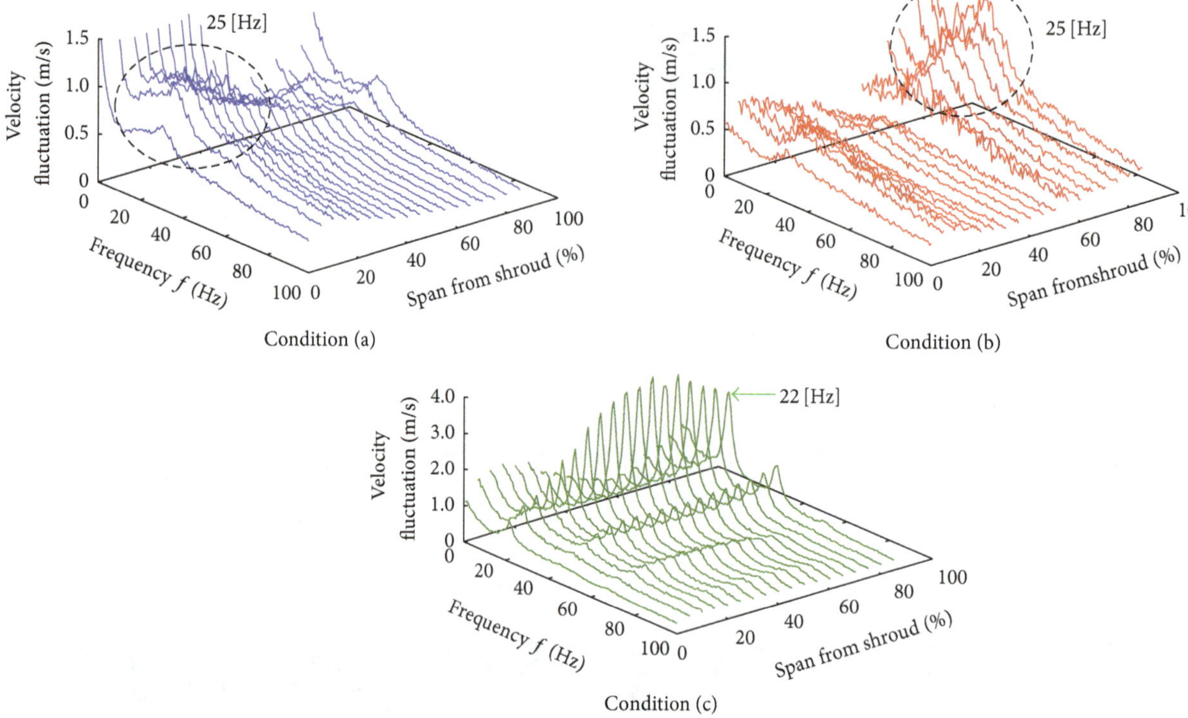

FIGURE 16: Distribution of velocity fluctuation in spanwise direction on three conditions (a)~(c).

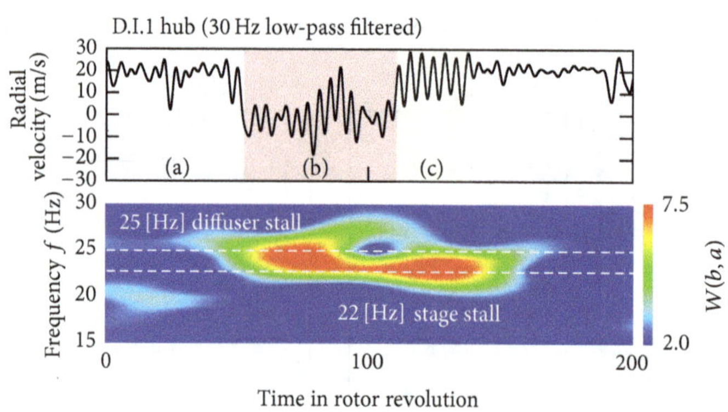

FIGURE 17: Contour map of the wavelet transform $W(a, b)$ for the low-pass filtered radial velocity traces on the hub side.

above. The CFD analysis was able to capture the impeller and diffuser stall cell; however the development from the diffuser stall to stage stall was not found at the flow coefficient $\phi = 0.10$. Therefore, the performance mismatch between experimental and CFD results occurred at $\phi = 0.10$ because the stage stall deteriorated the compressor performance. The detailed experimental and CFD analysis will be conducted in future studies to examine the relationship between the impeller and diffuser/stage stall cell.

In summary, the stage stall, which rotated within both impeller and diffuser passages, occurred instead of the diffuser stall as the mass flow rate was decreased ($\phi = 0.10$). The diffuser stall, which rotated on the shroud side, was suddenly shifted to the hub side. Then, the diffuser stall moved into the impeller passages and formed the stage stall. Therefore, the stage stall was caused by the development of the diffuser stall, which transferred from the shroud side to the hub side in the vaneless space and expanded to the impeller passages. In fact, the change in vertical structure from shifting the diffuser stall position was the key to understand the inception mechanism of stage stall. In future work, we will focus on the unsteady structure of diffuser stall before the occurrence of the stage stall by CFD analysis. In addition, we will reveal the clues of the stage stall getting back process to diffuser stall (conditions (c) to (a)).

5. Conclusions

The vortical structure of diffuser stall and the transition from diffuser stall to stage stall in a centrifugal compressor with

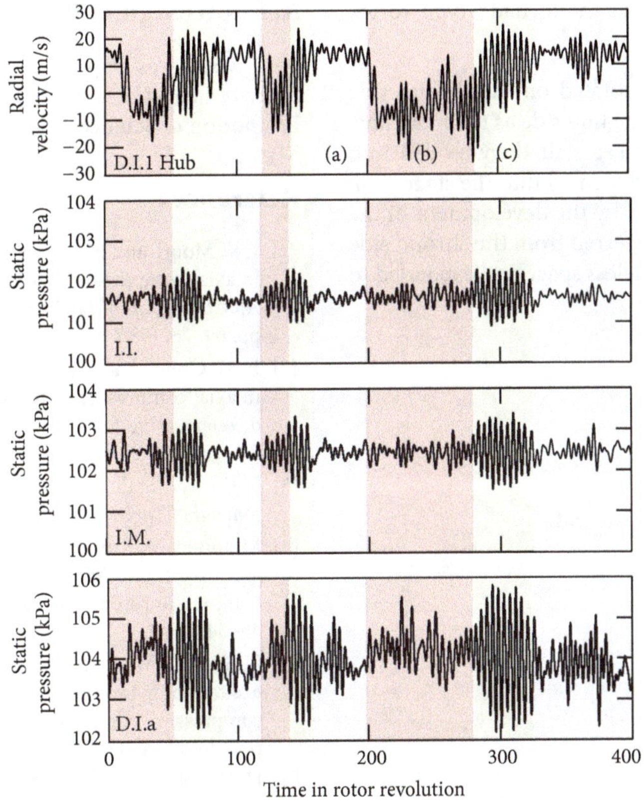

Figure 18: Radial velocity at diffuser inlet and static pressure traces at meridional direction ($\phi = 0.10$, 30 Hz low-pass filter).

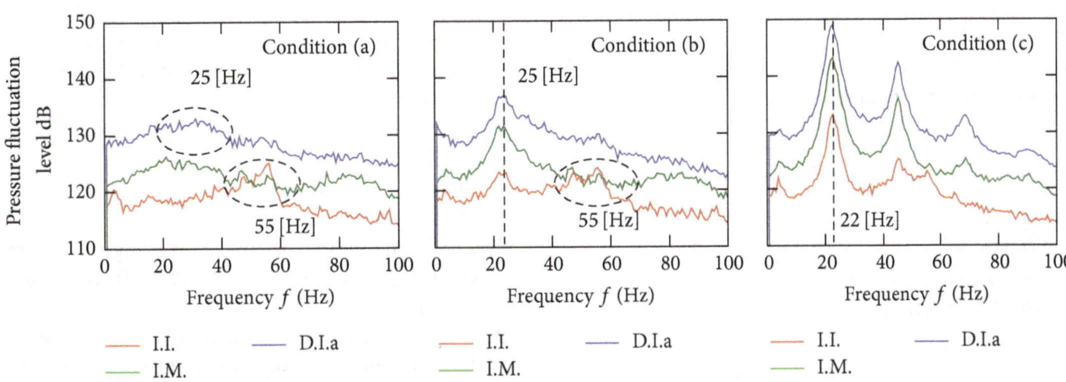

Figure 19: Power spectra of pressure fluctuations in impeller passage ($\phi = 0.10$).

a vaned diffuser were investigated by experimentation and through CFD analysis.

The results can be summarized as follows:

(i) In the tested compressor, the diffuser stall rotated at 25% of the impeller rotational speed during off-design flow operation. Both the experimental and numerical results revealed the rotation of the diffuser stall cell in the vaneless space on the shroud side of the flow passage. In addition, large fluctuations occurred at 22 Hz, and were caused by stage stall when the mass flow rate decreased. The stage stall expanded over the entire span in the vaneless space.

(ii) The vortical structure of the diffuser stall cell consisted of the tornado-type vortex, LEV, and throat blockage. The stall cell's typical vortical structure propagated to the succeeding diffuser passages.

(iii) The tornado-type separation was caused by the positive incidence angle to the diffuser vane near the shroud side. The LEV developed with the growth of the tornado-type vortex and extended to the succeeding diffuser vane. The developed LEV subsequently interacted with the succeeding diffuser leading edge and formed the throat area blockage. Finally, the tornado-type vortex and LEV were developed by the throat area blockage. The diffuser rotating stall was caused by the evolution of the tornado-type vortex and LEV.

(iv) The diffuser stall which rotated on the shroud side was suddenly shifted to the hub side as the mass flow rate decreased. The diffuser stall then moved into the impeller passages and formed into the stage stall. The stage stall was caused by the development of the diffuser stall, which transferred from the shroud side to the hub side in the vaneless space and expanded to the impeller passages.

Nomenclature

B: Diffuser passage height (m)
D: Diameter (m)
G: Mass flow rate (kg/s)
L_{id}: Distance between impeller exit and diffuser leading edge (m)
N: Rotational speed (min^{-1})
Q: Volume flow rate (m^3/s)
W: Wavelet coefficient
V: Number of diffuser vanes
Z: Number of impeller blades
f: Frequency (Hz)
t: Time (s)
t_{bp}: Blade passing time (s)
t^*: Nondimensional time ($= t/t_{\text{bp}}$)
p: Static pressure (Pa)
p_t: Total pressure (Pa)
u: Velocity (m/s)
u_r: Radial velocity (m/s)
u_t: Circumferential velocity at impeller exit (m/s).

Greek Letters

ρ: Air density (kg/m^3)
ϕ: Flow coefficient
ψ_t: Total pressure-rise coefficient
ψ^*: Mother wavelet function.

Subscripts

1: Impeller inlet
2: Impeller outlet
3: Diffuser leading edge
4: Diffuser trailing edge.

Abbreviations

D.E.: Diffuser exit
D.I.: Diffuser inlet
I.I.: Impeller inlet
I.M.: Impeller midsection
LEV: Leading edge vortex
ODV: Original wedge-type diffuser vane.

Conflicts of Interest

The authors declare that they have no conflicts of interest.

Acknowledgments

This work was supported by Grant-in-Aid for Research Activity Start-Up no. 16H07293 from Japan Society for the Promotion of Science.

References

[1] F. K. Moore and E. M. Greitzer, "A theory of post-stall transient in axial compressor systems: part 1 development of equations," *ASME Journal of Engineering for Gas Turbines Power*, vol. 108, pp. 69–76, 1986.

[2] E. M. Greitzer and F. K. Moore, "A theory of post-stall transient in axial compressor systems: part 2 application," *ASME Journal of Engineering for Gas Turbines and Power*, vol. 108, no. 2, pp. 231–239, 1986.

[3] I. J. Day, "Stall inception in axial flow compressors," *ASME Journal of Turbomachinery*, vol. 115, pp. 1-9, 1993.

[4] M. Inoue, M. Kuroumaru, T. Tanino, and M. Furukawa, "Propagation of multiple short-length-scale stall cells in an axial compressor rotor," *Journal of Turbomachinery*, vol. 122, no. 1, pp. 45–54, 2000.

[5] U. Haupt, M. Rautenberg, and A. N. Abdel-Hamid, "Blade excitation by broad-band pressure fluctuations in a centrifugal compressor," *ASME Journal of Turbomachinery*, vol. 110, no. 1, pp. 129–137, 1988.

[6] U. Haupt, U. Seidel, A. N. Abdel-Hamid, and M. Rautenberg, "Unsteady flow in a centrifugal compressor with different types of vaned diffusers," *ASME Journal of Turbomachinery*, vol. 110, no. 3, pp. 293–302, 1988.

[7] I. Tomita, S. Ibaraki, M. Furukawa, and K. Yamada, "The effect of tip leakage vortex for operating range enhancement of centrifugal compressor," *Journal of Turbomachinery*, vol. 135, no. 3, Article ID 051020, 2013.

[8] Y. Fukuda, Y. Takeyama, and Y. Ohta, "Characteristics of rotating instability in a centrifugal blower with shrouded impeller," *Transactions of the JSME*, vol. 80, no. 809, pp. 1–11, 2014 (Japanese).

[9] Z. S. Spakovszky, "Backward traveling rotating stall waves in centrifugal compressors," *Journal of Turbomachinery*, vol. 126, no. 1, pp. 1–12, 2004.

[10] J. N. Everitt and Z. S. Spakovszky, "An Investigation of Stall Inception in Centrifugal Compressor Vaned Diffuser," *Journal of Turbomachinery*, vol. 135, no. 1, Article ID 011025, 2012.

[11] Y. Bousquet, N. Binder, G. Dufour, X. Carbonneau, M. Roumeas, and I. Trebinjac, "Numerical simulation of stall inception mechanisms in a centrifugal compressor with vaned diffuser," *Journal of Turbomachinery*, vol. 138, no. 12, Article ID 121005, 2016.

[12] N. Fujisawa, S. Hara, Y. Ohta, and T. Goto, "Unsteady behavior of leading edge vortex and diffuser stall inception in a centrifugal compressor with vaned diffuser," in *Proceedings of the ASME FEDSM 2014*, 2014.

[13] N. Fujisawa, S. Ikezu, and Y. Ohta, "Structure of Diffuser Stall and Unsteady Vortices in a Centrifugal Compressor," in *Proceedings of the ASME Turbo Expo 2016*, 2016.

[14] E. Outa, Y. Ohta, D. Kato, and K. Chiba, "Two-dimensional study on evolution of deep rotating stall under uniform inlet conditions in an axial compressor cascades," in *Proceedings of the 14th International Society for Air Breathing Engines*, pp. 1–11, 1999.

[15] Y. Yoshida, H. Tsurusaki, Y. Murakami, and Y. Tsujimoto, "otating stalls in centrifugal impeller/vaned diffuser systems (1st Report)," *Transactions of the JSME*, vol. 56, no. 530, pp. 2991–2998, 1990 (Japanese).

[16] S. Mizuki and Y. Oosawa, "Unsteady flow within centrifugal compressor channels under rotating stall and surge," *Journal of Turbomachinery*, vol. 114, no. 2, pp. 312–320, 1992.

[17] X. Zheng and A. Liu, "Phenomenon and mechanism of two-regime-surge in a centrifugal compressor," *Journal of Turbomachinery*, vol. 137, no. 8, Article ID 081007, 2015.

[18] J. L. Steger and R. F. Warming, "Flux vector splitting of the inviscid gasdynamic equations with application to finite-difference methods," *Journal of Computational Physics*, vol. 40, no. 2, pp. 263–293, 1981.

[19] B. van Leer, "Towards the ultimate conservative difference scheme. V. A second-order sequel to Godunov's method," *Journal of Computational Physics*, vol. 32, no. 1, pp. 101–136, 1979.

[20] E. Shima, "A simple implicit scheme for structured/unstructured cfd," in *Proceedings of 29th Fluid Dynamic Conference*, pp. 325–328, 1997.

[21] P. R. Spalart, M.-H. Jou, M. Strelets, and S. R. Allmaras, "Comments on the feasibility of les for wings and on the hybrid rans/les approach, advances in DNS/LES," in *Proceedings of the First AFOSR International Conference on DNS/LES*, 1997.

[22] M. Strelets, "Detached Eddy simulation of massively separated flows," *AIAA Paper*, Article ID 2001-0879, 2001.

[23] S. Ikezu, N. Fujisawa, and Y. Ohta, "Characteristics of a Centrifugal Compressor with Vaned Diffuser," in *Proceedings of the 6th Asian Joint Workshop on Thermophysics and Fluid Science*, 2016.

[24] J. Jeong and F. Hussain, "On the identification of a vortex," *Journal of Fluid Mechanics*, vol. 285, pp. 69–94, 1995.

3

Theoretical Model for the Performance of Liquid Ring Pump based on the Actual Operating Cycle

Si Huang, Jing He, Xueqian Wang, and Guangqi Qiu

School of Mechanical and Automotive Engineering, South China University of Technology, Guangzhou 510640, China

Correspondence should be addressed to Si Huang; huangsi@scut.edu.cn

Academic Editor: Ryoichi Samuel Amano

Liquid ring pump is widely applied in many industry fields due to the advantages of isothermal compression process, simple structure, and liquid-sealing. Based on the actual operating cycle of "suction-compression-discharge-expansion," a universal theoretical model for performance of liquid ring pump was established in this study, to solve the problem that the theoretical models deviated from the actual performance in operating cycle. With the major geometric parameters and operating conditions of a liquid ring pump, the performance parameters such as the actual capacity for suction and discharge, shaft power, and global efficiency can be conveniently predicted by the proposed theoretical model, without the limitation of empiric range, performance data, or the detailed 3D geometry of pumps. The proposed theoretical model was verified by experimental performances of liquid ring pumps and could provide a feasible tool for the application of liquid ring pump.

1. Introduction

Liquid ring pump is a kind of rotary displacement machines, including liquid ring vacuum pump and liquid ring compressor. Owing to the advantages of isothermal compression process, simple structure, and liquid-sealing, liquid ring pump is widely applied in petroleum, chemical, power, metallurgy, pharmaceutical, and light industry and other fields, particularly suitable for the pumping of flammable, explosive, toxic, and harmful gases like hydrogen, oxygen, nitrogen, chlorine, ethylene, acetylene, propane, natural gas, and carbon dioxide. However, there are still some problems such as a higher power consumption, lower suction vacuum degree, and unstable operation in application of liquid ring pump. The theoretical result considerably deviates from the actual performance due to the lack of the fundamental study on the performance and restricts the improvement of performance and reliability of liquid ring pump. Therefore, it is of significant scientific and application values to carry out the fundamental study on the operating theory and performance of the liquid ring pump.

The operating principle of the liquid ring pump is illustrated in Figure 1. With the revolution of the impeller, the operating liquid is ejected outwards due to centrifugal forces and forms a liquid ring. Owing to the eccentricity of the impeller to the casing, a crescent-shaped chamber forms between the impeller hub and the liquid ring. The impeller blades divide the crescent-shaped chamber into several cells of different volumes. As the impeller runs, the cells in the range of the suction port become larger and suck the gas through the suction port. In the part opposite to the suction port, the gas is compressed in the reducing cells and emitted through the discharge port. Each revolution of the impeller corresponds to one operating cycle. Consequently, liquid ring pumps are the rotating machines with reciprocating pump characteristics and the ring liquid takes on the function of a reciprocating piston.

Since the first liquid ring pump was invented, many scholars have carried out a variety of work on the design and industrial application of liquid ring pumps [1–6]. Compared with the work on the other types of pumps and compressors, however, there were few reports of the fundamental studies on liquid ring pumps. Referring to ideal operating cycle theory of "suction-compression-discharge" in reciprocating compressor, Pfleiderer [7] derived an equation for theoretical suction capacity in liquid ring pump and developed a theoretical model for the compression ratio and the liquid ring

TABLE 1: Calculation models of gas volume flow and shaft power.

Model	q	N
Pfleiderer [7]	$q_{\text{th}} = \frac{1}{2}\mu\omega b r_2^2\left[(1-\alpha)^2 - \upsilon^2\right]$	
Prager [10]	$q_s = \frac{\lambda}{\lambda_{\max}} q_{s\max}$ For vacuum pumps: $\frac{\lambda}{\lambda_{\max}} = 0.147X^3 - 1.297X^2 + 0.150X + 1$ For compressors: $\frac{\lambda}{\lambda_{\max}} = 0.707X^3 - 1.527X^2 + 0.1782X + 1,$ $X = \frac{(p_d/p_s - 1)}{[(p_d/p_s)_{\max} - 1]}$	
Powle [11]	$q_d = \frac{q_{d\max} p_a \left(p^m - p_v^{m+1}/p\right)}{\left(p_a^{m+1} - p_v^{m+1}\right)}$	
Bodik [14]		$Au_2^3 + Bu_2$

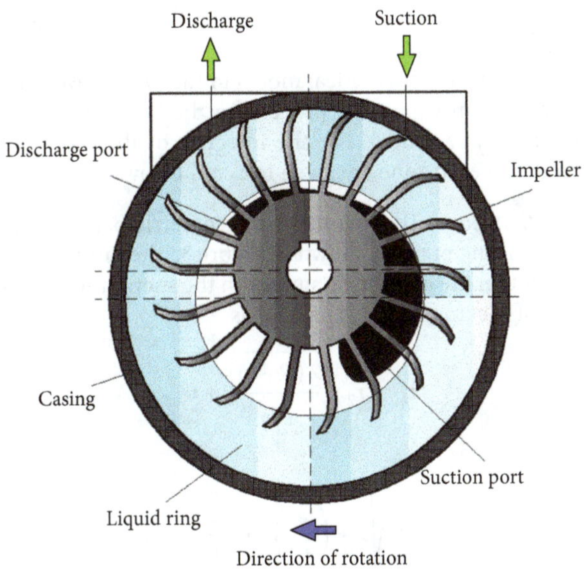

FIGURE 1: Operating principle of a liquid ring pump.

surface at each circumferential angle in the pump. Schulz [8] and Segebrecht [9], respectively, deduced similar equations for the theoretical suction capacity. So far, the theoretical model and the equations of theoretical suction capacity have been mainly used as reference for design of liquid ring pump. As shown in Table 1, the theoretical suction capacity, determined by the impeller rotational speed and the geometric parameters of liquid ring pump, remains constant regardless of the compression ratio of discharge pressure to suction pressure. This result obviously deviates from the actual situation, especially when the compression ratio is relatively high. To solve this problem, Prager [10] correlated an empirical formula for the ratio of the actual suction capacity to the maximum actual suction capacity, based on experimental performance data of liquid ring pump, but the validity of the statistical formula was restricted by the number of samples and the unknown maximum actual suction capacity. On the other hand, it was found from the actual applications that the minimum suction pressure of the liquid ring pump was close to the saturated vapor pressure of the operating liquid. Thus, Powle [11] developed an equation for the ratio of the actual discharge capacity to the maximum actual discharge capacity as a function of vapor pressure of the operating liquid and index of expansion process in the pump, but the maximum actual discharge capacity should be given in advance.

Concerning the power of liquid ring pump, due to the absorption and the removal of the heat by the operating liquid, the gas compression process in the liquid ring pump is generally regarded as isothermal one, so the formula for the isothermal compression power of the ideal gas is commonly used to calculate the effective power of the liquid ring pump [12, 13]. Bodik [14] deduced an approximate relation between the shaft power and rotational speed of liquid ring pump (Table 1). Besides the effective power, the shaft power includes the power consumed by volumetric losses, flow losses, and mechanical losses, but the coefficients in Bodik's power formula could not be determined. Recently, some investigators [15–18] used CFD numerical approach to simulate 3D flows and predicted performance of liquid ring pumps. Although numerical approach plays an important role for understanding internal flow field of liquid ring pump, it is still hard for practical application, due to the restriction of detailed 3D geometry of liquid ring pump, consumption of computation time, and the accuracy of gas-liquid two-phase flow model. In this paper, therefore, a convenient and universal theoretical model is proposed for the performance prediction of liquid ring pump based on the actual operating cycle, to obtain the rules of operating cycle performance and of power consumption in liquid ring pumps, without limitation of statistical range or the detailed 3D geometry.

2. Theoretical Model

2.1. Actual Operating Cycle of Liquid Ring Pump. The traditional theory [7] of liquid ring pump was based on the ideal operating cycle of "suction-compression-discharge,"

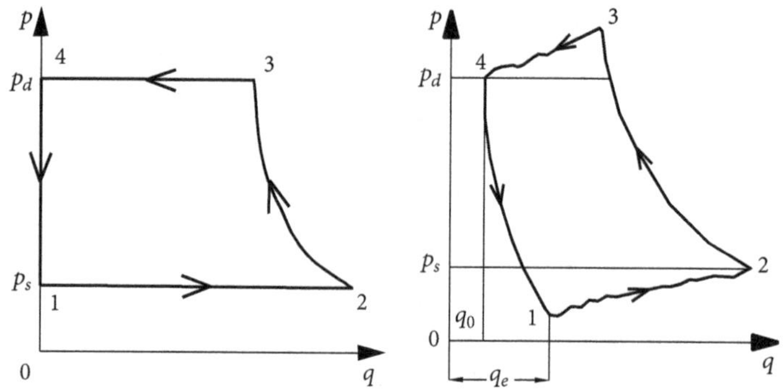

(a) Ideal (b) Actual

FIGURE 2: Diagram p-q of gas operating cycle in a liquid ring pump.

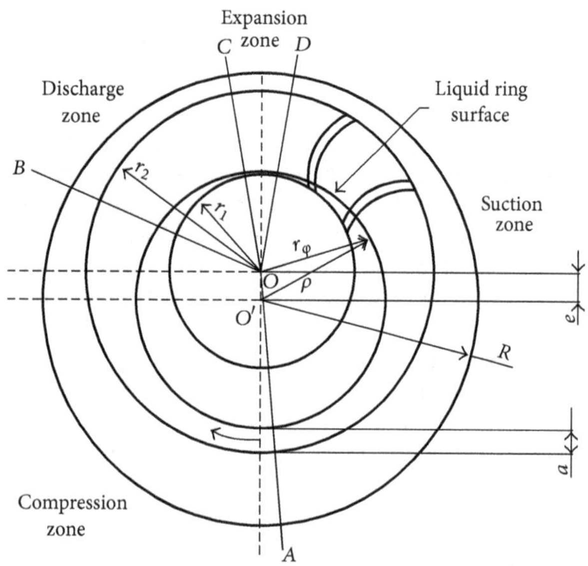

FIGURE 3: Geometric parameters in a liquid ring pump.

which was composed of three processes. During the suction process (Curves 1-2 shown in Figure 2(a)), the suction pressure p_s keeps constant and reaches the maximum suction capacity q_{th}. In the compression process (Curves 2-3), the gas volume reduces and the pressure increases to reach the discharge pressure p_d. In the discharge process (Curves 3-4), the discharge pressure p_d keeps constant and all the gas is discharged from the pump. The corresponding zones in a liquid ring pump are divided into the suction zone (Sector *DOA*), the compression zone (Sector *AOB*), and the discharge zone (Sector *BOC*) as shown in Figure 3. Like the actual operating principle of reciprocating compressor, however, some residual gas (capacity is q_o) is still available in the liquid ring pump after gas discharge and returns to the suction zone; the residual gas changes to q_e in capacity after a process of expansion (Curves 4-1 in Figure 2(b)) and goes to the next operating cycle. Thus, the actual operating cycle is composed of four processes: suction-compression-discharge-expansion, in which expansion process takes place at the expansion zone (Sector *COD*) shown in Figure 3. The actual suction capacity in a liquid ring pump is $q_s = q_{\text{th}} - q_e$ (if $p_s \le p_v$, $q_s = 0$). The residual gas and its expansion process were not taken into account in the theoretical model for ideal operating cycle of liquid ring pump, resulting in deviation between theoretical and practical results.

Therefore, the theoretical model in the present work is to supplement an expansion process on the basis of the ideal operating cycle model. For the integrity of the model, the formulas in each zone are given in detail below.

(1) Suction Zone. The definition of geometric parameters in the liquid ring pump is shown in Figure 3. The distance from any point on the liquid ring surface in the suction zone to the center of the impeller is expressed as

$$r_\varphi = r_2\sqrt{\frac{2(R-\rho)}{\mu r_2} + \upsilon^2}, \tag{1}$$

where

$$\rho = r_2\sqrt{1 + \left(\frac{e}{r_2}\right)^2 + 2\frac{e}{r_2}\cos\varphi}. \tag{2}$$

Theoretical suction capacity q_{th} is defined as the gas volume flow through the OA section assuming that the liquid ring pump inhales fully:

$$q_{\text{th}} = \frac{1}{2}\mu\omega b r_2^2\left[(1-\alpha)^2 - \upsilon^2\right]. \tag{3}$$

(2) Compression Zone. The distance from any point on the liquid ring surface in the compression zone to the center of the impeller is

$$r_\varphi = r_2\sqrt{\frac{(1-\alpha)^2 + (\sigma-1)\upsilon^2}{\sigma}}. \tag{4}$$

The compression ratio σ of the pressure at the circumferential angle φ to the suction pressure needs to solve an algebraic equation:

$$\frac{1}{\sigma^2} = A^2\left(1 - \frac{\sigma-1}{\varepsilon}\right), \tag{5}$$

where $A = 2(R-\rho)/\mu r_2[(1-\alpha)^2 - \upsilon^2]$; $\varepsilon = g\omega^2 r_2^2/2p_s$.

(3) Discharge Zone. The distance from any point on the liquid ring surface in the discharge zone to the center of the impeller is expressed as

$$r_\varphi = r_2\sqrt{v^2 + \frac{2(R-\rho)}{\mu r_2}\sqrt{\frac{1}{3}\left(1+\frac{1}{\varepsilon}\right)}}. \quad (6)$$

(4) Expansion Zone. The distance from any point on the liquid ring surface in the expansion zone to the center of the impeller is also calculated by (6). Therefore, the full shape of liquid ring surface in the liquid ring pump can be obtained according to (1), (4), and (6).

The capacity of gas returning to the suction zone of liquid ring pump is calculated by

$$q_0 = \mu\int_{r_1}^{r_{\varphi=2\pi}} \omega r b\, dr = \frac{1}{2}\mu\omega b\left(r_{\varphi=2\pi}^2 - r_1^2\right), \quad (7)$$

where $r_{\varphi=2\pi}$ is the distance from the liquid ring surface at the circumferential angle $\varphi = 2\pi$ to the center of the impeller. The gas capacity q_0 changes to q_e after a polytropic expansion process and then enters the next operating cycle:

$$q_e = q_0\left(\frac{p_d}{p_s}\right)^{1/m}. \quad (8)$$

Therefore, the actual suction capacity of liquid ring pump is expressed as

$$q_s = \begin{cases} 0 & p_s \le p_v \\ \frac{1}{2}\mu\omega b r_2^2\left[(1-\alpha)^2 - v^2\right] - q_0\left(\frac{p_d}{p_s}\right)^{1/m} & p_s > p_v. \end{cases} \quad (9)$$

Then the actual discharge capacity q_d can be calculated according to the isothermal compression process:

$$q_d = \frac{q_s p_s}{p_d}. \quad (10)$$

2.2. Shaft Power Model of Liquid Ring Pump. The effective power of the liquid ring pump is generally calculated according to the isothermal compression power. However, accurate prediction of shaft power N is essential for estimating global efficiency and studying how to reduce power consumption in a liquid ring pump. The shaft power is mainly composed of two parts: the total power consumption N_g of the gas and the friction loss N_l of the operating liquid. The total power consumption of the gas is the adiabatic compression work:

$$N_g = \frac{\kappa}{\kappa-1}p_s q_{\text{th}}\left[\left(\frac{p_d}{p_s}\right)^{(\kappa-1)/\kappa} - 1\right]. \quad (11)$$

The empirical formula of the liquid friction loss N_l in liquid ring pump for the turbulent flow was given by Prager [19]:

$$N_s = 0.708\frac{\rho}{2}\omega^3 r_2^5 \text{Re}^{-0.1732}, \quad (12)$$

$$\text{Re} = \frac{\omega r_2^2}{\nu}. \quad (13)$$

It is found that axial width b of the impeller is usually big in liquid ring pumps but not yet considered in Prager's equation (12). By referring to the formula of disk friction power in centrifugal pump [20], the equation (12) of liquid friction loss N_l can be corrected as

$$N_l = 0.708\frac{\rho}{2}\omega^3 r_2^5 \text{Re}^{-0.1732}\left(1 + f\frac{b}{r_2}\right). \quad (14)$$

Therefore, the shaft power of liquid ring pump is written as

$$\begin{aligned} N &= N_g + N_l \\ &= \frac{\kappa}{\kappa-1}p_s q_{\text{th}}\left[\left(\frac{p_d}{p_s}\right)^{(\kappa-1)/\kappa} - 1\right] \\ &\quad + 0.708\frac{\rho}{2}\omega^3 r_2^5 \text{Re}^{-0.1732}\left(1 + f\frac{b}{r_2}\right). \end{aligned} \quad (15)$$

The global efficiency η is finally estimated by the following expression [13]:

$$\eta = \frac{p_s q_s \ln(p_d/p_s)}{N}. \quad (16)$$

3. Performance Tests of Liquid Ring Pumps

To validate the feasibility of the proposed theoretical models on performance of liquid ring pumps, the performance tests of the liquid ring pumps were performed. Three types of single-stage double-acting liquid ring vacuum pumps (2BE1103, 2BE1253, and 2BE1353) were selected for performance tests. The design and operating parameters of the liquid ring vacuum pumps are listed in Table 2. Water and air were the operating liquid and gas in the tests, and room temperature was maintained at 28°C. The tests were performed under different rotational speed of the pumps by frequency control. A multiparameter data acquisition and processing system were used in the performance tests and the detailed measurement method was introduced in [21]. The experimental facility and the layout of measurement device are shown in Figure 4.

4. Comparisons of Performance between Test and Theoretical Models

4.1. Suction Capacity of Liquid Ring Pump. Figure 5 shows the comparisons between the tested suction capacity q_s values and theoretical ones from (3), Prager's equation, Powle's equation, and (9) ($m = 1.4$ for adiabatic expansion) under different rotational speeds, in which $q_{s\max}$ in Prager's equation and $q_{d\max}$ in Powle's equation are provided by (9). From the performance curves, it can be seen that the results of (9) under the adiabatic expansion process are close to the test results. The predicted suction capacities of the liquid ring vacuum pumps are within 15% relative error for all rotational speeds. The average relative errors of predicted values by different models are within 10% for three types of pumps at various rotational speeds as shown in Table 3. This

(a) Experimental apparatus

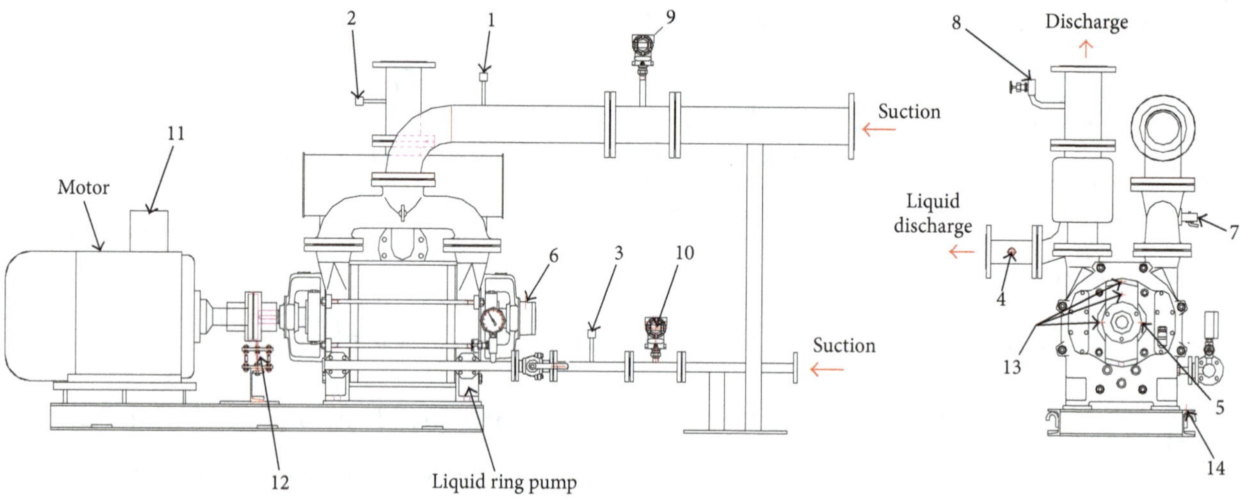

(b) Layout of monitoring points

FIGURE 4: Experimental facility. 1—suction temperature; 2—discharge temperature; 3—liquid suction temperature; 4—liquid discharge temperature; 5—drive bearing temperature; 6—nondriven bearing temperature; 7—suction pressure; 8—discharge pressure; 9—air capacity; 10—liquid capacity; 11—power; 12—rotational speed; 13—$X/Y/Z$-direction vibration; 14—substructure vibration.

TABLE 2: Parameters of liquid ring vacuum pump.

Category	Parameter	Unit	2BE1103	2BE1253	2BE1353
Casing	Radius R	m	0.1365	0.3145	0.407
Impeller	Rotational speed n	r/min	1170, 1300, 1450	590, 740, 920	372, 490, 590
	Axial width b	m	0.215	0.5	0.708
	Angle of blade β	Degree	40	42	45
	Impeller radius r_2	m	0.105	0.25	0.36
	Hub radius r_1	m	0.0535	0.125	0.177
	Eccentricity e	m	0.0136	0.0316	0.04
	Blade thickness δ	m	0.008	0.01	0.012
	Blade number Z		12	15	18

indicates that the expansion process of the residue gas is near the adiabatic one due to a shorter expansion time and path. In addition, Prager's and Powle's equations can also provide reasonable predictions of q_s, if $q_{s\,\max}$ and $q_{d\,\max}$ are calculated by (9).

4.2. Shaft Power of Liquid Ring Pump. Figure 6 shows the comparisons between the tested shaft power N values and theoretical ones by (15) in liquid ring pump under different rotational speeds. From Figure 6, it can be seen that the trends of the shaft power curves calculated by (15) agree well with the ones of the test results. The shaft power predictions of the liquid ring vacuum pumps are within 13% relative error for all rotational speeds. The average relative errors of predicted values by the proposed model are within 10% for three types of pumps at various rotational speeds as shown in Table 3. This means the effect of axial width b of the impeller should be taken into account for liquid friction loss in liquid ring

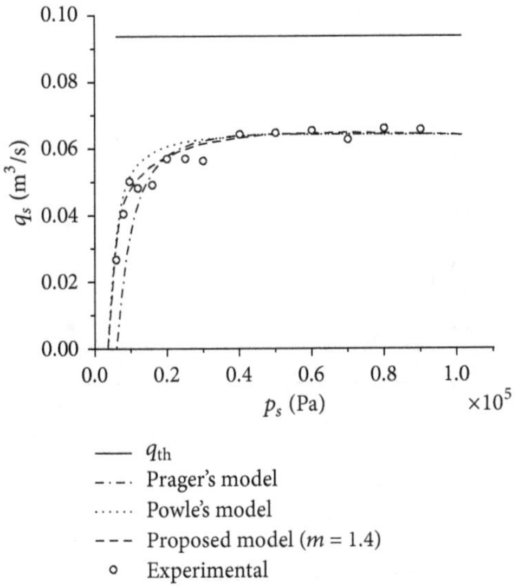

(a) n = 1170 r/min, type: 2BE1103

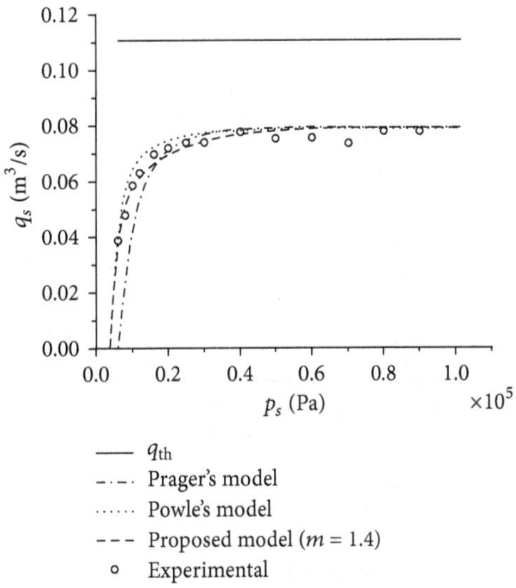

(b) n = 1300 r/min, type: 2BE1103

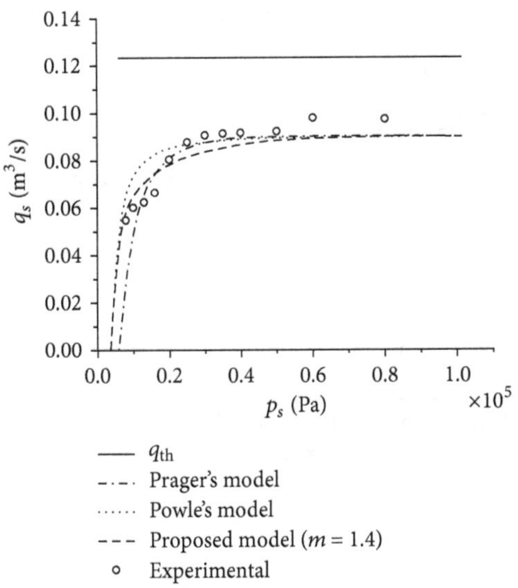

(c) n = 1450 r/min, type: 2BE1103

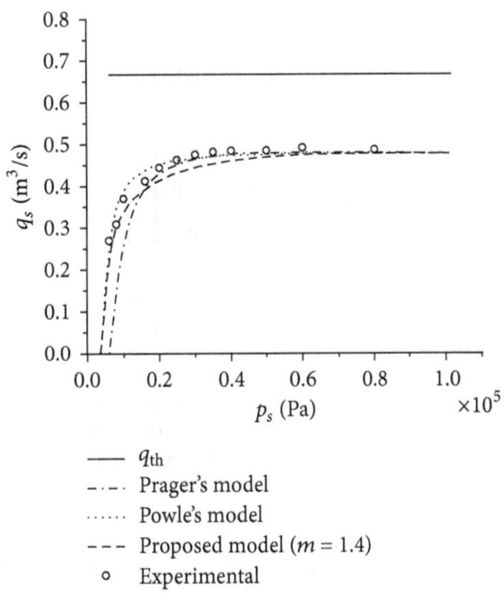

(d) n = 590 r/min, type: 2BE1253

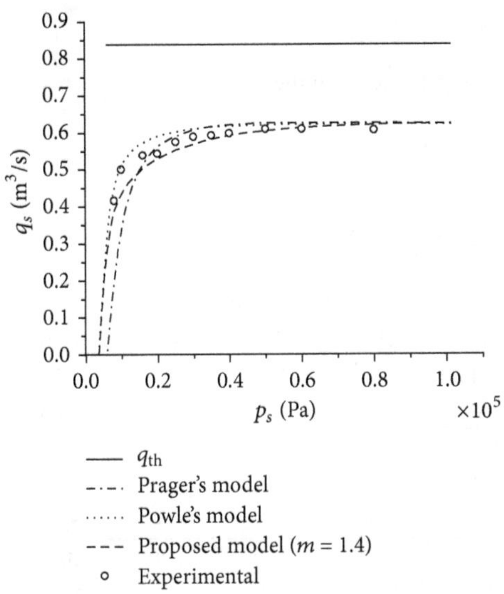

(e) n = 740 r/min, type: 2BE1253

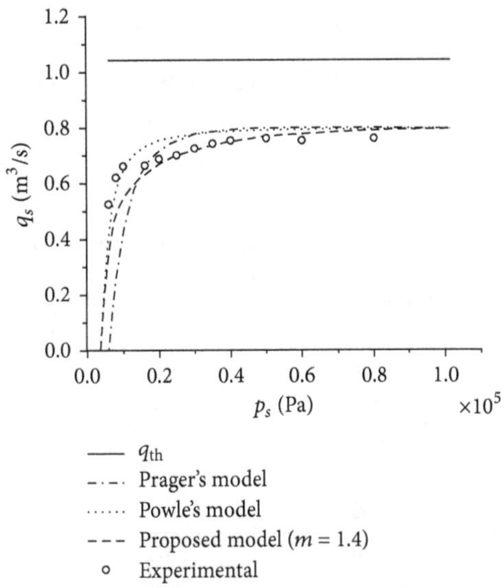

(f) n = 920 r/min, type: 2BE1253

Figure 5: Continued.

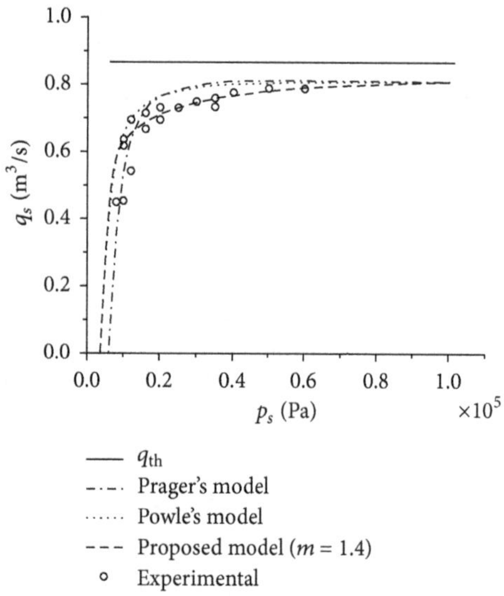

(g) n = 372 r/min, type: 2BE1353

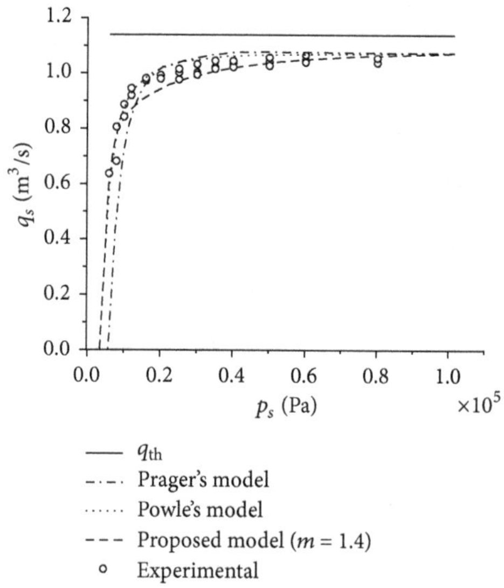

(h) n = 490 r/min, type: 2BE1353

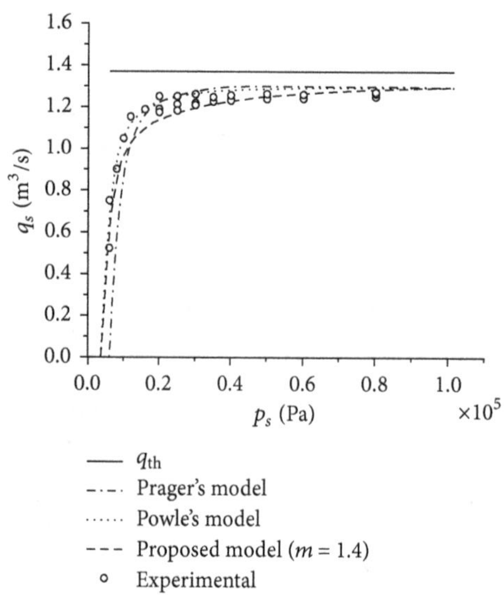

(i) n = 590 r/min, type: 2BE1353

FIGURE 5: Comparisons of q_s-p_s curves between the test and theoretical models.

TABLE 3: Average relative errors of predicted models.

Type of pump	Average relative errors (%)	Prager's model			Powle's model			Proposed model		
	n (r/min)	1170	1300	1450	1170	1300	1450	1170	1300	1450
2BE1103	Δq_s	5.23	4.25	5.41	4.68	5.36	10.06	1.77	1.83	5.32
	ΔN	—	—	—	—	—	—	6.43	4.87	6.69
	n (r/min)	590	740	920	590	740	920	590	740	920
2BE1253	Δq_s	4.17	7.13	9.38	4.27	5.02	5.80	3.34	2.99	4.94
	ΔN	—	—	—	—	—	—	6.40	10.53	7.62
	n (r/min)	372	490	590	372	490	590	372	490	590
2BE1353	Δq_s	8.85	7.38	7.07	12.95	4.01	4.52	8.66	1.88	1.40
	ΔN	—	—	—	—	—	—	5.02	4.66	4.46

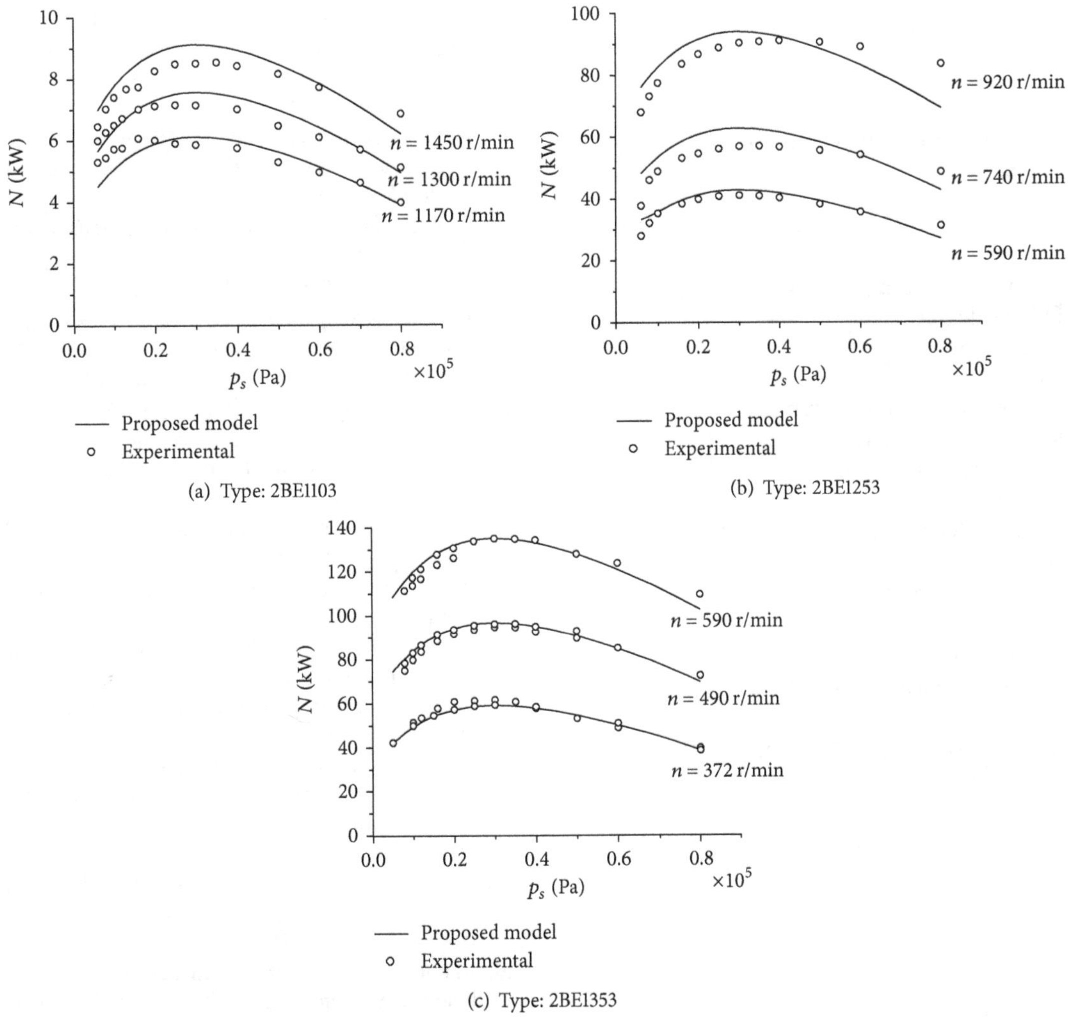

FIGURE 6: Comparisons of N-p_s curves between the test and theoretical model.

pump. More sophisticated models for shaft power of liquid ring pump are necessary to consider for higher predictive accuracy in further work.

5. Conclusions

Based on the actual operating cycle of "suction-compression-discharge-expansion," a universal theoretical model for performance of liquid ring pump was established and verified by performance test in this study. Some conclusions are summarized as follows:

(1) Considering expansion process of the residual gas in liquid ring pumps is an effective approach to solve the problem that the theoretical models deviate from the actual performance in operating cycle. The expansion process is likely near the adiabatic one due to a shorter expansion time and path.

(2) With the major geometric parameters and operating conditions of a liquid ring pump, the performance parameters such as the actual capacity for suction and discharge, shaft power, and global efficiency can be conveniently predicted using the proposed theoretical model, without the limitation of empiric range, performance data, or the detailed 3D geometry of pumps.

(3) The effect of axial width of the impeller should be taken into account for liquid friction loss in liquid ring pump model.

Nomenclature

A: Coefficient
B: Coefficient
N: Shaft power, W
R: Inner radius of casing, m
Re: Reynolds number
a: Immersion depth of the blades, m
b: Impeller width, m
e: Impeller eccentricity, $e < R - r_2$, m
f: Coefficient in friction power considering effect of impeller width, $f = 0.15$
m: Gas expansion index, $1 < m < \kappa$

n: Impeller rotational speed, r/min
p: Pressure, Pa
q: Volumetric capacity, m^3/s
r: Radius, m
u: Circumferential speed of impeller, m/s
Z: Blade number.

Greek Letters

α: Relative immersion depth of the blades
β: Angle of blade, degree
δ: Blade thickness, m
ε: Coefficient
η: Global efficiency of liquid ring pump
φ: Circumferential angle, degree
κ: Adiabatic index; for air, $\kappa = 1.4$
λ: Flow coefficient
μ: Coefficient considering the effect of blade thickness, $\mu = 0.68{\sim}0.85$
ν: Kinematic viscosity, m^2/s
ρ: Length, m; density, kg/m^3
σ: Compression ratio
ω: Angular speed of impeller, $\omega = 2\pi n/60$, rad/s
υ: Ratio of hub radius to impeller radius, $\upsilon = r_1/r_2$.

Subscripts

a: Atmospheric
d: Discharge
e: Expansion
g: Gas
l: Operating liquid
s: Suction
v: Saturated vapor
th: Theoretical
0: Residual gas
1: Impeller hub
2: Impeller.

Conflicts of Interest

The authors declare that they have no conflicts of interest.

References

[1] W. Burgmann and K. Göhler, "Modern vacuum pumps for the vacuum degassing of steel in small and large vacuum-degassing units," *Metallurgist*, vol. 57, no. 5-6, pp. 516–525, 2013.

[2] A. V. Kravtsov and O. A. Troshkin, "Enhancement of operational characteristics of liquid-piston compressors based on new designs," *Chemical and Petroleum Engineering*, vol. 46, no. 11-12, pp. 675-676, 2011.

[3] H. M. Yu, "Analysis on selection of water ring vacuum pumps in the chemical industry," *Applied Mechanics and Materials*, vol. 327, pp. 1435–1439, 2013.

[4] Y. M. Vertepov, V. N. Matsenko, and V. M. Antonov, "Advances in water-ring vacuum pumps and compressors," *Chemical and Petroleum Engineering*, vol. 33, no. 5, pp. 522-523, 1997.

[5] C. Schreiber and M. Kopf, "New application horizons for liquid ring vacuum pumps-extended capabilities in combinations with side channel blowers," *Vakuum in Forschung und Praxis*, vol. 26, no. 6, pp. 26–31, 2014.

[6] N. D. Karlsen-Davies and G. A. Aggidis, "Regenerative liquid ring pumps review and advances on design and performance," *Applied Energy*, vol. 164, pp. 815–825, 2016.

[7] C. Pfleiderer, *Centrifugal Pump for Liquids and Gases*, Berlin, Germany, 1961.

[8] H. Schulz, *The Pumps*, Berlin, Germany, 1977.

[9] U. Segebrecht, *Liquid Ring Vacuum Pumps and Liquid Ring Compressors: Technical Details and Fields of Application*, Verlag Moderne Industrie, Landsberg am Lech, Germany, 1994.

[10] R. Prager, "Operational Conditions and Application Field of Liquid-Ring Machines," in *Proceedings of the Third Conference on Fluid Mechanics and Fluid Machinery*, 1969.

[11] U. S. Powle and S. Kar, "Investigations on pumping speed and compression work of liquid ring vacuum pumps," *Vacuum*, vol. 33, no. 5, pp. 255–263, 1983.

[12] W. H. Faragallah, 1988, Liquid Ring Vacuum Pumps and Compressors, Gulf.

[13] H. Bannwarth, *Liquid Ring Vacuum Pumps, Compressors and Systems: Conventional and Hermetic Design*, John Wiley & Sons, Compressors and Systems: Conventional and Hermetic Design, 2006.

[14] I. S. Bodik and A. B. Tishchenko, "Effect of impeller circumferential velocity on the specific power demand of water-ring vacuum pumps," *Chemical and Petroleum Engineering*, vol. 24, no. 6, pp. 289-290, 1988.

[15] S. Huang, R. Zhi-Yong, Q. J. Deng, W. U. Tai-Zhong, and Z. H. Tan, "Numerical analysis of gas-liquid two-phase flow in liquid-ring vacuum pump," *Vacuum*, vol. 46, no. 2, pp. 49–52, 2009.

[16] K. Kakuda, Y. Ushiyama, S. Obara et al., "Flow simulations in a liquid ring pump using a particle method," *Computer Modeling in Engineering and Sciences*, vol. 66, no. 3, pp. 215–226, 2010.

[17] M. Radle and B. Shome, "Cavitation prediction in liquid ring pump for aircraft fuel systems by CFD approach," in *Proceedings of the SAE 2013 Aerotech Congress*, 2013.

[18] H. Ding, Y. Jiang, H. Wu, and J. Wang, "Two phase flow simulation of water ring vacuum pump using VOF model," in *Proceedings of the ASME/JSME/KSME 2015 Joint Fluids Engineering Conference*, Seoul, Republic of Korea, July 2015.

[19] R. Prager, *Investigation into Liquid Ring Vacuum Pumps and Compressors for the Determination of Performance Losses*, Dissertation, Dresden, Germany, 1968.

[20] A. Nemdili and D. H. Hellmann, "Investigations on fluid friction of rotational disks with and without modified outlet sections in real centrifugal pump casings," *Forschung im Ingenieurwesen*, vol. 71, no. 1, pp. 59–67, 2007.

[21] G. Q. Qiu, S. Huang, L. L. Zhu, Y. Chen, and J. He, "Performance monitoring analysis of liquid ring vacuum pumps," *Applied Mechanics and Materials*, vol. 853, pp. 463–467, 2017.

4

Rolling Bearing Fault Signal Extraction based on Stochastic Resonance-based Denoising and VMD

Xiaojiao Gu[1] and Changzheng Chen[1,2]

[1]*School of Mechanical Engineering, Shenyang University of Technology, Shenyang 110870, China*
[2]*Liaoning Engineering Center for Vibration and Noise Control, Shenyang 110870, China*

Correspondence should be addressed to Changzheng Chen; chencz6699@sina.com

Academic Editor: Hyeong Joon Ahn

Aiming at the difficulty of early fault vibration signal extraction of rolling bearing, a method of fault weak signal extraction based on variational mode decomposition (VMD) and quantum particle swarm optimization adaptive stochastic resonance (QPSO-SR) for denoising is proposed. Firstly, stochastic resonance parameters are optimized adaptively by using quantum particle swarm optimization algorithm according to the characteristics of the original fault vibration signal. The best stochastic resonance system parameters are output when the signal to noise ratio reaches the maximum value. Secondly, the original signal is processed by optimal stochastic resonance system for denoising. The influence of the noise interference and the impact component on the results is weakened. The amplitude of the fault signal is enhanced. Then the VMD method is used to decompose the denoised signal to realize the extraction of fault weak signals. The proposed method was applied in simulated fault signals and actual fault signals. The results show that the proposed method can reduce the effect of noise and improve the computational accuracy of VMD in noise background. It makes VMD more effective in the field of fault diagnosis. The proposed method is helpful to realize the accurate diagnosis of rolling bearing early fault.

1. Introduction

Rolling bearing is one of the most critical and easily damaged components in rotating machinery. Its running state is directly related to the performance of the whole mechanical system [1–3]. Therefore, it is significant to implement fault diagnosis for rolling bearings so as to prevent fatal malfunction of rotating machinery [4]. Vibration analysis is one of the most frequently used methods for health monitoring and fault diagnosis of rotating machinery. The vibration signal of mechanical equipment is rich in status information [5, 6]. When a rotating machine has incipient faults or works in a wicked environment, the useful fault signals are often submerged in strong background noise. It will seriously affect the accuracy of fault diagnosis. Therefore, the extraction of weak fault signal is a research hotspot in the field of signal processing [7].

In order to extract useful signals from complex original signals, many scholars have proposed a lot of effective weak signal extraction methods. One category is to extract weak signal from the perspective of denoising, such as wavelet transform, chaos theory, empirical mode decomposition (EMD), local mean decomposition (LMD), ensemble empirical mode decomposition (EEMD), and singular value decomposition (SVD). Under the influence of sampling frequency, the EEMD decomposition error is large. Meng and Xiang [8] proposed an improved EEMD which extracts the real intrinsic mode functions (IMF) by using the correlation coefficients between the original signal and IMFs. With this method, the pseudo low-frequency IMFs can be eliminated. Yi et al. [9] proposed the augmented quaternion singular spectrum analysis multichannel denoising method. This method has a better ability than multivariate EMD method in multisignal processing. Xie et al. [10] proposed an improved LMD based on extension characteristic wave method to eliminate the end effect. Yi et al. [11] proposed a convex optimization algorithm using nonconvex penalty functions based on SVD for extracting weak fault characteristics. Variational mode decomposition (VMD) is a new adaptive signal processing method proposed by Dragomiretskiy and Zosso in 2014 [12]. In this method, the frequency center and bandwidth of each component are determined by iteration searching for the optimal solution of the variational mode during the process of obtaining decomposition components. Therefore, the effective separation of the frequency domain can be realized adaptively. The existing study shows that in complicated signal decomposition the VMD has more advantage than EMD [13]. Compared with the recursive filtering mode of EEMD and LMD, VMD is a nonrecursive and variational mode decomposition. It has a solid theoretical basis. Its essence is a number of adaptive Wiener filter groups, which show better noise robustness. Through the reasonable control for the convergence condition, the influence of sampling on VMD is much smaller than EEMD and LMD. In terms of mode separation, the VMD can successfully separate two pure harmonic signals with similar frequencies. At present, the research results of the variational mode decomposition in the field of mechanical fault diagnosis show that VMD has better performance than LMD and EEMD in many aspects [14]. The fault feature extraction based on variational mode decomposition has important application value.

The other category is to use the noise to enhance the fault signal by means of stochastic resonance (SR). Because of its obvious advantages in weak signal detection, SR has become a hotspot in the field of signal processing in recent years [15, 16]. Han et al. [17] combined wavelet transform with stochastic resonance theory, which can detect several high frequency weak signals in strong noise background by adjusting the amplitude of wavelet multiscale decomposition. Rother et al. [18] combined stochastic resonance with EMD, which is applied to detect deviations in strip travel of hot strip mill. The results show that this method can reduce downtimes and extend the lifetime of components. Barbini et al. [19] focused on the use of SR in a two-dimensional system of gradient type for detection of weak signals submerged in Gaussian noise. The two-dimensional system is more accurate and more reliable than the traditional one-dimensional system. Castiglione et al. [20] used bistable stochastic resonance and Woods-Saxon stochastic resonance in bearing faults detection, respectively. Experimental data shows that each SR method has its own characteristic features. Gao et al. [21] constructed corresponding Duffing oscillator based on chaos theory to monitor weak signal, which reduces the SNR threshold compared with the traditional monitoring method. Lu et al. [22] enhanced stochastic resonance by full-wave signal construction, which surpass the traditional stochastic resonance in the field of fault signal processing and can be used in areas related to weak signal detection. Lin et al. [23] presented a new additive noise driving and aperiodic chirp signal stochastic dynamical system, which can improve the system performance of localization and moving target detection. Breen et al. [24] used the mechanical stochastic resonance to detect weak signals, which harvest weak periodic signals via the movement of an inverted pendulum. Shi et al. [25] combined analytical mode decomposition (AMD) and EEMD with adaptive stochastic resonance, which can improve the decomposition results of EEMD. This method has a good effect in the rolling bearing fault diagnosis. Xiang and Zhong [26] presented a two-step method of Duffing oscillator and stochastic resonance. And this method combined stochastic resonance with the artificial bee colony algorithm. It is favorable to detect weak signal from a strong noise environment. Liu et al. [27] combined wavelet reconstruction and stochastic resonance. The reconstruction scale was selected by comparing the wavelet entropy of each decomposition scale. This fault detection method is more accurate than the conventional wavelet reconstruction. The above studies which combine the stochastic resonance with the signal processing method or the optimization method show that the appropriate application of the stochastic resonance method can make the signal processing method have a better effect in fault diagnosis. Stochastic resonance has important application value in fault diagnosis under strong noise background.

Combining the respective advantages of VMD and SR, this study presents a weak signal extraction method of rolling bearing fault based on VMD and quantum particle swarm optimization (QPSO) adaptive stochastic resonance. First, the parameters of stochastic resonance system are optimized according to the original signal feature and quantum particle swarm optimization algorithm. Thereafter, the original signal is processed by the stochastic resonance so that the noise frequency component is weakened and the fault frequency component is enhanced. Lastly, the signal which is processed by stochastic resonance is decomposed by VMD method. The useful signal features are extracted. The rolling bearing measured signal analysis shows that the proposed method can discharge most of the interfering signal generated by the background noise so that the components obtained by the VMD can better reflect the fault signal information and enhance the useful signal amplitude. The detection results are more accurate and reliable.

The rest of this paper is organized as follows. Section 2 provides a brief description of the principal theory of VMD, SR, and QPSO. Section 3 introduces the proposed fault diagnosis process and discusses the simulation study to validate the performance of the proposed method. Section 4 discusses the case study to validate the practical application value of this method. Lastly, Section 5 presents the conclusion.

2. Basic Theory

2.1. Variational Mode Decomposition. VMD algorithm is a new nonrecursive variational mode signal decomposition method, which can decompose the complex input signal into a set of discrete mode components [28]. The implementation steps of the VMD algorithm are as follows.

Step 1. The Hilbert transform is performed for each mode function u_k to obtain its analytical signal, as follows:

$$\left[\delta(t) + \frac{j}{\pi t}\right] * u_k(t). \quad (1)$$

Step 2. The estimated center frequency $e^{j\omega_k t}$ of u_k is mixed. The spectrum of each mode is modulated to the corresponding base band, as follows:

$$\left[\left(\delta(t) + \frac{j}{\pi t}\right) * u_k(t)\right] * e^{-j\omega_k t}. \quad (2)$$

Step 3. Calculate the gradient square L^2 norm of the above demodulated signal and estimate the bandwidth of each mode signal. The variational problem is constructed as follows:

$$\min_{\{u_k\},\{\omega_k\}} \left\{ \sum_k \left\| \partial_t \left[\left(\delta(t) + \frac{j}{\pi t}\right) * u_k(t)\right] * e^{-j\omega_k t} \right\|^2 \right\}, \quad \text{s.t.} \quad \sum_k u_k = f, \quad (3)$$

where $\{u_k\} := \{u_1, \ldots, u_k\}$ represents K variational mode components. $\{\omega_k\} := \{\omega_1, \ldots, \omega_k\}$ represents the center frequency of each component. K is the number of variational mode components. t represents the time. ∂_t represents the partial derivative of t. $\delta(t)$ represents the impulse function. j is the square root of -1. ω represents the circular frequency. f represents the input signal.

Step 4. In order to transform the constraint variational problem into the unconstrained problem, the quadratic penalty factor α is introduced to ensure the accuracy of signal reconstruction in noise condition. The Lagrangian multiplier λ is introduced to ensure the stringency of the constraint condition. Through introducing the two parameters, the extended Lagrangian expression is obtained as follows:

$$\begin{aligned} L(\{u_k\}, \{\omega_k\}, \lambda) &:= \alpha \sum_k \left\| \partial_t \left[\left(\delta(t) + \frac{j}{\pi t}\right) * u_k(t)\right] * e^{-j\omega_k t} \right\|_2^2 \\ &+ \left\| f(t) - \sum_k u_k(t) \right\|_2^2 \\ &+ \left\langle \lambda(t), f(t) - \sum_k u_k(t) \right\rangle. \end{aligned} \quad (4)$$

Step 5. Solve the extended Lagrangian expression by using the alternate direction method of multipliers (ADMM). Specific steps are as follows:

(1) Initialize $\{\hat{u}_k^1\}$, $\{\omega_k^1\}$, $\hat{\lambda}^1$, n.

(2) Repeat $n = n + 1$.

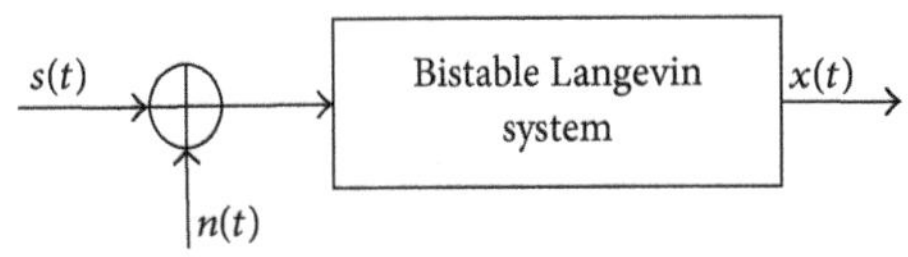

FIGURE 1: SR structure for the bistable Langevin system.

(3) Update u_k, ω_k, and λ, and do

$$\begin{aligned} \hat{u}_k^{n+1}(\omega) &\longleftarrow \frac{\hat{f}(\omega) - \sum_{i \neq k} \hat{u}_i(\omega) + \hat{\lambda}(\omega)/2}{1 + 2\alpha(\omega - \omega_k)^2}, \\ \omega_k^{n+1} &\longleftarrow \frac{\int_0^\infty \omega |\hat{u}_k(\omega)|^2 d\omega}{\int_0^\infty |\hat{u}_k(\omega)|^2 d\omega}, \\ \hat{\lambda}^{n+1}(\omega) &\longleftarrow \hat{\lambda}^n(\omega) + \tau \left[\hat{f}(\omega) - \sum_k \hat{u}_k^{n+1}(\omega)\right]. \end{aligned} \quad (5)$$

(4) Judge whether or not the following conditions are met. If met, the iteration is stopped; otherwise, return to execute Step (2).

$$\frac{\sum_k \left\|\hat{u}_k^{n+1} - \hat{u}_k^n\right\|_2^2}{\left\|\hat{u}_k^n\right\|_2^2} < e, \quad (6)$$

where e is the convergence stopped condition. k is the iteration number.

2.2. Stochastic Resonance. Benzi et al. [29] first introduced the mechanism of stochastic resonance in 1981 and used SR in climatic change. In the field of signal processing, stochastic resonance can transfer the energy of some noise signals to useful signals, which is beneficial to the identification of weak signals submerged in noise [30–32]. The noise signal is input to the bistable or multistable nonlinear system. The system performance can be optimized by adjusting the parameters so that the effect of stochastic resonance achieves the best. The equation of the bistable Langevin system is as follows:

$$\dot{x} = -\frac{dU(x)}{dx} + s(t) + n(t), \quad (7)$$

where $U(x)$ is the potential function.

$$U(x) = -\frac{1}{2}ax^2 + \frac{1}{4}bx^4, \quad (8)$$

where a and b are system parameters. The potential function (8) has one unstable equilibrium point and two stable equilibrium points. The height of the potential barrier is $a^2/(4b)$. $s(t) = \sum_{i=1}^n A_i \sin(2\pi f_i t)$ represents an input signal with amplitude A_i and frequency f_i. $n(t) = \sqrt{2D}\varepsilon(t)$ represents the Gaussian white noise signal with intensity D, mean 0, and variance 1. The bistable Langevin system structure is shown in Figure 1.

The stochastic resonance problem is equivalent to a multidimensional-continuous optimization problem [33].

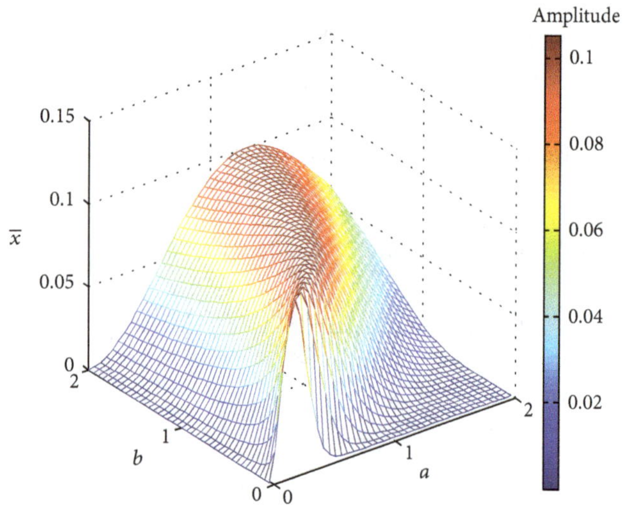

FIGURE 2: Influence of SR system parameters on amplitude.

The same set of input signals using different parameters of the system for stochastic resonance processing will produce different effect [34]. Setting $s(t) = A\sin(2\pi f_0 t)$, Figure 2 shows the response amplitude $\overline{x}$ changes with SR system parameters when $D = 0.2$, $f_0 = 0.1\,\text{Hz}$, and $A = 0.1$. It is shown that there exist optimal system parameters a, b to make the stochastic resonance performance optimal.

2.3. Quantum Particle Swarm Optimization. Quantum Particle Swarm Optimization was proposed on trajectory analysis of particle swarm optimization (PSO) and the quantum mechanism [35]. In the quantum space, the position and velocity of the particle cannot be determined simultaneously, so the wave function is used to describe the particle state [36]. The square of the wave function represents the probability of finding a particle at a particular position in the quantum space. By employing the Monte Carlo method, the position of a particle can be expressed as follows:

$$x(t) = P \pm \frac{L}{2}\ln\left(\frac{1}{v}\right), \tag{9}$$

where v is a random number between 0 and 1. L is the characteristic length of potential well which changes with the time as follows:

$$L(t+1) = 2\beta\left|\sum_{i=1}^{N}\frac{P_i}{N} - x(t)\right|, \tag{10}$$

where parameter β is the contraction expansion coefficient and is used to control the convergence speed of the particle. N is the number of particle swarms. P_i is the local optimal value of particle i. Finally, the position of the particle can be expressed as follows:

$$x(t+1) = P \pm \beta\left|\sum_{i=1}^{N}\frac{P_i}{N} - x(t)\right|\ln\left(\frac{1}{v}\right). \tag{11}$$

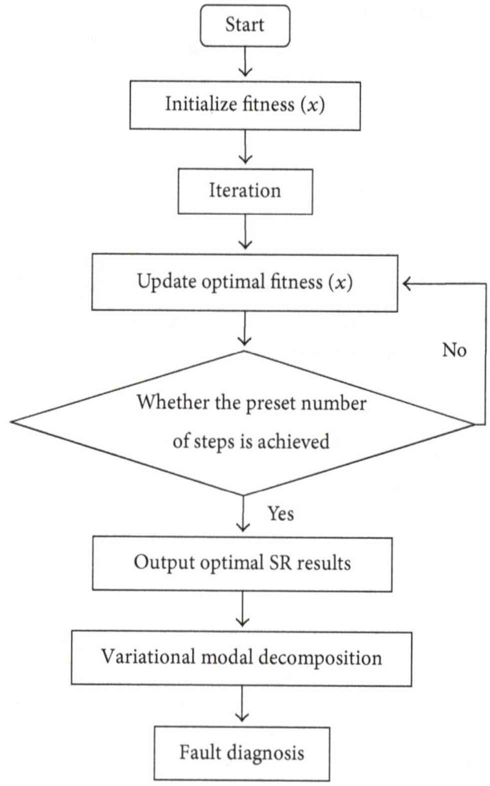

FIGURE 3: Flowchart of the proposed scheme.

3. Proposed Fault Identification Scheme

3.1. Fault Diagnosis Process. Different from the traditional fault identification method, this fault diagnosis process considers the noise feature of different input signal. In various moments, the fault features are varying degrees submerged in noise. Hence, fixed SR system parameters may be unable to suppress the noise and properly enhance the fault feature. This study proposes an adaptable and varying-scale stochastic resonance to overcome the shortcomings of the traditional method. The values of SR system parameters are optimized by using the quantum particle swarm optimization algorithm. The fitness of each particle is determined by the objective function. The performance of stochastic resonance system is mainly reflected in the signal to noise ratio (SNR) of the output signal [37]. Hence, the fitness function of QPSO algorithm is shown as follows:

$$\text{fitness}(x) = 10\log\frac{P_S}{P_N}, \tag{12}$$

where $P_S := 2|X[k_0]|^2$ is the active power of the signal. $P_N := \sum_{k=0}^{L-1}|X[k]|^2 - P_S$ is the active power of the noise. $|X[k]|$ is the discrete Fourier transform (DFT) of the sample sequence, which is defined according to the following formula:

$$X[k] = \sum_{t=0}^{L-1} x_t \exp\left(-j\frac{2\pi}{L}kt\right), \quad k = 0, 1, \ldots, L-1. \tag{13}$$

The procedures for this method of fault feature extraction are shown in Figure 3.

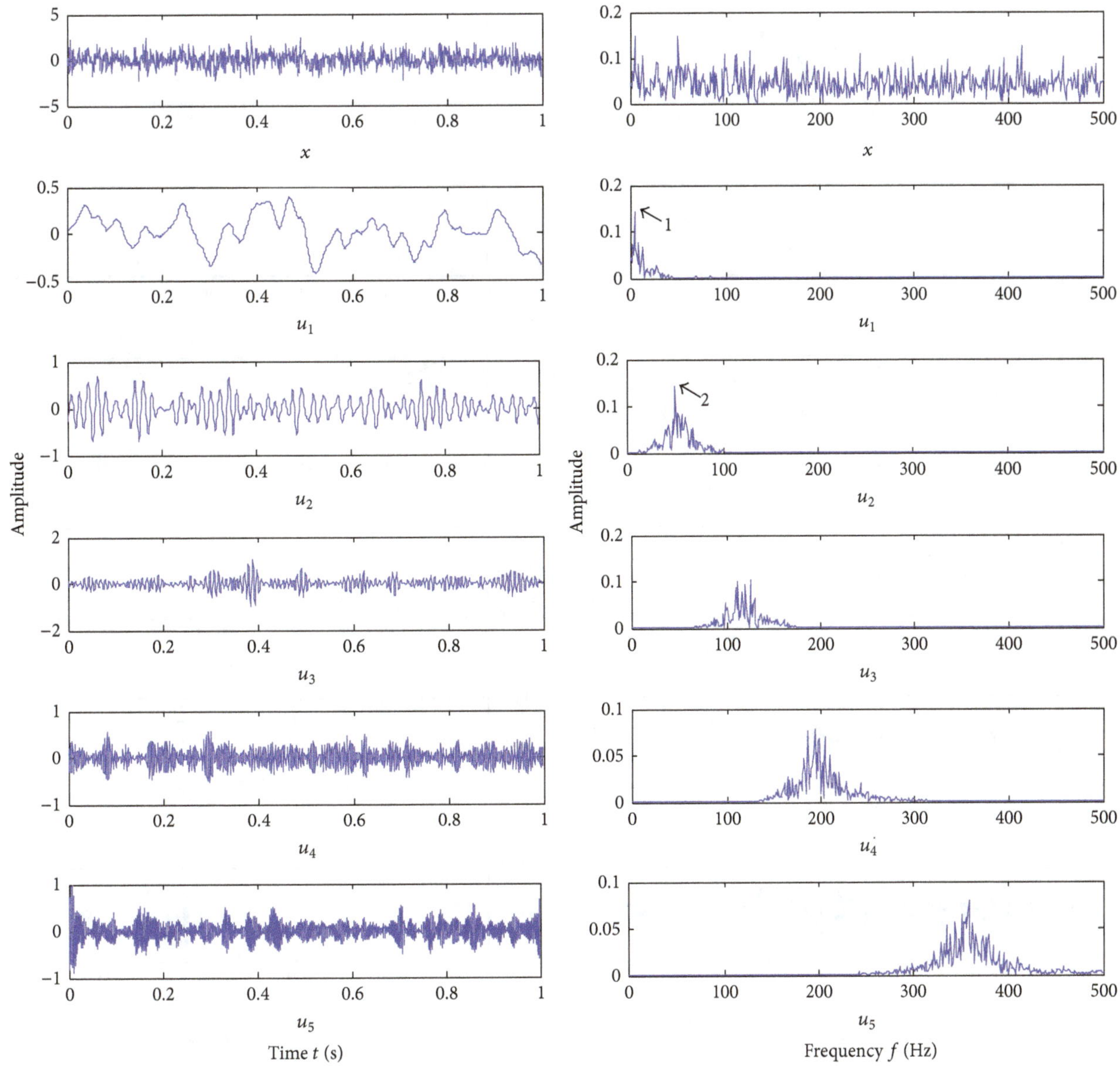

FIGURE 4: VMD analysis result of the simulated signal.

3.2. Simulation Signals Analysis. To demonstrate the differences between VMD and the improved VMD based on QPSO-SR, the synthesized signal which includes 5 Hz and 50 Hz sinusoidal signal and the Gauss white noise is used to simulate the fault signal. The analog signal expression is established as follows:

$$x(t) = 0.15\sin(2\pi \times 5t) + 0.15\sin(2\pi \times 50t) + n(t), \tag{14}$$

where $t := 0.001 : 0.001 : 1$ is the sample time points. $n(t)$ is the Gauss white noise. The intensity of the Gauss white noise is 0.8. For comparison, the simulated signal is first subjected to VMD decomposition. The quadratic penalty factor α selected the default value 2000. The bandwidth τ selected the default value 0. The number of variational mode components is determined by the method of observation center frequency [38]. Set the initial value of K to 2. The VMD is carried out with the number of variational mode components $K = K + 1$ and judged whether the center frequency of each mode component is close. When $K = 6$ began to show the phenomenon of overdecomposition, the optimal mode number K is 5. The variational mode components can be obtained through VMD algorithm. The VMD results and the Fast Fourier Transform (FFT) are shown in Figure 4.

From Figure 4, it can be seen that the fault information cannot be identified from time spectrum and frequency spectrum of the simulated signal. After the VMD algorithm, the amplitude of 5 Hz (represented by point 1) in u_1 FFT is 0.1416 and the amplitude of 50 Hz (represented by point 2) in u_2 FFT is 0.1439, showing the simulated fault information. And u_3, u_4, and u_5 show the noise signal information.

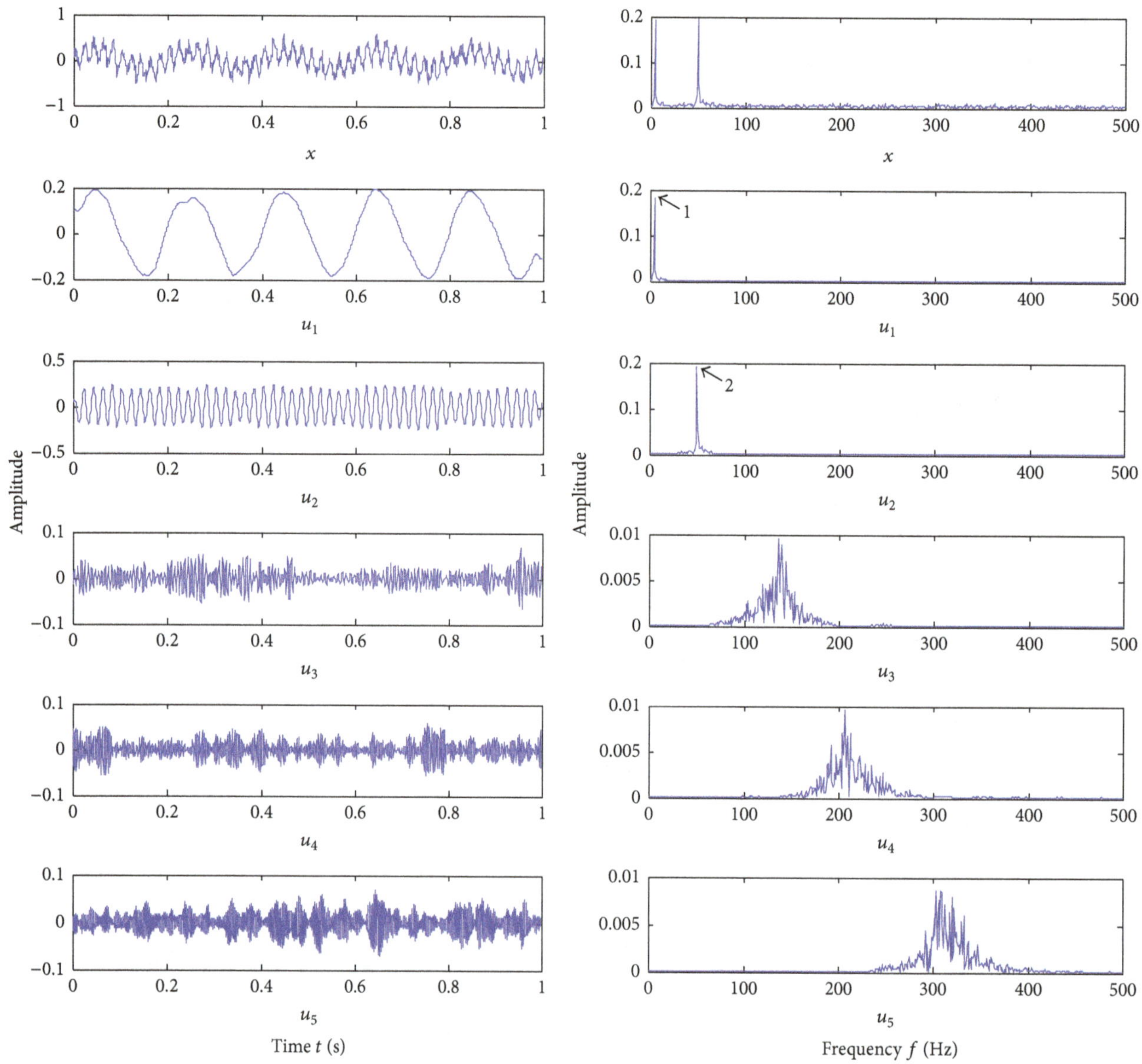

FIGURE 5: VMD analysis result of simulated signals based on QPSO-SR denoising.

However, u_1, u_2 mixed with the interference noise cannot accurately identify the simulated fault features.

In order to eliminate the effect of interference noise, the QPSO-SR method is used to denoise the simulated signal before VMD. The stochastic resonance system parameters a and b are optimized by quantum particle swarm optimization algorithm. The optimization range of system parameters is from 0.01 to 2. The optimization results show that SNR achieves the maximum value when $a = 0.05$ and $b = 0.17$. These system parameters are used to denoise the simulated signal. And then the signal processed by stochastic resonance is decomposed by VMD method. After QPSO-SR noise reduction, the result of VMD decomposition is shown in Figure 5.

From Figure 5, it can be seen that the interference noise in u_1 and u_2 is weakened. The amplitude of 5 Hz (represented by point 1) in u_1 FFT is increased from 0.1416 to 0.1820. The amplitude of 50 Hz (represented by point 2) in u_2 FFT is increased from 0.1439 to 0.1922. The amplitude of the fault information frequency is obviously enhanced and the useful information is accurately identified. The amplitude of the mode components u_3, u_4, and u_5 which represent the noise signal is weakened. After the noise reduction by QPSO-SR, a part of the noise energy can be transmitted to the useful signal, which is helpful for VMD to obtain the useful information more accurately.

4. Experimental Validations

The bearing vibration signals are supplied by the Shenyang University of Technology. The rolling bearing vibration signals were measured by QPZZ-II rotating machinery vibration fault experimental platform. The experimental platform is shown in Figure 6. The motor rotational speed is at 1720 r/min

TABLE 1: Geometry information of the tested rolling bearing.

Inside diameter (mm)	Outside diameter (mm)	Bearing thickness (mm)	Ball diameter (mm)	Pitch diameter (mm)
25.00	52.00	15.00	7.94	39.04

FIGURE 6: Experimental facility.

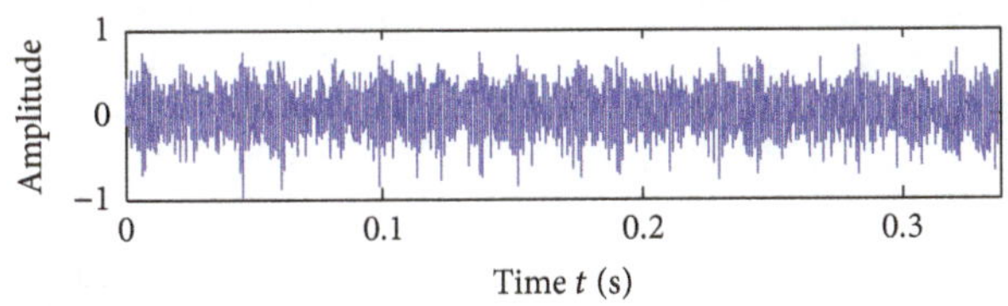

(a)

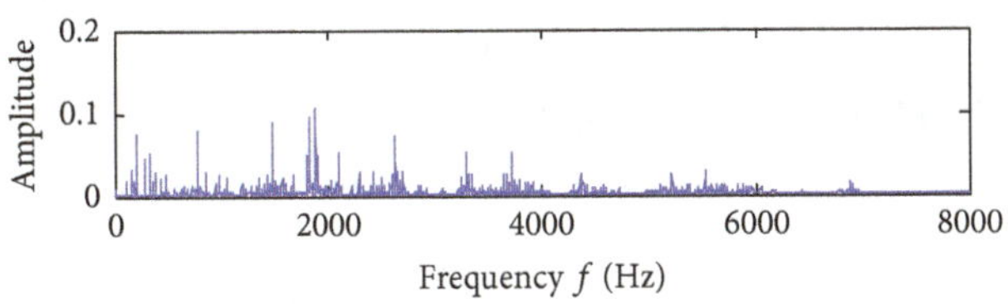

(b)

FIGURE 7: Raw vibration signal.

to 1797 r/min. The digital data used in this analysis were collected at 16,000 samples per second and the sampling number is 5468. A 0.2 mm wide and 0.1 mm deep groove was wire-cut on the surface of the bearing inner ring to simulate early fault. The detailed information of the tested rolling bearing is presented in Table 1. The fault feature frequency of the tested rolling bearing is presented in Table 2.

The time domain spectrum and frequency domain spectrum of rolling bearing vibration signal with inner ring fault are shown in Figure 7. The SNR of raw vibration signal is −15.3097 dB. It can be seen from the waveform that there are obvious interference noise and shock in the rolling bearing vibration signal and the frequency component is rich. Therefore, the fault information cannot be obtained from the spectrum directly.

The raw vibration signal is decomposed by VMD. The decomposition which is obtained by VMD is processed by

TABLE 2: Fault feature frequency of the tested rolling bearing.

Inner ring (Hz)	Outer ring (Hz)	Rolling element (Hz)	Cage element (Hz)
$5.4152f_r$	$3.5848f_r$	$4.7135f_r$	$0.39828f_r$

f_r = 29.95 Hz.

TABLE 3: The SNR of the output signal.

a	b	$\mathrm{SNR}_{\mathrm{out}}$ (dB)
0.01	0.07	3.8942
0.03	0.06	1.8319
0.05	0.16	3.0896
0.10	0.12	2.5896
1.19	1.97	5.0046

using Envelope analysis and fast Fourier transform algorithm. The VMD results and the FFT are shown in Figure 8. In contrast, replace VMD with EEMD and perform the same processing on the raw signal. The EEMD results and the FFT are shown in Figure 9. The EEMD decomposition result reveals that the phenomena of mode mixing make the early fault indistinguishable. The VMD decomposition result reveals that the VMD method solves the phenomena of mode mixing but the component still contains noise disturbance and shock. VMD algorithm has not realized the potential that was expected of it. Characteristic frequency 162 Hz (represented by point 1), double frequency 324 Hz (represented by point 2), and third frequency 486 Hz (represented by point 3) of bearing inner ring fault are not obvious in the decomposition result (see Figure 8). This shows that it is difficult to ensure the accuracy of the fault diagnosis in the practical engineering if we use the VMD algorithm independently.

The traditional stochastic resonance method is employed to process the raw vibration signal. SR system parameters are selected according to experience. The parameter a takes on 0.01, 0.03, 0.05, 0.10, and 1.19, respectively. The parameter b takes on 0.07, 0.06, 0.16, 0.12, and 1.97, respectively. The raw vibration signal is subjected to SR processing by the above five sets of parameters, respectively. The SNR of the output signal is shown in Table 3.

The proposed fault weak signal extraction method based on VMD and QPSO-SR denoising is employed to process the raw vibration signal. Firstly, the raw signal is preprocessed with QPSO-SR method. According to the raw signal characteristics, the quantum particle swarm optimization algorithm is used to optimize the SR system parameters. The signal is

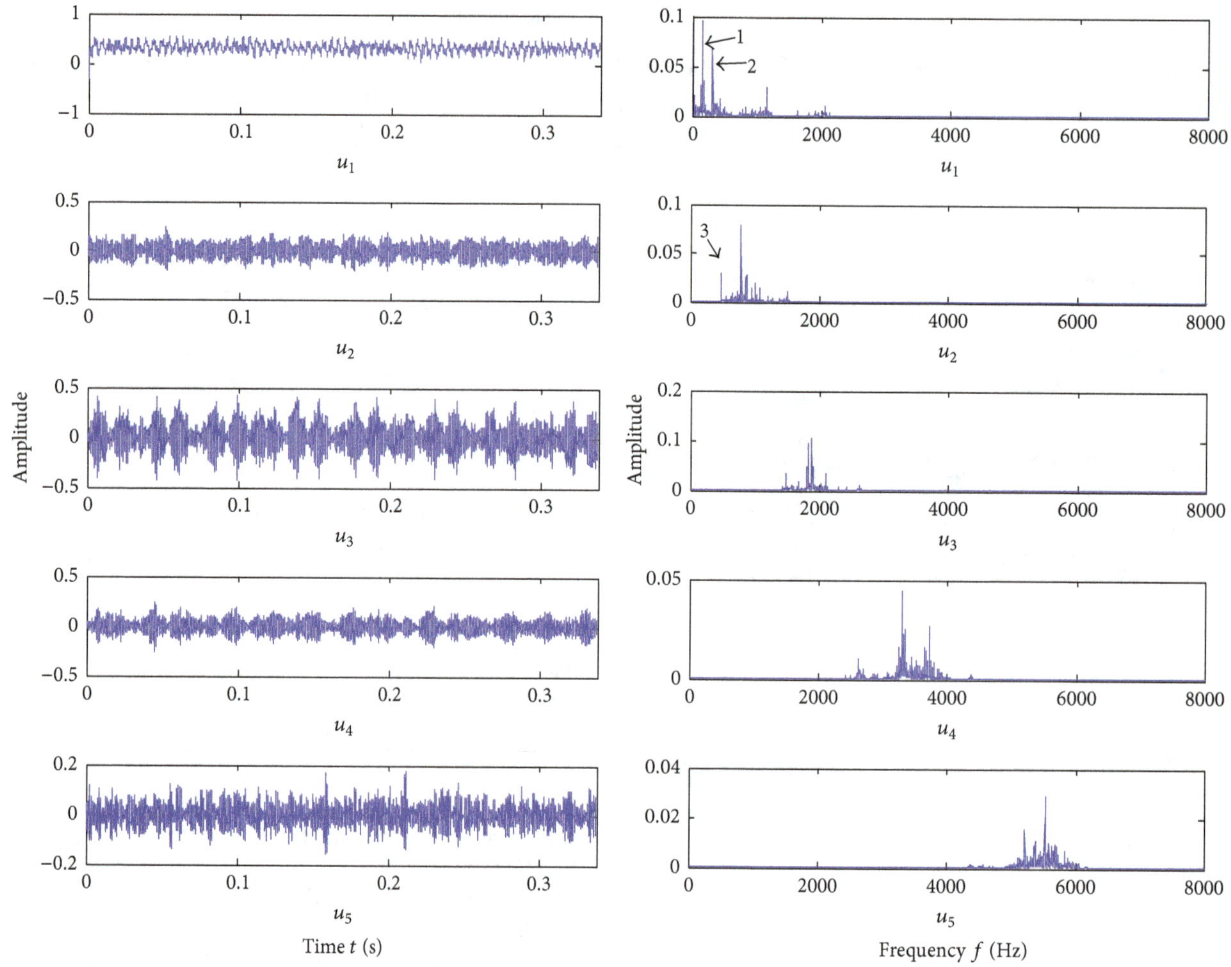

FIGURE 8: VMD analysis result of the rolling bearing vibration signal.

subjected to SR processing based on the optimal parameters. The SNR of the output signal is 7.6465 dB. Evidently, the processed signals have been well denoised with the aid of QPSO-SR method. The frequency domain information exchange method is used to overcome the limitation of SR to large parameter signal. Liu et al. [39] described the specific calculation process of the frequency domain information exchange method in detail. The signal is subjected to VMD after denoising. Center frequency corresponding to different K is shown in Table 4. When K is 6, overdecomposition began to happen, so K was chosen as 5. The time domain and frequency domain waveforms of the results are shown in Figure 10.

From Figure 10, it can be seen that the VMD results are less affected by interference noise. Characteristic frequency (represented by point 1), double frequency (represented by point 2), and third frequency (represented by point 3) of bearing inner ring fault are obvious in the result. The fault feature is highlighted and the noise component is weakened. The bearing fault type can be analyzed intuitively and effectively. Compared with the original signal which has not been processed by QPSO-SR method, the shock wave of the processed signal is obviously reduced in time domain waveform. The energy distribution of the processed signal spectrum is also more concentrated in the bearing fault frequency band. This proofs that a portion of the noise energy can be passed to the useful signal by the appropriate SR method, which is helpful for VMD to obtain the useful information more accurately.

TABLE 4: Center frequency corresponding to different K.

K	Center frequency (Hz)					
2	1767	4784	—	—	—	—
3	1767	2789	4784	—	—	—
4	824	1767	2789	4784	—	—
5	824	1767	2789	3641	4784	—
6	824	1767	2789	3641	4668	4784

5. Conclusion

This study presents a novel weak fault extraction method. This method includes QPSO-SR method and variational

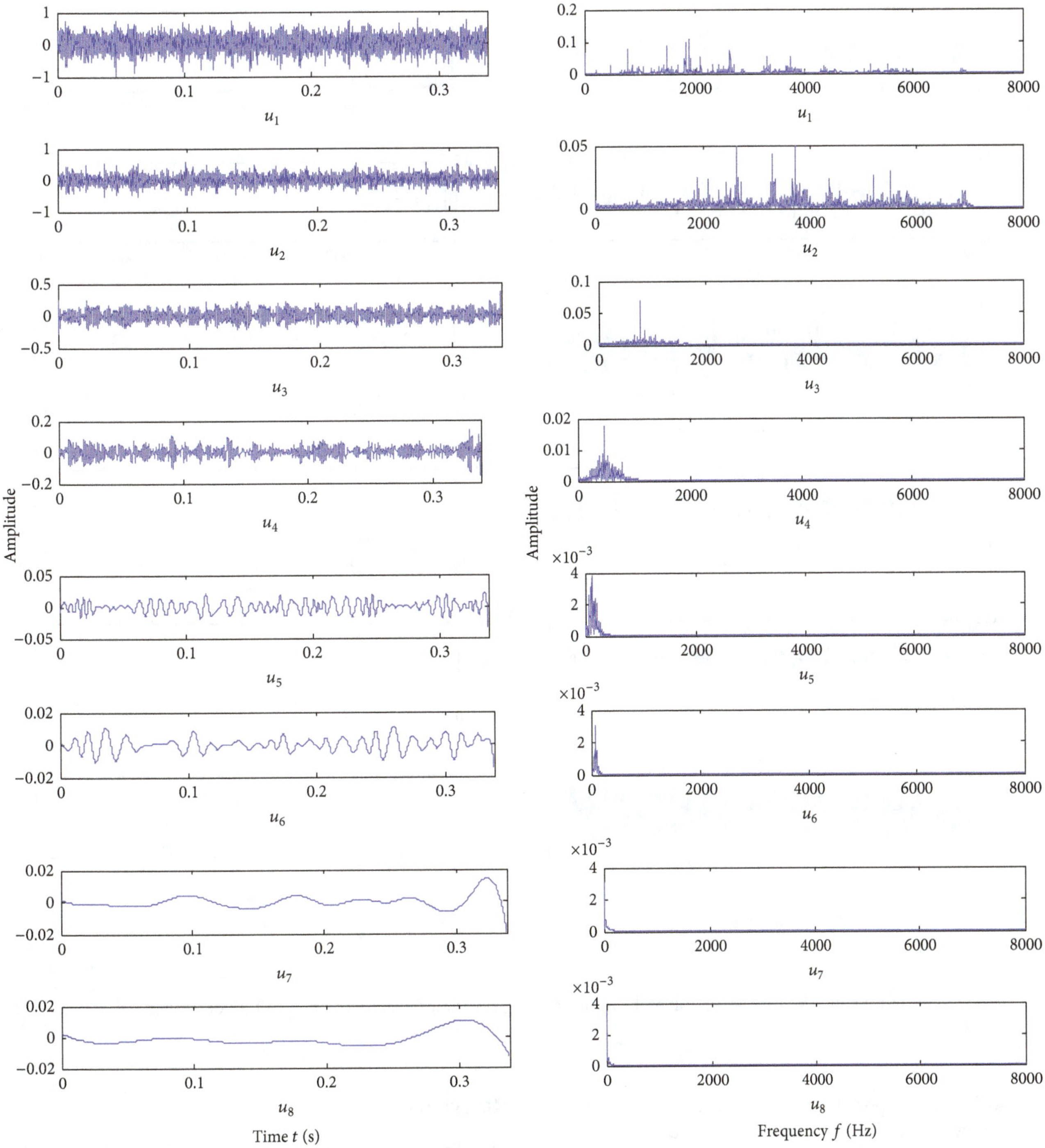

FIGURE 9: EEMD analysis result of the rolling bearing vibration signal.

mode decomposition. QPSO-SR method is an adaptive denoising method that can transmit a part of the noise energy to the weak fault signal. First, QPSO-SR method is proposed to enhance the fault characteristic signal and reduce the influence from the interference noise and the shock. According to the characteristics of the original fault vibration signal, the parameters of SR system are optimized by QPSO algorithm dynamically. Thereafter, the optimal SR system is used for denosing and the processed signal is decomposed by VMD method. The failure feature information can be extracted through the FFT of the VMD component. The results obtained in the investigation of the simulated fault signal and the rolling bearing data set show that the proposed scheme avoids the influence of the interference noise and

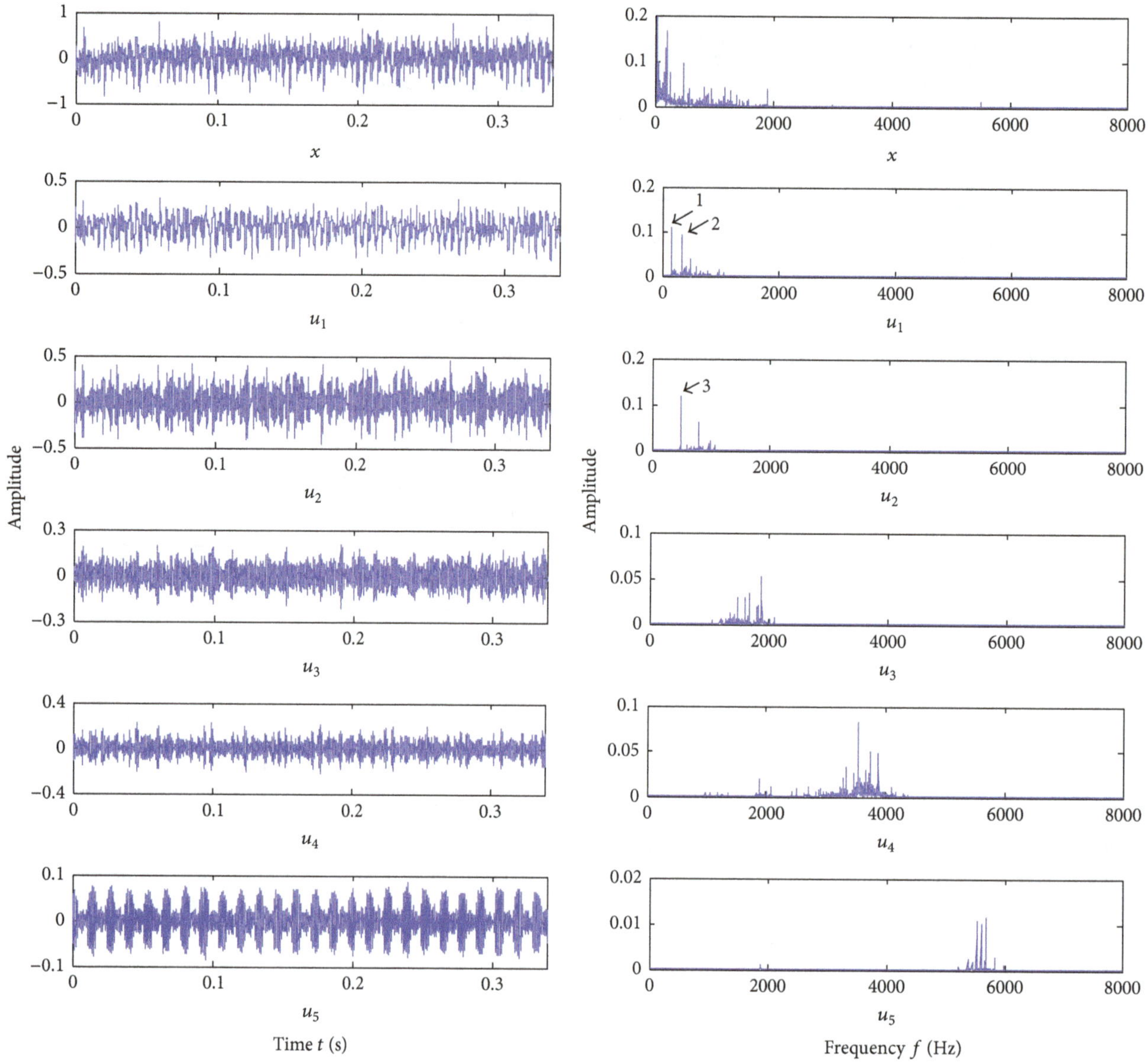

FIGURE 10: VMD analysis result of the rolling bearing vibration signal based on QPSO-SR denoising.

the shock. The proposed scheme reaches a higher degree of accuracy in machinery fault diagnosis compared with previous VMD method.

Conflicts of Interest

The authors declare that they have no conflicts of interest.

Acknowledgments

This research was financially supported by the National Natural Science Foundation of China (Grants nos. 51675350 and 51575361) and the Doctoral Research Initial Fund of Liaoning Province (201601154).

References

[1] J. Shuai, C. Shen, and Z. Zhu, "Adaptive Morphological Feature Extraction and Support Vector Regressive Classification for Bearing Fault Diagnosis," *International Journal of Rotating Machinery*, vol. 2017, Article ID 2384184, 2017.

[2] C. Peeters, P. Guillaume, and J. Helsen, "A comparison of cepstral editing methods as signal pre-processing techniques for vibration-based bearing fault detection," *Mechanical Systems and Signal Processing*, vol. 91, pp. 354–381, 2017.

[3] L. Barbini, A. P. Ompusunggu, A. J. Hillis, J. L. du Bois, and A. Bartic, "Phase editing as a signal pre-processing step for automated bearing fault detection," *Mechanical Systems and Signal Processing*, vol. 91, pp. 407–421, 2017.

[4] A. Kumar and R. Kumar, "Least Square Fitting for Adaptive Wavelet Generation and Automatic Prediction of Defect Size

in the Bearing Using Levenberg–Marquardt Backpropagation," *Journal of Nondestructive Evaluation*, vol. 36, no. 1, article no. 7, 2017.

[5] D. P. Jena and S. N. Panigrahi, "Automatic gear and bearing fault localization using vibration and acoustic signals," *Applied Acoustics*, vol. 98, pp. 20–33, 2015.

[6] L. L. Xia, L. Wen, H. M. Liu, Y. Chu, and J. J. Tai, "Fault diagnosis and pattern recognition of power transformer based on improved PSO-BPNN," *Journal ofShenyang University of Technology*, vol. 38, no. 6, pp. 606–611, 2016.

[7] C. Mishra, A. K. Samantaray, and G. Chakraborty, "Bond graph modeling and experimental verification of a novel scheme for fault diagnosis of rolling element bearings in special operating conditions," *Journal of Sound and Vibration*, vol. 377, pp. 302–330, 2016.

[8] L. J. Meng and J. W. Xiang, "Rolling bearing fault diagnosis based on second generation wavelet denoising and improved EEMD," *Applied Mechanics and Materials*, vol. 556-562, pp. 2677–2680, 2014.

[9] C. Yi, Y. Lv, Z. Dang, H. Xiao, and X. Yu, "Quaternion singular spectrum analysis using convex optimization and its application to fault diagnosis of rolling bearing," *Measurement*, vol. 103, pp. 321–332, 2017.

[10] L. Xie, X. Lang, J. Chen, A. Horch, and H. Su, "Time-varying oscillation detector based on improved LMD and robust Lempel-Ziv complexity," *Control Engineering Practice*, vol. 51, pp. 48–57, 2016.

[11] C. Yi, Y. Lv, M. Ge, H. Xiao, and X. Yu, "Tensor singular spectrum decomposition algorithm based on permutation entropy for rolling bearing fault diagnosis," *Entropy*, vol. 19, no. 4, article no. 139, 2017.

[12] K. Dragomiretskiy and D. Zosso, "Variational mode decomposition," *IEEE Transactions on Signal Processing*, vol. 62, no. 3, pp. 531–544, 2014.

[13] C. Yi, Y. Lv, and Z. Dang, "A fault diagnosis scheme for rolling bearing based on particle swarm optimization in variational mode decomposition," *Shock and Vibration*, vol. 2016, Article ID 9372691, 10 pages, 2016.

[14] T. Dutta, U. Satija, B. Ramkumar, and M. S. Manikandan, "A novel method for automatic modulation classification under non-Gaussian noise based on variational mode decomposition," in *Proceedings of the 22nd National Conference on Communication, NCC 2016*, ind, March 2016.

[15] Y. J. W. Ngouongo, G. D. Kenmoe, and T. C. Kofane, "Effect of coupling on stochastic resonance and stochastic antiresonance processes in a unidirectionally N-coupled systems in periodic sinusoidal potential," *Physica A: Statistical Mechanics and its Applications*, vol. 472, pp. 25–31, 2017.

[16] Z.-H. Lai and Y.-G. Leng, "Weak-signal detection based on the stochastic resonance of bistable Duffing oscillator and its application in incipient fault diagnosis," *Mechanical Systems and Signal Processing*, vol. 81, pp. 60–74, 2016.

[17] D. Han, P. li, S. An, and P. Shi, "Multi-frequency weak signal detection based on wavelet transform and parameter compensation band-pass multi-stable stochastic resonance," *Mechanical Systems and Signal Processing*, vol. 70-71, pp. 995–1010, 2016.

[18] A. Rother, M. Jelali, and D. Söffker, "A brief review and a first application of time-frequency-based analysis methods for monitoring of strip rolling mills," *Journal of Process Control*, vol. 35, pp. 65–79, 2015.

[19] L. Barbini, M. O. T. Cole, A. J. Hillis, and J. L. Du Bois, "Weak signal detection based on two dimensional stochastic resonance," in *Proceedings of the 23rd European Signal Processing Conference, EUSIPCO 2015*, pp. 2147–2151, Nice, Franc, September 2015.

[20] R. Castiglione, L. Garibaldi, S. Marchesiello, and R. Serra, "Stochastic Resonance algorithms to enhance damage detection in bearing faults," *Matec Web of Conferences*, vol. 20, p. 07005, 2015.

[21] S. L. Gao, S. C. Zhong, K. Wei, and H. Ma, "Weak singal detection based on chaos and stochastic resonance," *Acta Physica Sinica*, vol. 61, no. 18, Article ID 180501, 2012.

[22] S. Lu, Q. He, H. Zhang, and F. Kong, "Rotating machine fault diagnosis through enhanced stochastic resonance by full-wave signal construction," *Mechanical Systems and Signal Processing*, vol. 85, pp. 82–97, 2017.

[23] L.-f. Lin, L. Yu, H. Wang, and S. Zhong, "Parameter-adjusted stochastic resonance system for the aperiodic echo chirp signal in optimal FrFT domain," *Communications in Nonlinear Science and Numerical Simulation*, vol. 43, pp. 171–181, 2017.

[24] B. J. Breen, J. G. Rix, S. J. Ross et al., "Harvesting wind energy to detect weak signals using mechanical stochastic resonance," *Physical Review E: Statistical, Nonlinear, and Soft Matter Physics*, vol. 94, no. 6, Article ID 062205, 2016.

[25] P. Shi, C. Su, and D. Han, "Fault diagnosis of rotating machinery based on adaptive stochastic resonance and AMD-EEMD," *Shock and Vibration*, vol. 2016, Article ID 9278581, 11 pages, 2016.

[26] J. Xiang and Y. Zhong, "A two-step method using Duffing oscillator and stochastic resonance to detect mechanical faults," in *Proceedings of the 2016 IEEE International Instrumentation and Measurement Technology Conference, I2MTC 2016*, Taipei, Taiwan, May 2016.

[27] W.-X. Liu, Y.-J. Wang, X. Liu, and M.-J. Zhang, "Weak thruster fault detection for AUV based on stochastic resonance and wavelet reconstruction," *Journal of Central South University*, vol. 23, no. 11, pp. 2883–2895, 2016.

[28] Z. Li, J. Chen, Y. Zi, and J. Pan, "Independence-oriented VMD to identify fault feature for wheel set bearing fault diagnosis of high speed locomotive," *Mechanical Systems and Signal Processing*, vol. 85, pp. 512–529, 2017.

[29] R. Benzi, A. Sutera, and A. Vulpiani, "The mechanism of stochastic resonance," *Journal of Physics A: Mathematical and General*, vol. 14, no. 11, pp. L453–L457, 1981.

[30] H. Zhang, Y. Zheng, and F. Kong, "Weak impulsive signals detection based on step-varying asymmetric stochastic resonance," *Proceedings of the Institution of Mechanical Engineers, Part C: Journal of Mechanical Engineering Science*, vol. 231, no. 2, pp. 242–262, 2017.

[31] M. Uzuntarla, J. J. Torres, P. So, M. Ozer, and E. Barreto, "Double inverse stochastic resonance with dynamic synapses," *Physical Review E: Statistical, Nonlinear, and Soft Matter Physics*, vol. 95, no. 1, Article ID 012404, 2017.

[32] P. Zhou, S. Lu, F. Liu, Y. Liu, G. Li, and J. Zhao, "Novel synthetic index-based adaptive stochastic resonance method and its application in bearing fault diagnosis," *Journal of Sound and Vibration*, vol. 391, pp. 194–210, 2017.

[33] P. Addesso, V. Pierro, and G. Filatrella, "Interplay between detection strategies and stochastic resonance properties," *Communications in Nonlinear Science and Numerical Simulation*, vol. 30, no. 1-3, pp. 15–31, 2016.

[34] Z. Qiao, Y. Lei, J. Lin, and F. Jia, "An adaptive unsaturated bistable stochastic resonance method and its application in

mechanical fault diagnosis," *Mechanical Systems and Signal Processing*, vol. 84, pp. 731–746, 2017.

[35] . Shuyuan Yang, . Min Wang, and . Licheng jiao, "A quantum particle swarm optimization," in *Proceedings of the 2004 Congress on Evolutionary Computation*, pp. 320–324, Portland, OR, USA.

[36] W. Fang, J. Sun, Y. Ding, X. Wu, and W. Xu, "A review of quantum-behaved particle swarm optimization," *IETE Technical Review*, vol. 27, no. 4, pp. 336–347, 2010.

[37] D. Yang, H. Li, Y. Hu, J. Zhao, H. Xiao, and Y. Lan, "Vibration condition monitoring system for wind turbine bearings based on noise suppression with multi-point data fusion," *Journal of Renewable Energy*, vol. 92, pp. 104–116, 2016.

[38] A. A. Abdoos, "A new intelligent method based on combination of VMD and ELM for short term wind power forecasting," *Neurocomputing*, vol. 203, pp. 111–120, 2016.

[39] J. J. Liu, Y. G. Leng, Z. H Lai, and T. Dan, "Stochastic resonance based on frequency domain information exchange," *Acta Physica Sinica*, vol. 65, no. 22, Article ID 220501, 2016.

5

Aerothermodynamic Effects and Modeling of the Tangential Curvature of Guide Vanes in an Axial Turbine Stage

Tzong-Hann Shieh

Department of Aerospace and Systems Engineering, Feng Chia University, No. 100, Wenhwa Rd., Seatwen, Taichung 40724, Taiwan

Correspondence should be addressed to Tzong-Hann Shieh; thshieh@fcu.edu.tw

Academic Editor: Jechin Han

By tangential curvature of the stacking line of the profiles guide vanes can be designed, which have on both ends an obtuse angle between suction side and sidewall. This configuration, according to literature, is capable of reducing secondary loss. This type of vanes develops considerable radial components of the blade force and effects a displacement of the meridional flow towards both sidewalls. In this paper we work with a finite-volume-code for computations of the three-dimensional Reynolds averaged Navier-Stokes equations for an axial turbine stage with radial and two types of tangentially curved guide vanes. With computational results, mathematical formulations are developed for a new flow model of deflection of such blades that are formally compatible with the assumption of a rotation-symmetrical flow and with the existing throughflow codes, in order to predict the deflection angle over the blade height for the tangential leaned and curved blades.

1. Introduction

The classic design method for the blading of axial turbines, applied for many decades, stipulates that the profiles constituting the blade body are stacked in radial direction. Such stacked profiles can be different forms in the case of twisted blades. When such blading profiles are arranged with the profile centers of inertia on a radial line, the stacking line, bending stress by centrifugal force is avoided in rotor blades. The normal of the surface elements of blades designed accordingly is directed essentially to be tangential-axial, so that the pressure forces of the blades have no or quite small radial components; these are usually neglected in the aerodynamic calculation. Consequently only the radial pressure gradient and the centrifugal force resulting from the circumferential component of the flow velocity are acting in radial direction on the fluid particles passing through the turbine stage. Minor additional radial forces may result from streamline curvature and inclination. They enter as important parts into the equation of radial equilibrium, which is one of the basic equations for the fluid-dynamic design of axial turbomachines.

The classic blade design results in difficulties, particularly in the case of long blades as in the LP-part of condensing steam-turbines (e.g., [1]), because the large radial pressure gradient in the gap between stator and rotor causes a considerable variation of the degree of reaction over the blade length (e.g., 0% at the hub and 80% at the tip), which complicates blade design and reduces efficiency.

Consequently the idea was born by Deich et al. [2], to produce additional blade forces in radial inward direction by circumferential inclination of the stator blades in order to counteract the centrifugal forces and so to reduce the radial pressure gradient. The idea proved successful and caused extensive research on different blade forms with nonradial surface elements (e.g., [3–6]).

It was found in addition that a blade inclination in the sidewall area, which produces an obtuse angle between blade suction side and sidewall, can reduce secondary loss, while an acute angle will increase it. A pure inclination of the stacking line in circumferential direction *(lean)* will therefore create a positive effect on secondary loss at one end of the blade and a negative effect at the other end. Such inclined blade has been discussed by several authors; see, for example, Deich et al. [2, 7–9]; Han et al. [10, 11]; Harrison [12]; Hourmouziadis and Hübner [13]; Suslov and Filippov [14]; Wang et al. [15], an experiment research, and Deich et al. [2, 7–9]; Hourmouziadis and Hübner [13]; Pioske and Gallus [16];

Smith Jr. and Yeh [5]; Suslov and Filippov [14], a numerical research, as well as the references therein.

In order to realize the loss reducing obtuse angle between suction side and sidewall on both ends of the blade a curvature *(bow)* of the stacking line of the profiles is necessary. For stator blades this can be realized without mechanical trouble. This design appears favourable especially for short blades, as in height pressure (HP) and middle pressure (MP), part of steam-turbines, since there secondary loss together with clearance loss amounts to 20–40% of the total aerodynamic loss, while a reduction of the radial pressure gradient, not realizable with this type of blade, is not important. The research work of the flow in axial turbine stages with curved stator blades is therefore an important research topic for actual turbomachinery and aeroengines. And some results of the curvature blades can be found in Deich et al. [7–9]; Filippov and Wang [17, 18]; Wanjin et al. [11]; Harrison [12]; Hourmouziadis and Hübner [13]; Jiang et al. [19]; Vogt and Zippel [20]; Wang and Zheng [21]; Wolf and Romanov [22], an experiment research, and Filippov et al. [18, 23]; Hourmouziadis and Hübner [13]; Jiang et al. [19]; Pioske and Gallus [16]; Suslov and Filippov [14]; Wang and Zheng [21], a numerical research, where the curvature form of such researches is however mostly a simple curvature (circle) form.

Experimental researches and numerical computations are available as tools to obtain a detailed knowledge of the 3D-flow field and to understand the effect of geometric modifications. Simulation offers the opportunity of a very detailed insight in the complete flow field which is hardly accessible by experimental means. In addition it is ideally suitable for systematic research of geometry variations, as necessary in the present case. However, the quality of simulation results will be influenced by numerical iteration procedure, turbulence model, mesh-structure, wall-flow model, and so forth in a way that is difficult to control. It is therefore highly recommendable to validate the results of simulation with the aid of experimental data, as has been possible in the present case.

The object of the present paper is the numerical simulation of the flow in an axial turbine stage equipped with several variations of tangentially curved stator blades by using a finite-volume-code for the 3D Reynolds averaged Navier-Stokes equations. The research work is centered on

(i) effects of blade curvature on the flow in the stator,

(ii) effects of the stator blade curvature on the flow in the rotor (which remained unchanged in geometry),

(iii) effects of the blade curvature on stage characteristic,

(iv) new modeling of the radial blade force and the deflection angle over the blade height for the tangential leaned and curved blades that are formally compatible with the assumption of a rotation-symmetrical flow and with the existing throughflow codes.

Definition of Curved Blade. The research is restricted to stator blades curved of both ends as follows. The stacking line of the profiles is curved in circumferential direction as, for example, in Figure 1.

The curvature is defined by the angle ε between the stacking line and the sidewall at the blade end and by the extension h^k of the nonradial part of the stacking line. These values can be different at both ends of the blade.

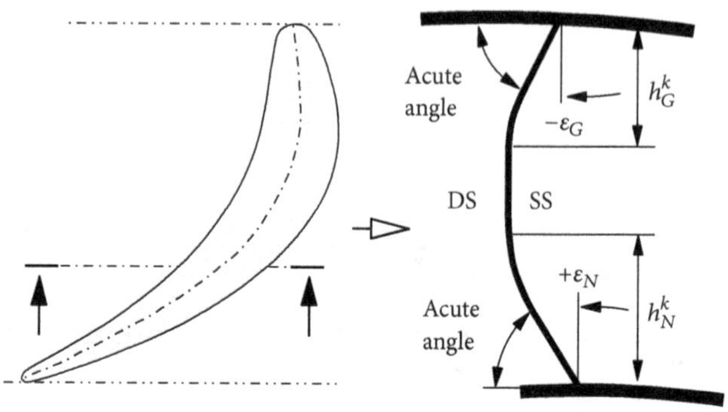

FIGURE 1: Definition of guide vanes curved on both ends by its stacking line in S_3-plane.

(i) Positive Tangential Blade Curvature. If an acute angle between the pressure side DS and the sidewall is realized, the curvature is defined as positive.

(ii) Negative Tangential Blade Curvature. A blade with negative curvature shows acute angles between suction side SS and sidewall.

2. Computational Methods

The applied numerical procedure solves the 3D-time-dependent Reynolds averaged Navier-Stokes equations with the finite volume method and comprises different turbulence models. Some validations showed a very good quality of the results. For turbomachinery as well as aeroengine applications the program contains a procedure for circumferential averaging of flow parameters in planes between blade rows, for example, between stators and rotor, in order to make a steady-state calculation feasible in both blade rows. A suitable averaging plane was provided in the grid design. The calculation used the *Baldwin-Lomax* turbulence model [24] and a three-level full multigrid method for convergence acceleration. Our simulation results showed a difference of mass flow in the inlet and outlet plane of less than about 0.02%. Details on the computational methods of this code are available in some literatures, for example, [25–30].

2.1. Governing Equations. The applied numeric procedure regards the so-called Reynolds averaged Navier-Stokes *(RANS)* equations as the physical basic model. Their hyperbolic system of conservation equations applied for mass, impulse, and energy depends on velocity, pressure, and enthalpy. They can be mostly written as a recapitulatory form

$$\frac{\partial U}{\partial t} + \underline{\nabla} \cdot \underline{F} = Q. \tag{1}$$

Thereby $U = (\rho, \rho\underline{v}, \rho E)^T$ denotes the conservative variables, $\underline{F}$ the matrices of the fluxes, and Q the source terms column-vector. $\underline{F}$ is actually combined from the matrices of the frictionless flux (Euler) $\underline{F}^{\mathrm{Eu}}$ and the dissipative flux (Navier-Stokes) $\underline{F}^{\mathrm{NS}}$. It can be rewritten as the following compact form further into the following general conservative form [25, 27, 29, 30]:

$$\frac{\partial}{\partial t}\begin{bmatrix}\rho \\ \rho\underline{v} \\ \rho E\end{bmatrix} + \underline{\nabla}\cdot\begin{bmatrix}\rho\underline{v} \\ \rho\underline{v}\otimes\underline{v} + p\underline{\underline{I}} - \underline{\underline{\tau}} \\ \rho\underline{v}H - \underline{\underline{\tau}}\cdot\underline{v} - k\underline{\nabla}T\end{bmatrix} = \begin{bmatrix}0 \\ \rho\underline{f}_e \\ W_f + q_H\end{bmatrix}, \quad (2)$$

where $\rho, \underline{v}, E, H, \underline{\underline{I}}, \underline{\underline{\tau}}, \underline{f}_e, W_f, q_H$ denote the density, the velocity vector, the total energy, the total enthalpy, the 3×3 unit matrix tensor, the viscous shear stress tensor, the external force vector, the column-vector of characteristic variables, and the source term, whereby the velocity vector has components u, v, w. According to the character of flow we can differentiate in each case according to the hyperbolic, parabolic, and elliptical type of equation. The time-dependent Navier-Stokes equation possesses a hyperbolic-parabolic character within the temporal-spatial ranges, because the steady Navier-Stokes equations are a mixed type in the spatial domain. From it we obtain the elliptical-parabolic type which describes subsonic flows and the hyperbolic-parabolic type supersonic flows. For the flows with high Reynolds numbers the system of the conservative equations causes convection obtained in most flow domains.

2.2. Method of the Spatial Discretization. The spatial fluxes of the applied basic equations of (1) and (2) are actually discretized using the finite volume scheme. The finite volumes scheme has been usually spread using numeric technology in particular for flow problems, because the integral form of the conservative equations can be interpreted by direct discretization for a volume cell of the grids with the structured as well as unstructured grids very flexibly and easily [25, 27, 29, 30].

Under the condition that the temporal derivatives are exchangeable with the spatial integrations (i.e., the integrands are constantly differentiable in t), the integral forms can be normally deduced from the above differential expression (1). For this reason the following integral form of the conservative equations is given by

$$\frac{\partial}{\partial t}\int_V U\,dV + \oint_{\partial V}\underline{F}\cdot d\underline{S} = \int_V Q\,dV; \quad (3)$$

therefore it is formulated for any volume V with the control surface $S = \partial V$. The discretization in the spatial is realized with an approximation of the volume V by elementary cells, which contain, respectively, an element ΔS of control surface S. Then the above integral equation (3) can be discretized and rewritten as follows:

$$\frac{\partial}{\partial t}\int_V U\,dV + \sum_{\text{faces}}\underline{F}\Delta\underline{S} = \int_V Q\,dV, \quad (4)$$

whereby $\underline{F}\Delta\underline{S}$ denotes the flux where they are combined, respectively, from frictionless $\underline{F}^{\mathrm{Eu}}\Delta\underline{S}$ and dissipative $\underline{F}^{\mathrm{NS}}\Delta\underline{S}$ fluxes.

2.2.1. Numerical Fluxes. The frictionless flux $\underline{F}^{\mathrm{Eu}}\Delta\underline{S}$ and the dissipative flux $\underline{F}^{\mathrm{NS}}\Delta\underline{S}$ are discretized, for example, in each case by the *upwind scheme* and an application of *Gauss theorems*. The treatment of the two fluxes will be not implemented here; their detailed description is represented in [25, 27, 29–35]. The physical fluxes are mainly discretized in the approximation of the *central scheme* as follows:

$$[\underline{F}\cdot\Delta\underline{S}]_{i+(1/2)} = \frac{1}{2}[\underline{F}_i + \underline{F}_{i+1}]\cdot\Delta\underline{S}. \quad (5)$$

This treatment for the numeric flux does not lead actually to a stable scheme, because it does not give sufficient numeric dissipation. Therefore one needs an artificial dissipation term, in order to eliminate odd-even oscillations because of the central discretization *(high frequency oscillations)* and in order to capture shocks without oscillation. The usual formulation prosecutes the type of the dissipation from Jameson et al. [31]; it is actually formulated as sum of conservative variable derivation, respectively, in second order with nonlinear coefficients for *shock capturing* and in linear fourth order for the *background oscillations*. In this way the numeric flux (5) transferred into the form [25, 27, 29, 30]

$$\underline{F}^{\star}_{i+1/2}\cdot\Delta\underline{S} = \frac{1}{2}[\underline{F}_i + \underline{F}_{i+1}]\cdot\Delta\underline{S} - d_{i+1/2}, \quad (6)$$

where $d_{i+(1/2)}$ is the artificial dissipation. This term from Jameson et al. is defined as follows in one-dimensional form [31]:

$$\begin{aligned} &d^j_{i+1/2} \\ &= \epsilon^{(2)}_{i+1/2}(u_{i+1} - u_i) \\ &\quad - \epsilon^{(4)}_{i+1/2}[(u_{i+2} - u_{i+1}) - 2(u_{i+1} - u_i) + (u_i - u_{i+1})]. \end{aligned} \quad (7)$$

The scalar coefficients ϵ are defined in the following way

$$\begin{aligned} \epsilon^{(2)} &= \kappa^{(2)}|a_c|\phi, \\ \epsilon^{(4)} &= \max\left[0, \left(\kappa^{(4)}|a_c| - \epsilon^{(2)}\right)\right] \end{aligned} \quad (8)$$

and connected with the variables of pressure and temperature

$$\phi_i = \max\left(\left|\frac{p_{i+1} - 2p_i + p_{i-1}}{p_{i+1} + 2p_i + p_{i-1}}\right|, \left|\frac{T_{i+1} - 2T_i + T_{i-1}}{T_{i+1} + 2T_i + T_{i-1}}\right|\right). \quad (9)$$

The pressure quantities are only used for liquids.

For the Navier-Stokes equation, the friction-affected dissipation is normally not sufficient in the solution and therefore is the above term of the artificial dissipation necessarily by the numeric solution of the Navier-Stokes equation. The unexpected difficulty with this approximation hence comes from the fact that the solution and the convergence can be strongly dependent on the coefficients [25, 27]. This approximation will be used with either explicit or implicitly temporal integration.

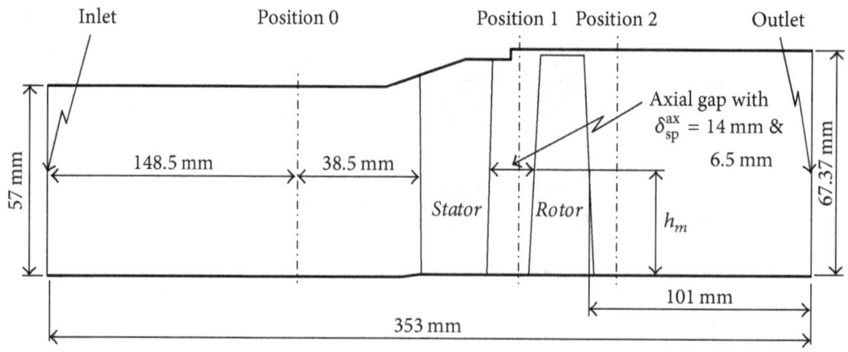

FIGURE 2: Meridional plane of the experimental and computational domain for the applied axial turbine.

2.3. Turbulence Model. Since there is still no universal and general turbulence model, the turbulence model will govern the results simulated [24, 25, 27]. In this study, Baldwin-Lomax turbulence model, which is suitable for the wider class of interference model to achieve enough accurate and reliable results for a turbomachinery, is chosen for the fully turbulence condition. The added blended wall function implemented in the Baldwin-Lomax turbulence model leads to the boundary layer at the solid walls with further refinement by y^+ value with $y^+ \approx 1$.

3. Computational Conditions

As a starting point data from an experimental research of a single stage air turbine with different types of curved guide vanes were made available to us by the department for power, nuclear turbine construction, and aircraft engines of State St. Petersburg Technical University (Russia). They comprised stage characteristics as well as detailed flow measurements with aerodynamic probes. Comparison of these data with the results of numerical simulation of the same geometries, reported in [36], showed good agreement and served as validation of the simulation technique chosen. A number of stator configurations with different types of bow was derived from the geometry of the experimental turbine and investigated numerically.

3.1. Single Stage Experimental Air Turbine. Figure 2 shows the meridional section of the turbine, the test results of which have been used for the validation of the numerical simulation procedure. For the measuring planes positions 0, 1, and 2 data of probe measurements were available. The extension of the meridional section in Figure 2 shows, too, the field of the numerical calculation.

3.2. Geometrical Data of the Researched Blades. The experimental investigations were made for a radial guide vane $Basis_{2A}$ and a curved guide vane Bow_{2B}, both combined with the same rotor. Characteristic data of the center-line profiles are given in Table 1. For the numerical research 8 different stator blade rows have been calculated with a systematic variation of curvature by tangential displacement of the unchanged profiles of the radial stator $Basis_{2A}$. Results are given here for the basis blading and the two most characteristic bow stators. Data describing the curvature are shown in Table 2; the parameters describing the curvature are ε and the relative extension $h_{\text{bez}}^k = h^k/h$ according to Figure 1. Figure 3 shows a 3D representation of the three guide vanes.

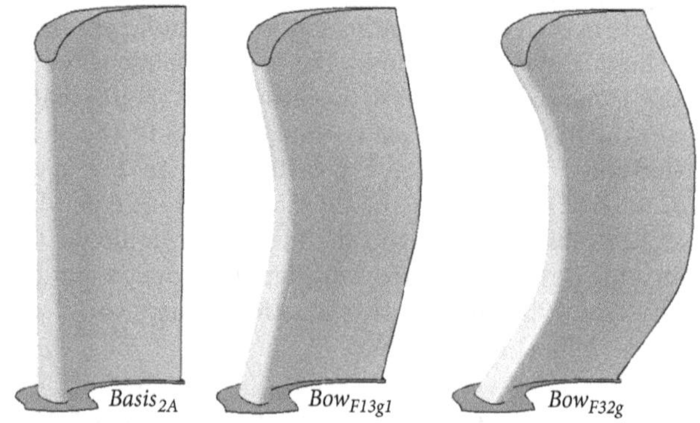

FIGURE 3: Schematic three-dimensional representation of the radial and the curvature guide vanes.

3.3. Computational Mesh. The calculation mesh for the numerical simulation is identical in form as well as in cell number for all investigated blading. Table 3 gives the cell numbers in three coordinate directions (i circumferential, j radial, and k axial direction). The entire computation area was divided into three zones in order to realize a rather orthogonal mesh. Near the blade profiles an O-mesh was used, for the inlet channel an H-mesh, and for the outlet area of the stator an I-mesh with H-mesh extension in order to adapt the wake of the blade. In the outlet channel of the rotor only an H-mesh was applied. A total of 973191 nodes resulted for the complete calculation area.

Figure 4 shows the grid configuration for the applied guide vanes and rotor blades. The interfaces of the two domains for stator and rotor are the only overlapped region and there is only one accepted virtual translating direction along the interfaces; that is, it is unacceptable to translate across the interfaces. Considering the complex geometry of this system, for accuracy of the prediction, the regions near interfaces and boundary layers of the stator and rotor, finer

TABLE 1: Geometrical data of the center line profiles.

Notation	Unit	$Basis_{2A}$	Bow_{2B}	Rotor
Blade height	[mm]	65.6	65.6	66.87
Middle diameters	[mm]	363.98	363.98	364.24
Blade chord	[mm]	31.159	32.101	20.26
Max. camber	[mm]	7.047	5.254	6.063
Stagger-angle	[deg.]	43.00	36.00	57.20
Inlet angle	[deg.]	80.00	90.00	42.00
Outlet angle	[deg.]	15.37	13.95	21.33
Rel. pitch	[−]	0.4898	0.4745	0.7433
Blade number	[−]	74	74	76
δ_{sp}	[mm]	—	—	0.5

TABLE 2: Geometrical data of the investigated guide vanes.

Notation	Abbreviation	ε_N	ε_G	$h^k_{\text{bez},N}$	$h^k_{\text{bez},G}$
Unit	[−]	[deg.]	[deg.]	[−]	[−]
1	$Basis_{2A}$	0.0	0.0	—	—
2	Bow_{F13g1}	+13.2	−12.2	0.453	0.438
3	Bow_{F32g}	+32.0	−29.0	0.409	0.374

TABLE 3: The mesh distributions of the turbine stage.

Notation	H-mesh (Inlet) $[i \times j \times k]$	O-mesh (Blade) $[i \times j \times k]$	I + H/H-mesh (Outlet) $[i \times j \times k]$
le	33 × 57 × 33	33 × 57 × 217	33 × 57 × 49
la	41 × 57 × 17	41 × 57 × 177	41 × 57 × 25

structured mesh was employed in the present numerical computation; on the other side, regions far from stator and rotor are meshed with coarser structured mesh.

This mesh configuration considers the calculation and comparison of turbulence model and combines it with the near wall modified function with low Reynolds numbers in order to effectively offset the interference that occurs in the internal flow field to improve the near wall computational grid's deficiencies.

Therefore, the distance of the first cell layer from the blade surface was 0.001 mm; the added blended wall function implemented in the turbulence model leads to the boundary layer at the solid walls with further refinement by y^+ value; this gave a value of the dimensionless wall-distance $y^+ \approx 1$.

The computation of entire mesh used a three-level full multigrid method for convergence acceleration. Our simulation results for such a mesh configuration showed a difference of mass flow in the inlet and outlet plane of less than about 0.02%. This validation showed a very good quality of the applied mesh configuration.

3.4. Operating Conditions of the Turbine Stage. The numerical calculation of the flow in the turbine stage was realized for its optimum operating point deduced from the experimental data, all relevant data for inlet and outlet are summarized in Table 4. The total-static pressure ratio $\Pi \equiv p^{\text{aus}}_{2,s} / p^{\text{ein}}_{0,t} = 0.7$.

The applied 3D calculations of total stage efficiency with axial gap 6.5 mm were performed for two different ranges of the run number ν.

(i) Run Number ν in the Range 0.3 to 0.6. In this range of the run number, the pressure ratio $\Pi = 0.7$ remained constant, so the enthalpy change remained also constant. But the rotational speed was changed for different operating points. The applied data are shown in Table 5.

(ii) Run Number ν in the Range 0.6 to 0.9. In this range of the run number, the rotational speed was no longer changed, but the reduced rotational speed was maintained at a constant value 4000. However, on the contrary, the pressure ratio was changed by various static pressures on entry. This data is shown in Table 6. Therefore, the enthalpy or the mass flow is also changed.

4. Results and Discuss

4.1. Validation of Numerical Results. For the operating points in both Tables 5 and 6, the turbine stage was computed with a real axial gap 6.5 mm, the results are shown in Figure 5. It shows that the stage efficiency is a function of the run number ν. The run number ν has been divided into two ranges for the computations of the stage efficiency. In the range of the run number between 0.3 and 0.6, the stage efficiency of the radial blade Basis2A agrees very well with the experimental data. However, the computations of this radial blade in the range between 0.6 and 0.9 also show a pretty good consistency outside the range between 0.65 and 0.85 since in this range only

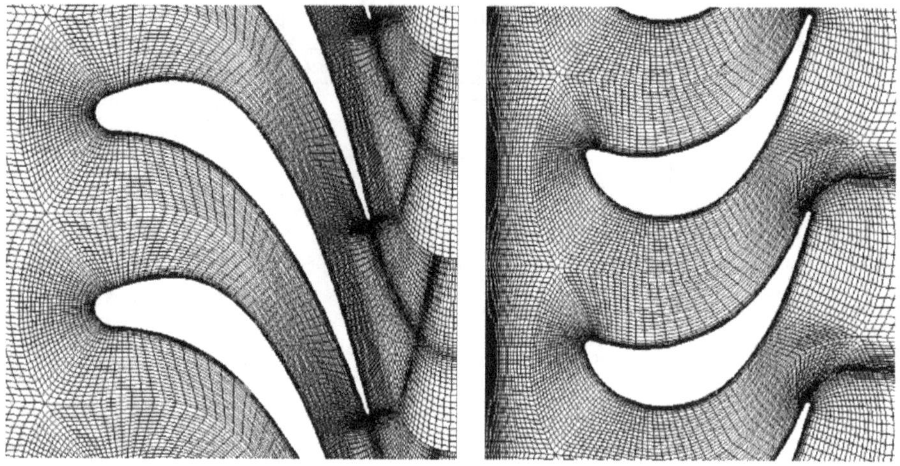

FIGURE 4: Configuration of the calculating mesh in the middle section of the radial blades (mesh shown in (b) (rotor) and mesh shown in (a) (stator)).

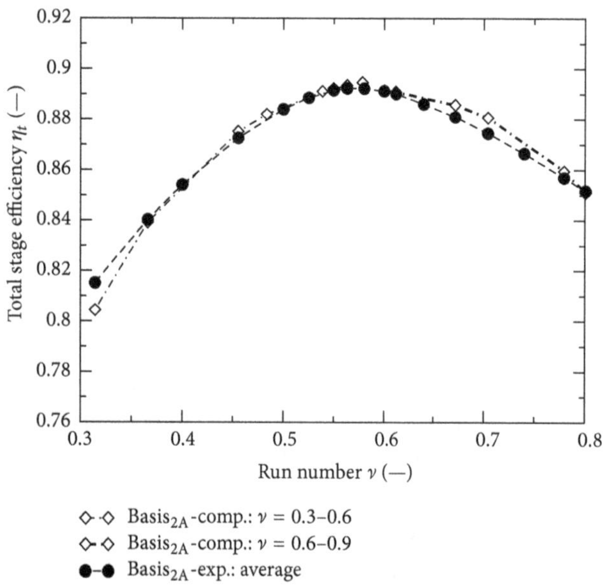

FIGURE 5: Total stage efficiency as a function of the run number ν for the radial blade Basis2A with real axial gap 6.5 mm.

TABLE 4: The aerodynamic data in the optimal operation point.

Notation	Unit	Value
$p_{0,t}^{ein}$	[Pa]	142857
$p_{0,s}^{ein}$	[Pa]	140812.5
$p_{2,s}^{aus}$	[Pa]	100000
$T_{0,t}^{ein}$	[K]	323.1
$T_{0,s}^{ein}$	[K]	321.772
Tu	[%]	$1\cdots2\%$
h/S_{Bi}	[—]	2.1^{le}
c_z^{ein}	[m/s]	51.657
c_r^{ein}	[m/s]	0.0
c_φ^{ein}	[m/s]	0.0
$\dot{m}$	[kg/s]	4.93
n	[rpm]	7190
R	[J/(kgK)]	287.14
ρ	[kg/m^3]	1.53

a deviation of about 0.3 occurs. The accuracy of presented numerical results can therefore be validated by this comparison result clearly.

4.2. Effect on the Meridional Streamlines. The form of meridional streamlines is essentially determined by the radial blade forces acting on the flow, mainly the centrifugal force c_φ^2/r and the radial pressure gradient $\partial p/\partial r$. In the familiar stage design according to the free vortex law (approximately it is realized in the $Basis_{2A}$-stage), these are balanced so that no radial displacement of streamlines occurs, as in Figure 6.

The type of bow applied in two guide vanes presented here will create additional radial blade forces acting towards the hub and the casing in the sidewall regions. The balance of radial forces and the radial flow displacement will vary to extent in the circumferential direction too and then a 3D-flow field will be formed really. A simplified presentation is shown in Figure 6; a circumferential averaged flow over the pitch is calculated and applied to the calculation of meridional streamlines in this figure; a similar result would be achieved by a 2D-calculation of the flow field.

By the comparison with both curved guide vanes of the meridional flow in Figure 6, it shows that the radial displacement of streamlines towards the guide vane ends has the same effect of bow. The bow-effect for the vane Bow_{F32g}, whose inclination (bow) angles ε have more than double the value of the vane Bow_{F13q1}, is more distinct. The radial deflection tapers off in the gap between stator and rotor and is concentrated in the outlet part of guide vanes, where the radial blade force has its full effect. Consequently, the deflection of

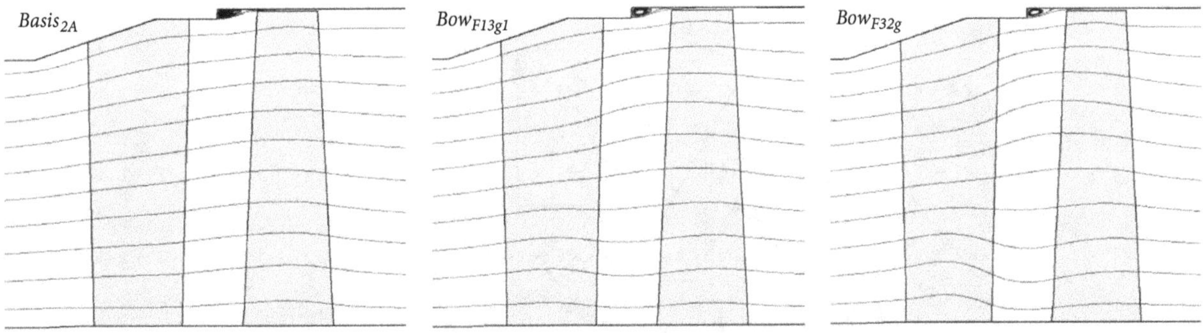

FIGURE 6: Circumferentially averaged meridional streamlines for radial and bow stators.

TABLE 5: The operating points of rotating speed as a variable in the range of run number ν between 0.3 and 0.6.

ν [−]	0.314	0.366	0.455	0.484	0.539	0.550	0.563	0.578
n [rpm]	4113	4790	5955	6326	7046	7190	7366	7567

TABLE 6: The operating points of total pressure at the inlet as a variable in the range of run number ν between 0.3 and 0.6.

ν [−]	0.600	0.612	0.672	0.704	0.778	0.800
$p_{0,t}^{\text{ein}}$ [Pa]	134543	132957	126460	123753	118911	117797

streamlines towards the sidewall will cause a deceleration of the axial flow velocity in the middle part of guide vanes, which can be seen under the increased distance of streamlines compared with the radial guide vane $Basis_{2A}$; such an effect for the velocity distribution over the blade height is important.

4.3. Radial Blade Forces Produced by Blade Curvature

4.3.1. Treatment and Modeling of the Radial Blade Force. Blading designed according to standard principles with radial stacking line produces blade forces acting essentially in circumferential and axial directions, and the radial component remains very small. In addition, the inclined and curved guide vanes develop the radial blade force, which can attain the same magnitude as the circumferential and axial components and therefore enter into the radial equilibrium as important terms, which influences the flow field.

A schematic representation of three-dimensional blade force and an element angle of inclined or curved guide vane are represented as a definition of balance direction in Figure 7 at the blade trailing edge. It shows that on the flow the reaction blade forces of guide vane are exerted on the pressure side $\underline{f}_s^{DS}$ and the suction side $\underline{f}_s^{SS}$, respectively. They act on the blade surface perpendicularly and have different values on both sides.

The local blade force $\underline{f}_s$ is defined by the inclined surface element of the blade and its axial and radial length elements Δz_z and Δh_r can be calculated from the local static pressure p_s at the blade surface in the following form:

$$\underline{f}_s = p_s \Delta A_E = p_s \frac{\Delta z_z}{\cos\nu} \frac{\Delta h_r}{\cos\varepsilon}, \tag{10}$$

where ε is the inclination angle against the radial direction and the deflection angle ν against the axial direction of the surface element. In addition, the element length along the blade contour Δz_E and the element height along the inclined blade height Δh_E can be computed by the forms $\Delta z_E = \Delta z_z / \cos\nu$ and $\Delta h_E = \Delta h_r / \cos\varepsilon$, respectively.

The local blade force $\underline{f}_s$ in Figure 7 can be divided into two components of blade force $\underline{f}_{s,1}$ in the S_1-plane (blade to blade) and $\underline{f}_{s,3}$ in the S_3-plane. The blade force in the S_1-plane arises from the blade force $\underline{f}_s$ with the inclined angle ε_s in the following form:

$$\underline{f}_{s,1} = \underline{f}_s \cos\varepsilon_s; \tag{11}$$

thereby the inclined angle ε_s is determined over the tangent through the radial blade force $\underline{f}_{r,s}$ and the blade force in the S_1 plane $\underline{f}_{s,1}$

$$\begin{aligned}\varepsilon_s &= \arctan\left(\frac{\underline{f}_{r,s}}{\underline{f}_{s,1}}\right) = \arctan\left(\frac{\underline{f}_{\varphi,s}\tan\varepsilon}{\underline{f}_{\varphi,s}/\cos\nu}\right) \\ &= \arctan(\tan\varepsilon\cos\nu).\end{aligned} \tag{12}$$

Both blade forces in the S_1 and S_3-planes can be further divided into three coordinate directions, $\underline{f}_{\varphi,s}$, $\underline{f}_{r,s}$, and $\underline{f}_{z,s}$ describing the blade forces in circumferential, radial, and axial directions, respectively. By the blade force in the S_1-plane, the circumferential blade force can be also indicated in the form

$$\underline{f}_{\varphi,s} = \pm \underline{f}_{s,1} \cos\nu; \tag{13}$$

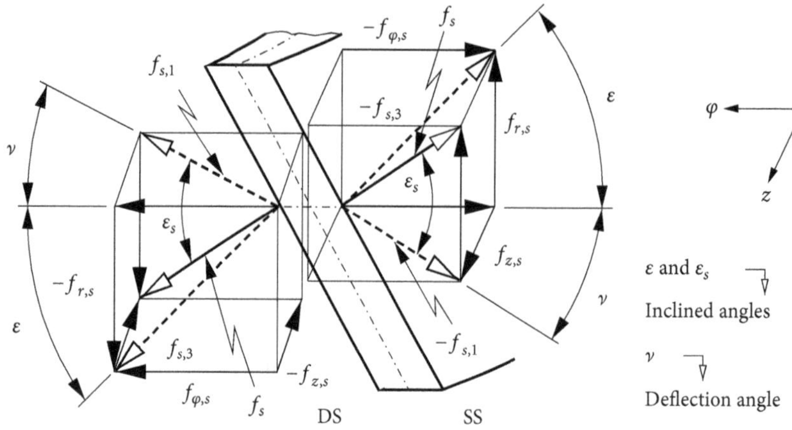

FIGURE 7: Schematic representation of 3D-blade forces and angles on the surface element of inclined or curved guide vane as a definition of the force direction (DS = pressure side; SS = suction side).

with this definition the radial blade force in the S_3-plane arises

$$\underline{f}_{r,s} = \pm \underline{f}_{\varphi,s} \tan \varepsilon = f\left(\varepsilon, \underline{f}_{\varphi,s}\right). \quad (14)$$

The local radial blade force $\underline{f}_{r,s}$ can be therefore obtained by the following new general form:

$$\underline{f}_{r,s} = \left[\left[p_s \frac{\Delta z_z \; \Delta h_r}{\cos \nu \; \cos \varepsilon} \cos\left[\arctan\left(\tan \varepsilon \cos \nu\right)\right]\right] \cdot \cos \nu\right] \tan \varepsilon. \quad (15)$$

4.3.2. Distribution of the Resultant Radial Blade Force. The distribution of the radial blade force over the entire blade height must be described in detail for the effect of the blade curvature. The total radial blade force $\tilde{\underline{f}}_{r,s}$ of a profile section with the elementary height dh results from the integration over all surface elements on this profile

$$\tilde{\underline{f}}_{r,s} = -\sum_{i=1}^{m} \left(\underline{f}_{r,s}^{\mathrm{DS}}\right)_i + \sum_{j=1}^{n} \left(\underline{f}_{r,s}^{\mathrm{SS}}\right)_j, \quad (16)$$

where the indices i, j are the ith and jth surface element of the blade on the pressure (DS) and suction (SS) sides, respectively. The total radial blade force is the sum of the pressure side part and the suction side part; it is always directed from the pressure side to the suction side. The distribution of the total radial blade force per length-element of the blade is shown in Figure 8 for the radial guide vane (assuming no radial blade force) and both curved guide vanes. From Figure 8 we find the following phenomena:

(i) The region for the radial blade force is about equal to the region of blade zones for bow (ca. 40% of blade height on both ends).

(ii) The radial blade force is increasing in the inclination (curved) angle ε as well as on the same blade in direction of the increasing bow as from blade to blade with increased bow.

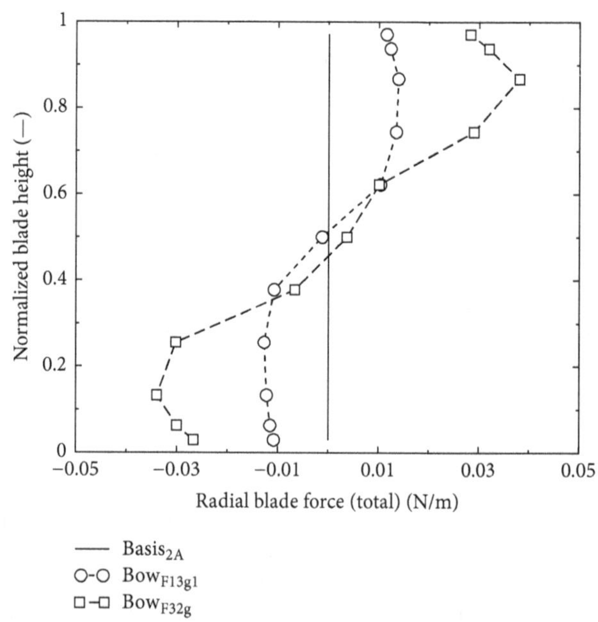

FIGURE 8: Distributions of the resultant radial blade force for the guide vanes over the blade height.

(iii) The radial blade force tapers off in the proximity of sidewall, even if the inclination (curved) angle increases there. This is probably due to reduced aerodynamic loading of profiles there.

The distributions and magnitude of the radial blade force correspond directly with the deflection of meridional streamlines in Figure 6, introducing the radial blade force in the equation of radial equilibrium; their form can be calculated quantitatively.

4.4. Effect of the Blade Curvature on the Static Pressure Distribution at the Profiles. The aerodynamic blade forces and their components result from the static pressure on the profile and the geometric orientation of its surface elements. The pressure distribution itself is, also, of great importance for the flow over

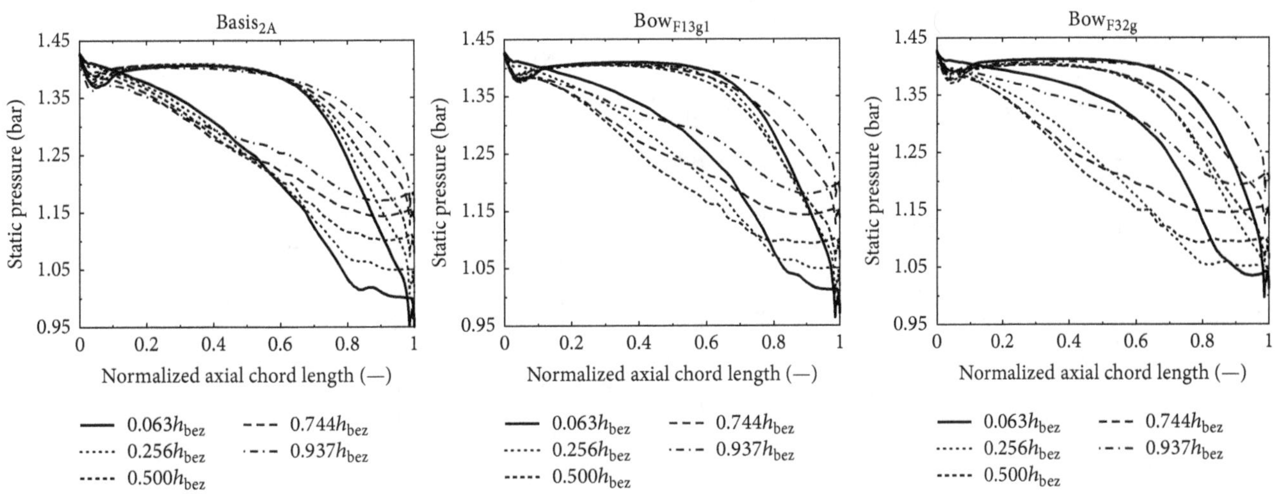

FIGURE 9: Pressure distributions over the blade profile of the stator at the 0.063 (thick line) 0.256, 0.5, 0.744, and 0.937 normalized blade height.

the profile, especially the boundary layer and its development. Boundary layer thickness, transition, and eventually separation depend to a great deal on the pressure distribution. It is of great interest therefore how the inclination and curvature of a blade influence the pressure distribution.

4.4.1. Aerodynamic Load of the Blade Profiles. If the static pressure distribution over a profile is presented over the axial chord, as in Figure 9, the area circumscribed by the pressure side and the suction side curve is directly proportional to the circumferential (or working) total force of the blade profile; the difference between the two curves at an axial station describes the local aerodynamic loading and the form, especially of the suction side curve, is important for the boundary layer development and hence for the profile loss.

Figure 9 shows the static pressure distributions over the profile sections at different height positions for the three guide vanes treated here. Two curves are near the sidewalls (0.063 and 0.937 normalized blade height), two are in the middle of the bow zones (0.256 and 0.744), and one is in the middle of the blades (0.5). It is obvious that, because of the radial pressure gradient between stator and rotor, the outlet pressure increases with increasing blade height, as visible in Figure 9 for all three stators. Noteworthy is the decrease in aerodynamic loading (circumscribed area) towards the sidewall with increasing bow, especially in the hub area. Finally remarkable and quite unexpected is the fact that the pressure distributions of the profiles close to the sidewalls of the curved vanes are distinctly deformed, so that the aerodynamic loading is not only reduced, but also displaced towards the trailing edge and acceleration on the suction side is delayed. This effect seems, too, to be proportional to the degree of curvature.

4.4.2. Pressure Distribution on the Suction Side. The variation of the pressure distribution, especially on the suction side, by the blade curvature is shown in Figure 9 for several distinct profile sections. It can also be demonstrated in a continuous way by a graph of the pressure distribution on the suction side over the blade height as shown in Figure 10 for the three vanes. The isobaric lines of the radial basis vane extend essentially in radial direction over the blade height; the peaks near the trailing edge are a result of a slight overspeed on the profile. In direct proximity of the sidewalls they bow a little downstream. This somewhat more distinct effect near the casing is probably due to the conical contour.

Blade curvature now effects a clear displacement of the isobars in downstream direction near both sidewalls, more pronounced by stronger blade curvature, and resulting from the downstream movement of the blade loading described in Section 4.4.1. While the local inclination of the blade produces for the bulk of the flow an additional force directed towards the sidewalls, as shown before, it creates, somewhat unexpected, near the suction surface of the blade a component of the pressure gradient, which is directed towards the middle of the blade. As is well-known the low energy fluid of the secondary flow accumulates in the corner between suction side and sidewall. This static pressure component directed towards the center of the blade might be able to transport part of the low energy fluid out of the suction corner and might be assisted by the obtuse angle between sidewall and suction side resulting from the blade inclination. Here is a reason for the repeatedly observed reduction of secondary loss by bowing of blades.

4.5. Effect of the Curvature on the Vane Outlet Angle. Since the inclination of the blade changes the aerodynamic loading as well as the circumferential component of the blade force the turning of the flow (e.g., [37]), that is, the outlet angle of the vane, will be influenced, too (e.g., [38]). Figure 11 shows the distribution of the flow outlet angle, averaged over one pitch, over the blade height. The angle of the radial blade is equal to the value according to the sine-rule (effective angle) over most of the blade height. The curved blades exhibit clear underturning over 20% of the blade length at both ends, intensified by higher blade inclination, which is certainly due

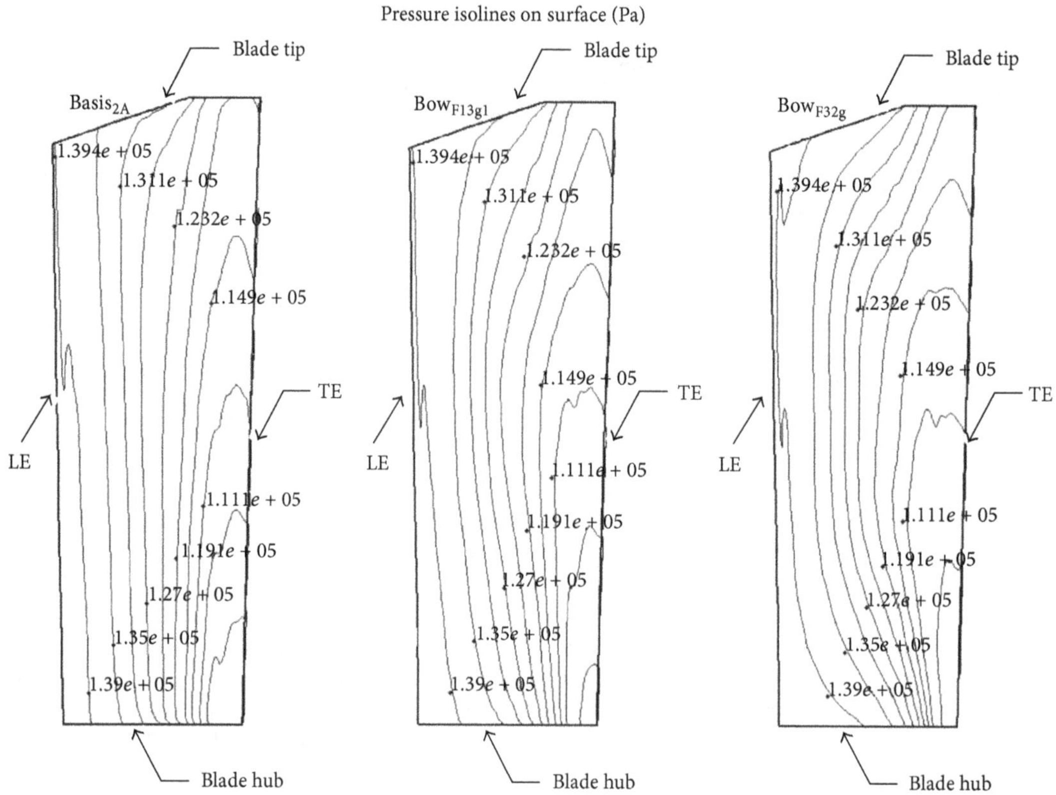

FIGURE 10: Pressure distributions on the suction side of the blade surface, where the unit of the pressure isolines on the blade surface is in Pascal. Therefore, LE and TE denote the leading edge and the trailing edge, respectively.

to the reduced blade loading in these areas. Not so obvious is the explanation for the overturning by 2 to 4 deg. in the middle zone. Probably it can be attributed to the deceleration of the axial flow velocity in this area because of the deflection of the flow towards the sidewalls.

4.6. Effect of the Blade Curvature on the Loss Coefficient of the Vanes. As explained initially the advantage gained by use of partly or completely inclined vanes is the reduction of the radial pressure gradient and hence the gradient of the degree of reaction especially for long blades. The interest in real bow vanes in turbines accrues mainly from the possibility of reducing the flow losses. Therefore the research of the influence of vane curvature on aerodynamic loss is of central importance. The vane loss coefficient was defined according to Traupel [39] as

$$\zeta^{\text{le}} = 1 - \frac{c_1^2/2}{\Delta h_s^{\text{le}} + c_0^2/2}. \tag{17}$$

The isentropic enthalpy drop Δh_s^{le} of the stator is

$$\Delta h_s^{\text{le}} = \frac{\kappa}{\kappa - 1} R T_{0,s} \left[1 - \left(\frac{p_{1,s}}{p_{0,s}} \right)^{(\kappa-1)/\kappa} \right]. \tag{18}$$

The loss coefficient at a sufficient number of stations along the streamtubes of constant % blade height was determined from the results of the 3D-simulation with circumferentially

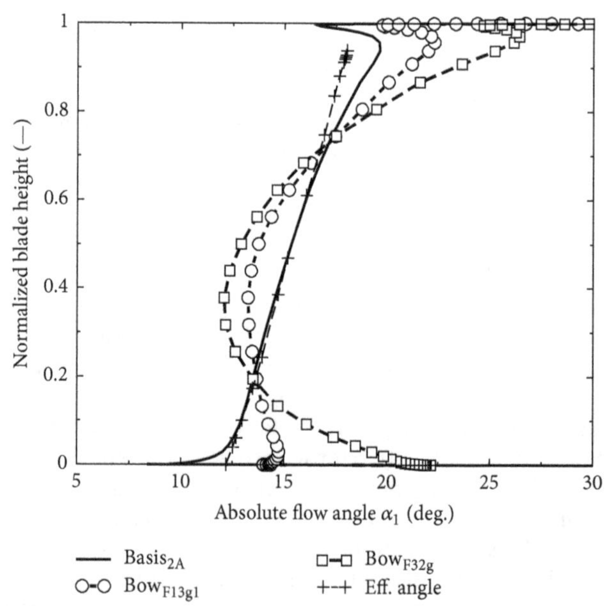

FIGURE 11: Circumference averaged exit flow angle of the stators over the channel.

averaged values of the parameters and is presented for the three different vane types in Figure 12, whereby $p_{0,s}$ is a function of the radius in (18) and all parameters are used in this connection along the streamlines over blade height in

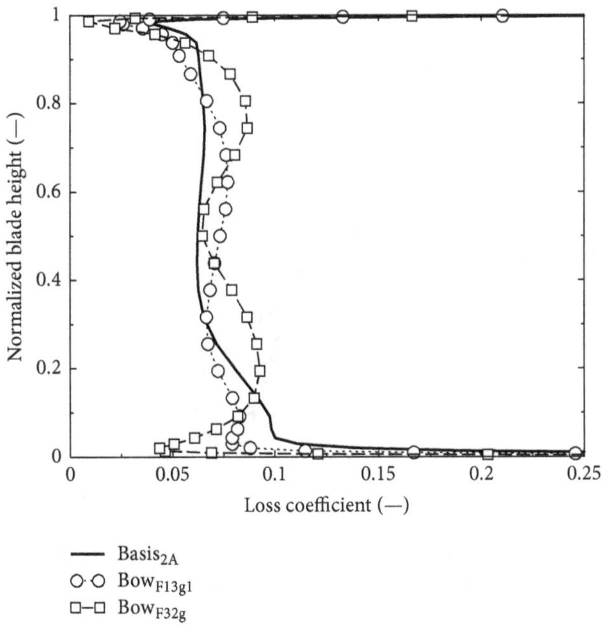

FIGURE 12: Circumference averaged loss distributions of the stators over the channel height.

each case for stator and rotor. The curves are not simple to interpret, but it is obvious that the vane with moderate bow has lower losses near both sidewalls and a certain increase in the middle zone, compared with the radial vane. The vane with high bow exhibits a sharp reduction of the loss coefficient in direct proximity of the sidewalls, but then a considerable increase around 0.2 and 0.8 normalized blade length, while in the middle of the blade the loss is equal to that of the radial blade.

The changes of the loss distribution and magnitude will be due to a combination of influences, as different development of profile boundary layer because of different pressure distribution, oblique movement of boundary layer on inclined parts of the blade, different secondary flow, and so forth. Finally the well-known deficiencies of Navier-Stokes-solvers in calculating dissipation correctly must not be forgotten.

The integrated loss of the three stators is calculated by the presented curves of the illustrations according to

$$\overline{\zeta}_{\text{int}} = \frac{1}{h_{\text{bez}}} \int_0^{h_{\text{bez}}} \zeta\left(h_{\text{bez}}\right) dh_{\text{bez}} \tag{19}$$

and their differences according to

$$\Delta\overline{\zeta}_{\text{rech}} = \frac{\overline{\zeta}_{\text{rech}} - \overline{\zeta}_{\text{rech}}^{\text{Basis}_{2A}}}{\overline{\zeta}_{\text{rech}}^{\text{Basis}_{2A}}}. \tag{20}$$

The values are presented in Table 7. Such results are identical to Figure 6 for the effects of the radial displacement of the meridional streamlines for radial and curved blades. It turns out that a moderate bow effects a reduction in total loss, while the high bow vane exhibits higher loss than the radial vane.

TABLE 7: Integrated loss coefficient of all investigated vanes.

Notation	$\overline{\zeta}_{\text{rech}}$	$\Delta\overline{\zeta}_{\text{rech}}$ [%]
$Basis_{2A}$	0.0767	—
Bow_{F13g1}	0.0744	−2.943
Bow_{F32g}	0.0782	+2.005

TABLE 8: Integrated loss coefficient of the rotor for the calculated blading.

Notation	$\overline{\zeta}_{\text{rech}}$	$\Delta\overline{\zeta}_{\text{rech}}$ [%]
$Basis_{2A}$	0.0814	—
Bow_{F13g1}	0.0771	−5.346
Bow_{F32g}	0.0728	−10.531

5. Effects of the Vane Curvature on the Rotor Flow

Any modification of the stator of a turbine stage, which changes the flow field at its outlet, will directly influence the flow entering the rotor and therefore its aerodynamic effectivity (e.g., [40–45]). In our case the curvature of the stator vanes changes the outlet angle and the axial flow velocity; together they define in the velocity triangle the relative inlet flow angle for the rotor. The difference between aerodynamic inlet angle and the geometric inlet angle of the rotor blade, the incidence angle, is of central importance for the aerodynamic action of the rotor blade profiles. High positive inlet angles cause an aerodynamic overload of the profile and high loss. Negative incidence angle may result in adverse profile pressure distribution and increased loss, too.

Figure 13, showing the rotor incidence angles for the three cases, makes it clear that the basis stage, which had been conceived for the experiments at the department for power, nuclear turbine construction, and aircraft engines of State St. Petersburg Technical University, had already not been designed in an optimum way, since in the top half of the rotor blade a considerable negative incidence prevails. The loss distribution, presented in Figure 14, shows however that this design seemed not to have a negative consequence. The influence of the vane bow on rotor incidence in Figure 13 results directly from the variation of the vane outlet angle α_1. Therefore its values move towards negative in the two sidewall areas and towards positive on the middle of the blade. Interesting, though not quite easy to explain, is the reaction of the rotor loss coefficient on the bow of the stators. The remarkable reduction of the loss coefficient in the lower half of the blade is not directly attributable to the variation of the incidence. Probably it results from a reduction of profile loss as well as secondary loss and would require a detailed study of the loss characteristics of the rotor profiles in function of the incidence. Values of the integrated loss coefficient of the rotor for the three cases in question are given in Table 8.

As could be expected from Figure 14 they show a distinct reduction of loss for the cases with bow vanes. Finally the rotor outlet angle β_2 is presented in Figure 15. Since the outlet angle of turbine blade profiles is within a certain range

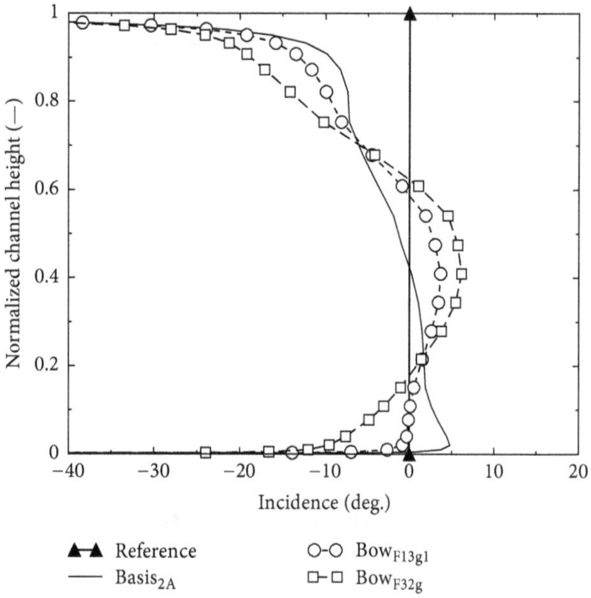

Figure 13: Circumference averaged incidence of the rotor over the channel height.

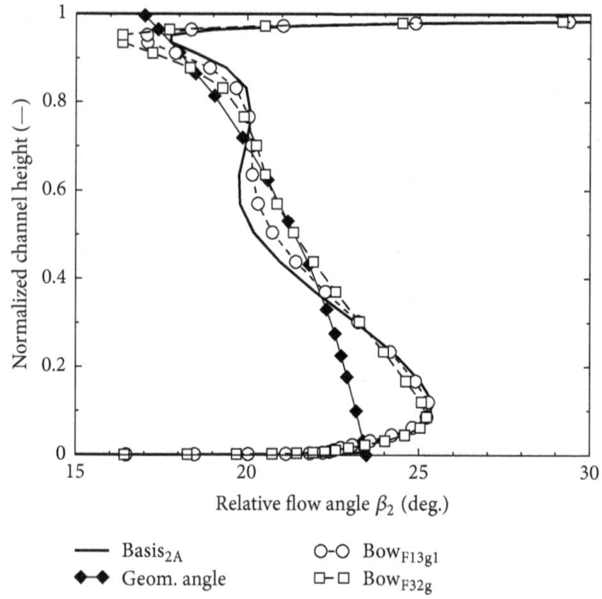

Figure 15: Circumference averaged relative exit flow angle β_2 of the rotor in position 2 (MP2).

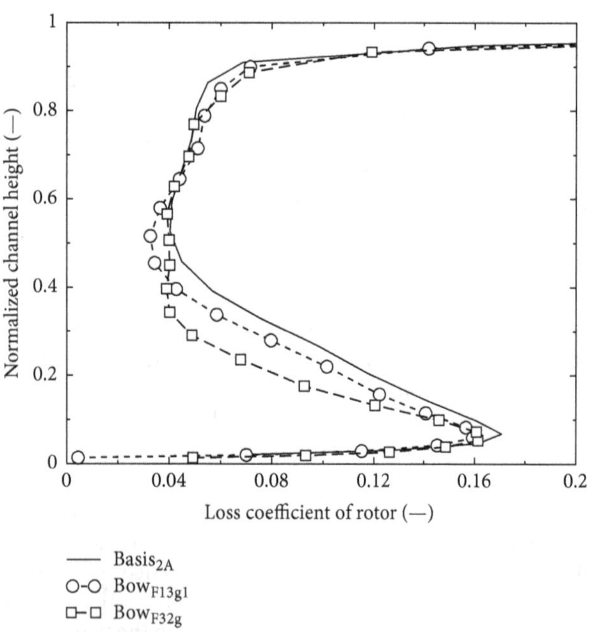

Figure 14: Circumference averaged loss coefficient of the rotor over the channel height.

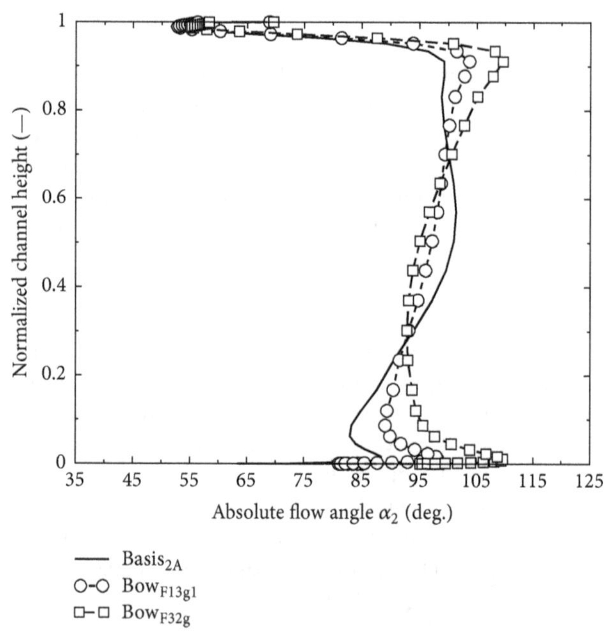

Figure 16: Circumference averaged absolute exit flow angle α_2 over the channel height in position 2 (MP2).

independent of the inflow angle, it could be expected that the β_2-distribution is little influenced by the bow of the vanes. It shows the familiar underturn near the wall and the overturn directly at the wall, both due to secondary flow, in a pronounced way near the hub and less clear near the tip, where directly at the wall the clearance flow predominates. The influence of the vane bow is small. The absolute flow angle α_2 behind the rotor is defined not only by β_2, but via the velocity triangle by the axial velocity, too. This is, as we have seen, influenced considerably by the curvature of the vanes and this influence is still sensible behind the rotor. Therefore, as Figure 16 shows, the flow angle α_2 can be changed by 10 deg. and more through the vane bow, even if β_2 differs by one or two deg. only for the different cases.

6. 2D-Modeling of the Flow Deflection

In preliminary design and optimization of turbines (e.g., [46]) the two-dimensional flow calculation *(throughflow method)* assuming rotation-symmetric flow is still widely in use and of high practical importance, since the blades cannot be introduced directly into the calculation models for flow deflection and loss, representing the chosen blading, has to be used. For radial blading the familiar *sine-rule* is quite appropriate to describe the exit flow angle α_1 of the cascade in a simple form

$$\alpha_1 = \arcsin\left(\frac{a_1}{t_1}\right), \tag{21}$$

where a_1 is the exit channel width and t the blade pitch. We denote the exit angle and so calculated the effective angle.

6.1. Correction Function for the Exit Flow Angle of Curved Blade. Tangential curvature will change the exit flow angle of a blade over the blade height in a way mainly influenced by the local tangential inclination angle ε. In a general way a correction function for the effective flow angle α_1 can be defined, which renders the exit flow angle $\tilde{\alpha}_1$ of the curved blade

$$\tilde{\alpha}_1(h_{\text{bez}}) = f_\varepsilon^\alpha(h_{\text{bez}})\,\alpha_1(h_{\text{bez}}). \tag{22}$$

This function will depend on the local inclination ε and the effective flow angle α_1 and varies over the blade height.

6.1.1. Correction Function Based on Experimental Data. Experimental investigations were made at the department for power, nuclear turbine construction, and aircraft engines of State St. Petersburg Technical University for the guide vane Bow_{2B} (not treated here in detail), which is inclined in the lower half and curved towards radial near the casing. The relevant data in our definition are $\varepsilon_N = +13.2°$, $\varepsilon_G = 0°$, and $h^k_{\text{bez},N} = 1.0$. Because of the moderate curvature a rather simple correction function is proposed

$$f_\varepsilon^\alpha = 1 + k\tan\varepsilon. \tag{23}$$

From the experimental data a value for $k = 0.435$ is deducted. A validation is given in Section 6.2.

6.1.2. Correction Function for Vanes with High Curvature. The numerically investigated vanes with high curvature on both ends produce high underturning in the sidewall zones, which cannot be described by a correction approximately proportional to ε as in (23). A nonlinear version for the correction parameter k has been devised, which makes use of several additional parameters

$$k = \frac{k_k - 1}{\alpha_k - \alpha_k^{\text{eff}}}\left(\alpha_k - \alpha_1\right)^n, \tag{24}$$

where α_1 is again the sine-rule flow angle along the blade height, α_k^{eff} the effective flow angle at the extreme blade end (hub or tip). and α_k an assumed maximum value of the exit flow angle of the curved blade at its end.

The value α_k^{eff} depends on the geometry of the hub profile cascade and is invariable (e.g., 13° for the vane $Basis_{2A}$ and the curved vanes deduced from it). The two coefficients α_k and k_k have to be chosen suitably for the different curvatures.

TABLE 9: Values of the empirical coefficients in (25) for the curvature blading (a = general).

Notation	Bow_{2B}	Bow_{F32g}	K
n	3	3	k
α_k^{eff}	13	13	k
α_k'	20	15.35	a
k_k'	1.195	1.085	a

The exponent n will take care of the nonlinear character of the angle distribution; a value $n = 3$ seems appropriate. From the results of experimental investigations of turbine stages with curved vanes it seems that an optimal angle of tangential inclination of approximately 20° exists. We assume therefore that the value of α_k can be chosen up to a maximum of 20° only; similarly the coefficient k_k should not be assumed higher than 1.2. If (24) is inserted in (23) we receive a new general formulation for the exit flow angle of curved vanes

$$\tilde{\alpha}_1 = \left[1 + \left(\frac{k_k - 1}{\alpha_k - \alpha_k^{\text{eff}}}\left(\alpha_k - \alpha_1\right)^n\right)\tan\varepsilon\right]\alpha_1. \tag{25}$$

6.2. Validation of the Correction Function. Since the numerically investigated vanes have been developed by tangential displacement of the unvaried profiles of the radial basis stator their effective exit flow angle distribution is identical with that of the stator $Basis_{2A}$. The geometry and hence the effective flow angle distribution of the stator Bow_{2B}, for which experimental data are available, are quite different. The exit flow angle distribution of the two curved stators Bow_{2B} and Bow_{F32g} has been calculated with the correction function described above. The values of the geometrical and empirical coefficients that have been used are shown in Table 9. Further validations of other blade types from the numerical simulation are carried out in this investigation, respectively (not treated here in detail).

Figure 17 shows the results for the vane Bow_{2B}, which comprises inclination mainly in the lower half. The resulting exit flow angle differs sensibly in this region from the sine-rule value; the corrected angle according to (25) and the coefficients in Table 9 agree well with the experimental and the values of the 3D-computation. Because of the moderate degree of inclination the correction in the simplified form of (23) with a constant value of $k = 0.435$ would render good results.

Figure 18 shows results for the vane Bow_{F32g} with considerable curvature on both ends. The corrected angles according to (25) with the coefficients in Table 9 represent the results of the 3D-simulation very well in both sidewall regions. Since the correction model cannot take care of the redistribution of the aerodynamic blade load into the middle of the blade the reduction of the flow angle in that region will not be correctly described (see Figure 11). The validation of this correlation is tested at present likewise for the throughflow calculation by the TU St. Petersburg as well as industry in St. Petersburg.

Because the sine-rule is valid for two-dimensional cascade flow, the exit flow angle distribution will show some deviations from these values in the sidewall zones of radial

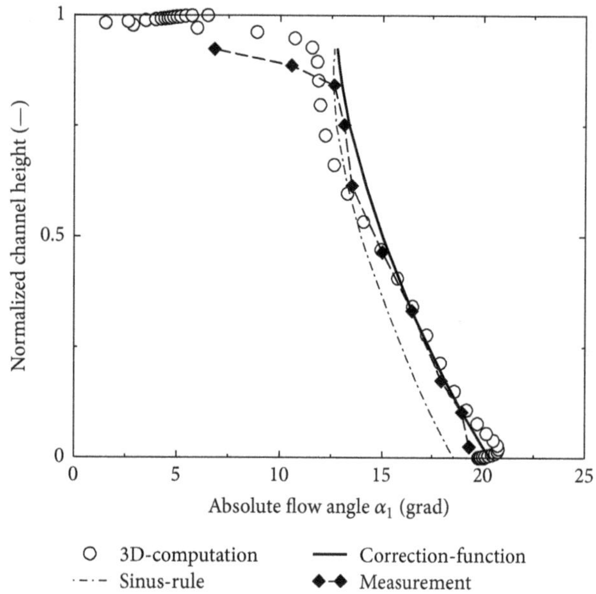

FIGURE 17: Validation of the deflection model of (25) over the blade height for the curvature blade Bow_{2B}.

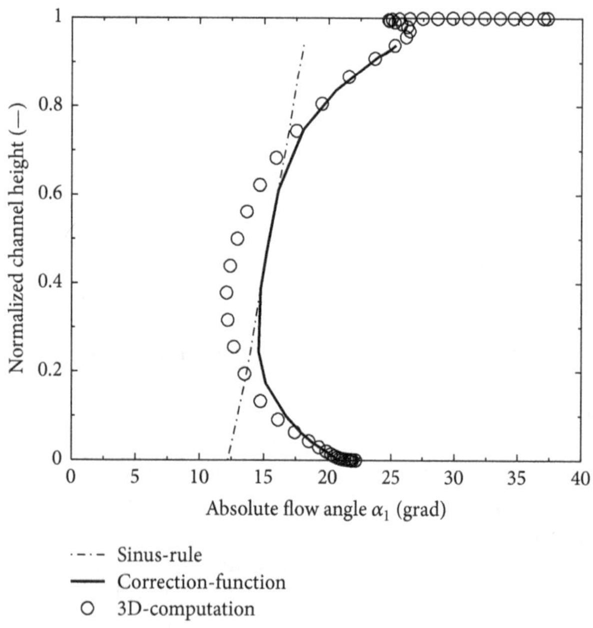

FIGURE 18: Validation of the deflection model of (25) over the blade height for the curvature blade Bow_{F32g}.

blades, too. Secondary flow will cause underturning when approaching the sidewall and overturning in its direct proximity; radial clearance will cause considerable underturning. These deviations from the sine-rule values can be calculated by specific flow models, described, for example, in [47], and then introduced in our correction procedure instead of the effective flow angle α_1 for improved modeling. In Figure 17 these influences are visible to some extent. A distinct tangential curvature, however, will render flow angle modifications, which surpass the secondary flow influence considerably, as is clearly visible in Figure 18.

7. Conclusions

Results of numerical 3D flow simulations of an axial turbine stage with radial and two different curved guide vanes are presented. The curved guide vanes are developed by tangential displacement of the profiles of the radial vane. The simulation procedure was validated with the aid of experimental data, not presented here.

The guide vanes with curvature on both ends form an obtuse angle between suction side and sidewall. They generate radial pressure forces, which displace the fluid flow towards the sidewalls. The axial velocity is increased there as well as the exit flow angle. The pressure distribution of the blade profiles is deformed, so that the aerodynamic loading is shifted backwards. This produces a radial component of the pressure gradient on the suction side directed towards the middle of the blade. This phenomenon probably promotes a displacement of the secondary flow and a reduction of the secondary loss. All effects are intensified by increased blade inclination and extended area of the curvature.

The modification of vane exit angle and axial velocity distribution will cause changes of the incidence to the rotor up to 10° and more, so that adaption of the rotor geometry, when applying curved stators, seems necessary.

In order to render the flow calculation of curved stators feasible by the widespread 2D-methods *(through flow method)* a radial force model and an exit angle model for curved stators were developed. It departs from the familiar sine-rule and comprises a correction function, which depends mainly on the local inclination angle ε of the vane. General coefficients permit the adaption to different blade curvatures. Further research by numerical simulation, finally validated by experiments, which is extended to other profile types, profile modifications in the inclined zone, other types of curvature, and so forth, will probably lead to optimum forms of bow vanes with reduced secondary loss and increased efficiency.

Nomenclature

Symbols

a: Exit channel width
c: Flow velocity in abs. system
$\underline{f}$: Vector of the force
h_s: Specific enthalpy
h: Blade and channel height
Δh_s: Change of the isentropic spe. enthalpy
i: Incidence
k: Coefficient
$\dot{m}$: Mass flow rate
n: Speed in normal direction
p: Pressure
r: Radius
R: Gas constant
S: Blade chord length
T: Temperature
Tu: Turbulence degree
α: Flow angle in the abs. system

β: Flow angle in the rel. system
Δ: Difference
δ: Curving angles of meridian streamlines
δ_{sp}: Gap width of tip clearance
ε: Curvature angle of blade stacking line
ζ: Loss coefficient
κ: Specific heat ratio
ν: Angle of the meridional stream surface
ρ: Flow density
Π: Pressure ratio of turbine stage.

Indices

0, 1, 2: Exp. measured positions
aus: Outlet of the stage
ax: Axial
bez: Normalized
Bi: Bitangent
DS: Pressure side
ein: Inlet of the stage mesh
eff: Effective according to sine-rule
G: At the casing
int: Integration
is: Isentropic
k: Curvature zone and coefficient
la: Rotor
le: Stator
m: Meridional comp. and middle value
N: At the hub
r: Radial component
rech: Computation
s: Static
sp: Gap losses or gap
SS: Suction side
t: Total
z: z-axial component
φ: Circumference component.

Conflicts of Interest

The author declares that they have no conflicts of interest.

Acknowledgments

This study described here was supported by the Ministry of Science and Technology and the National Science Council (MOST104-2221-E-035-052; MOST103-2632-E-035-001-MY3; MOST103-2221-E-035-064; NSC102-2221-E-035-029; NSC101-2221-E-035-010) and was completed due to the support of Professor W. Riess and Professor J. Seume of the Institute for Turbomachinery and Fluid Dynamics of Leibniz University of Hannover. The author is very grateful for this aid and wants to thank their continuously encouragement and cooperation as well as their helpful discussions to this work for the writing. The basic blade data and experimental results were completed and available due to the support of Professor Lapschin of the Department for Power, Nuclear Turbine Construction and Aircraft Engines of State St. Petersburg Technical University.

References

[1] J. A. MacDonald, "Increasing Steam Turbine Power Generation Efficiency," June 2003 issue of Energy-Tech magazine (2003).

[2] M. E. Deich, A. V. Gubarev, G. A. Filippov, and Z. Q. Wang, "A new method of profiling the guide vane cascades of stages with small ratios of diameter to length," *Thermal Engineering*, no. 8 (in Russian), pp. 42–46, 1962, Associated Electrical Industries, Research Laboratory, Manchester, Translation No. 3277.

[3] A. S. Leyzerovich, *Steam Turbine for Modern Fossil-Fuel Power Plants*, Taylor & Francis, UK, 2008.

[4] M. Deckers and E. W. Pfitzinger, "The exploitation of advanced blading technologies for the design of highly efficient steam turbines," in *Proceedings of the 6th International Charles Parsons Turbine Conference*, Dublin, Ireland, 2003.

[5] L. H. Smith Jr. and H. Yeh, "Sweep and Dihedral Effects in Axial-Flow Turbomachinery. Trans. of the ASME," *Journal of Basic Engineering*, vol. 85, pp. 401–406, 1963.

[6] E. Y. Watanabe, Y. Tanaka, T. Nakano et al., "Development of New High Efficiency Steam Turbine," *Mitsubishi Heavy Industries Technical Review*, vol. 40, no. 4, pp. 1–6, 2003.

[7] M. E. Deich and B. M. Trojanovskij, *Investigation and Design of Axial Turbine Stages (Untersuchung und Berechnung Axialer Turbinenstufen)*, VEB Verlag Technik, Berlin, Germany, 1973.

[8] M. E. Deich, B. M. Troyanovskii, and G. A. Filippov, "An effective way of improving the efficiency of turbine stages," *Thermal Engineering*, vol. 37, no. 10, pp. 520–523, 1990.

[9] M. E. Deich, *Gas Dynamics of Turbine Blade Rows, Edited by G.A. Filippov*, Energoatomizdat, Moscow, 1996.

[10] W. Han, Z. Wang, and W. Xu, "An experimental investigation into the influence of blade leaning on the losses downstream of annular cascades with a small diameter-height ratio," in *Proceedings of the International Gas Turbine and Aeroengine Congress and Exposition (GT '88)*, pp. 1–9, June 1988, ASME Paper, 88-GT-19 (1988).

[11] H. Wanjin, W. Zhongqi, T. Chunqing, S. Hong, and Z. Mochun, "Effects of leaning and curving of blades with high turning angles on the aerodynamic characteristics of turbine rectangular cascades," *Journal of Turbomachinery*, vol. 116, no. 3, pp. 417–424, 1994.

[12] S. Harrison, "Influence of blade lean on turbine losses," *Journal of Turbomachinery*, vol. 114, no. 1, pp. 184–190, 1992.

[13] J. Hourmouziadis and N. Hübner, "3-D Design of Turbine Airfoils," ASME Paper 85-GT-188, 1985, pp. 1-7.

[14] D. I. Suslov and G. A. Filippov, "An approximate method for calculating and profiling cascades of curved vanes," *Thermal Engineering*, vol. 42, no. 3, pp. 259–264, 1995.

[15] Z. Wang, W. Xu, W. Han, and B. Jie, "An experimental investigation into the reasons of reducing secondary flow losses by using leaned blades in rectangular turbine cascades with incidence angle," in *Proceedings of the ASME 1988 International Gas Turbine and Aeroengine Congress and Exposition (GT '88)*, June 1988.

[16] C. Pioske and H. E. Gallus, "Dreidimensionale Turbinenbeschaufelung," *MTZ Motortechnische Zeitschrift*, pp. 358–362, 1, Band 58, Heft 6 (1997).

[17] G. A. Filippov and Z. Q. Wang, "Die Berechnung der rotationssymmetrischen Strömung für die letzte Stufe der turbomaschinen," *Publikation des Moskauer Energetisches Institut (MEI)*, no. 47, 1963.

[18] G. A. Filippov and Z. Q. Wang, "The effect of flow twisting on the characteristics of guide rows," *Thermal Engineering (Teploenergetika)*, no. 5, pp. 69–73, 1964.

[19] H. D. Jiang, Y. L. Lu, F. Z. Zhou, and S. J. Wang, "Study of a subsonic combind-leaning turbine guide vane with tip endwall contouring," *Journal of Aerospace Power*, vol. 8, no. 1, pp. 41–44, 1993.

[20] H.-F. Vogt and M. Zippel, "Sekundärströmungen in turbinengittern mit geraden und gekrümmten schaufeln: visualisierung im ebenen wasserkanal," *Forschung im Ingenieurwesen/Engineering Research*, vol. 62, no. 9, pp. 247–253, 1996.

[21] Z. Wang and Y. Zheng, "Research status and development of the bowed-twisted blade for turbomachines," *Engineering Science*, vol. 2, pp. 40–48, 2000.

[22] R. Wolf and K. Romanov, *Steam Turbines: Siemens Reactive Blading - Designed for Highest Efficiency and Minimal Performance Degradation*, 2014, Russia Power 2014 (March 2003), pp. 1-19.

[23] G. Filippov, V. Gribin, A. Tischenko, I. Gavrilov, and V. Tischenko, "Experimental studies of polydispersed wet steam flows in a turbine blade cascade," *Proceedings of the Institution of Mechanical Engineers, Part A: Journal of Power and Energy*, vol. 228, no. 2, pp. 168–177, 2014.

[24] B. Baldwin and H. Lomax, "Thin layer approximation and algebraic model for separated turbulent flow," in *Proceedings of the 16th Aerospace Sciences Meeting*, vol. 78-0257, AIAA, 1978.

[25] Ch. Hirsch, *Numerical Computational of Internal and External Flows, Fundamentals of Computational Fluid Dynamics*, Butterworth-Heinemann, 2nd edition, 2007.

[26] Ch. Hirsch et al., "An Integrated CFD System for 3D Turbomachinery Applications," ADARD-CP-510, 1992.

[27] Ch. Hirsch, *CFD Methodology and Validation for Turbomachinery Flows. AGARD Lecture Series on 'Turbomachinery Design Using CFD'*, 1994, pp. 4.1-4.17, May to June.

[28] S. Kang and C. Hirsch, "Numerical simulation of three-dimensional viscous flow in a linear compressor cascade with tip clearance," *Journal of Turbomachinery*, vol. 118, no. 3, pp. 492–502, 1996.

[29] D. Kröner, *Numerical Schemes for Conservation Laws*, Wiley, Teubner, 1997.

[30] J. W. Slooff and W. Schmidt, "Computational Aerodynamics Based on the Euler Equations," AGARD-AG-325, 1994.

[31] A. Jameson, W. Schmidt, and E. Turkel, "Numerical simulation of the euler equations by finite volume methods using runge-kutta time stepping schemes," in *Proceedings of the AIAA 5th Computational Fluid Dynamics Conference*, AIAA Paper 81-1259 (1981).

[32] T.-H. Shieh and M.-R. Li, "Numeric treatment of contact discontinuity with multi-gases," *Journal of Computational and Applied Mathematics*, vol. 230, no. 2, pp. 656–673, 2009.

[33] T.-H. Shieh, T.-M. Liou, M.-R. Li, C.-H. Liu, and W.-J. Wu, "Analysis on numerical results for stage separation with different exhaust holes," *International Communications in Heat and Mass Transfer*, vol. 36, no. 4, pp. 342–345, 2009.

[34] T.-H. Shieh, M.-R. Li, Y.-T. Li, and M.-C. Chen, "A comparative study of flux limiters using new numerical methods in unsteady supersonic flows," *Numerical Heat Transfer, Part B: Fundamentals*, vol. 67, no. 2, pp. 135–160, 2015.

[35] T.-H. Shieh and M.-C. Chen, "Study of high-order-accurate limiters for time-dependent contact discontinuity and shock capturing," *Numerical Heat Transfer, Part B: Fundamentals*, vol. 70, no. 1, pp. 56–79, 2016.

[36] T. H. Shieh, *Untersuchung von Axialturbinen-Beschaufelungen mit dreidimensionalen Gestaltungselementen [Ph.D. dissertation]*, Universität Hannover, 2003.

[37] D. Wang, H. Ding, and J. Zhong, "The influence of tailboard on the exit flow fields of compressor cascade with curved blade," in *Proceedings of the China Engineering Thermal Physics Society (HMA'00)*, pp. 344–348, 2000.

[38] H. Jujii, T. Kimura, and K. Segawa, "A high-efficiency steam turbine utilizing optimized reaction blades—an application to the kwangyang combined-cycle power plant of K-power of Republic of Korea," *Hitachi Review*, vol. 56, no. 4, pp. 104–108, 2007.

[39] W. Traupel, *Thermische Turbomaschinen*, vol. 1, Springer-Verlag, Berlin, Germany, 2000.

[40] Z.-M. Feng, H.-B. Gu, J.-D. Zhang, and Y. Lu, "Different three-dimensional blades aerodynamic performance research comparison," *Information Technology Journal*, vol. 12, no. 11, pp. 2219–2224, 2013.

[41] Z. M. Feng, H. B. Gu, and J. D. Zhang, "Curved rotor blade cascades aerodynamic performance at off-design incidence in axial steam turbine," *International Journal of Digital Content Technology and its Applications*, vol. 7, no. 7, pp. 1086–1093, 2013.

[42] M. Murugan, A. Ghoshal, F. Xu et al., "Articulating turbine rotor blade concept for improved off-design performance of gas turbine engines," in *Proceedings of ASME 2016 Conference on Smart Materials, Adaptive Structures and Intelligent Systems*, vol. 1, 2016, Paper No. SMASIS2016-9045, pp. V001T04A004.

[43] K. Segawa et al., "Development of a highly loaded rotor blade for steam turbines (2nd report, performance verification)," *JSME International Journal, Series B*, vol. 45, no. 4, 2002.

[44] K. Segawa et al., "A high performance optimized reaction blade for high pressure steam turbines," in *Proceedings of ASME Power Conference*, pp. 307–314, March 2004.

[45] B. Mischo, *Axial Turbine Rotor Aero-thermal Blade Tip Performance Improvement through Flow Control [Ph.D. dissertation]*, TU Kaiserslautern, 2008.

[46] P. Lampart and L. HirT, "Complex multidisciplinary optimization of turbine blading systems," *Archives of Mechanics*, vol. 64, no. 2, pp. 153–175, 2012.

[47] M. Petrovic, *Berechnung der Meridianströmung in mehrstufigen Axialturbinen bei Nenn- und Teillastbetrieb [Ph.D. dissertation]*, Universität Hannover, VDI Fortschritt-Berichte, Netherlands, 1995.

6

Numerical Investigation of Fluid Flow and Performance Prediction in a Fluid Coupling using Large Eddy Simulation

Wei Cai,[1] Yuan Li,[1] Xingzhong Li,[2] and Chunbao Liu[1]

[1] *School of Mechanical Science and Engineering, Jilin University, Changchun 130022, China*
[2] *State Key Laboratory of Automotive Simulation and Control, Jilin University, Changchun, China*

Correspondence should be addressed to Chunbao Liu; liuanbc@126.com

Academic Editor: Funazaki Ken-ichi

Large eddy simulation (LES) with various subgrid-scale (SGS) models was introduced to numerically calculate the transient flow of the hydraulic coupling. By using LES, the study aimed to advance description ability of internal flow and performance prediction. The CFD results were verified by experimental data. For the purpose of the description of the flow field, six subgrid-scale models for LES were employed to depict the flow field; the distribution structure of flow field was legible. Moreover, the flow mechanism was analyzed using 3D vortex structures, and those showed that DSL and KET captured abundant vortex structures and provided a relatively moderate eddy viscosity in the chamber. The predicted values of the braking torque for hydraulic coupling were compared with experimental data. The comparison results were compared with several simulation models, such as SAS and RKE, and SSTKW models. Those comparison results showed that the SGS models, especially DSL and KET, were applicable to obtain the more accurate predicted results than SAS and RKE, and SSTKW models. Clearly, the predicted results of LES with DSL and KET were far more accurate than the previous studies. The performance prediction was significantly improved.

1. Introduction

A fluid coupling or hydraulic coupling is a hydrodynamic device commonly used in applications requiring smooth torque transmission. It is called coupling for short, which consists of an input rotor (pump impeller) and a matching output rotor (turbine runner), facing each other with a small gap without mechanical contact. It enables transmitting rotating mechanical power, where it requires variable speed operation and controlled start-up without shock loading of the power transmission system. Fluid couplings can also act as hydrodynamic brakes, named hydraulic retarder, dissipating rotational energy as heat through frictional forces.

The transient flow field in fluid coupling is very representative. The rotor and stator impellers comprised the enclosed fluid chamber, also involved in rotor-stator interaction. The chamber could be partially filled during operation. A number of literatures had demonstrated how complex and unsteady the flow was in fluid coupling (see, for example, McKinnon et al. [1] and Hampel et al. [2, 3]). Actually, there were unsteady and turbulent flows with intense secondary flows in the chamber, which were very difficult to calculate because of the lack of understanding of unsteady turbulent flows. In this study, a numerical investigation was presented using large eddy simulation (LES) to study the flow circulation during energy transmission and predict the external performances. Bai et al. utilized standard k-ε turbulence model as closure of Reynolds averaged Navier-Stokes (RANS) equations to simulated fluid flow in fluid coupling [4]. Huitenga and Mitra developed a code in the body-fitted rotating coordinates to solve RANS equations with k-ε turbulence model [5, 6]. Luo et al. employed SIMPLEC algorithm and SST k-ω turbulence model to simulate the steady state flows [7]. They also used PISO algorithm and RNG k-ω turbulence model to calculate the gas-liquid two-phase flows in fluid coupling [8]. Hur et al. applied the realizable k-ε turbulence model together with the VOF method to investigate two-phase flow of water fluid coupling [9]. Obviously, the RANS method was dominant. However, it was well known that this method had theoretical defects, such as the overprediction of viscosity and the failure of describing the dynamic pressure fluctuation. Those defects definitely resulted in prediction errors both in fluid flow and in external performances.

Compared with RANS, LES had essentially theoretical advantage, which directly resolved large-scale eddies. It also could alleviate the serious area of modeling uncertainty and was much less model dependent than RANS. LES had been successfully applied in many industrial flows, either in compressible flow, such as compressors [10] and gas turbine [11], or in incompressible flow, such as Francis turbine [12]. Basically, those publications proved that LES was more accurate in flow calculation. Hence, the motivation of present study came from the current situation of turbulence models applied in computation of fluid coupling which were relatively backward. In this paper, a simplified hydraulic coupling, removing the inlet and outlet, was chosen to carry out the investigation. By using LES, the study aimed to advance description ability of internal flow, especially unsteady flow phenomena. LES simulation results were verified by experimental data. Moreover, the performance prediction was improved. It should be noted that accurate prediction has significant value. And the accuracy was good enough, which would reach a level permitting design solely from computational results instead of experiments.

2. Numerical Simulation

2.1. SGS Models. The filtering governing equations for fluid coupling simulations can be written as in the following form:

$$\frac{\partial \overline{u_i}}{\partial x_i} = 0, \tag{1}$$

$$\frac{\partial \overline{u_i}}{\partial t} + \frac{\partial \left(\overline{u_i}\,\overline{u_j}\right)}{\partial x_j} = -\frac{1}{\rho}\frac{\partial \overline{p}}{\partial x_i} + \mu \frac{\partial^2 \overline{u_i}}{\partial {x_j}^2} - \frac{\partial \tau_{ij}}{\partial x_j}. \tag{2}$$

The equations are unclosed because of the introduction of τ_{ij} and require closure by structuring the model. The SGS stress tensor in LES is formulated using the concept of eddy viscosity.

$$\tau_{ij} - \frac{1}{3}\delta_{ij}\tau_{kk} = -2\mu_t \overline{S_{ij}}, \tag{3}$$

$$\overline{S_{ij}} = \frac{1}{2}\left(\frac{\partial \overline{u_i}}{\partial x_j} + \frac{\partial \overline{u_j}}{\partial x_i}\right). \tag{4}$$

The eddy viscosity μ_t, which can be obtained from the SGS model in (3), represents the SGS eddy viscosity. Thus, (3) and (4) and the specific SGS model lead to the closure of (2).

In LES, eddies in the turbulence are divided into large-scale and small-scale eddies by the filtering way, which characterize the vortex of energy and dissipation, respectively. Afterwards, large eddies are resolved directly by the instantaneous Navier-Stokes equations, while the small eddies are modeled by virtue of subgrid-scale (SGS) models. SGS models including Smagorinsky-Lilly (SL) model, Dynamic Smagorinsky-Lilly (DSL) model, Wall-Adapting Local Eddy Viscosity (WALE) model, Algebraic Wall-Modeled LES (WMLES) model, Algebraic WMLES *S*-Omega (WMLES *S*-Ω) model, and Dynamic Kinetic Energy Transport (KET) model were involved in this study, which were introduced in Table 1.

2.2. Computational Setup and Solution. In this paper, three-dimensional models of the hydraulic coupling are shown in Figure 1(a).

The complex geometry of hydraulic elements was accurately represented and the computational meshes were appropriately distributed with ANSYS ICEM to generate the structured hexahedral grid of the entire hydraulic coupling. The intensive type grid was used to refine the boundary layer. Mesh-point clustering was performed near the boundary layer and cell deformity was checked with ANSYS ICEM. Figure 1(b) shows the generation process of the structured grid of the entire passage. The detailed properties of CFD model were summarized in Table 2. Furthermore, the dependence on grid was tested by High Performance Computer (HPC) (2.60 GHz, 8 × 2 CPU) under different grid numbers based on the pump rotational speed of 600 rpm. The calculation results are shown in Figure 1(c).

3. Internal Fluid Flow

3.1. Reynolds Number Configuration. Reynolds number (Re) was introduced to evaluate the flow condition. It was defined as the ratio of inertia force to viscous force:

$$\mathrm{Re} = \frac{\rho v L}{\mu}, \tag{5}$$

where L was the characteristic length and could be calculated as

$$L = \frac{4A}{p}, \tag{6}$$

where A was the sectional area for the flow passage; p was the wetted perimeter length.

Three annular sections perpendicular to the interface along the radial direction were equally extracted out for appropriately studying the flow conditions. Figure 2 describes the distributions of Re for three axial sections in different turbulence models and R85, R120, and R155, respectively, represented different positions. The position was a ring surface in the annular sections and its radius was R. The value of R was 85 mm, 120 mm, and 155 mm, respectively.

Re ranged from 0 and 60000, suggesting that the flow within the turbine was mainly in a sufficient turbulent flow conditions. WALE and WMLES models allowed the locally larger Re values at R120 and R155 and the corresponding gradient change was distorted according to the working principle of hydraulic coupling. The flow at the outer ring was more intense and corresponding gradient changes were more obvious. However, WALE and WMLES provided a relatively large Re values in the near-wall region and laminar flows were transited into turbulence earlier. Although WMLES S-Ω, compared with WALE and WMLES, had the smaller Re values in the near-wall region, the Re value in the mainstream

TABLE 1: The equations of the six SGS models.

SGS models	Equations	Note
SL [13, 14]	$\lvert\bar{S}\rvert = \sqrt{2\bar{S}_{ij}\bar{S}_{ij}}$ $L_s = \min(\kappa d, C_s\Delta)$ $\Delta = V^{1/3}$ $\mu_t = \rho L_s^2 \lvert\bar{S}\rvert$	L_s is the mixing length for subgrid scales; κ is the von Kármán constant; d is the distance to the closest wall; C_s is the Smagorinsky constant, 0.1; Δ is the local grid scale.
DSL [15, 16]	$L_{ij} = T_{ij} - \widehat{\tau_{ij}} = \widehat{\bar{\rho}\tilde{u}_i\tilde{u}_j} - \frac{1}{\hat{\bar{\rho}}}\left(\widehat{\bar{\rho}\tilde{u}_i}\,\widehat{\bar{\rho}\tilde{u}_j}\right)$ $\tau_{ij} = -2C\bar{\rho}\Delta^2 \lvert\tilde{S}\rvert\left(\tilde{S}_{ij} - \frac{1}{3}\tilde{S}_{kk}\delta_{ij}\right)$ $T_{ij} = -2C\hat{\bar{\rho}}\hat{\Delta}^2 \lvert\hat{\tilde{S}}\rvert\left(\hat{\tilde{S}}_{ij} - \frac{1}{3}\widehat{\tilde{S}_{kk}}\delta_{ij}\right)$ $C_D M_{ij} = L_{ij} - \frac{1}{3}L_{kk}\delta_{ij}$ $M_{ij} = -2\left(\hat{\Delta}^2\hat{\bar{\rho}}\lvert\hat{\tilde{S}}\rvert\hat{\tilde{S}}_{ij} - \Delta^2\widehat{\bar{\rho}\lvert\tilde{S}\rvert\tilde{S}_{ij}}\right)$	The new filter width $\hat{\Delta}$ is equal to two times the grid filter width Δ; T_{ij} is the SGS stress tensor with $\hat{\Delta}$; L_{ij} is the added value of SGS stress tensor with two times consecutive filtering compared to the filtering only with $\hat{\Delta}$.
WALE [14, 16–18]	$\mu_t = \rho L_s^2 \frac{\left(S_{ij}^d S_{ij}^d\right)^{3/2}}{\left(\bar{S}_{ij}\bar{S}_{ij}\right)^{5/2} + \left(S_{ij}^d S_{ij}^d\right)^{5/4}}$ $L_s = \min\left(\kappa d, C_w V^{1/3}\right)$ $S_{ij}^d = \frac{1}{2}\left(\bar{g}_{ij}^2 + \bar{g}_{ji}^2\right) - \frac{1}{3}\delta_{ij}\bar{g}_{kk}^2,$ $\bar{g}_{ij} = \frac{\partial\bar{u}_i}{\partial x_j}$	$C_w = 0.325$.
WMLES [19, 20]	$\nu_t = \min\left[(\kappa d_w)^2, \left(C_{smag}\Delta\right)^2\right]\cdot S\cdot\left\{1 - \exp\left[-\left(\frac{y^+}{25}\right)^3\right]\right\}$ $\Delta = \min\left(\max\left(C_w\cdot d_w; C_w\cdot h_{max}, h_{wn}\right); h_{max}\right)$	d_w is the wall distance; S is the strain rate; $\kappa = 0.41$, $C_w = 0.15$, and $C_{smag} = 0.2$ are constants; y^+ is the normal to the wall inner scaling. h_{max} is the maximum edge length for a rectilinear hexahedral cell; h_{wn} is the wall-normal grid spacing.
WMLES S-Ω	$\mu_t = (C\Delta)^2 \lvert S - \Omega\rvert\left\{1 - \exp\left[-\left(\frac{y^+}{25}\right)^3\right]\right\}$	
KET [14, 21]	$k_{sgs} = \frac{1}{2}\left(\overline{u_k^2} - \bar{u}_k^2\right)$ $\mu_t = C_k\rho k_{sgs}^{1/2}\Delta_f$ $\rho\frac{\partial\bar{k}_{sgs}}{\partial t} + \rho\frac{\partial\bar{u}_j\bar{k}_{sgs}}{\partial x_j} = -\tau_{ij}\frac{\partial\bar{u}_i}{\partial x_j} - C_\varepsilon\rho\frac{k_{sgs}^{3/2}}{\Delta_f} + \frac{\partial}{\partial x_j}\left(\frac{u_t}{\sigma_k}\frac{\partial k_{sgs}}{\partial x_j}\right)$ $\tau_{ij} - \frac{2}{3}\rho k_{sgs}\delta_{ij} = -2C_k\rho k_{sgs}^{1/2}\Delta_f\bar{S}_{ij}$	C_k is model coefficient; K_{sgs} is the subgrid-scale kinetic energy. μ_t is the subgrid-scale eddy viscosity; Δ_f is the filter size computed from $\Delta_f = V^{1/3}$. σ_k is 1.0.

TABLE 2: CFD model description.

Analysis type	Transient state
Subgrid-scale models	SL, DSL, WALE, WMLES, WMLES S-Ω, and KET
Pressure-velocity coupling	SIMPLEC
Transient formulation	Second-order implicit
Momentum	Bounded central differencing
Pump status	Varied from 200 rpm to 600 rpm
Turbine status	Stationary
Viscosity and density	0.0258 Pa·s, 860 kg/m^3

was larger than that in other models. SL, DSL, and KET could simulate a relatively small Re values in the near-wall region and a relatively moderate Re values in the mainstream.

3.2. Vorticity Distribution. *Vorticity* was a physical parameter to indicate the vortex motion and represent the energy transportation process. The generation of vortex was caused by the velocity gradient in the flow field. Its development, diffusion, and decline were caused by the viscous interaction in the flow field. The value of vorticity was two times that of the rigid rotation angular velocity of fluid micelle.

Figure 3 shows the vorticity distribution patterns of the mid chord R120 in the flow field obtained with all subgrid-scale models. The large vorticity values were concentrated in the near-wall area in the turbine blade and the low vorticity values were concentrated in middle cascade area, especially in the pump wheel. The vorticity ranged from 0 to 4600 1/s^2. In the braking process, the pump impeller stirred oil to convert the mechanical energy into kinetic energy of working oil. Then the oil flowed into the turbines and impacted the turbine blades, thus producing strong relative effects between the oil and the turbine blade surface. Then a large number of

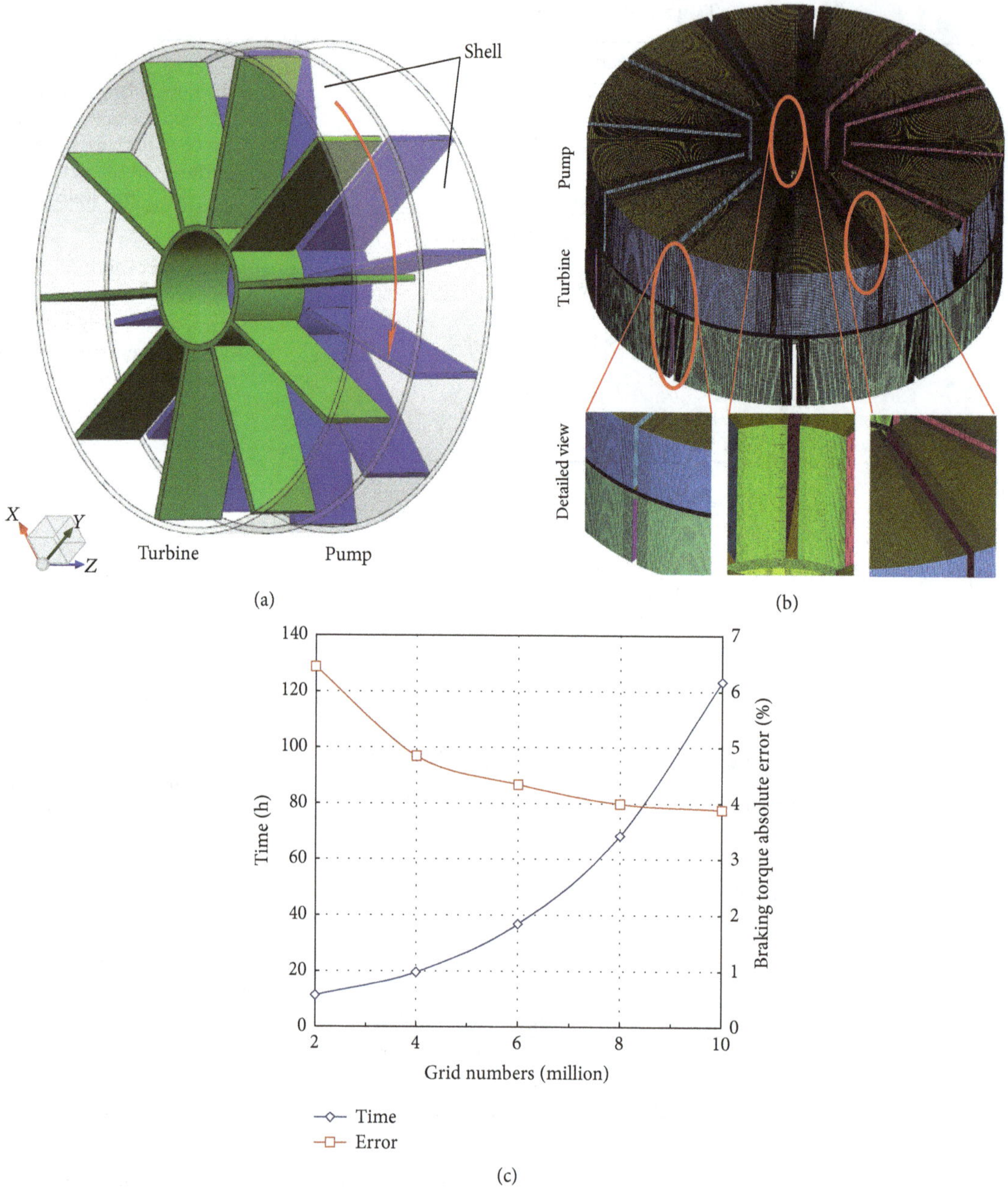

FIGURE 1: Hydraulic coupling. (a) 3D model; (b) grid layout; (c) grid independence test conditions.

vortexes were generated. In the SL model, vortexes formed inside the turbine were scattered and the vortex distribution near the wall was thin. In DSL, the adopted dynamic change C_d led to ample and continuous vorticity in turbine, which could improve the capture ability of vortex effectively. High vorticity value distribution of WALE was scattered and the predicted value of WALE was larger than that of other models. The remaining three SGS models could well capture the abundant structure of vorticity distribution, and the vorticity in the turbine also showed the continuous distribution. In conclusion, in the large eddy simulation, WMLES, WMLES S-Ω, and KET models showed more advantages in analyzing the vorticity distribution in the flow field and the distribution structures were abundant and continuous.

3.3. Changes in Rothalpy. Figure 4 describes the changes of *Rothalpy* from the inlet to the outlet on the nodal surface of turbine under different subgrid models. The nondimensional distance x could be defined as

$$x = \frac{S}{L} \tag{7}$$

L was the length of flow streamline on turbine's surface. If a point cut the flow streamline into two segments, then

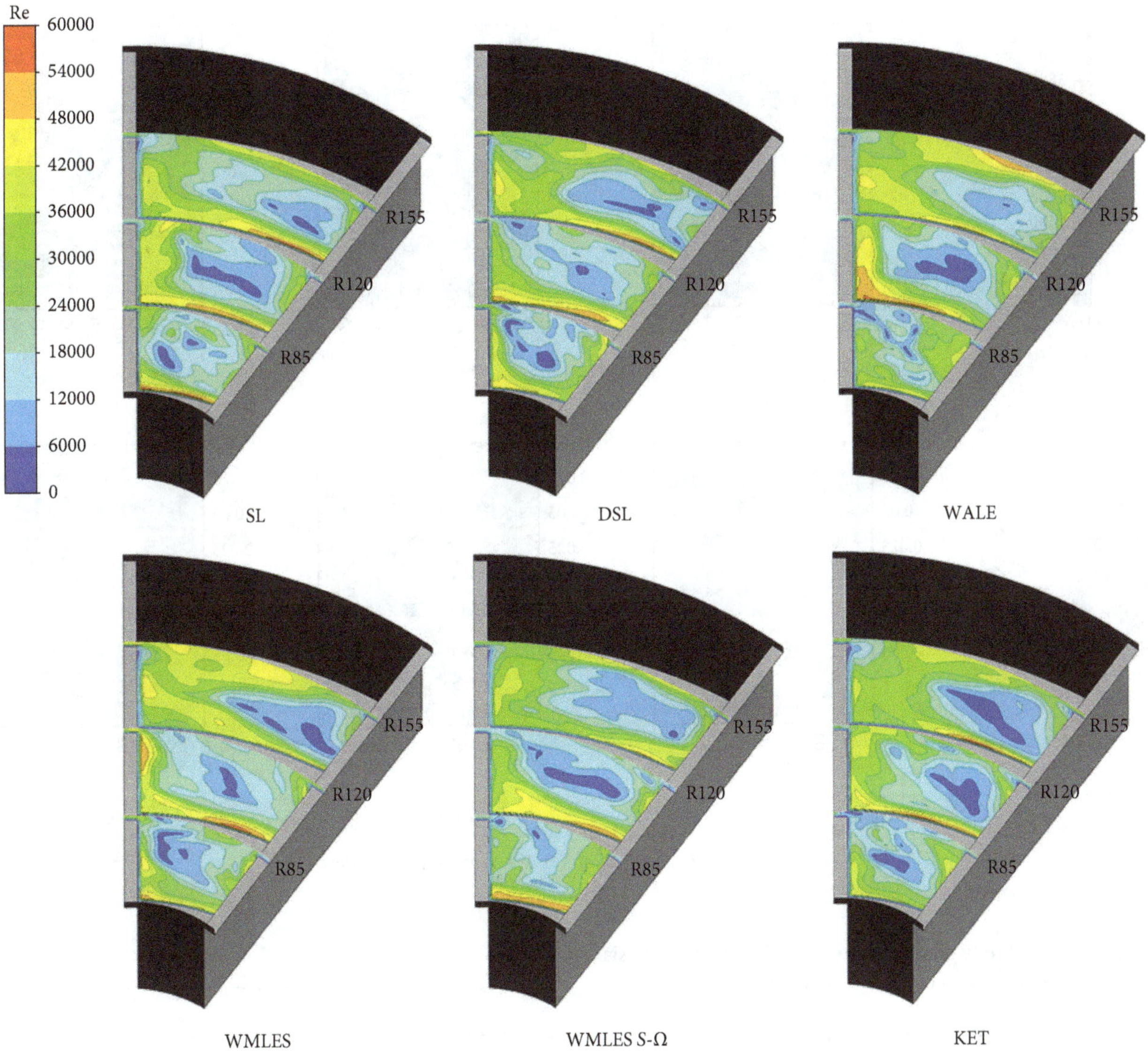

Figure 2: Re distribution on radial sections.

the part approaching the turbine's inlet was marked as part one. S was the length of part one. "0" and "1" represented, respectively, the locations of inlet and outlet of turbine. The x coordinates of A, B, C, and D point were 0.04, 0.60, 0.73, and 0.96, respectively. "A" and "B" in Figure 4 represent the leading and trailing edges, respectively. On the whole, *Rothalpy* value was decreased gradually with the increase in dimensionless distance, suggesting that the hydraulic loss was generated when oil impacted the turbine blades. In Figure 4, Point A shows that the oil flowed from the nonblade area into the blade edge. *Rothalpy* was reduced because the leading edge of blades hampered the working oil from flowing into the turbine. Correspondingly, hydraulic loss behind Point B referred to the diffusion loss generated when oil flowed into the nonblade area. Additionally, the regional largest changes of *Rothalpy* appeared in Points C and D.

Among various models, the SL model allowed the largest change of *Rothalpy*, 31.4%; WMLES and WALE provided almost the consistent change, 28.7%; WMLES S-Ω gave the change of 26.3%. The changes in the results of DSL and KET were the smallest, 24.9%. The phenomenon could be interpreted as follows: *Rothalpy* reflected the interaction strength between oil and the blades to a certain extent, indicating that the more intense interactions and viscosity dissipation led to the more obvious *Rothalpy* changes and the larger brake torque. These tendencies were consistent with that of the predicted braking torque obtained with various models.

4. Flow Mechanism Analysis

Q-Criterion was introduced to capture the vortex structure in the flow field. The Q-Criterion was defined as the second

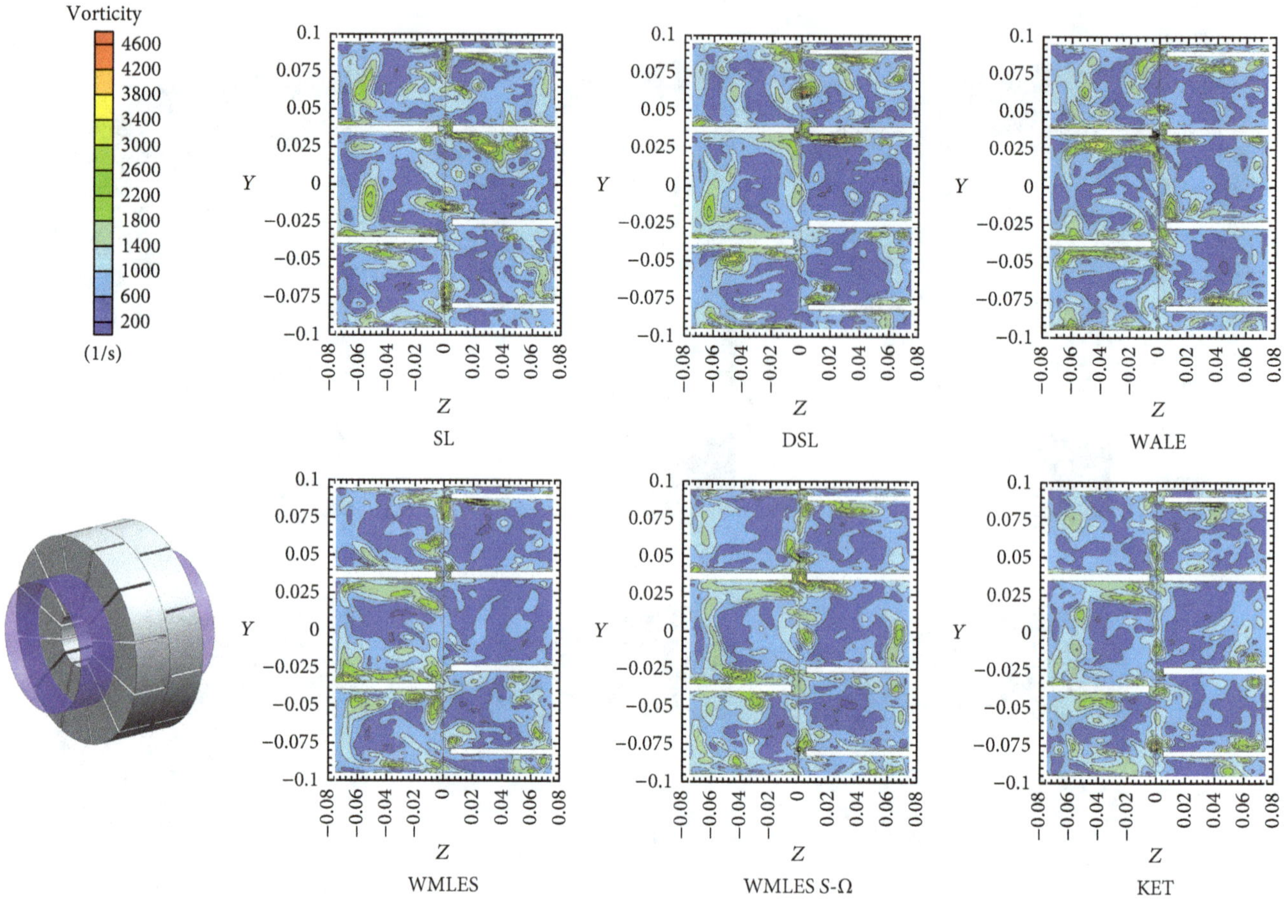

FIGURE 3: *Vorticity* distribution for different subgrid models.

invariant of the velocity gradient tensor. Under the hypothesis of incompressible flow, Q could be written as

$$Q = \frac{1}{2}\left(\|\Omega\|^2 - \|S\|^2\right). \quad (8)$$

$\|\|$ represented the 2-norm of tensor, where S and Ω were the symmetric and antisymmetric portions of the velocity gradient tensor, respectively. The Q value showed the degree that the rotational rate $\|\Omega\|^2$ overcame the strain rate $\|S\|^2$; therefore, positive Q value corresponded to dominant eddy.

As a comparison, the vortex structures of the SAS and RKE, and SSTKW models were also calculated and presented to highlight the ability of capturing the vortex structures for those SGS models. Figure 5 reveals the ability to capture the vortex structures for SGS models within the turbine under $Q = 8.7 \times 10^5\ 1/s^2$. Because the effect of all turbulent eddies was mimicked in the turbulence model of RANS, the effect of most turbulence eddies was directly simulated in the LES. It was evident that those SGS captured more vortex structures than SAS and RANS. Meanwhile, SAS and SSTKW provided the higher value of eddy viscosity ratio than SGS. Among the SGS models, DSL and WMLES S-Ω captured the most abundant vortex structures, followed by KET. SL provided a relatively larger eddy viscosity compared to other SGS models, followed by WMLES and WALE. WMLES S-Ω provided the smallest eddy viscosity. As a result, DSL and KET captured abundant vortex structures and provided a relatively moderate eddy viscosity.

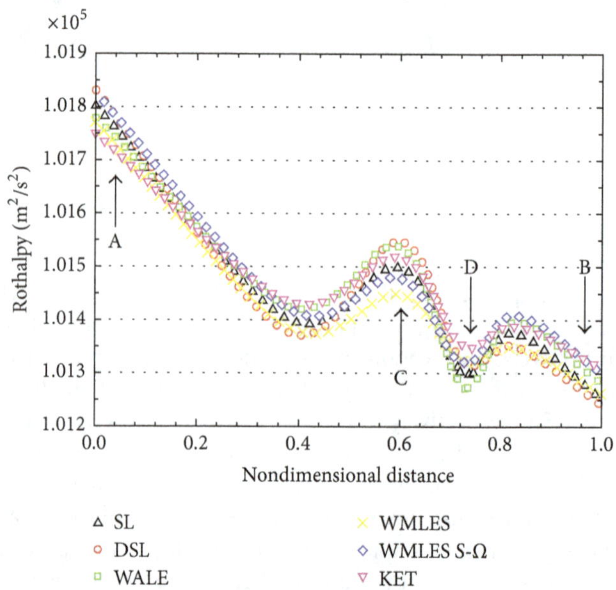

FIGURE 4: Changes in *Rothalpy* for different subgrid models.

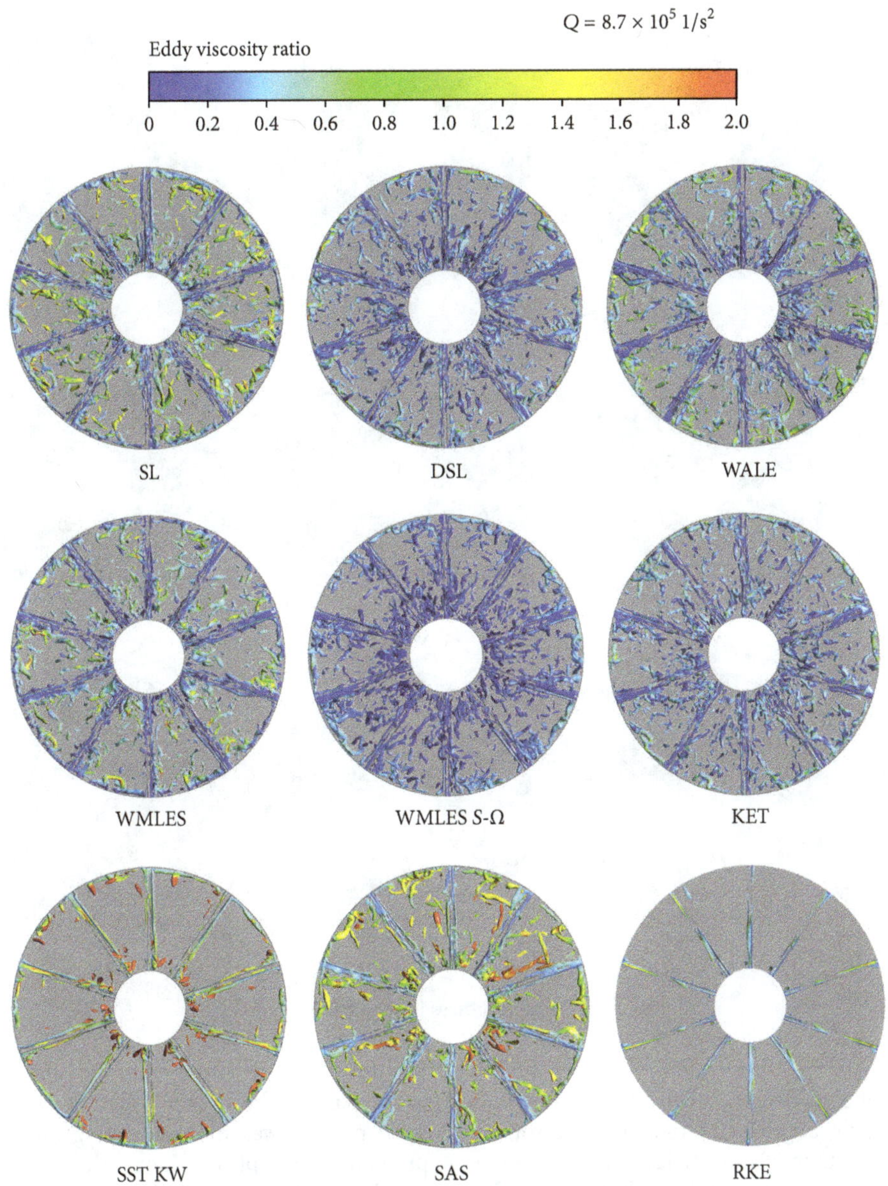

FIGURE 5: *Vortex* structures within turbine.

Figure 6 shows the vortex structures as a whole from front view of pump. The vortexes were concentrated in the region which was from the hub to the one-third blade height, where the kinetic energy of fluid transmission took place under 600 rpm. If the engineer wanted to advance the performance through flow field control, boundary control, just like the work in [5, 6], the position obtained could be significant, which inspired directly and avoided the attempts. The models all gave rich vortexes and were hardly told the differences through qualitative analysis, except SSTKW, RKE, and SAS.

Figure 7 shows the vortex structures in interior of the pump and turbine. Thus we found that there were several eddy structures along the wall of flow field, and it was diverse under different models. Fluid impacted on the impeller directly, so there was a linear flow eddy in the blade of the pump and turbine. Linear flow was divided into two sections, namely, the straight flow near the upper portion of the outer ring and the high-speed linear flow in the foot of the inner ring.

5. Performance Prediction and Verification

The comparison between the experimental and simulated results is plotted in Figure 8(a). To facilitate the comparison of the prediction ability of braking performance of hydraulic coupling among various models, we introduced the braking torque error defined as

$$E_S = \frac{|T_E - T_S|}{T_E} \times 100\%, \tag{9}$$

where T_E was the experimental torque value; T_S represented the torque value of simulation prediction; E_S was the absolute

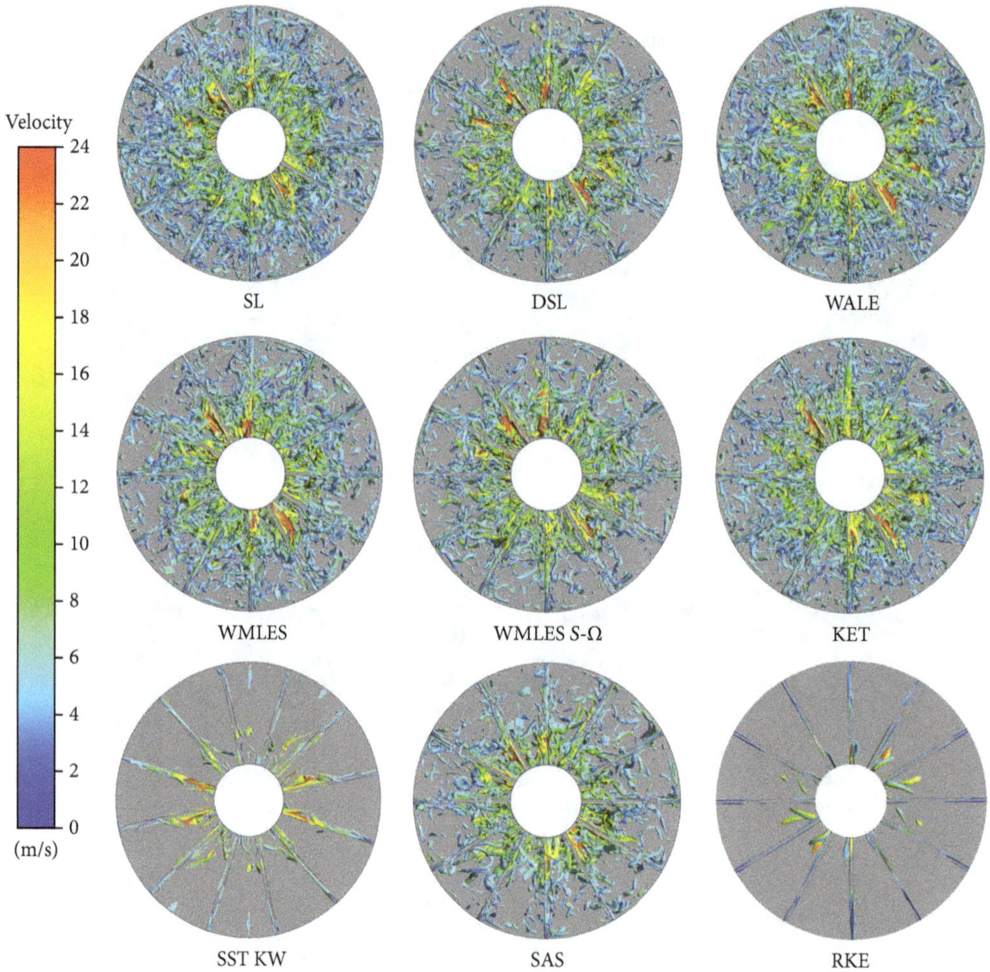

FIGURE 6: *Vortex* structures in whole flow channel.

error of braking torque prediction. Figure 8(b) shows the prediction errors of each subgrid-scale model. The errors of DSL and KET were less than 5% under low speed, and the DSL was lower, but the error of SL was the highest. Except the SL, all the models provided the errors below 4% under high speed. Moreover, with the increase in the speed, except the SL, all the models allowed the decreasing errors below 5%. The results were compared with the previous researchers: in the study of Sun et al. [22], with the standard k-ε model, the error was within 7.5%~15%. Fortunately, a lower error was acquired in He et al.'s research [23], and the error was close to but more than 6% under the pump rotational speed of 600 rpm by using standard k-ε model. However, in the research of Song et al. [24], the error was just less than 10% with the realizable k-ε model. The results of these explained that the error of subgrid model was more accurate than the former researcher's model.

Therefore, at the low speed, due to the influences of other factors in the flow field, the weak centrifugal force, and the insufficient flow, the predicted values were relatively large. With the increase in the rotating speed, the proportion of the centrifugal force increased and the flow developed fully. Therefore, other factors could be ignored and the prediction condition was closer to the ideal condition. In conclusion, at low speed, the higher order of DSL and KET subgrid-scale models was more advantageous to the performance prediction. Except the SL one, the other models showed the consistent prediction abilities for the performance at high speed.

Furthermore, the SAS and RKE, and SSTKW models were also calculated. SAS was a new class of URANS models which had been developed that could provide a LES-like behavior in unsteady flow regions. Meanwhile, they were more compatible with LES formulations than standard RANS models. RKE and SSTKW were typical RANS models. As shown in Figure 8(b), the error of those three is larger than that of the LES subgrid-scale models. The results showed that prediction accuracy of LES was higher than RANS and SAS.

Within the boundary layer, the velocity near the wall compared to the free flow speed would be decreased to zero quickly. Therefore, the velocity gradient was large. Especially in the viscous sublayer, the viscosity played the main role. With the increase of the distance to the wall, the velocity gradient decreased. Meanwhile, the viscosity effect reduced, but the inertia effect increased, and the flow finally developed into the full turbulence. Therefore, the interpretation of boundary layer transition was an important standard of a

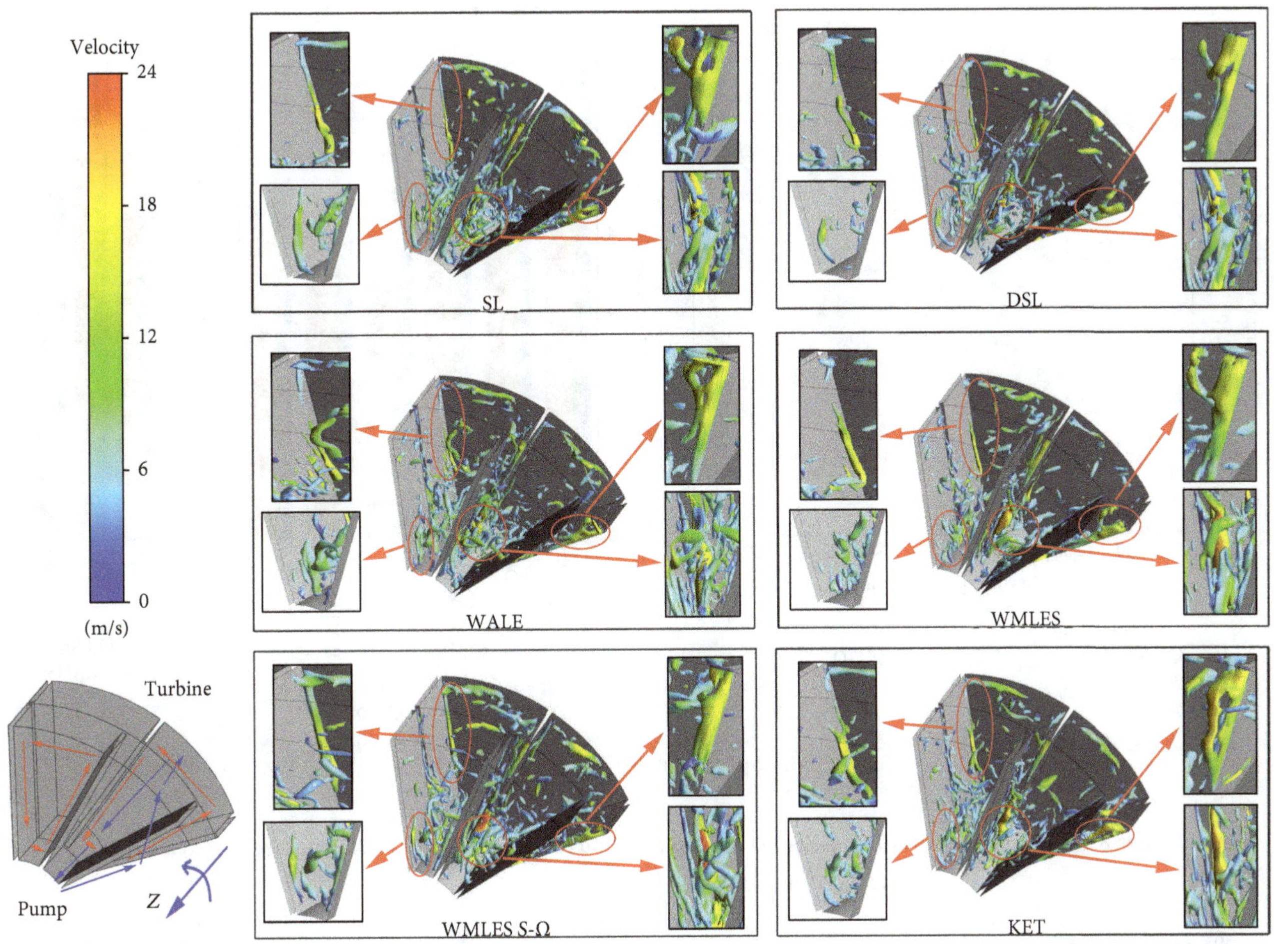

FIGURE 7: The *vortex* structures and flow field in the interior.

turbulence model. The reasons for improvement were mainly including their capacity of treatment for near-wall with fine grid. The near-wall treatments consisted of applying wall functions and resolving the viscous sublayer. Within the boundary layers, the most important region was the viscous sublayer, immediately adjacent to the wall and the log law region, slightly further away from the wall. Usually the near-wall mesh should be fine enough to be able to resolve the viscous sublayer, typically with the first near-wall node placed at. In this case, the SGS models could resolve effectively the viscous sublayer.

Figure 9 shows the analytical ability to the boundary layer of various grids models in the turbine pressure surface, where the RANS and SAS fail to get the data. DSL, WMLES, WMLES S-Ω, and KET could solve the viscous sublayer well. In addition, the interpretations of DSL and KET in the viscous sublayer were richer, especially at $y^+ \approx 1$. The results indicated that the dynamic stress solution of subgrid-scale model in the viscous sublayer was more advantageous. The locations of the transition layer in six models were similar. In other locations, according to the comparison results of the u^+ value under the same value of y^+ of various models, the WMLES allowed the largest predicted value and the SL allowed the smallest predicted value. With the increase of y^+ value, the u^+ values of six models tended to be consistent because the *N-S* equations were adopted to solve the transient in the mainstream zone.

In conclusion, DSL and KET performed better than other models in solving the boundary layer. Moreover, DSL and KET allowed the better predicted value of the braking torque of the hydraulic coupling under low speed conditions.

6. Conclusions

In order to study different ability of six subgrid-scale models to predict the flow field and performance of a fluid coupling, some studies, such as Re distribution, vorticity distribution, changes in Rothalpy, flow mechanism analysis, the error of braking torque prediction, and analytical ability to the boundary layer, had been conducted. Some conclusions had been drawn.

In the descriptions of the Re distribution, SL, DSL, and KET could simulate a relatively small Re values in the near-wall region and a relatively moderate Re values in the main-stream. In conclusion, in the large eddy simulation, WMLES, WMLES S-Ω, and KET models showed more advantages in analyzing the vorticity distribution in the flow field and the distribution structures were abundant and continuous.

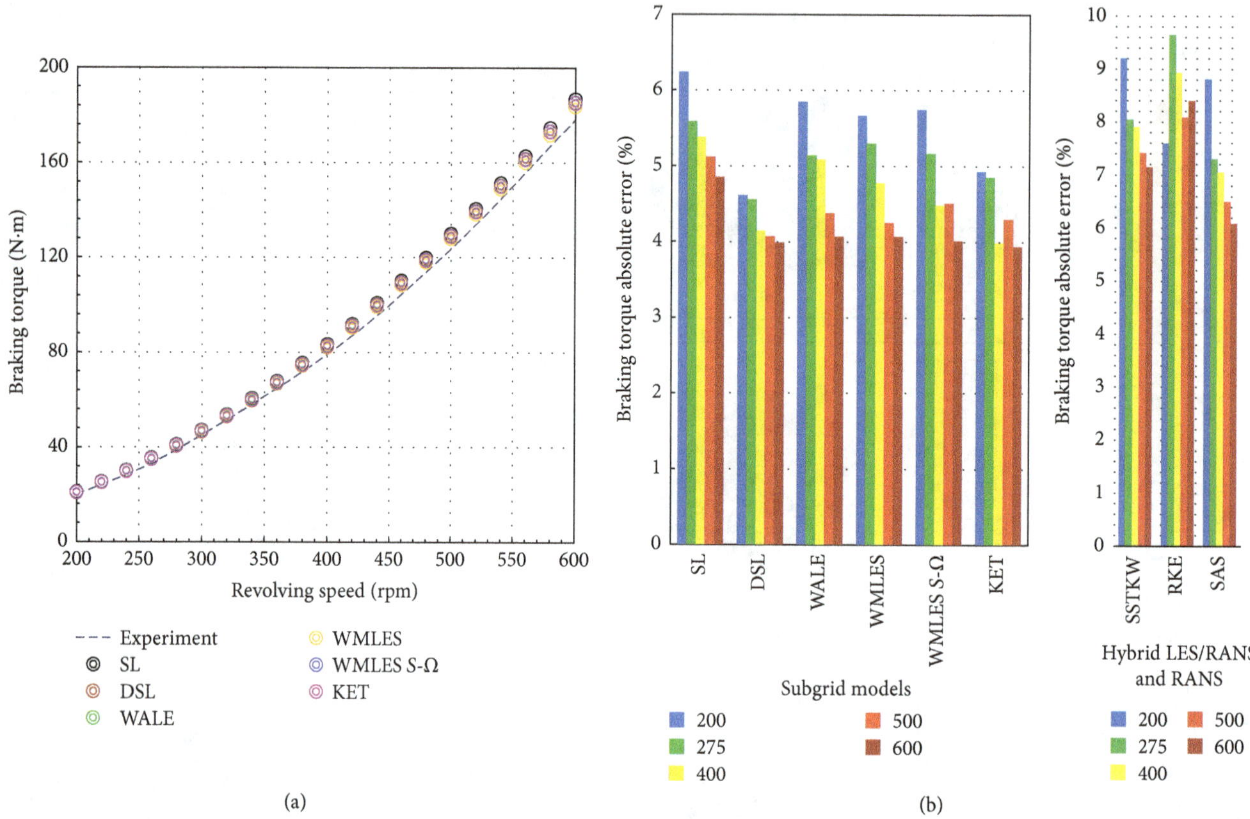

FIGURE 8: Braking torque calculated results. (a) Experimental and calculated results. (b) Braking torque absolute error.

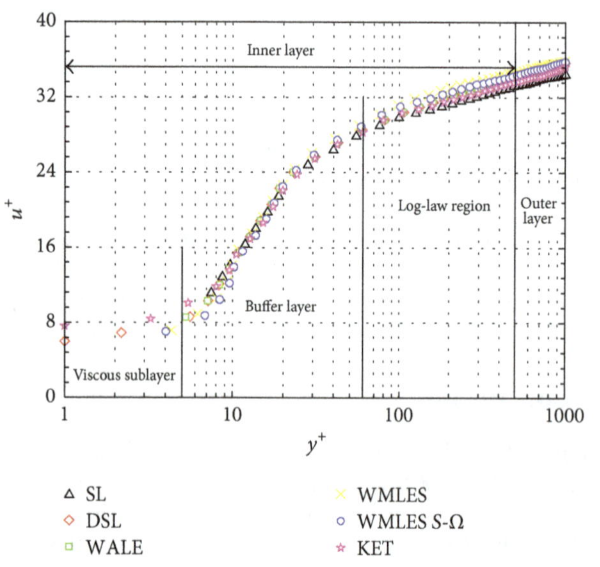

FIGURE 9: Boundary layer profiles near the pressure surface for turbine.

To some degree, *Rothalpy* changes reflected the strength of the interaction between oil and blade surface. The more intense interaction and the more viscosity dissipation meant the more obvious *Rothalpy* changes. Correspondingly, brake torque also was greater.

As we all know, the effect of all turbulent eddies was mimicked in the turbulence model of RANS, but the effect of most turbulence eddies was directly simulated in the LES. As a result, DSL and KET captured abundant vortex structures and provided a relatively moderate eddy viscosity within turbine. When the vortex structures as a whole were observed from front view of pump, the vortexes were concentrated in the region which was from the hub to the one-third blade height, where the kinetic energy of fluid transmission took place under 600 rpm. The models all gave rich vortexes and were hardly told the differences through qualitative analysis, except SSTKW, RS, and SAS. Because fluid impacted on the impeller directly, a linear flow eddy appeared in the blade of the pump and turbine.

In the prediction of braking torque, DSL and KET showed the smallest errors (less than 5%) at low speed, while the errors of other models were more than 5.6%. With the increase in rotational speed, the prediction errors of all the subgrid-scale models except SL model gradually maintained at 4%. DSL, WMLES, WMLES S-Ω, and KET may solve the viscous sublayer well. In addition, the interpretations of DSL and KET in the viscous sublayer were richer, especially at $y^+ \approx 1$. It meant that DSL and KET performed better than other models in solving the boundary layer. It led to the smallest errors showed by DSL and KET. Moreover, RKE, SSTKW, and SAS were also calculated. These results showed that SGS models were more accurate by contrast. Moreover, the predicted results of DSL and KET were far more accurate than the former study. Therefore, in the prediction of braking torque, we could choose DSL or KET subgrid-scale model to obtain the predicted results.

In conclusion, the six SGS models, especially DSL and KET, were applicable to obtain the more accurate predicted results and depict the flow field. The study not only provided the detailed study method for the application of large eddy simulation in hydraulic coupling, but also proposed many opinions on the model selection.

Conflicts of Interest

The authors declared that there were no conflicts of interest regarding the publication of this paper.

Acknowledgments

This work was also supported by the National Natural Science Foundation of China (Grant no. 51675219).

References

[1] C. N. McKinnon, D. Brennen, and C. E. Brennen, "Hydraulic analysis of a reversible fluid coupling," *Journal of Fluids Engineering, Transactions of the ASME*, vol. 123, no. 2, pp. 249–255, 2001.

[2] U. Hampel, D. Hoppe, K.-H. Diele et al., "Application of gamma tomography to the measurement of fluid distributions in a hydrodynamic coupling," *Flow Measurement and Instrumentation*, vol. 16, no. 2-3, pp. 85–90, 2005.

[3] U. Hampel, D. Hoppe, A. Bieberle et al., "Measurement of fluid distributions in a rotating fluid coupling using high resolution gamma ray tomography," *Journal of Fluids Engineering, Transactions of the ASME*, vol. 130, no. 9, pp. 0914021–0914028, 2008.

[4] L. Bai, M. Fiebig, and N. K. Mitra, "Numerical analysis of turbulent flow in fluid couplings," *Journal of Fluids Engineering, Transactions of the ASME*, vol. 119, no. 3, pp. 569–576, 1997.

[5] H. Huitenga and N. K. Mitra, "Improving startup behavior of fluid couplings through modification of runner geometry: part I—fluid flow analysis and proposed improvement," *Journal of Fluids Engineering, Transactions of the ASME*, vol. 122, no. 4, pp. 683–688, 2000.

[6] H. Huitenga and N. K. Mitra, "Improving startup behavior of fluid couplings through modification of runner geometry: part II—modification of runner geometry and its effects on the operation characteristics," *Journal of Fluids Engineering, Transactions of the ASME*, vol. 122, no. 4, pp. 689–693, 2000.

[7] Y. Luo, L. H. Feng, S. H. Liu, T. J. Chen, and H. G. Fan, "Numerical comparisons of the performance of a hydraulic coupling with different pump rotational speeds," in *Proceedings of the 6th International Conference on Pumps and Fans with Compressors and Wind Turbines (ICPF '13)*, vol. 52, no. 7, IOP Publishing, September 2013.

[8] Y. Luo, Z. G. Zuo, S. H. Liu, H. G. Fan, and W. L. Zhuge, "Numerical simulation of the two-phase flows in a hydraulic coupling by solving VOF model," in *Proceedings of the 6th International Conference on Pumps and Fans with Compressors and Wind Turbines (ICPF '13)*, vol. 52, no. 7, IOP Publishing, September 2013.

[9] N. Hur, M. Kwak, W. J. Lee, M. Moshfeghi, C.-S. Chang, and N.-W. Kang, "Unsteady flow analysis of a two-phase hydraulic coupling," in *Proceedings of the International Conference of Numerical Analysis and Applied Mathematics 2015 (ICNAAM '15)*, vol. 1738, no. 1, AIP Publishing, September 2015.

[10] N. Gourdain, F. Sicot, F. Duchaine, and L. Gicquel, "Large eddy simulation of flows in industrial compressors: a path from 2015 to 2035," *Philosophical Transactions of the Royal Society A: Mathematical, Physical and Engineering Sciences*, vol. 372, no. 2022, Article ID 20130323, 2014.

[11] F. Duchaine, N. Maheu, V. Moureau, G. Balarac, and S. Moreau, "Large-Eddy simulation and conjugate heat transfer around a low-mach turbine blade," *Journal of Turbomachinery*, vol. 136, no. 5, Article ID 051015, 2013.

[12] C. Trivedi, M. J. Cervantes, and O. G. Dahlhaug, "Experimental and numerical studies of a high-head Francis turbine: a review of the Francis-99 test case," *Energies*, vol. 9, no. 2, pp. 1–24, 2016.

[13] B. Wu, "Large eddy simulation of mechanical mixing in anaerobic digesters," *Biotechnology and Bioengineering*, vol. 109, no. 3, pp. 804–812, 2012.

[14] T. Lu, Y. Wang, and P. Wang, "Large eddy simulation with three kinds of sub-grid scale model on temperature fluctuation of hot and cold fluids mixing in a tee," *Applied Mechanics and Materials*, vol. 152-154, pp. 1307–1312, 2012.

[15] M. Jahoda, M. Moštěk, A. Kukuková, and V. Machoň, "CFD modelling of liquid homogenization in stirred tanks with one and two impellers using large eddy simulation," *Chemical Engineering Research and Design*, vol. 85, no. 5 A, pp. 616–625, 2007.

[16] Z. Li, G. Song, Y. Bao, and Z. Gao, "Stereo-PIV experiments and large eddy simulations of flow fields in stirred tanks with Rushton and curved-Blade turbines," *AIChE Journal*, vol. 59, no. 10, pp. 3986–4003, 2013.

[17] G. Lodato, L. Vervisch, and P. Domingo, "A compressible wall-adapting similarity mixed model for large-eddy simulation of the impinging round jet," *Physics of Fluids*, vol. 21, no. 3, Article ID 035102, 2009.

[18] B. Ji, X.-W. Luo, X.-X. Peng, and Y.-L. Wu, "Three-dimensional large eddy simulation and vorticity analysis of unsteady cavitating flow around a twisted hydrofoil," *Journal of Hydrodynamics*, vol. 25, no. 4, pp. 510–519, 2013.

[19] R. Menter Florian, *Turbulence Modeling for Engineering Flows*, Ansys Inc, 2011.

[20] S. Patil and D. Tafti, "Large-eddy simulation with zonal near wall treatment of flow and heat transfer in a ribbed duct for the internal cooling of turbine blades," *Journal of Turbomachinery*, vol. 135, no. 3, Article ID 031006, 2013.

[21] M. Ricci, L. Patruno, S. de Miranda, and F. Ubertini, "Effects of low incoming turbulence on the flow around a 5:1 rectangular cylinder at non-null-attack angle," *Mathematical Problems in Engineering*, vol. 2016, Article ID 2302340, 12 pages, 2016.

[22] Z. Sun, J. Chew, N. Fomison, and D. Edwards, "Analysis of fluid flow and heat transfer in industrial fluid couplings," *Proceedings of the Institution of Mechanical Engineers, Part C: Journal of Mechanical Engineering Science*, vol. 223, no. 9, pp. 2049–2062, 2009.

[23] Y. He, W. Ma, and C. Liu, "Numerical simulation and characteristic calculation of hydrodynamic coupling," *Transactions of the Chinese Society of Agricultural Machinery*, vol. 40, no. 5, pp. 24–28, 2009.

[24] B. Song, J. Lv, S. Guo, X. Zhang, and F. Kong, "Simulation and characteristic analysis on flow field of fluid couplings during braking," *Machine Design & Research*, vol. 27, no. 1, pp. 26–30, 2011.

7

A Comparison of Off-Grid-Pumped Hydro Storage and Grid-Tied Options for an IRSOFC-HAWT Power Generator

Mahdi Majidniya,[1] Kobra Gharali,[1] and Kaamran Raahemifar[2]

[1]*Department of Mechanical Engineering, University of Tehran, Tehran, Iran*
[2]*Department of Electrical & Computer Engineering, Ryerson University, Toronto, ON, Canada*

Correspondence should be addressed to Kobra Gharali; kgharali@ut.ac.ir

Academic Editor: Yuanzheng Li

An Internal Reforming Solid Oxide Fuel Cell (IRSOFC) is modeled thermodynamically; a Horizontal Axis Wind Turbine (HAWT) is designed; the combined IRSOFC-HAWT system should create a reliable source of electricity for the demand of a village located in Manjil, Iran. The output power of HAWT is unstable, but by controlling the fuel rate for the IRSOFC it is possible to have a stable power output from the combined system. When the electricity demand is over the peak or the wind speed is low/unstable/significantly high, the generated power may not be sufficient. To solve this problem, two scenarios are considered: connecting to the grid or using a Pumped Hydro Storage (PHS). For the second scenario, the extra produced electricity is saved when the production is more than demand and can be used when the extra power is needed. The economic analysis is done based on the economic conditions in Iran. The results will show a period of return about 9.5 and 13 years with the levelized cost of electricity about 0.0747 and 0.0882 \$/kWh for the first and second scenarios, respectively. Furthermore, effects of some parameters such as the electricity price and the real interest rate are discussed.

1. Introduction

Making a balance between energy productions and energy demands is a challenging issue for Renewable Energy (RE) systems. Energy production from RE sources is affected by low efficiencies, high capital costs, unreliable energy sources, and environmental impacts. Recently, combined systems are introduced to create efficient and economical power sources and mostly to overcome the unreliability of RE systems such as wind turbines. Based on the local conditions and the main energy source, the way that systems are combined varies. One of the options is combining the systems with a fuel cell. A fuel cell is a Renewable Energy source with controllable output electricity. A fuel cell has a short startup time. It is possible to add or remove cells immediately from the power cycle without any trouble. An SOFC is a common type of fuel cells for power plant applications. Recent studies include an SOFC in the power generation systems. In the following, some of the recent combinations of SOFC with other systems are discussed.

Soheyli et al. [1] analyzed a combined system in a hypothetical hotel in Kermanshah, Iran. The final optimized system was a combination of 10 wind turbines, 430 PV modules, 11 SOFCs, 106 batteries, and 2 heat storage tanks. Their results showed that the fuel consumption and pollution were decreased 263 and 353 times, respectively. Fathy [2] studied a system including photovoltaic modules, wind turbines, and fuel cells in Egypt. Obara et al. [3] used a numerical analysis to stabilize photovoltaic cyclic fluctuations by using a governor-free control of an SOFC and a flywheel inertia system. Hosseini et al. [4] analyzed a combination of photovoltaic-electrolyzer, SOFC, and a heat recovery unit for a residential area. The system supported an absorption chiller and a heat recovery steam generator using the excess heat of an SOFC. They conducted an energy and exergy analysis of the system. Vigneysh and Kumarappan [5] studied a combination of photovoltaic cells, an SOFC, and a battery energy storage system. For making a balance between power generation and consumption, a fuzzy logic method was applied. The SOFC was a backup generator. It reached its

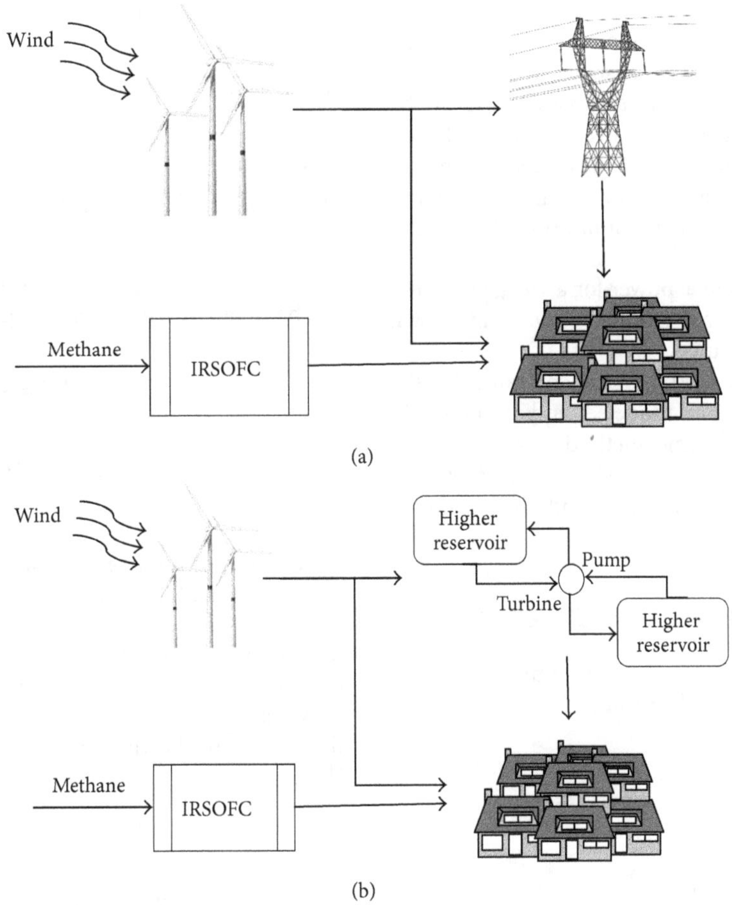

FIGURE 1: Schematic of the system: (a) first scenario and (b) second scenario.

rated power when the charge of the battery energy storage system was at the minimum limit. Tenfen and Finardi [6] presented a mathematical model to manage the energy from a microgrid. They modeled a micro turbine and an SOFC and combined them with a battery bank, photovoltaic cells, and a wind turbine. Akikur et al. [7] developed a mathematical model for a reversible SOFC combined with photovoltaic cells. To produce 100 kW of electricity, they investigated thermodynamically and economically three modes: a solar-SOFC mode for low solar radiation, a solar-solid oxide steam electrolyzer mode for high solar radiation, and an SOFC mode at night. Bai et al. [8] modeled and controlled an SOFC with photovoltaic cells. They used two PI controllers using an artificial bee colony.

It was shown that the combination of an SOFC with a wind turbine or photovoltaic cells is a state-of-the-art topic. In this study, a PHS is also added to a combined system of a HAWT and an IRSOFC. The system can run on grid or off grid with the aid of PHS as an energy storage component. The results are compared based on the economic and thermodynamic analysis to study the feasibility of each scenario; this comparison has not been done before based on the knowledge of the authors. In addition, the systems should fulfill the power demand of a village located in Manjil, Iran. The parameters are chosen based on the local economic and environmental conditions, making this study unique. It should be noticed that the system is sized based on monthly average wind speed and the needed electricity. That means the wind speed may not always match the needed electricity. For solving this problem, the two following scenarios are introduced.

The First Scenario (Figure 1(a)). The system is connected to the grid. It is possible to sell and buy electricity from the grid. The electricity can be sold to the grid when the produced electricity is more than demand. The electricity can be bought from the grid when the production cannot meet the demand. It is assumed that the prices of the selling and buying electricity are the same.

The Second Scenario (Figure 1(b)). The system is not connected to the grid. The extra electricity production of the wind turbine is saved by a Pumped Hydro Storage (PHS) system and can be released when it is needed. A PHS is a system that contains a pump, a turbine, and two reservoirs, a higher reservoir and a lower one. During high electricity production, the PHS pumps the water from the lower reservoir to the higher one to save the energy. When the electricity is needed, the water goes from the higher reservoir to the lower one through a turbine. Because of the pump and the turbine efficiencies and the other losses, some of the energy will be lost during these processes. To make up this energy, the size of the HAWT has been increased. The size of the PHS is assumed to be equal to the maximum needed electricity minus maximum production of the IRSOFC.

The case study is to produce power for a village with 100 houses located in Manjil, Iran. An economic analysis will show the possibility of the study.

Based on two scenarios, the main components, the IRSOFC and the HAWT, are modeled and designed in Sections 2 and 3. The economic method is described in Section 4. After validating the IRSOFC and the HAWT in Section 5, the results of the modeling and the economic analysis are discussed in Section 6.

2. Thermodynamic Modeling of IRSOFC

The modeling of components is done by home codes. The assumptions of the fuel cell modeling are

(i) ideal gases,

(ii) linear variation of enthalpy,

(iii) working in a steady condition,

(iv) being isolated with no heat exchange,

(v) no contact resistance,

(vi) ignoring radiation heat transfer between the gas canal and the solid container,

(vii) tubular cells and stack.

2.1. Direct Internal Reformer. For a fuel cell consuming natural gas, an internal or an external reformer is needed. Since an external reformer needs a cooling system and is less cost-effective than an internal one, the internal reformers are preferred [9].

In the internal reforming process, the occurring reactions are very endothermic, while the heat is provided by the fuel cell. The reactions in this process are the steam reforming reaction ($x \Rightarrow CH_4 + H_2O \rightarrow CO + 3H_2$) and the gas water shifting reaction ($y \Rightarrow CO + H_2O \rightarrow CO_2 + H_2$). According to these reactions, natural gas transforms to hydrogen in the fuel cell. Hydrogen takes part in an electrochemical reaction ($z \Rightarrow H_2 + (1/2)O_2 \rightarrow H_2O$). Here, x, y, and z are the molar rate progress of reforming, shifting, and electrochemical reactions, respectively [10]. This molar rate progress is calculated based on the relation among the equation constants, the temperature, and the pressure of the gasses.

TABLE 1: IRSOFC resistivity constants [12].

	A (Ω·cm)	B (K)	δ (cm)
Anode	0.00298	1392	0.01
Cathode	0.00811	−600	0.19
Electrolyte	0.0294	−10350	0.004
Internal connections	0.1256	−4690	0.0085

2.2. Reversible Voltage. The reversible voltage (E_{rev}) of the cell can be found by using Nernst equation:

$$E_{\text{rev}} = E^{\circ} + \Delta E = \frac{-\Delta G^{\circ}}{n_e F} + \frac{-\Delta G}{n_e F} = \frac{-\Delta G^{\circ}}{2F} + \frac{RT}{2F} \ln \frac{p_{H_2} p_{O_2}^{0.5}}{p_{H_2O}}. \quad (1)$$

Because of the irreversibilities, the voltage of a real fuel cell is less than that of a reversible one. These irreversibilities are called overpotentials. Overpotentials can be divided into three categories: Ohmic (η_{Ohm}), activation (η_{Act}), and concentration (η_{Conc}). The magnitude of these overpotentials grows as current density increases, which results in reducing the voltage of the cell. The cell voltage can be described as ([11])

$$E = E_{\text{rev}} - (\eta_{\text{Act}} + \eta_{\text{Ohm}} + \eta_{\text{Conc}}). \quad (2)$$

2.2.1. Ohmic Overpotential. Ohmic overpotential (η_{Ohm}) can be calculated as

$$\eta_{\text{Ohm}} = ir, \quad (3)$$

where r is the resistance:

$$r = \delta \rho. \quad (4)$$

δ is the thickness (Table 1) of components and ρ is the resistivity:

$$\rho = A \exp\left(\frac{B}{T}\right). \quad (5)$$

A and B are resistivity constants (Table 1).

2.2.2. Activation Overpotential. Butler–Volmer equation [13] describes

$$i = i_0 \left\{ \exp\left(\beta \frac{n_e F \eta_{\text{Act}}}{RT}\right) - \exp\left(-(1-\beta) \frac{n_e F \eta_{\text{Act}}}{RT}\right) \right\}. \quad (6)$$

β is the heat transfer coefficient and i_0 is the exchange current density. The heat transfer coefficient is the ratio of the exchange in the potential drop which results in changing the rate of constants. The value of β for a fuel cell is 0.5. For $\beta = 0.5$, (6) can be

$$i = 2 i_0 \sinh\left(\frac{n_e F \eta_{\text{Act}}}{2RT}\right). \quad (7)$$

TABLE 2: Values of activation overpotential constants [11].

		Cathode	Anode
γ	A/m^2	1.49×10^{10}	2.13×10^{8}
E_{Act}	kJ/mol	110	160

Therefore,

$$\eta_{Act,An} = \frac{2RT}{n_e F}\sinh^{-1}\left(\frac{i}{2i_{0,An}}\right),$$
$$\eta_{Act,Ca} = \frac{2RT}{n_e F}\sinh^{-1}\left(\frac{i}{2i_{0,Ca}}\right). \quad (8)$$

Equation (7) should be applied to both anode and cathode [13].

$$i_{0,An} = \gamma_{An}\left(\frac{p_{H_2}}{p_{ref}}\right)\left(\frac{p_{H_2O}}{p_{ref}}\right)\exp\left(-\frac{E_{Act,An}}{RT}\right),$$
$$i_{0,Ca} = \gamma_{Ca}\left(\frac{p_{O_2}}{p_{ref}}\right)^{0.25}\exp\left(-\frac{E_{Act,Ca}}{RT}\right) \quad (9)$$

γ (preexponential factor) like the activation energy (E_{Act}) depends on electrode materials [11] (Table 2).

2.2.3. Concentration Overpotential. Concentration overpotential can be calculated as

$$\eta_{Conc} = \frac{RT}{n_e F}\left(1 - \frac{i}{i_l}\right), \quad (10)$$

where i_l is the limiting current density. Considerable growth of concentration overpotential will be observed if the current density meets its limit.

3. HAWT Modeling

A home code for designing a HAWT is written based on Blade Element Momentum (BEM) method. For a rotor of a HAWT with the radius (R), the blade should be divided into some sections or elements. Each section (i) has a distance r from the hub of the rotor.

For each element, the local chord of the blade ($c(r)$) can be calculated from Schmitz formulation [14]:

$$c(r) = \frac{8\pi r}{NC_L}\left(1 - \cos\left(\frac{2}{3}\tan^{-1}\left(\frac{R}{r\lambda_r}\right)\right)\right). \quad (11)$$

λ_r is the local tip speed ratio, N is the number of blades, and C_L is the lift coefficient. The power coefficient (C_p) of the turbine is determined from the following equation [15]:

$$C_p = \frac{8}{\lambda_R N}\sum_{i=k}^{N} F_i \sin^2\varphi_i\left(\cos\varphi_i - \lambda_{ri}\sin\varphi_i\right) \cdot \left(\sin\varphi_i + \lambda_{ri}\cos\varphi_i\right)\left[1 - \frac{C_D}{C_L}\cos\varphi_i\right]\lambda_{ri}^2, \quad (12)$$

where F is Prandtl's correction factor, φ is the angle between the resultant velocity and the rotor plane, and C_D is the drag coefficient.

3.1. Corrections. In (12), F is Prandtl's correction factor. This factor is calculated as [16]

$$F = \frac{2}{\pi}\cos^{-1}\left(e^{-f}\right),$$
$$f = \frac{N}{2}\frac{R-r}{r\sin\varphi} = \frac{N}{2}\frac{1-r/R}{(r/R)\sin\varphi}. \quad (13)$$

Another correction applied to the modeling is based on the Reynolds number. For the variable range of the Reynolds number, Hernandez and Crespo [17] suggested a correction in which the lift coefficient remains unchanged and the drag coefficient is corrected:

$$C_D = C_{D,ref}\left(\frac{Re_{ref}}{Re}\right)^{0.2}. \quad (14)$$

Re_{ref} is the Reynolds number in which the drag coefficient ($C_{D,ref}$) of the blade is computed and Re is the Reynolds number based on the working conditions.

Glauert correction is also applied in this modeling to calculate the induced velocities.

4. Economic Modeling

Economic analysis is developed based on the Annualized Cost of System (ACS). In this method, revenues and costs are converted to a uniform annuity payment. ACS is the combination of annualized capital cost (C_{acap}), annualized replacement cost (C_{arep}), annualized maintenance cost (C_{amain}), and annualized operating cost (C_{aope}) [18].

$$\begin{aligned} \text{ACS} &= C_{acap}(\text{total instrument}) \\ &+ C_{arep}(\text{total instrument}) \\ &+ C_{amain}(\text{total instrument}) \\ &+ C_{aope}(\text{labor cost} + \text{fuel cost} + \text{insurance cost}). \end{aligned} \quad (15)$$

Annualized capital cost of the system is

$$C_{acap} = C_{cap} \cdot \text{CRF}\left(i, Y_{proj}\right),$$
$$\text{CRF}\left(i, Y_{proj}\right) = \frac{i(1+i)^{Y_{proj}}}{(1+i)^{Y_{proj}-1}}. \quad (16)$$

C_{cap} is the capital cost of each component in terms of US dollar and Y_{proj} is the lifetime of each component in terms of years. The rate of payback (CRF) calculates the present value of the annual cost (a set of annual homogeneous cash flows). The annual real interest rate (i) is

$$i = \frac{j-f}{1+f}. \quad (17)$$

In (17), f is annual inflation rate and j is the nominal interest rate. For an initial cost of equipment, see Table 3.

TABLE 3: Initial capital costs of instruments ([18–20]).

Component	Cost
Compressor	$C_{\text{comp}} = 91562\left(\frac{\dot{W}_{\text{comp}}}{445}\right)^{0.67}$
Heat exchanger	$C_{\text{HX}} = 8500 + 409\,(A_{\text{HX}})^{0.85}$
IRSOFC	$C_{\text{SOFC}} = (n_{\text{cells}}\pi D_{\text{cell}} L_{\text{cell}})(2.96T_{\text{cell}} - 1907)$
IRSOFC auxiliary component	$C_{\text{aux,SOFC}} = 0.1C_{\text{SOFC}}$
Inverter	$C_{\text{inv}} = 10^5\left(\frac{\dot{W}_{\text{cell}}}{500}\right)^{0.7}$
GT	$C_{\text{GT}} = \dot{W}_{\text{GT}}\left(-98.328\ln\left(\dot{W}_{\text{GT}}\right) + 1318.5\right)$
Combustion chamber	$C_{\text{cc}} = \left(1 + \exp\left(0.018T_{\text{GT}} - 26.4\right)\right) \times \left(\frac{46.08\dot{m}_{\text{cc}}}{0.995 - P_{\text{GT}}/P_{\text{cc}}}\right)$
HAWT	1300 ($/kW)
PHS	2000 ($/kW)

The annual replacement cost for this project is considered to be zero; that means the lifetime of the components is equal to the lifetime of the project. For each element, 5% of the initial cost is considered for the cost of the maintenance [18].

The annual costs of labor, fuel, and insurance are considered as the annual operating costs [18].

4.1. Net Present Value (NPV). The NPV in the engineering economics is one of the standard methods to evaluate the economic plans:

$$\text{NPV} = \frac{\text{ACS}}{\text{CRF}\left(i, Y_{\text{proj}}\right)}. \quad (18)$$

NPV is the present value of installing and operating the system over its lifetime in a project. The higher the amount of NPV is, the more profitable the project is. If NPV is negative, the project will lose money and it is infeasible in terms of economics [18].

4.2. Levelized Cost of Product (LCOP). The LCOP is the average cost of total usable power production capacity of the system [18]:

$$\text{LCOP} = \frac{\text{ACS}}{\text{SOPC}}. \quad (19)$$

The SOPC is Summation of Product Costs in a year.

4.3. Other Economic Parameters. Since the LCOP is based on total cost in the lifetime of the project, the Prime Cost (PC) is used based on the market price [18]:

$$\text{PC} = \frac{\text{OFC}}{\text{VOP}}. \quad (20)$$

The OFC or Operating Flow Costs include operating cost, fuel cost, labor cost, and the insurance cost for a year. The VOP is the Volume of Product in a year.

The period of return (POR) is the ratio of the capital costs to the Net Annual Benefit (NAB). Projects with a shorter POR are more attractive for the investment compared to plans which have longer periods. The POR means the total time to earn the entire initial capital by the profit [18]:

$$\text{POR} = \frac{C_{\text{Cap}}}{\text{NAB}}. \quad (21)$$

C_{Cap} is the capital cost. The NAB is the subtraction of income taxes from Annual Benefit (AB). AB can be calculated as [18]

$$\text{AB} = \text{SOPC} - \text{OFC}. \quad (22)$$

5. Validation

5.1. IRSOFC. The modeled fuel cell is validated with a model proposed by Pirkandi et al. [10] (Figures 2(a)–2(d)). The fuel cell was run based on Pirkandi et al.'s [10] condition which was a tubular fuel cell at 1273 K and 100 kPa with 0.85 of fuel utilization factor and 9000 A/m^2 of limiting current density. The fuel composition was 97% methane, 1.5% carbon dioxide, and 1.5% nitrogen. The air composition was 21% oxygen and 79% nitrogen. In Figure 2, the variation of total voltage and overpotentials are compared with the results of Pirkandi et al. [10]. Since the compared curves are close, the current model is valid and reliable. Increasing the current density (i) enriches the overpotential values (Figures 2(a)–2(c)), resulting in decreasing the cell voltage (Figure 2(d)).

5.2. HAWT. The home code is validated by the experimental work of Khalfallah and Koliub [21] for a three-bladed HAWT. The current home code was run for the referred conditions. The output powers from the home code were compared with the results of Khalfallah and Koliub [21] from cut-in wind speed to the designed wind speed, where the maximum power is accessible. As it is shown in Figure 3, the overall agreement between two curves is visible.

6. Results and Discussion

The required power for the targeted village is 320 kW, but, in this study, the generated power of 347 kW is considered.

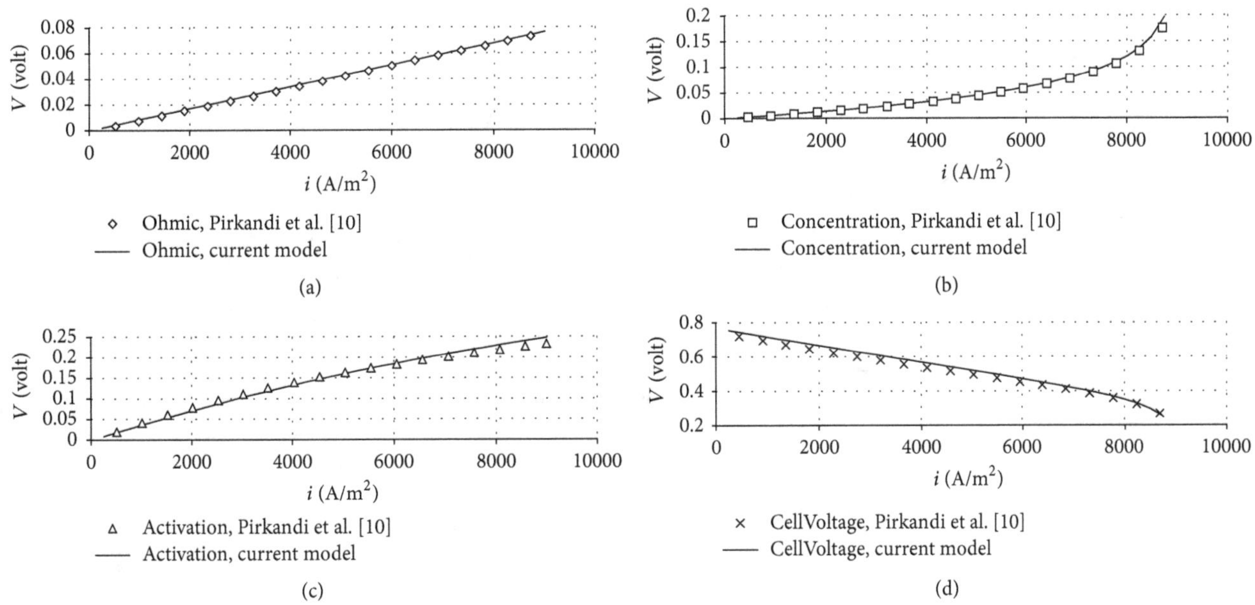

FIGURE 2: IRSOFC validation: (a) ohmic overpotential, (b) concentration overpotential, (c) activation overpotential, and (d) cell voltage.

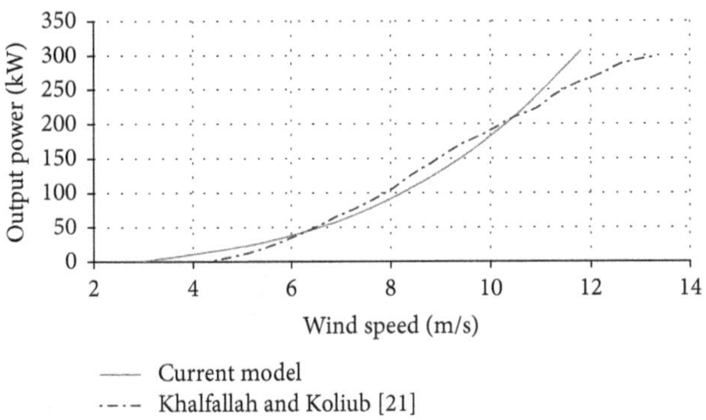

FIGURE 3: Validation of HAWT.

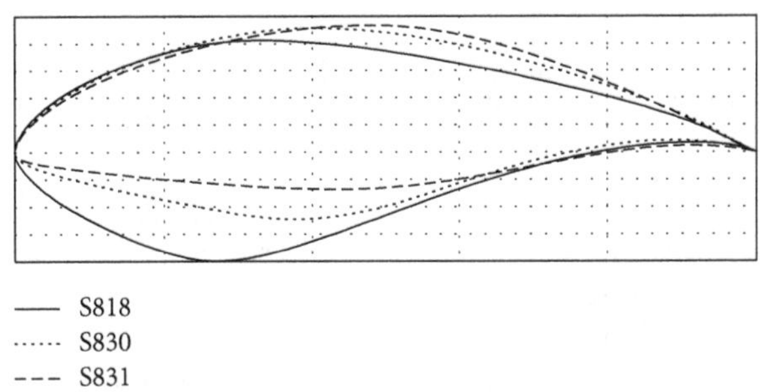

FIGURE 4: Selected airfoils.

TABLE 4: Details of the modeled HAWT.

	First scenario	Second scenario
Tip speed ratio	6	6
Number of blades	3	3
Rotor diameter (m)	58.14	66.14
Blade radius (m)	26.67	30.34
Hub height (m)	50	50
Airfoils	S818, S830, S831	S818, S830, S831

Then, based on 347 kW power which is assumed to be the targeted power demand, an IRSOFC and a HAWT are modeled.

6.1. Designed HAWT. Based on 10-year NASA records [22], the monthly averaged wind speeds at 50 meters above the surface of the earth for Manjil, north of Iran, were extracted. According to these data, a stall regulated wind turbine with nominal wind speed at 7.9 m/s is chosen. For the selection of airfoils, the blade is divided into three parts. An airfoil from the S-serious NREL families is selected for each part.

The details of the design of the wind turbine for two scenarios are provided in Table 4. As it was mentioned in the second scenario (off-grid IRSOFC-HAWT-PHS) the size of HAWT will increase because of the PHS efficiency and energy loss. The rated power of the HAWT for the first scenario is 347 kW. For the second scenario, the rated power is increased about 30%.

The chosen airfoil profiles were shown in Figure 4. The S818 airfoil, a thin airfoil, is used at the tip to absorb the maximum power. From the tip to the root, the airfoils become thicker. S830 and S831 are the airfoils used for the mid and the root of the blade, respectively. The variations of blade pitch (φ) and the chord during the blade are shown in Figure 5. The behaviors of both scenarios are almost the same.

After designing the HAWT and based on the wind speed variations, monthly power generation from the HAWT for both scenarios is shown in Figure 6. Because of the low wind speed from September to February, the power production of the HAWT has the lowest values in these months. In July and August, the power reaches its maximum values as the wind is so strong.

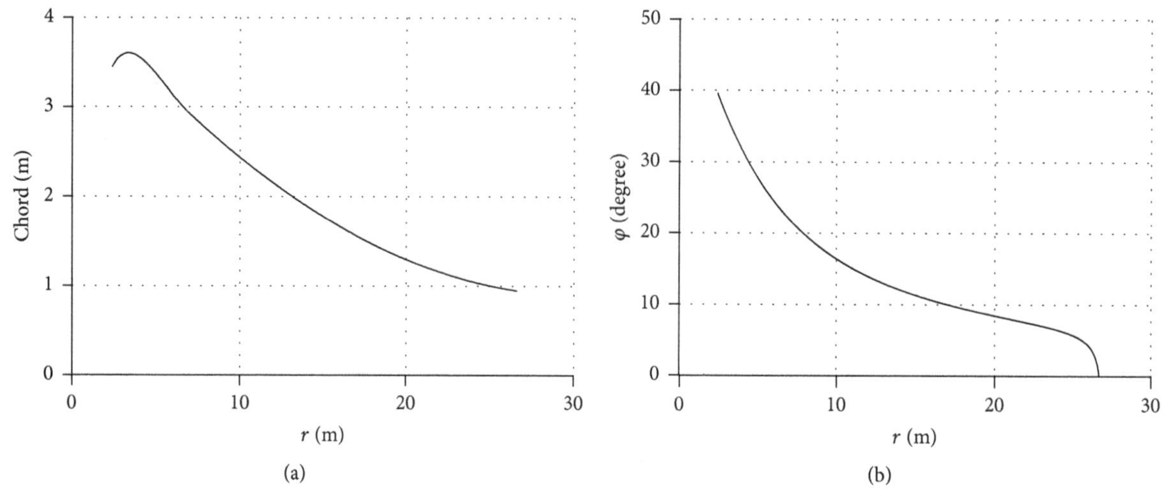

FIGURE 5: Variation of (a) chord and (b) φ during the blade.

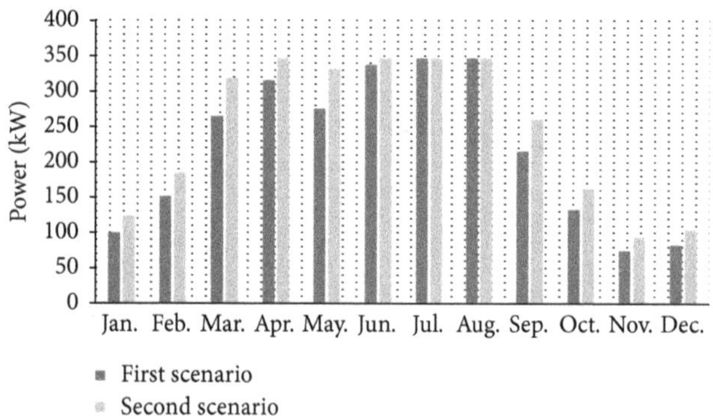

FIGURE 6: Monthly power generation from the HAWT.

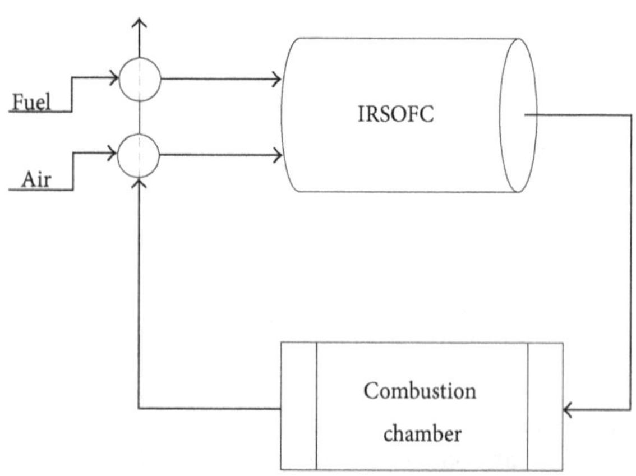

FIGURE 7: A schematic of the IRSOFC cycle.

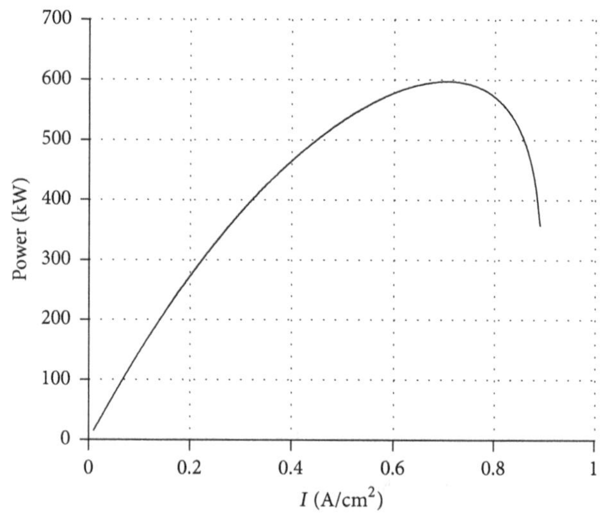

FIGURE 8: Variation of IRSOFC power production versus current density (first scenario).

6.2. IRSOFC Model. IRSOFC works at 1 bar pressure. The fuel that is fed to the IRSOFC is methane. A schematic of it is shown in Figure 7. The details of the IRSOFC inputs are provided in Table 5.

According to the power generated by the wind turbine during a year (Figure 6), the rest of the needed power should be generated by the IRSOFC. The results of IRSOFC modeling for both scenarios are shown in Table 6. The fuel flow rate in Table 6 is based on a full load operation. The IRSOFC does not work under full load all the time but using full load fuel flow rate for the whole year in economic analysis will provide a safety margin that the village will always have stable electricity.

In the second scenario, the size of the PHS equals 94 kW, which is the difference between the power demand and the IRSOFC capacity. Power and efficiency variation versus current density for the first scenario are shown in Figures 8 and 9, respectively. There is no significant difference between the results of two scenarios. As it is shown, increasing current density (up to 0.65 A/cm^2) increases the power but decreases the efficiency of the system because the fuel consumption strongly depends on the current density. Thus, the current density of IRSOFC is set to 2000 A/m^2.

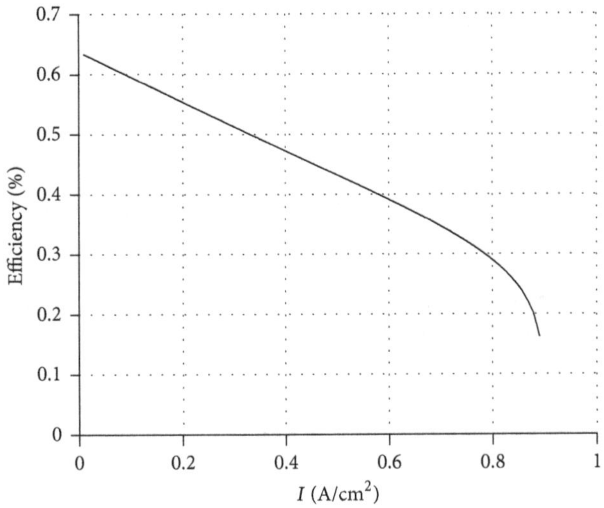

FIGURE 9: Variation of IRSOFC efficiency versus current density (first scenario).

TABLE 5: Input parameters for IRSOFC system.

Component efficiency		
Combustion chamber	95	%
Heat exchanger	80	%
Inverter	96.5	%
Pressure drop		
Heat exchanger	1	%
SOFC	2	%
Combustion chamber	1	%
SOFC input data		
Area	0.0834	m^2
Recirculation ratio	60	%
Limiting current density	9000	A/m^2
Current density	2000	A/m^2
Fuel utilization factor	85	%
Air utilization factor	25	%

TABLE 6: Results of IRSOFC modeling.

		First scenario	Second scenario
Electrical efficiency	%	55.3	55.3
Number of cells		2554	2371
SOFC power production	kW	272.5	253.0
Fuel flow rate	kg/s	0.0099	0.0091

6.3. *Economic Study.* It is assumed that government is going to run this project. The goal is to compare the current results with those from the conventional methods to produce electricity. Therefore, the electricity price and the fuel price are considered based on the prices for the government. For inputs and conditions, see Table 7.

Because of the size of the systems, grid connection cost (first scenario), and the PHS cost (second scenario), the economic results from two scenarios are different. The wind farm price is divided into 69% for the turbine, 12% for the grid connection, 8% for the foundation, 3% for the internal grid, 3% for the installation, and 5% for the rest [23]. For the second scenario, the cost of grid connection is replaced by the PHS cost. The results of the economic analysis are shown in Table 8.

TABLE 7: Economic parameter.

Number of labors	3	
Labor's salary	500	\$/month
Project lifetime	20	Year
Price of products	0.0879	\$/kWh
Interest rate	22	%
Annual inflation rate	15.6	%
Tax	8	%
Fuel price	0.23	$\$/m^3$

TABLE 8: Economic results.

		First scenario	Second scenario
ACS	\$	2.2709×10^5	2.6833×10^5
C_{acap}	\$	1.0919×10^5	1.3206×10^5
C_{arep}	\$	0	0
C_{amain}	\$	6.5046×10^4	7.8674×10^4
C_{aope}	\$	5.2854×10^4	5.7591×10^4
LCOP	\$/kWh	0.0747	0.0882
NPV	\$	2.7056×10^6	3.1970×10^6
AB	\$	1.4941×10^5	1.3104×10^5
PC	\$/kWh	0.0388	0.0448
POR	Year	9.4645	13.0515

All economic parameters are dominated by the capital cost which is the highest in the second scenario. The POR and the PC for the second scenario are about 38% and 15% higher than those of the first one, respectively. Thus, because of the capital cost of the PHS system, the first scenario is more economical than the second scenario.

In both projects, PC is lower than the price of electricity (POP); in the first scenario, PC is 126% lower than POP and, in the second scenario, this number is 96%, which is the advantage of using renewable energies with low energy costs.

Figure 10 shows the effects of the real interest rate on economic parameters. Increasing the real interest rate decreases the NPV and increases the LCOP and the ACS of the system (Figure 10). The slopes of the curves for the second scenario are higher than those of the first scenario. The real interest rate for the current project is 0.0554 (Figure 11), which is a high value; the detailed values are provided in Table 7 for Iran. That means, in countries with the low values of the real interest rate, the project will be more profitable in particular for the second scenario. For example, 10% lessening the real interest rate can improve all parameters about 2% (Figure 10). As it is seen, the economic parameters are not too sensitive to real interest rate.

The value of POP for this project is 0.0879 (Table 7), which is shown in Figure 11. For this project, the period of return (POR) is 9.46 years for the first scenario and 13.05 years for the second one. In Figure 11, the slopes of the curves at the

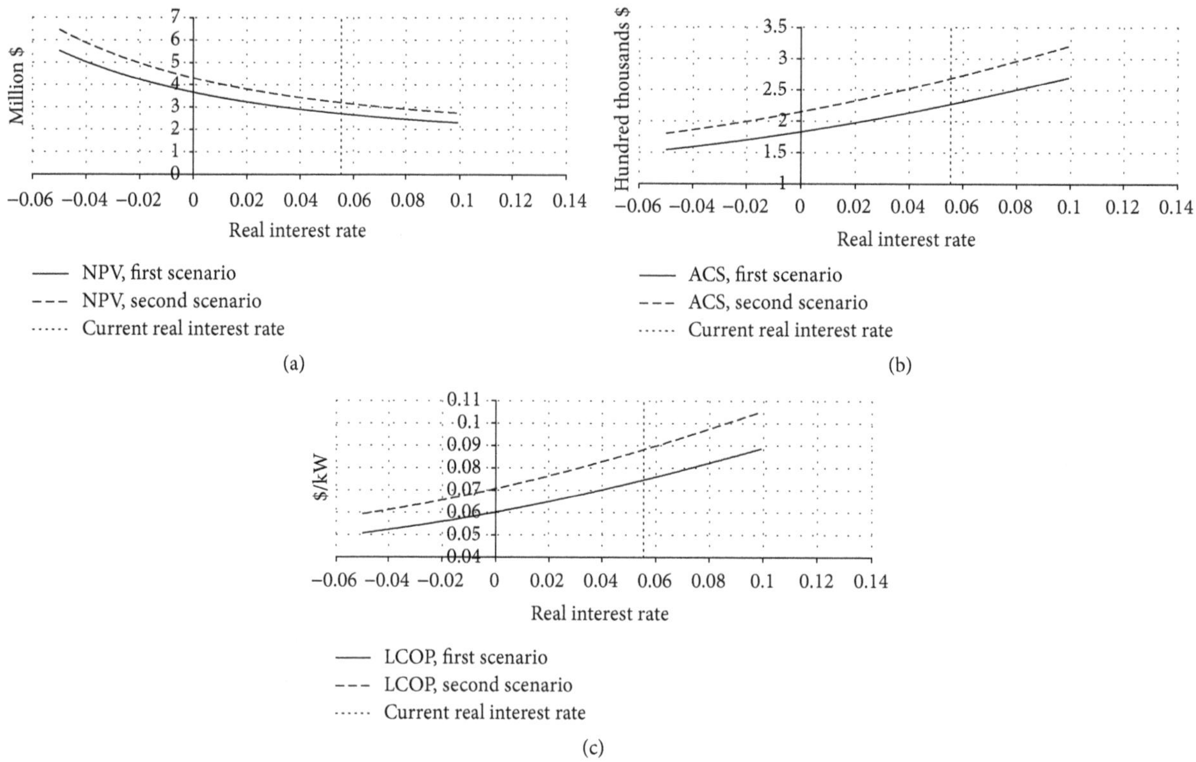

FIGURE 10: Variation of (a) NPV, (b) ACS, and (c) LCOP versus real interest rate.

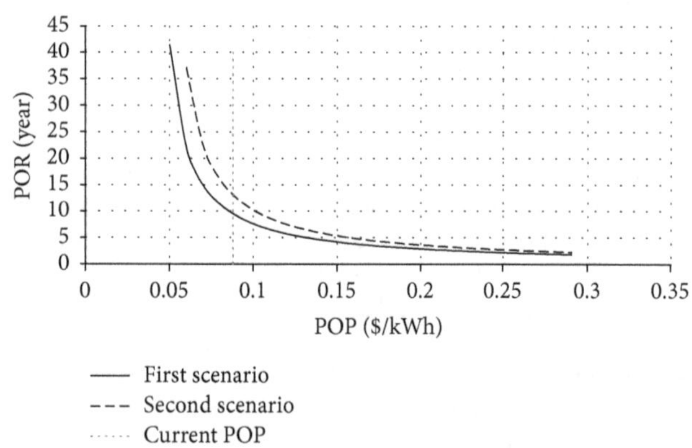

FIGURE 11: Variation of POR versus POP.

beginning are very high; then, small changes in the POP can make significant changes in the POR. For the second scenario, about 15% increase in POP decreases the POR under 10 years making the project more economical.

7. Conclusion

For a combined IRSOFC-HAWT system, thermodynamic and economic analyses were done based on two scenarios: an on-grid-IRSOFC-HAWT system and an off-grid-IRSOFC-HAWT system with a PHS. Based on the thermodynamic analysis, the best working condition for the IRSOFC was chosen and then a proper HAWT was designed to generate the demand power of a village in Manjil, Iran. The high capital cost of the PHS dominated the economic parameters and made the second scenario the worst one from the economic point of view. The POR for the second scenario is about 3 years longer compared to that of the first scenario which was 9.46 years. Although the second scenario is less economical, for the case of having no access to the grid, this scenario makes the project possible. If the price of the electricity or the POP was increased only 15%, around 30% of the years of the POR would be reduced, making the second scenario more economical. In both projects, PC is lower than the price of electricity (POP), which is the advantage of using renewable energies with low energy costs.

It is suggested for the future work of this study that the waste heat of the system be used directly for the heating and cooling purposes, making both scenarios more profitable.

Nomenclature

A: Area [m^2]
AB: Annual Benefit [$]
A, B: Constants [—]
ACS: Annualized Cost of System [$]
c: Chord [m]
C: Cost [$]
C_D: Drag coefficient [—]
C_L: Lift coefficient [—]
C_p: Power coefficient [—]
CRF: Capital Recovery Factor [—]
D: Diameter [m]
E: Cell voltage [V]

E_{Act}: Activation energy [kJ/mol]
f: Annual inflation rate [—]
F: Prandtl's correction factor [—]
F: Faraday constant [C]
i_0: Exchange current density [A/m^2]
i: Current density [A/m^2]
i: Real interest rate [—]
I: Current density [A/cm^2]
i_l: Limiting current density [A/m^2]
j: Nominal interest rate [—]
L: Length [m]
LCOP: Levelized Cost of Product [$/kW]
$\dot{m}$: Mass flowrate [kg/s]
n: Number of cells [—]
N: Number of blades [—]
NAB: Net Annual Benefit [$]
n_e: Number of electrons [—]
NPV: Net Present Value [$]
OFC: Operating Flow Cost [$]
p: Partial pressure [—]
P: Pressure [bar]
PC: Prime Cost [$/kW]
POP: Price of Product [$/kW]
POR: Period of return [year]
r: Resistance [Ω]
r: Radius [m]
R: Universal gas constant [J/mol]
R: Radius [m]
Re: Reynolds number [—]
SOPC: Summation of Product Cost in a year [$]
T: Temperature [K]
VOP: Volume of Product [kW]
$\dot{W}$: Power [Watts]
x, y, z: Molar rate [mol/s]
Y: Year [—].

Greek Letters

β: Heat transfer coefficient [—]
γ: Preexponential factor [A/m^2]
δ: Thickness [cm]
ΔG°: Gibbs free energy changes at standard condition [—]
η: Overpotential [V]
λ_r: Tip speed ratio at radius r [—]
ρ: Resistivity [Ω·cm]
φ: Angle between the resultant velocity and the rotor plane [degree].

Subscript

acap: Annualized capital cost
Act: Activation
amain: Annualized maintenance cost
An: Anode
aope: Annualized operating cost
arep: Annualized replacement cost
Ca: Cathode
Cap: Capital
cc: Combustion chamber
comp: Compressor
Conc: Concentration
GT: Gas turbine
HX: Heat exchanger
i: Section number of the blade
inv: Inverter
proj: Project
ref: Reference
rep: Replacement
rev: Reversible.

Conflicts of Interest

The authors declare that they have no conflicts of interest.

References

[1] S. Soheyli, M. H. Shafiei Mayam, and M. Mehrjoo, "Modeling a novel CCHP system including solar and wind renewable energy resources and sizing by a CC-MOPSO algorithm," *Applied Energy*, vol. 184, pp. 375–395, 2016.

[2] A. Fathy, "A reliable methodology based on mine blast optimization algorithm for optimal sizing of hybrid PV-wind-FC system for remote area in Egypt," *Renewable Energy*, vol. 95, pp. 367–380, 2016.

[3] S. Obara, J. R. Morel Rios, and M. Okada, "Control of cyclic fluctuations in solid oxide fuel cell cogeneration accompanied by photovoltaics," *Energy*, vol. 91, pp. 994–1008, 2015.

[4] M. Hosseini, I. Dincer, and M. A. Rosen, "Hybrid solar-fuel cell combined heat and power systems for residential applications: Energy and exergy analyses," *Journal of Power Sources*, vol. 221, pp. 372–380, 2013.

[5] T. Vigneysh and N. Kumarappan, "Autonomous operation and control of photovoltaic/solid oxide fuel cell/battery energy storage based microgrid using fuzzy logic controller," *International Journal of Hydrogen Energy*, vol. 41, no. 3, pp. 1877–1891, 2016.

[6] D. Tenfen and E. C. Finardi, "A mixed integer linear programming model for the energy management problem of microgrids," *Electric Power Systems Research*, vol. 122, pp. 19–28, 2015.

[7] R. K. Akikur, R. Saidur, H. W. Ping, and K. R. Ullah, "Performance analysis of a co-generation system using solar energy and SOFC technology," *Energy Conversion and Management*, vol. 79, pp. 415–430, 2014.

[8] W. Bai, M. R. Abedi, and K. Y. Lee, "Distributed generation system control strategies with PV and fuel cell in microgrid operation," *Control Engineering Practice*, vol. 53, pp. 184–193, 2016.

[9] S. H. Chan, H. K. Ho, and Y. Tian, "Modelling of simple hybrid solid oxide fuel cell and gas turbine power plant," *Journal of Power Sources*, vol. 109, no. 1, pp. 111–120, 2002.

[10] J. Pirkandi, M. Ghassemi, M. H. Hamedi, and R. Mohammadi, "Electrochemical and thermodynamic modeling of a CHP system using tubular solid oxide fuel cell (SOFC-CHP)," *Journal of Cleaner Production*, vol. 29-30, pp. 151–162, 2012.

[11] J. Larminie and A. Dicks, *Fuel Cell Systems Explained*, John Wiley & Sons, West Sussex, UK, 1st edition, 2003.

[12] S. H. Chan, K. A. Khor, and Z. T. Xia, "Complete polarization model of a solid oxide fuel cell and its sensitivity to the change of cell component thickness," *Journal of Power Sources*, vol. 93, no. 1-2, pp. 130–140, 2001.

[13] P. Costamagna and K. Honegger, "Modeling of solid oxide heat exchanger integrated stacks and simulation at high fuel utilization," *Journal of the Electrochemical Society*, vol. 145, no. 11, pp. 3995–4007, 1998.

[14] R. Gasch and J. Twele, Eds., *Windkraftanlagen: Grundlagen, Entwurf, Planung und Betrieb*, Springer-Verlag, 2010.

[15] J. Manwell, *Wind Energy Explained: Theory, Design and Application*, 1st edition, 2002.

[16] M. Hansen, *Aerodynamics of Wind Turbines*, Earthscan, Sterling, VA, USA, 1st edition, 2008.

[17] J. Hernandez and A. Crespo, "Aerodynamic calculation of the performance of horizontal axis wind turbines and comparison with experimental results," *Wind Engineering*, vol. 11, no. 4, pp. 177–187, 1987.

[18] M. Meratizaman, S. Monadizadeh, and M. Amidpour, "Introduction of an efficient small-scale freshwater-power generation cycle (SOFC-GT-MED), simulation, parametric study and economic assessment," *Desalination*, vol. 351, pp. 43–58, 2014.

[19] M. Gökçek and M. S. Genç, "Evaluation of electricity generation and energy cost of wind energy conversion systems (WECSs) in Central Turkey," *Applied Energy*, vol. 86, no. 12, pp. 2731–2739, 2009.

[20] P. Du and N. Lu, *Energy Storage for Smart Grids: Planning and Operation for Renewable and Variable Energy Resources (VERs)*, Academic Press, 2014.

[21] M. G. Khalfallah and A. M. Koliub, "Effect of dust on the performance of wind turbines," *Desalination*, vol. 209, no. 1-3, pp. 209–220, 2007.

[22] ASDC — Processing, archiving, and distributing Earth science data at the NASA Langley Research Center, https://eosweb.larc.nasa.gov/ [Accessed 5 April 2017].

[23] H. Auer, M. Stadler, G. Resch, T. Schuster, J. Twidell, and D. J. Swider, "Pushing A Least Cost Integration Of Green Electricity Into The European Grid (GreenNet)-Cost and Technical Constraints of RES-E Grid Integration," Tech. Rep., Energy Economics Group, 2006.

8

Active Flow Control in a Radial Vaned Diffuser for Surge Margin Improvement: A Multislot Suction Strategy

Aurélien Marsan,[1] Isabelle Trébinjac,[2] Stéphane Moreau,[1] and Sylvain Coste[3]

[1]*Department of Mechanical Engineering, Université de Sherbrooke, Sherbrooke, QC, Canada J1K 2R1*
[2]*LMFA, Ecole Centrale de Lyon, 36 Avenue Guy de Collongue, 69134 Ecully, France*
[3]*Safran Helicopter Engines, Avenue Joseph Szydlowski, 64510 Bordes, France*

Correspondence should be addressed to Aurélien Marsan; aurelien.marsan@usherbrooke.ca

Academic Editor: Yutaka Ohta

This work is the final step of a research project that aims at evaluating the possibility of delaying the surge of a centrifugal compressor stage using a boundary-layer suction technique. It is based on Reynolds-Averaged Navier-Stokes numerical simulations. Boundary-layer suction is applied within the radial vaned diffuser. Previous work has shown the necessity to take into account the unsteady behavior of the flow when designing the active flow control technique. In this paper, a multislot strategy is designed according to the characteristics of the unsteady pressure field. Its implementation results in a significant increase of the stable operating range predicted by the unsteady RANS numerical model. A hub-corner separation still exists further downstream in the diffuser passage but does not compromise the stability of the compressor stage.

1. Introduction

The stable operating range of centrifugal compressors is limited toward low mass flow rate by the onset of flow instabilities, either rotating stall or surge. This limitation is detrimental to the acceleration rate of the compressor that is a key performance parameter in particular for turboshaft engines and can also prevent operation at maximum efficiency which may lay at or close to the surge line. It is then of primary concern for turboshaft designers to better understand the surge inception process in centrifugal compressors and to develop design or control techniques in order to delay the surge.

In this context, the radial vaned diffuser appears to be a key component for surge inception in centrifugal compressors operating at high rotational speed. This was already pointed out by Came [1] in 1976 and has been confirmed by several studies since, but at this time the impeller was the most critical element of the stage from a performance point-of-view and the improvement efforts have then mostly focused on it.

In 1975 for a centrifugal compressor for automotive engine, Amann et al. [2] attributed the occurrence of rotating stall and surge to the radial vaned diffuser.

In 1994, Hunziker and Gyarmathy [3] have worked on a single-stage centrifugal compressor equipped with a radial vaned diffuser made of circular-arc vanes whose stagger angle could be adjusted. Maximum rotational speed is 22,000 rpm and maximum pressure ratio is about 2,0. By analyzing the stability in various operating conditions and configurations using the criteria about the slope of the characteristics proposed by Greitzer in [4], the authors have concluded that the diffuser has a destabilizing influence at high rotational speed and for low mass flow rate which may explain the onset of surge. In particular, the diffuser entry zone is expected to play a major role since it has a stabilizing effect over the whole operating line but switches to a destabilizing effect at low mass flow rate.

In 2001, Wernet et al. [5] investigated the surge inception in the NASA-CC3 centrifugal compressor equipped with the wedge vane diffuser using digital particle imaging velocimetry. The design rotational speed is 21,789 rpm for a design pressure ratio of about 4. They have shown that the flow reversal at surge starts within the diffuser and then propagates upstream until forward flow is reestablished.

More recently, in 2010, Everitt and Spakovszky [6, 7] studied the onset of instabilities in a high-speed preproduction centrifugal compressor stage used for large diesel engines. The design pressure ratio is 5 and the impeller tip Mach number is greater than one at design speed. The radial vaned diffuser is made of airfoil vanes. Using the strategy first investigated by Hill IV [8] based on an isolated diffuser numerical model, the authors identify the growth of a flow separation at the leading-edge of the radial vaned diffuser as one of the necessary conditions for the formation of short-wavelength spike stall precursors.

In 2014, Bousquet et al. [9] studied a 2.5 pressure ratio subsonic centrifugal compressor stage designed by Liebherr-Aerospace Toulouse SAS and integrated it in an air-conditioning system. It is composed of a backswept splittered unshrouded impeller and a radial vaned diffuser made of wedge blades. Design rotational speed is 38000 rpm. Based on the results of full-annulus 3D unsteady RANS simulations, the authors observed the growth of a flow separation at the leading-edge of the diffuser vane when moving the operating point toward the surge limit, in a similar manner to that previously reported by Everitt and Spakovszky [6, 7]. For that particular test case, however, this flow separation does not lead to an accumulation of vortical structure in the vaneless diffuser and the operating point remains stable until the flow separates at the leading-edge of the impeller blades [10].

Then, Buffaz et al. [11, 12] studied the onset of instabilities in a supersonic centrifugal compressor stage designed and built by Safran Helicopter Engines (SHE), both experimentally and numerically. The compressor was mounted on a 1 MW test rig at the LMFA, École Centrale de Lyon, France, and detailed unsteady pressure measurements in the diffuser entry zone were performed. A full-annulus simulation of the surge inception was also conducted. The growth of a boundary-layer separation on the suction side of the diffuser vanes when moving the operating point toward the surge limit was again observed. This separation degenerates into a rotating stall finally leading to surge.

Other numerical investigations were conducted by Benichou and Trébinjac [13, 14] on another transonic centrifugal compressor designed and built by Safran Helicopter Engines (SHE). The test case is composed of a backswept unshrouded impeller and a splittered vaned diffuser. Pressure ratio is about 4 at a rotational speed of 40,000 rpm. The analysis of the flow field predicted by URANS simulations for an operating point near surge suggested that a boundary-layer separation on the suction side of the diffuser vanes was responsible for the surge inception.

Finally, the surge inception scheme proposed by Everitt and Spakovszky and based on the growth of a separation at the leading-edge of the diffuser vanes was found by Fujisawa et al. [15] in a centrifugal compressor of a turbocharger for marine diesel engines operating at low-speed. The study also showed the interaction between the diffuser leading-edge separation vortex and the tip leakage flow from the impeller.

Given the importance of the flow separation in the surge inception process for centrifugal compressor stages enlightened by these studies, the use of an active control technique aiming at controlling the flow separation within the diffuser is expected to be an effective way for delaying the surge. In particular, the boundary-layer suction technique has proven its effectiveness in case of axial compressors in order to either delay the surge or increase the blade loading [16–20].

In the next section, previous control strategies are reviewed.

2. Previous Control Strategies

Came [1] already mentioned in 1976 the possibility of designing efficient diffusers of smaller radial extent using boundary-layer suction. Still, there is no evidence yet in the literature of the interest of the boundary-layer suction in case of centrifugal compressors. The following aims at giving an overview of the other flow control techniques that have been applied to centrifugal compressors equipped with a radial vaned diffuser.

Botros and Henderson [21] give an extensive technology assessment of the surge control techniques for centrifugal compressors developed until 1993. The authors report the common use of 1D control technique based on the opening of a recycle valve when the surge is detected or on active control of the throttling element. Design methods are also given, as, for example, the use of a backswept impeller. The reduction of the diffuser exit width or specific designs of the diffuser inlet are also mentioned as means to influence the surge flow rate. Finally, a pioneering casing treatment developed by Amann et al. [2] in 1975 is reported. This treatment is applied to a centrifugal compressor whose surge is triggered by the occurrence of rotating stall within the radial vaned diffuser. It consists in a circumferential slot put at the impeller trailing edge and connected to an annular chamber. A significant decrease of the surge flow rate is reported. The effectiveness of the control is attributed to the uniformization of the pressure distribution over the circumference. Another technique relying on the same principle was proposed by Raw [22] in 1986. The so-called "porous throat diffuser" technique is based on the linking of the diffuser throats through slots connected to an azimuthal chamber. Again, the stabilizing effect is explained by the uniformization of the flow between all the diffuser passages, which may prevent the stall of any one diffuser vane before another one and then avoid the occurrence of rotating stall.

Later, Nelson et al. [23] managed to experimentally reduce the surge mass flow rate of an axicentrifugal stage by 1% using air injection into the throats of the diffuser passages.

Skoch also investigated the effectiveness of the air injection technique applied to the NASA CC3 compressor [24, 25] on the basis of the work of Spakovszky [26]. Injector nozzles were inserted first at the shroud of the vaneless diffuser space. It was found that the mere obstruction created by the injector nozzles at the shroud was sufficient to significantly decrease the surge mass flow rate and concluded that the benefit in diffuser stability came from the reduction of the incidence on the diffuser vanes. This conclusion was supported by hub-side injection experiments, whose conclusions were similar. The effects of the air injection technique in the NASA CC3 test case have been recently investigated numerically by Halawa et al. [27].

Other publications can be found in patents. A device is suggested by Schönenborn [28] based on boundary-layer suction on the suction face of the diffuser vanes. He also suggests that the bled flow could be used for the cooling of the turbine blades, for example. Another is proposed by Leblanc [29], the fluid being bled from the suction face of the diffuser vanes and then reinjected upstream in the vaneless diffuser space.

No detailed assessment of the potential of the boundary-layer suction technique applied to centrifugal compressor has been done yet. This is the objective of the present study.

3. Test Case

The test case is a centrifugal compressor stage designed and built by Safran Helicopter Engines (SHE), composed of a backswept splittered unshrouded impeller and a radial vaned diffuser. It is used as a rear compression stage in a helicopter engine. The flow at the impeller outlet is transonic. This compressor is installed in a 400 kW test rig of the DAEP Laboratory at the Institut Supérieur de l' Aéronautique et de l'Espace (ISAE), Toulouse, France. A complete description of the experimental test rig is available in [30].

For the compressor operating at its nominal rotational speed, previous work based on steady-state numerical calculations has shown the growth of corner stall within the semivaneless space of the diffuser when moving the operating point of the compressor toward the low mass flow rate [31]. This corner stall was supposed to limit the stability of the compressor stage and was also predicted by unsteady numerical simulations [32]. A boundary-layer suction technique had been foreseen and then studied in order to control the corner stall. According to steady-state numerical calculations relying on the mixing-plane approach, the suction led to a significant increase of the stable operating range of the compressor. But these promising results were challenged by unsteady numerical simulations [30]. The detailed analysis of the flow in the diffuser with suction has shown that the hub-corner separation was not completely removed by the suction, contrary to the prediction of the steady-state numerical model: a corner separation still existed in the diffuser downstream of the location of the initial boundary-layer separation, and the stable operating range was not extended. This result confirms the high influence of the unsteadiness on the flow in the diffuser entry zone, especially in centrifugal compressors with small radial gap between the impeller and the vaned diffuser [33]. The scrolling of the impeller blades in front of the diffuser vanes generates intense high pressure waves, in the same manner as already observed in another transonic Safran Helicopter Engines centrifugal compressor [34]. These high pressure waves are reinforced when crossing the diffuser throat. This generates intense instantaneous adverse pressure gradients that cannot be predicted by the steady-state numerical model and are supposed to provoke the boundary-layer separation. Figure 1 shows the map of the maxima of the instantaneous adverse pressure gradient in the diffuser over one period.

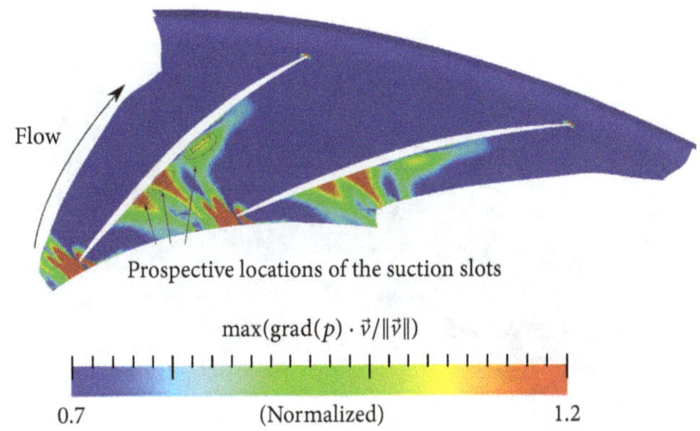

Figure 1: Maxima of instantaneous adverse pressure gradient, without control. Prospective locations of the suction slots.

Given that conclusion, the control strategy was adapted in order to take into account the unsteadiness. A multislot suction strategy was proposed but not evaluated. This strategy relies on three suction slots distributed along the diffuser suction side on the hub surface, at the locations of the maxima of the instantaneous adverse pressure gradient, as shown in Figure 1. The paper presents the results obtained by this multislot suction strategy.

4. Numerical Setup

Numerical simulations have been performed with the elsA software, developed by ONERA, the French aerospace laboratory [35]. This code solves the Reynolds-Averaged Navier-Stokes equations with a cell-centered finite volume method applied on multiblock-structured meshes [31, 32]. The flow is considered fully turbulent and the effect of turbulence on the mean flow field is modeled thanks to the two variables (k-l) turbulence model of Smith. The time integration is performed with a backward Euler scheme with implicit operators. Spatial discretization is achieved with Roe's second-order scheme and Harten's correction.

The mesh is the same as for the previous studies. It has been generated using the commercial meshing software Autogrid from Numeca. It counts 6,07 million points for one impeller-diffuser passage. The first cell height on the walls is 1 micrometer, which corresponds to a dimensionless wall distance y^+ approximately equal to 1 along all solid surfaces. The impeller tip clearance is included in the numerical model and is meshed with a C-H topology. Filets between the endwalls and the blades are included neither in the impeller nor in the radial vaned diffuser. Previous studies have shown that mesh independence in the diffuser is reached.

Unsteady numerical simulations of the flow within the compressor stage have been performed using the phase-lagged assumption which is based on the Tyler and Sofrin relation and is also known as the chorochronic approach [36–38].

In a previous publication [32], experimental unsteady pressure signals available in the diffuser entry zone have

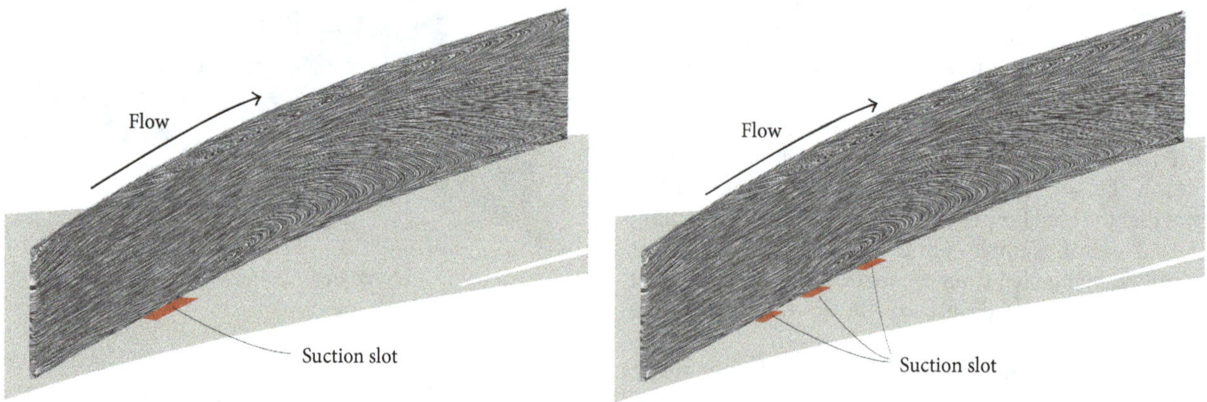

(a) Single-slot strategy

(b) Three-slot strategy

FIGURE 2: Suction strategies.

been compared with the unsteady numerical results (without suction). The comparison has ensured the capability of the numerical model at describing the flow in this compressor, even at near surge operating points.

For the present study, the numerical parameters have been kept the same except for the time step which is defined as

$$\delta t = \frac{2\pi}{\omega N_R N_S Nqo}, \quad (1)$$

where N_R and N_S are the numbers of rotor and stator blades and ω is the rotational speed and Nqo is a parameter. For the base case, the Nqo parameter has been progressively doubled in order to check the time step convergence, starting from $Nqo = 20$ and ending at $Nqo = 160$. Regarding the convergence, Nqo was set to 60 for the single-slot suction case, as a compromise between the time step convergence and the computational time, and 80 for the multislot suction case.

The exit static pressure was set by prescribing a ratio between the value of the static pressure and the mass flow rate at the mesh exit plane. This allows computing the flow at a mass flow rate lower than the peak of the performance curve, which is necessary in order to reach the stability limit.

The boundary-layer suction was modeled by a surface mass-flux boundary-condition. The results obtained with this simplified model had been previously compared with those obtained by including the entire slot within the numerical model, using the chimera technique [39]. The relevance of the boundary-condition approach for analyzing the suction effect on the diffuser flow has been demonstrated: the boundary limit condition technique cannot describe the possible recirculation zone within the suction slot, but its effect on the main flow is consistent with the effect predicted using the chimera technique [31] and is more appropriate for the present study that does not deal with the important issue of designing an optimal suction slot.

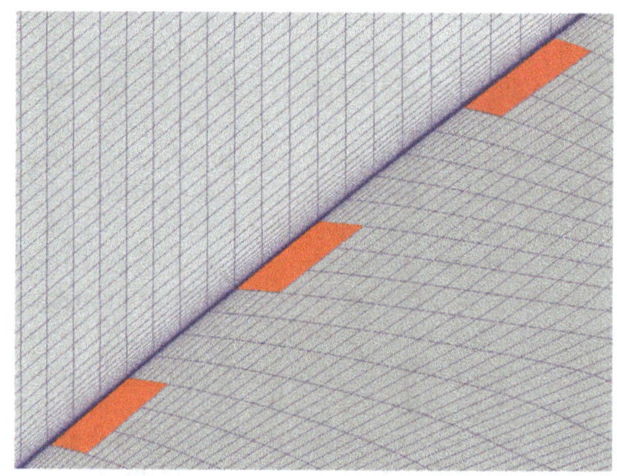

FIGURE 3: Numerical model of the three-slot suction effect by a boundary limit condition.

5. Control Strategy

Figure 2 shows the two suction strategies that are compared here. The single-slot strategy is the one that has already been studied and did not reach any improvement of the surge margin according to the time-dependent numerical simulations [30]. The three-slot strategy is the one which is under the scope of the present paper, the slots being located at the maxima of the instantaneous adverse pressure gradient.

Mass flow rate removed through the suction slots was chosen equal to 1 percent of the surge mass flow rate of the compressor without suction, which is consistent with the previous study, and the surface mass-flux was prescribed homogeneous all over the aspiration surface. In the case of the three-slot suction strategy, the surfaces of the three suction slots are slightly different from each other. The effect of suction is indeed modeled by a boundary limit condition and the possible area of the slots is constrained by the areas of the faces of the wall mesh (Figure 3). 0.32 percent of the surge mass flow rate is then removed through the first upstream slot, 0.33 percent through the second one, and 0.35 percent through the last downstream (the total being still equal to 1 percent of the surge mass flow rate).

6. Results

Given the strong unsteady interaction between the impeller and the diffuser for lowest mass flow rate operating points, the phase-lagged numerical calculation has required more than 25 impeller revolutions in order to reach the convergence for the last converged operating point with the three-slot suction strategy.

The limit of stability was searched by progressively increasing the prescribed ratio between the pressure and the mass flow rate at the outlet. It is important to notice that the limit of stability of the numerical model toward low mass flow rates has been reached for the base and the single-slot cases, but not for the three-slot case. Numerically stable operating points with the three-slot suction strategy could then exist at mass flow rate lower than the last plotted operating point.

7. Performance

Figure 4(a) shows the total-to-static pressure ratio of the compressor stage, plotted as a function of the standard mass flow rate at the impeller inlet. As indicated in the figure, no stable operating point has been found for the base case and the single-slot case below $\dot{m}_R^{\text{std}} \simeq 0.97$. On the contrary, it has been possible to reach the convergence for a much lower mass flow rate with the three-slot suction strategy. Both suction techniques result in an increase of the total-to-static pressure ratio of the compressor stage.

In order to analyze more precisely the effect of the suction on the performance of the compressor stage, Figures 4(b) and 4(c) show the main performance coefficients for the impeller and the radial vaned diffuser. Figure 4(b) shows the total-to-static pressure ratio of the impeller plotted as a function of the diffuser inlet standard mass flow rate. Figure 4(c) shows the static pressure recovery coefficient in the different parts of the diffuser, also plotted as a function of the standard mass flow rate at the diffuser inlet. All contributions are calculated with respect to the kinetic energy at the vaneless diffuser inlet, that is, the impeller trailing edge, so that they can be added in order to find the static pressure recovery coefficient of the whole vaned diffuser.

The total-to-static pressure ratio of the impeller reaches a maximum at a mass flow rate which is close to the stability limit of the base case and the single-slot suction case. The three-slot suction strategy allows reaching a stable operating point for a mass flow rate lower than that of the total-to-static pressure ratio maximum. Indeed, the control should stabilize the operation of the diffuser.

The origin of the saturation of the total-to-static pressure rise of the impeller is not yet identified. It could be caused by a flow separation within the impeller or by a particular behavior of the tip leakage vortex, for example. Some future study should aim at clarifying the flow pattern within the impeller. Given the fact that the limit of stability of the base case is reached at the peak of the total-to-static pressure ratio of the impeller, the surge inception could be possibly explained by an interaction with a particular flow structure within the impeller which leads to the destabilizing of the diffuser hub-corner separation. Such an interaction between boundary-layer separation in the diffuser and the flow structure at the impeller outlet has already been observed [40].

The effect of the suction on the diffuser static pressure recovery process is visible in Figure 4(c). The C_p of the semivaneless diffuser space is greatly increased by the control, with both the single-slot and the three-slot suction strategies. According to previous works which have identified the entry zone of the diffuser as a critical zone for the onset of instabilities [3, 41], the fact that the slope is steeper (negative slope) should have a stabilizing effect on the diffuser work. But the single-slot suction strategy does not improve the overall operating range despite the increase of C_p in the semivaneless space. The only observation of the slopes to check a stability is thus proved to be insufficient.

Figure 5 shows the blockage coefficients calculated at the impeller trailing edge, the diffuser leading-edge, the diffuser throat and the diffuser trailing edge, near hub (between 0% and 30% of span-height), Figure 5(a), in the mid-span region (between 30% and 70% of span-height), Figure 5(b), and near shroud (between 70% and 100% of span-height), Figure 5(c). Blockage coefficient in a section S is defined as

$$B_S = 1 - \frac{\int_S \rho \mathbf{v} \cdot \mathbf{n} \, dS}{\left|\rho \mathbf{v} \cdot \mathbf{n}\right|_{\max}}. \tag{2}$$

These coefficient are plotted for several operating points, as a function of the standard mass flow rate at the diffuser inlet.

In all sections except at the diffuser trailing edge, the blockage coefficients are very similar between the single-slot and the three-slot suction cases. In particular, there is no difference between the blockage curves at the diffuser throat. Finally, the only difference between the single-slot and the three-slot strategies is visible on the blockage coefficient at the diffuser trailing edge. The blockage decreases in the near hub region with the three-slot suction strategy. On the contrary, it increases in the near shroud region. This balance between the near hub and the near shroud blockage has been previously explained by the growth of a shroud-corner separation on the suction side of the diffuser vanes when the hub-corner separation is controlled [30].

8. Flow Structure

Figure 6 compares the skin-friction pattern on the suction side of the diffuser vanes for the three test cases. The compared operating points are those labeled in Figure 4(c) and have similar diffuser inlet standard mass flow rates, that is, similar diffuser inlet flow field conditions. The location of the suction slots are also shown. A hub-corner separation still exists with the three-slot strategy downstream of the last suction slot. Compared with the single-slot case, it is however translated downstream, and the control of the hub-corner separation is more effective. It now takes place far downstream of the diffuser throat, which could explain the

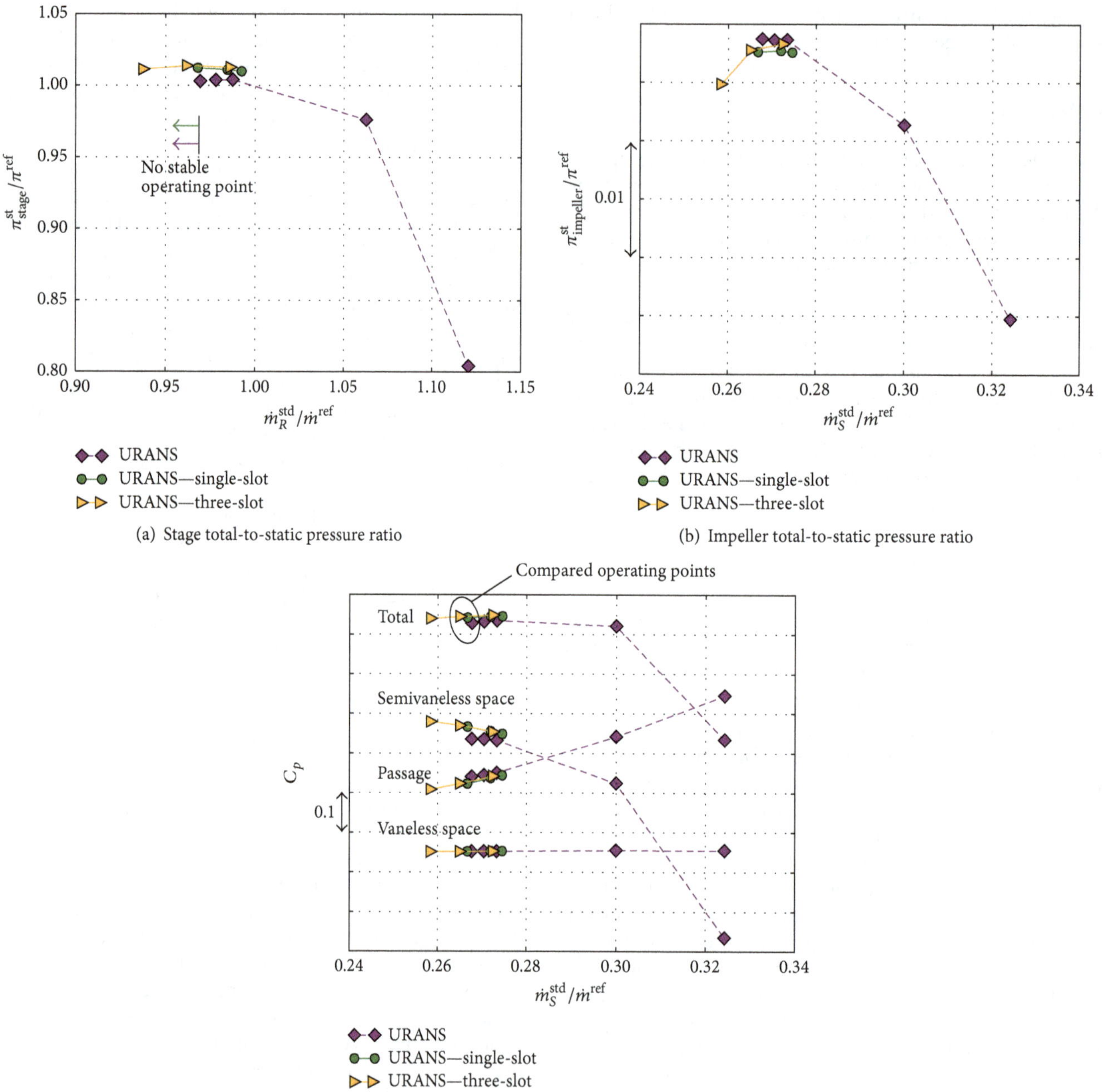

(a) Stage total-to-static pressure ratio

(b) Impeller total-to-static pressure ratio

(c) Diffuser static pressure recovery coefficient

FIGURE 4: Performance of the compressor.

stabilizing effect of the three-slot suction case. As a consequence, the shroud-corner separation develops more. This explains the increase of the blockage near shroud previously mentioned, seen in Figure 5(c).

With the single-slot suction, the new location of the hub-corner separation downstream of the suction slot had been explained by the existence of intense instantaneous adverse pressure gradients [30]. In order to challenge that criterion for boundary-layer separation, Figure 7 shows the fields of the maximum of the instantaneous adverse pressure gradient in a plane located at 10 percent of span-height for the three-slot suction case. The location of the main saddle point of the skin-friction pattern in the corner is indicated by a black square. Here, the field of the maximum of instantaneous adverse pressure gradient is not modified by the three-slot suction effect, and the maxima are still located at the suction slots. The separation yet occurs downstream of the third maximum limit. A perspective of the present study could be to extend the third suction slot in order to cover the entire third maximum of instantaneous adverse pressure gradient, and verify if it could lead to a complete removal of the hub-corner separation.

Figure 8 shows a general overview of the flow structure in the radial vaned diffuser with three-dimensional streamlines calculated from time-averaged flow field. The three-dimensional streamlines are colored by the value of the Mach

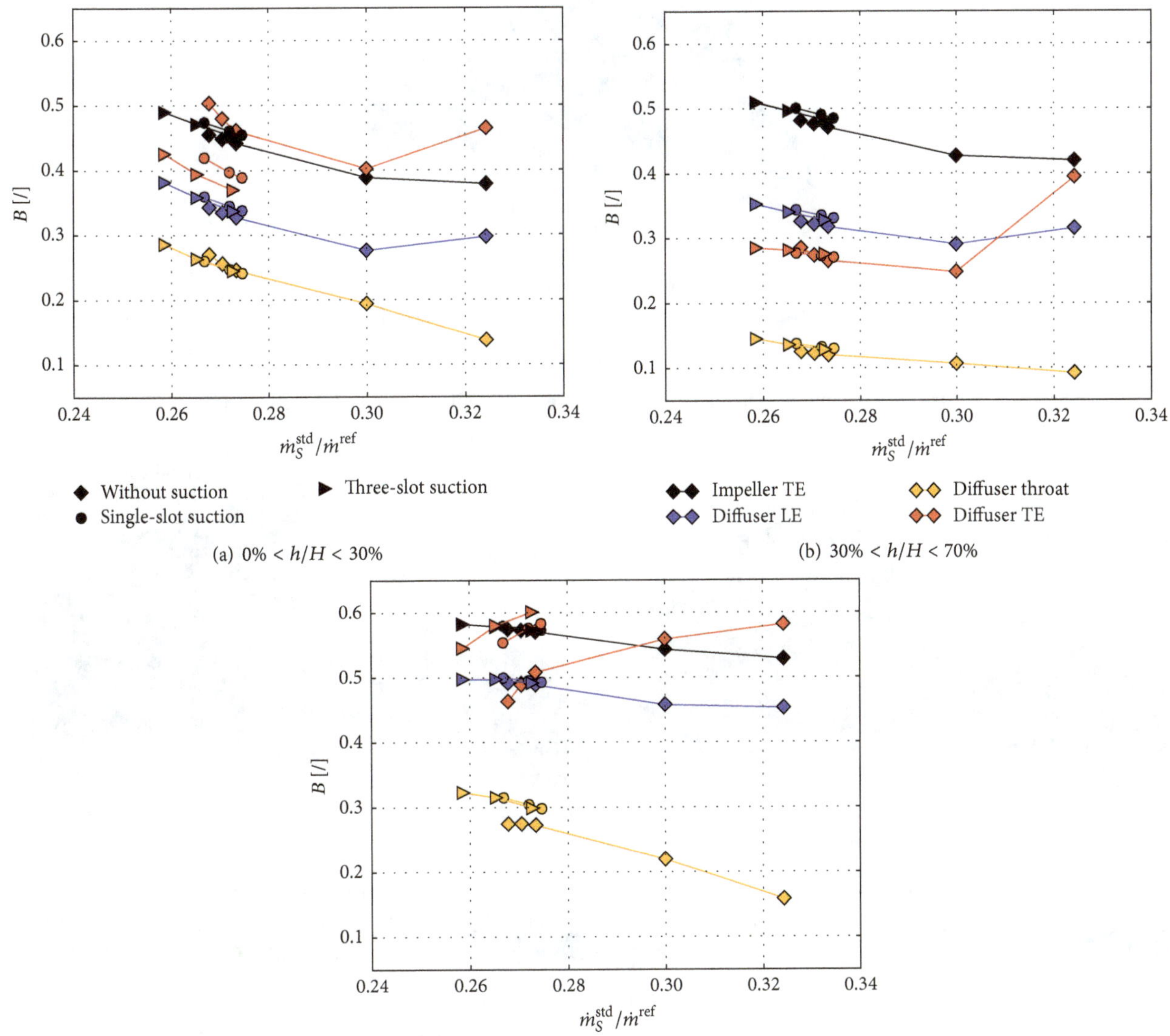

(a) $0\% < h/H < 30\%$

(b) $30\% < h/H < 70\%$

(c) $70\% < h/H < 100\%$

FIGURE 5: Blockage coefficients in the diffuser.

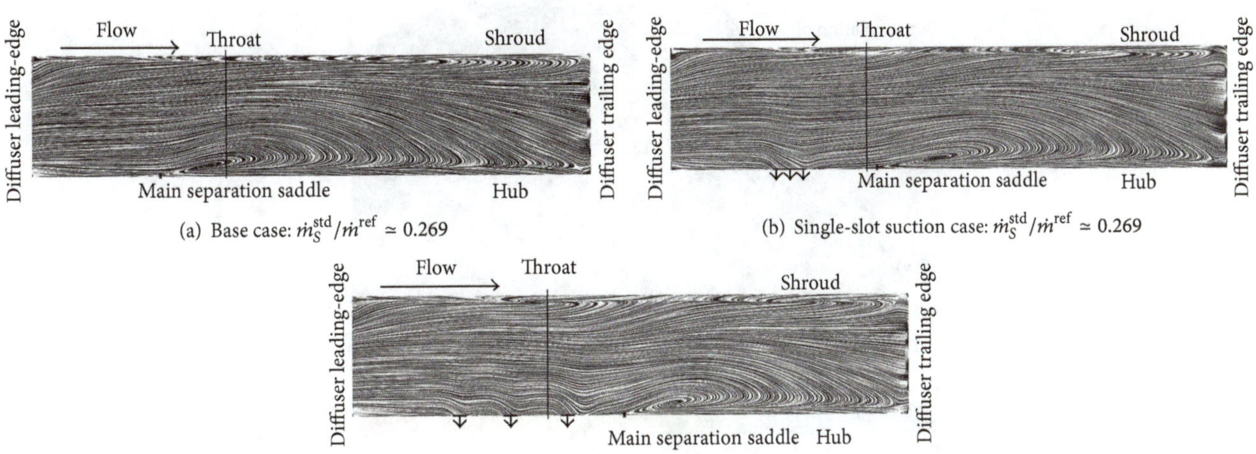

(a) Base case: $\dot{m}_S^{std}/\dot{m}^{ref} \simeq 0.269$

(b) Single-slot suction case: $\dot{m}_S^{std}/\dot{m}^{ref} \simeq 0.269$

(c) Three-slot suction case: $\dot{m}_S^{std}/\dot{m}^{ref} \simeq 0.268$

FIGURE 6: Skin-friction on the diffuser suction side.

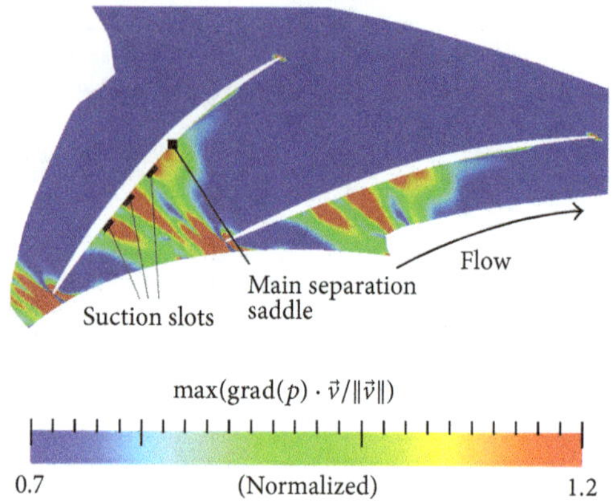

FIGURE 7: Maximum of instantaneous adverse pressure gradient—three-slot suction case.

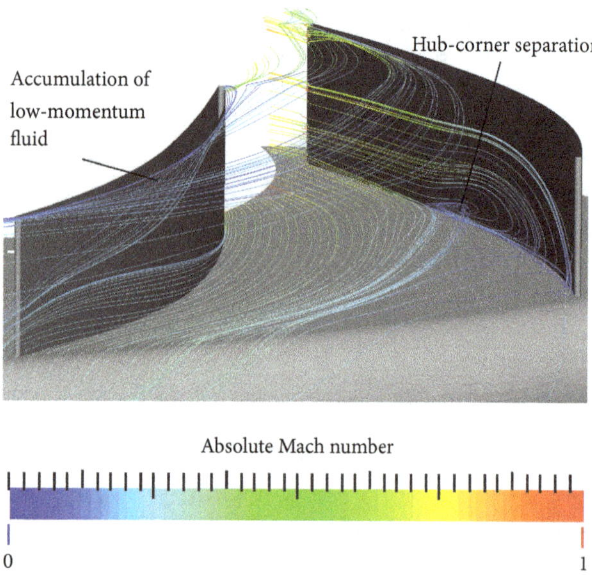

(a) Base case: $\dot{m}_S^{std}/\dot{m}^{ref} \simeq 0.268$

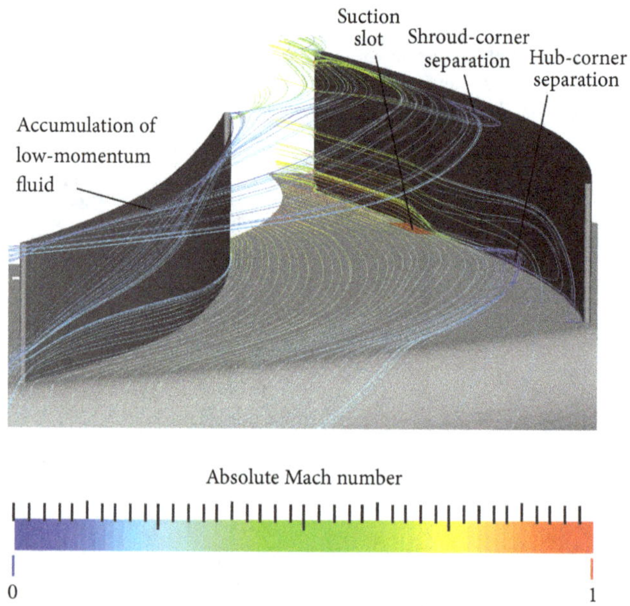

(b) Single-slot suction case: $\dot{m}_S^{std}/\dot{m}^{ref} \simeq 0.268$

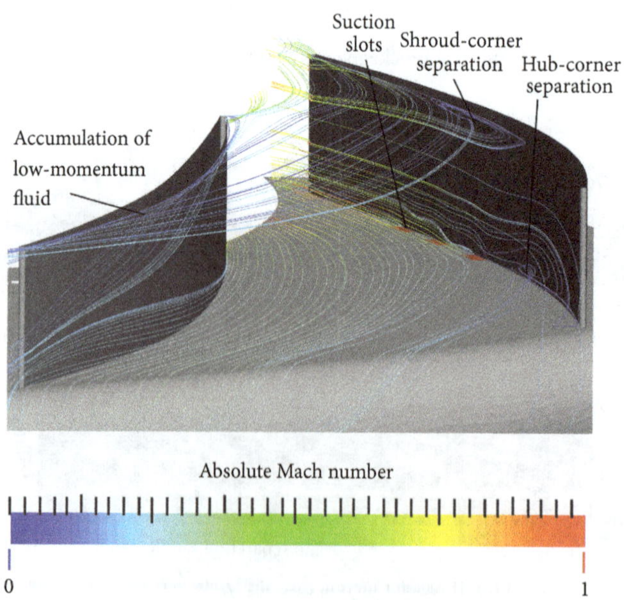

(c) Trislot suction case: $\dot{m}_S^{std}/\dot{m}^{ref} \simeq 0.269$

FIGURE 8: Three-dimensional streamlines.

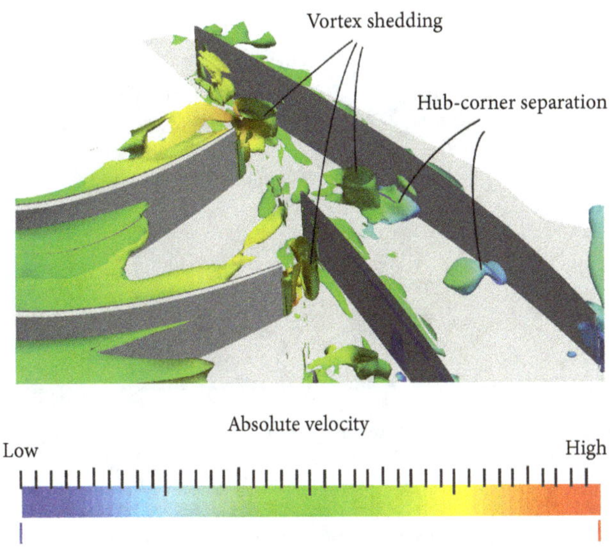

(a) Base case: $\dot{m}_S^{\text{std}}/\dot{m}^{\text{ref}} \simeq 0.269$

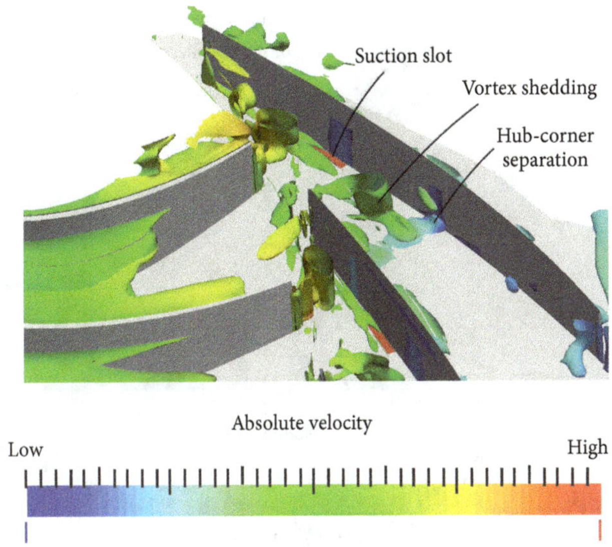

(b) Single-slot suction case: $\dot{m}_S^{\text{std}}/\dot{m}^{\text{ref}} \simeq 0.269$

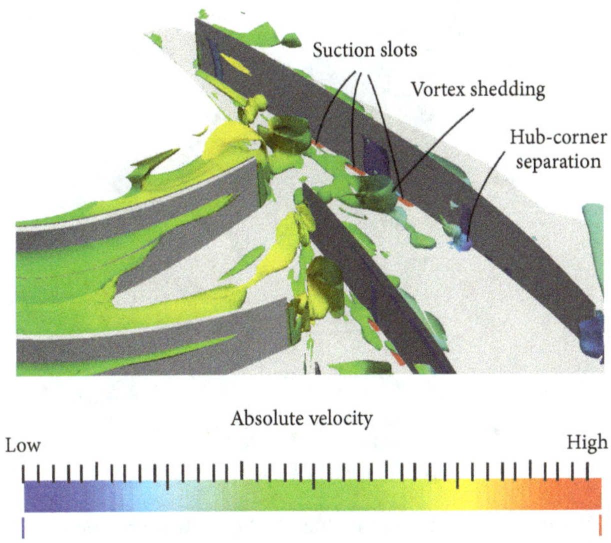

(c) Three-slot suction case: $\dot{m}_S^{\text{std}}/\dot{m}^{\text{ref}} \simeq 0.268$

FIGURE 9: Isocontours of Q-criterion in the diffuser colored by velocity: $Q = 30$.

number along the streamlines. It stresses the translation of the hub-corner separation downstream under the effect of the suction and the growth of the shroud-corner separation, as previously mentioned.

In order to give more insight on the effect of suction, Figure 9 shows isocontours of Q-criterion in the diffuser computed from instantaneous flow fields for the three cases without and with control. The Q-criterion is a local measure of the excess of rotation rate relative to the strain rate [42]. It allows identifying the vortices as flow regions with positive second invariant of the velocity gradient tensor [43]. For clarity, isocontours are not plotted in the near-tip region, above $h/H = 80\%$.

The impeller vortex shedding and the hub-corner separation vortex are both visible in the three figures. For the base case, Figure 9(a), these two vortices merge together, which is expected to notably increase the size of the hub-corner separation vortex. With the single-slot control strategy, Figure 9(b), they still merge together but their interaction is decreased as the hub-corner separation is deflected. With the three-slot suction strategy, the interaction is barely visible given the chosen value for the isocontour.

This analysis stresses that the hub-corner separation development does not only result from the growth of the boundary-layers in the diffuser. It is also promoted by the interaction between the hub-corner separation vortex and other vortices like the impeller trailing edge vortex.

The vorticity transport equation that simplifies as

$$\frac{\partial \Omega_i}{\partial t} + U_j \frac{\partial \Omega_i}{\partial x_j} = \underbrace{\Omega_j \frac{\partial U_i}{\partial x_j}}_{A} + \nu \frac{\partial^2 \Omega_i}{\partial x_j^2} \tag{3}$$

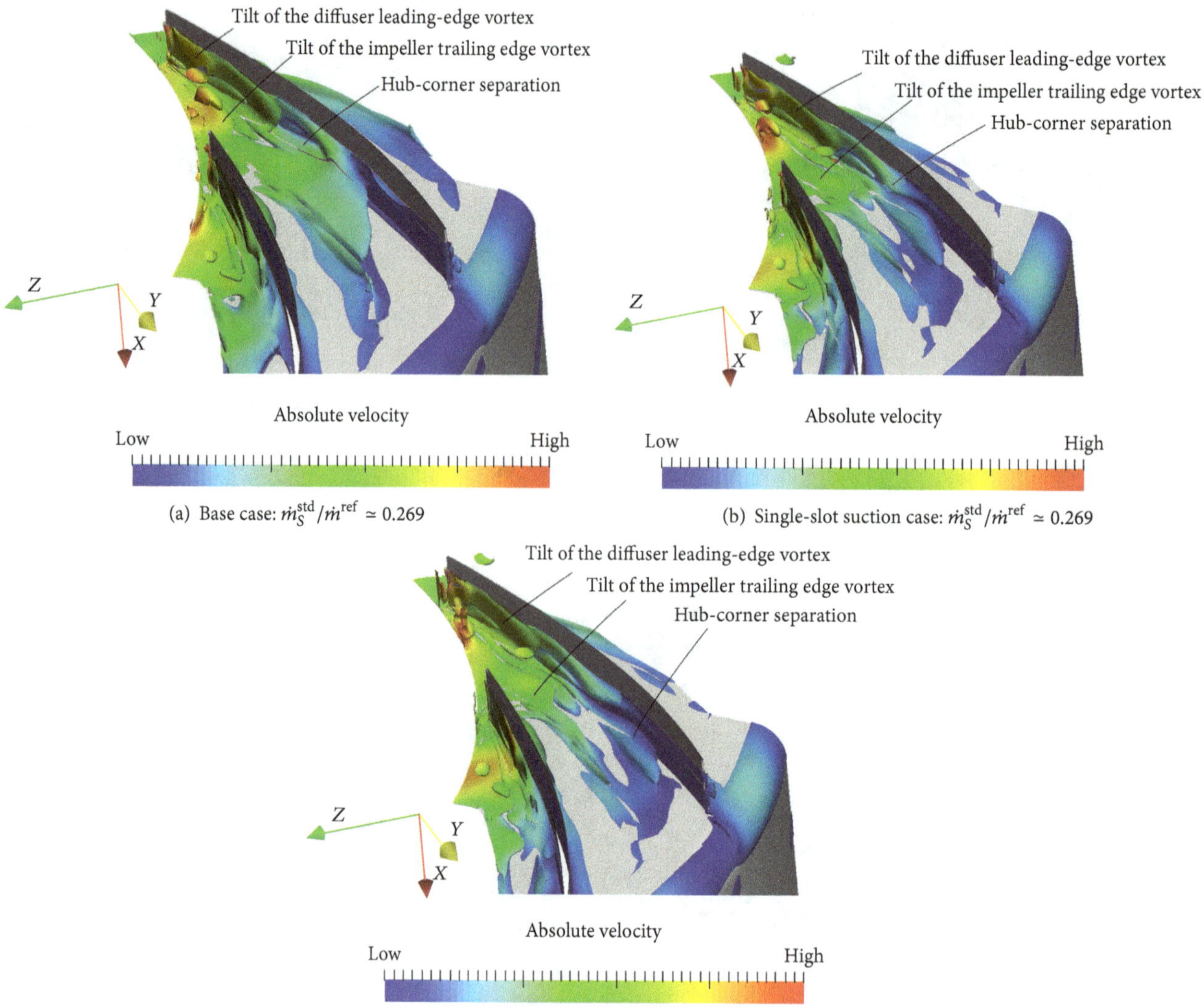

(a) Base case: $\dot{m}_S^{\mathrm{std}}/\dot{m}^{\mathrm{ref}} \simeq 0.269$

(b) Single-slot suction case: $\dot{m}_S^{\mathrm{std}}/\dot{m}^{\mathrm{ref}} \simeq 0.269$

(c) Three-slot suction case: $\dot{m}_S^{\mathrm{std}}/\dot{m}^{\mathrm{ref}} \simeq 0.268$

FIGURE 10: Isocontours of tilt of the Ω_x vorticity component.

for incompressible and isotropic fluids subject to conservative body forces gives more understanding about the feeding of the hub-corner vortex by other vortices. In this equation, the term A stands for the increase of the vorticity due to the tilt of vortices under velocity gradients. In case of a radial diffuser, the spanwise component of the vorticity can then feed the blade-to-blade vorticity under the effect of the spanwise velocity gradient. The impeller trailing edge vortex can then tilt and feed the hub-corner separation vortex thanks to the terms $\Omega_x(\partial U_y/\partial x)$ and $\Omega_x(\partial U_z/\partial x)$, x being the spanwise axis. In order to illustrate this phenomenon and to highlight the interaction between the impeller trailing edge vortex and the hub-corner separation vortex, Figure 10 shows isocontours of $\sqrt{(\Omega_x(\partial U_y/\partial x))^2 + (\Omega_x(\partial U_z/\partial x))^2}$ for the three cases.

In Figure 10(a), a large structure is indeed visible near the hub surface and in the middle of the diffuser passage. It corresponds to the tilt of the foot of the impeller trailing edge vortex that flows into the hub-corner separation and contributes to its growth. The hub-corner separation is also clearly visible in the figure, since the hub-corner separation vortex captures the boundary-layer on the diffuser suction side. It then redirects the Ω_x vorticity component of the boundary-layer toward the blade-to-blade direction.

Another structure is also visible on the suction side of the diffuser vane. It corresponds to the tilt of the Ω_x vorticity generated at the diffuser leading-edge due to the high incidence on the diffuser vanes at near surge operating point. Then, the diffuser leading-edge vortex also contributes to the growth of the hub-corner separation.

Under the effect of the suction, the hub-corner separation growth is slowed down and the interaction with the impeller vortex shedding is decreased. In Figure 10(b), the tilt of the impeller trailing edge vortex is less than for the base case. With the three-slot suction case, in Figure 10(c), the interaction is further decreased, and this may be the explanation of the increase of stable operating range when the hub-corner separation is pushed downstream.

However, the role played by the diffuser leading-edge vortex is unchanged under the effect of the suction, and it now appears to be the main visible interaction that feeds the hub-corner separation. Further work should aim at gaining more understanding of the role played by this diffuser leading-edge vortex and its consequences for the surge inception. Controlling the development or the trajectory of this vortex could lead to a complete removal of the hub-corner separation.

9. Conclusions and Outlooks

According to the URANS numerical model, and at the nominal rotational speed, the three-slot suction strategy allows a significant increase of the centrifugal compressor operating range toward low mass flow rate. This demonstrates the interest of the boundary-layer suction technique in order to improve the performance of centrifugal compressor stages. It also confirms that the hub-corner separation in the diffuser is involved in the surge inception for the present test case.

The single-slot suction strategy allows delaying the hub-corner separation growth, and the three-slot strategy delays it further. As a consequence, the interaction between the hub-corner separation vortex and the impeller trailing edge vortex shedding is reduced, which is expected to be the key feature of the control of the surge. With the three-slot strategy, this interaction is almost cancelled, and this may explain the increase of the stable operating range.

The analysis has also revealed the interaction between the hub-corner separation and the diffuser leading-edge separation vortex, which should be taken into consideration in order to achieve a complete removal of the hub-corner separation. The leading-edge separation may be controlled by suction applied to the diffuser suction side, or by other control techniques as vortex generator, for example.

Finally, the interest of the multislot control strategy should also be challenged at other rotational speeds, and the issue of the optimal shapes of the suction slots should be addressed, prior to conducting experiments.

Conflicts of Interest

The authors declare that there are no conflicts of interest regarding the publication of this paper.

Acknowledgments

The authors would like to thank Turbomeca, Safran group, for providing the test case and supporting the present study. Thanks also are due to the ISAE laboratory at Université de Toulouse, where the experimental test rig is set up. Thanks are due to ONERA, which develops the elsA software and provides support regarding the numerical simulations.

References

[1] P. Came, "The current state of research and design in high pressure ratio centrifugal compressors," DTIC Document, 1976.

[2] C. Amann, G. Nordenson, and G. Skellenger, "Casing modification for increasing the surge margin of a centrifugal compressor in an automotive turbine engine," *ASME*, vol. 97, no. 3, pp. 329–335, 1975.

[3] R. Hunziker and G. Gyarmathy, "Operational stability of a centrifugal compressor and its dependence on the characteristics of the subcomponents," *Journal of Turbomachinery*, vol. 116, no. 2, pp. 250–259, 1994.

[4] E. Greitzer, "The stability of pumping systems," *Journal of Fluids Engineering*, vol. 103, pp. 193–242, 1981.

[5] M. P. Wernet, M. M. Bright, and G. J. Skoch, "An investigation of surge in a high-speed centrifugal compressor using digital PIV," *Journal of Turbomachinery*, vol. 123, no. 2, pp. 418–428, 2001.

[6] J. N. Everitt, *Investigation of stall inception in centrifugal compressors using isolated diffuser simulations [M.S. thesis]*, MIT Dept. Aero/Astro, 2010.

[7] J. N. Everitt and Z. S. Spakovszky, "An investigation of stall inception in centrifugal compressor vaned diffuser," *Journal of Turbomachinery*, vol. 135, Article ID 011025, 2012.

[8] R. A. Hill IV, *Simulation of spike stall inception in a radial vanted diffuser [M.S. thesis]*, Massachusetts Institute of Technology, 2007.

[9] Y. Bousquet, X. Carbonneau, G. Dufour, N. Binder, and I. Trebinjac, "Analysis of the unsteady flow field in a centrifugal compressor from peak efficiency to near stall with full-annulus simulations," *International Journal of Rotating Machinery*, vol. 2014, Article ID 729629, 11 pages, 2014.

[10] Y. Bousquet, N. Binder, G. Dufour, X. Carbonneau, M. Roumeas, and I. Trebinjac, "Numerical simulation of stall inception mechanisms in a centrifugal compressor with vaned diffuser," *Journal of Turbomachinery*, vol. 138, no. 12, Article ID 121005, 2016.

[11] N. Buffaz and I. Trébinjac, "Aerodynamic instabilities in transonic centrifugal compressor," *Mechanics and Industry*, vol. 15, no. 3, pp. 191–196, 2014.

[12] I. Trébinjac, E. Benichou, and N. Buffaz, "Full-annulus simulation of the surge inception in a transonic centrifugal compressor," *Journal of Thermal Science*, vol. 24, no. 5, pp. 442–451, 2015.

[13] E. Benichou and I. Trébinjac, "Comparison of steady and unsteady flows in a transonic radial vaned diffuser," *Journal of Turbomachinery*, vol. 138, no. 12, Article ID 121002, 2016.

[14] E. Benichou and I. Trébinjac, "Numerical analysis of an alternate stall in a radial vaned diffuser," in *ASME Turbo Expo 2016: Turbomachinery Technical Conference and Exposition*, American Society of Mechanical Engineers, 2016.

[15] N. Fujisawa, S. Ikezu, and Y. Ohta, "Structure of diffuser stall and unsteady vortices in a centrifugal compressor with vaned diffuser," in *ASME Turbo Expo 2016: Turbomachinery Technical Conference and Exposition*, American Society of Mechanical Engineers, 2016.

[16] A. Merchant, J. L. Kerrebrock, J. J. Adamczyk, and E. Braunsheidel, "Experimental investigation of a high pressure ratio aspirated fan stage," *American Society of Mechanical Engineers*, pp. 493–502, 2004.

[17] B. J. Schuler, J. L. Kerrebrock, and A. Merchant, "Experimental investigation of a transonic aspirated compressor," *Journal of Turbomachinery*, vol. 127, no. 2, pp. 340–348, 2005.

[18] S. A. Gbadebo, N. A. Cumpsty, and T. P. Hynes, "Control of three-dimensional separations in axial compressors by tailored boundary layer suction," *Journal of Turbomachinery*, vol. 130, no. 1, Article ID 011004, 2008.

[19] J. L. Kerrebrock, A. H. Epstein, A. A. Merchant et al., "Design and test of an aspirated counter-rotating fan," *Journal of Turbomachinery*, vol. 130, no. 2, Article ID 021004, 2008.

[20] A. Godard, F. Bario, S. Burguburu, and F. Lebuf, "Experimental and numerical study of a subsonic aspirated cascade," in *ASME Turbo Expo 2012: Turbine Technical Conference and Exposition*, pp. 253–267, American Society of Mechanical Engineers, 2012.

[21] K. K. Botros and J. F. Henderson, "Developments in centrifugal compressor surge control - a technology assessment," *Journal of Turbomachinery*, vol. 116, no. 2, pp. 240–249, 1994.

[22] J. Raw, "Surge margin enhancement by a porous throat diffuser," *Canadian Aeronautics and Space Journal*, vol. 32, no. 1, pp. 54–61, 1986.

[23] E. B. Nelson, J. D. Paduano, and A. H. Epstein, "Active stabilization of surge in an axicentrifugal turboshaft engine," *Journal of Turbomachinery*, vol. 122, no. 3, pp. 485-493, 2000.

[24] G. J. Skoch, "Experimental investigation of centrifugal compressor stabilization techniques," *Journal of Turbomachinery*, vol. 125, no. 4, pp. 704–713, 2003.

[25] G. J. Skoch, "Experimental investigation of diffuser hub injection to improve centrifugal compressor stability," *Journal of Turbomachinery*, vol. 127, no. 1, pp. 107–117, 2005.

[26] Z. S. Spakovszky, "Backward traveling rotating stall waves in centrifugal compressors," *Journal of Turbomachinery*, vol. 126, no. 1, pp. 1–12, 2004.

[27] T. Halawa, M. S. Gadala, M. Alqaradawi, and O. Badr, "Optimization of the efficiency of stall control using air injection for centrifugal compressors," *Journal of Engineering for Gas Turbines and Power*, vol. 137, no. 7, Article ID 072604, 2015.

[28] H. Schönenborn, "Removal of cooling air on the suction side of a diffuser vane of a radial compressor stage of gas turbines," US Patent 6,210,104, April 2001.

[29] A. Leblanc, "Diffuser with enhanced surge margin," US Patent 8,556,573, October 2013.

[30] A. Marsan, I. Trébinjac, S. Coste, and G. Leroy, "Influence of unsteadiness on the control of a hub-corner separation within a radial vaned diffuser," *Journal of Turbomachinery*, vol. 137, no. 2, 12 pages, 2015.

[31] A. Marsan, I. Trébinjac, S. Coste, and G. Leroy, "Study and control of a radial vaned diffuser stall," *International Journal of Rotating Machinery*, vol. 2012, Article ID 549048, 12 pages, 2012.

[32] A. Marsan, I. Trébinjac, S. Coste, and G. Leroy, "Temporal behaviour of a corner separation in a radial vaned diffuser of a centrifugal compressor operating near surge," *Journal of Thermal Science*, vol. 22, no. 6, pp. 555–564, 2013.

[33] I. Trébinjac, P. Kulisa, N. Bulot, and N. Rochuon, "Effect of unsteadiness on the performance of a transonic centrifugal compressor stage," *Journal of Turbomachinery*, vol. 131, no. 4, Article ID 041011, 9 pages, 2009.

[34] N. Bulot and I. Trébinjac, "Effect of the unsteadiness on the diffuser flow in a transonic centrifugal compressor stage," *International Journal of Rotating Machinery*, vol. 2009, Article ID 932593, 11 pages, 2009.

[35] L. Cambier and J. P. Veuillot, "Status of the elsA CFD software for flow simulation and multidisciplinary applications," in *Proceedings of the 46th AIAA Aerospace Science Meeting and Exhibit*, vol. 664, Reno, Nev, USA, 2008.

[36] J. I. Erdos, E. Alzner, and W. McNally, "Numerical solution of periodic transonic flow through a fan stage," *AIAA Journal*, vol. 15, no. 11, pp. 1559–1568, 1977.

[37] J. M. Tyler and T. G. Sofrin, "Axial flow compressor noise studies," *SAE International*, vol. 70, Article ID 620532, 24 pages, 1962.

[38] G. A. Gerolymos, G. J. Michon, and J. Neubauer, "Analysis and application of chorochronic periodicity in turbomachinery rotor/stator interaction computations," *Journal of Propulsion and Power*, vol. 18, no. 6, pp. 1139–1152, 2002.

[39] L. Castillon, G. Billonnet, S. Péron, and C. Benoit, "Numerical simulations of technological effects encountered on turbomachinery configurations with the chimera technique," in *Proceedings of 27th International Congress of the Aeronautical Sciences*, Nice, France, September 2010.

[40] Y. Bousquet, X. Carbonneau, G. Dufour, N. Binder, and I. Trebinjac, "Analysis of the unsteady flow field in a centrifugal compressor from peak efficiency to near stall with full-annulus simulations," *International Journal of Rotating Machinery*, vol. 2014, Article ID e729629, 11 pages, 2014.

[41] Z. S. Spakovszky and C. H. Roduner, "Spike and modal stall inception in an advanced turbocharger centrifugal compressor," *Journal of Turbomachinery*, vol. 131, no. 3, Article ID 031012, 2009.

[42] P. Chakraborty, S. Balachandar, and R. J. Adrian, "On the relationships between local vortex identification schemes," *Journal of Fluid Mechanics*, vol. 535, pp. 189–214, 2005.

[43] J. C. Hunt, A. Wray, and P. Moin, "Eddies, streams, and convergence zones in turbulent flows," Tech. Rep. CTR-S88, Center for Turbulence Research, 1988.

9

Variable Ratio Hydrostatic Transmission Simulator for Optimal Wind Power Drivetrains

Jose M. Garcia-Bravo,[1] Ivo N. Ayala-Garcia,[2] and Juan L. Cepeda-Aguilar[3]

[1] *Purdue University, West Lafayette, IN, USA*
[2] *Universidad Tecnológica de Querétaro, Santiago de Querétaro, QRO, Mexico*
[3] *Universidad de los Andes, Bogotá, Colombia*

Correspondence should be addressed to Jose M. Garcia-Bravo; jmgarcia@purdue.edu

Academic Editor: Lei Chen

This work presents a hydromechanical transmission coupled to an electric AC motor and DC generator to simulate a wind power turbine drive train. The goal of this project was to demonstrate and simulate the ability of a hydrostatic variable ratio system to produce constant electric power at varying wind speeds. The experimental results show that the system can maintain a constant voltage when a 40% variation in input speed is produced. An accompanying computer simulation of the system was built and experimentally validated showing a discrete error no larger than 12%. Both the simulation and the experimental results show that the electrical power output can be regulated further if an energy storage device is used to absorb voltage spikes produced by abrupt changes in wind speed or wind direction.

1. Introduction

Wind power generators are a technology that contributes to sustainable energy and low impact to the environment. However, some of the challenges that the current wind power industry face are high installation and maintenance costs and reliability issues [1–4]. To be able to reduce the cost of wind power energy in general, the wind power industry must invest resources in developing wind power technologies that decrease the total cost of the produced energy. This can be done if wind power turbines operate efficiently at broader wind speed ranges, while maintaining optimum electrical generator shaft speeds, so they can generate more electric power regardless of the wind speed. There is a great impact on the turbine reliability when major components of the wind power turbine like the gearbox or the generator fail because this creates an extended mean time to repair (MTTR) [4–6]. Therefore, reducing failure and maintenance of these major components reduces the cost of operation of the wind power turbine, which in turn reduces the cost of energy itself.

The power coefficient C_p in (1) is a parameter used for measuring the efficiency of the wind energy captured by the turbine; this coefficient represents the ratio of the mechanical power output of the turbine to the power input from the wind.

$$C_p = \frac{\eta_d}{\eta_m \eta_g}. \tag{1}$$

η_d is the aerodynamic efficiency of the turbine and η_m and η_g are the electric generator efficiency and the gearbox efficiency, respectively [7]. The maximum theoretical achievable C_p is 59%; this condition, which is derived from fluid mechanics principles is known as the Betz limit [7]. The direct consequence of this limit is that no more than 59% of the available power in the wind can be captured.

The tip speed ratio λ (see (2)) is parameter used to define the ratio between the tangential velocity of the blade and the wind speed. The family of plots shown in Figure 1 illustrates the effect of the tip speed ratio on the power coefficient C_p for various pitch angles of the same blade.

$$\lambda = \text{TSR} = \frac{\omega R}{V} = \frac{2\pi R N}{V}, \tag{2}$$

where ω is the rotational speed of the blades, R is the radius of the wind turbine, and V is the wind speed. As

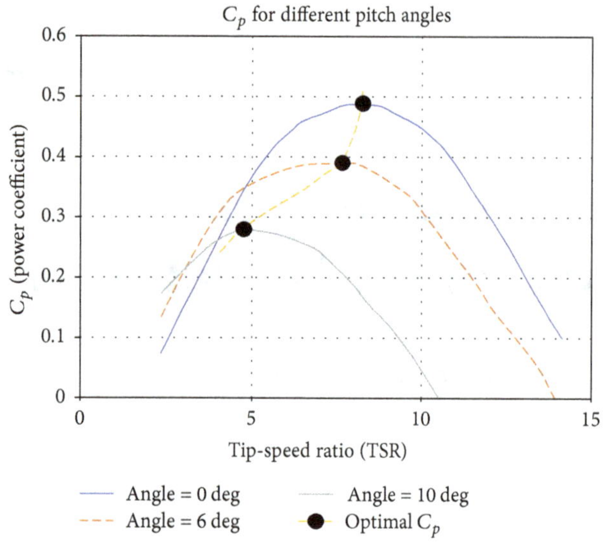

FIGURE 1: Power coefficient versus tip speed ratio [8].

the wind speed varies, the pitch angle of the blade needs to be adjusted to operate at the optimal C_p. Inevitably, when the wind speed varies, the shaft speed of the generator also changes because the rotor's shaft is directly (direct drive) or indirectly (gearbox) coupled to the generator's shaft. That means that the electric generator will produce electric power at variable frequencies that need to be conditioned using power electronics to match the constant frequency standard of the electric grid. The challenge of using this approach is the increased complexity, cost, weight, and size of the wind turbine components and it reduces its efficiency and reliability. Additionally, most common current electric generators operate more efficiently at their constant rated shaft speed, which is normally a single value between 1800 and 2200 RPM. This is particularly true, for field induced generators, but it also affects permanent magnet motors because the amount of power produced is proportional to the input shaft speed.

Various studies [9–15] have proposed the use of hydrodynamic, hydrostatic, or mechanical continuously variable transmissions (CVT) instead of using fixed or variable speed gearboxes; the objectives of incorporating a CVT to replace the gearbox are to improve the overall reliability of the wind turbine, reduce the weight of the nacelle components, reduce down-time, and significantly reduce the cost of installation and operation. The use of power electronics systems used to convert the electric output of the generator with the grid is known [11] to reduce the captured wind power in the 93-94% range for wind speeds between 6 and 11 m/s.

When replacing the fixed ratio mechanical gearbox or the power electronics systems in a wind power drive train with a CVT, the overall efficiency of the turbine will be increased because the generator's shaft will turn at its optimal speed while maintaining an ideal tip speed ratio λ to achieve a maximum power coefficient C_p [7]. While the hydraulic hydrostatic transmission will introduce a power loss due to its overall efficiency of approximately 86–93%, the cost, reliability, and maintenance benefits of implementing this solution would offset the ~1–12% loss in energetic efficiency.

The hydrostatic transmission will be the focus of this project because it is a technology that has been proven to be reliable and cost effective. Additionally, hydrostatic transmissions allow novel and flexible designs where the generator is placed at the floor level for easier, less costly, and safer maintenance. A hydrostatic transmission is a variable speed rotary drive that consists of one or more positive displacement hydraulic pumps connected to one or more positive displacement hydraulic motors. The main function of the hydrostatic transmission is to convert mechanical power into hydraulic power (pressure and flow) and after transmitting it through piping or hoses convert it back into mechanical power to be used in an application. A hydrostatic transmission fitted with a fixed or a variable displacement (flow) pump and a fix or a variable displacement motor allows the output shaft to deliver a nearly constant power at varying speeds, similar to the requirement for wind power turbines.

The work by Thul et al. [9] provided theoretical evidence that these types of transmissions would dampen the large torque impulses created from wind gusts and turbulence, which translates into lower loads on the bearings of the generator, therefore increasing their life. Preliminary simulation results from Thul et al. [9] and Yang et al. [12] revealed that a wind power turbine fitted with pitch control and or a continuously variable transmission could operate at an optimal power coefficient even at speeds close to the cut in speed (minimum operating speed); this would allow capturing more energy from the wind at these lower speeds and capturing more power at the turbine's rated speed.

The main challenge for using hydrostatic transmissions in wind power generation is the lack of pumps and motors that can match the power requirements of current mid-sized (250 kW to 750 kW) and large sized wind power turbines (750 kW or more); most of the larger displacement pumps are rated for a little less than 250 kW. Researchers investigating the use of hydrostatic transmissions for wind power use have recurred to three alternatives: (1) limiting the wind power capacity to less than 250 kW [9], (2) using more than one pump and more than one motor to meet the flow requirements of larger size wind turbines [11], and (3) designing hydrostatic transmissions with an experimental large displacement pump. Wind power turbines larger than 150 kW are typically fitted with three phase alternating current (AC) generators at 690 V; this current is then sent to a transformer to increase the voltage several thousand volts up to 30,000 volts, making it up for transmission in high voltage lines [16].

The aim of this project was to design, build, and test a small scale experimental bench system with the flexibility to simulate various wind speed profiles. At the same time, the principal objective was to integrate control algorithms for measuring and demonstrating a variable ratio hydrostatic transmission.

The design of the simulator encompasses two associated fields: hydraulics and electrical systems. The first system is involved in the transmission of the mechanical energy from the wind turbine into the electrical generator by using a hydrostatic transmission. The second system, is in charge of

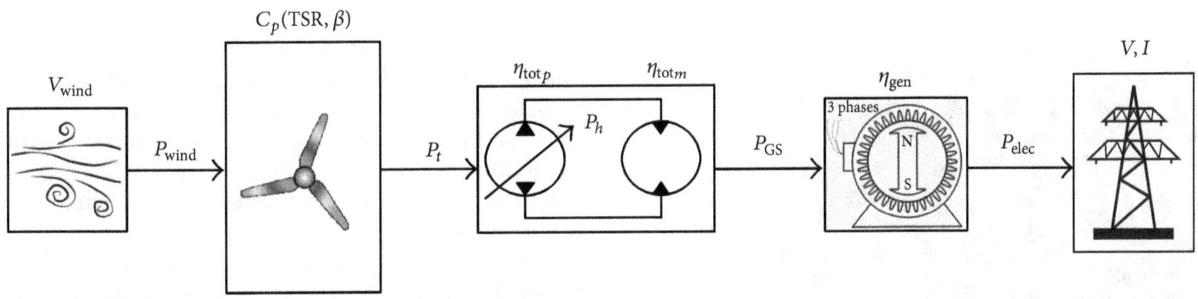

FIGURE 2: Power flow through the system.

controlling the different gear ratio of the system to maintain a constant voltage output and also adjusting the simulated wind speed.

2. Theoretical Approach

This study section focuses on the flow of power through the system, as it is depicted in Figure 2. The magnitude of the power available in the wind P_{wind} is quantified using the expression in

$$P_{\text{wind}} = \frac{1}{2}\rho A V^3, \quad (3)$$

where ρ is the density of the air at the altitude at the wind turbine is location, A is the transversal area that is being passed by the wind, in this case the area swept by the blades while it is spinning, and V is the velocity of the wind.

The power that the turbine extracts from the wind (P_t), as it was mentioned in the background section, is calculated using the power coefficient using (3):

$$P_t = C_p(\text{TSR}, \beta) P_{\text{wind}} = \frac{1}{2} C_p \rho A V^3. \quad (4)$$

The power extracted by the turbine is equal to the power on the shaft, if the rotor is assumed to be rigid (no torsion).

$$P_t = \omega_{\text{tS}} T_{\text{aero}}. \quad (5)$$

Then, the mechanical power from the rigid shaft is converted into hydraulic power by a hydrostatic pump.

$$P_h = \eta_{P_{\text{tot}}} P_t. \quad (6)$$

And the hydraulic power is defined as

$$P_h = \Delta p Q. \quad (7)$$

The total efficiency of the pump ($\eta_{P_{\text{tot}}}$) is the product between the volumetric (η_{vol}) and the mechanical (η_{mec}) efficiencies of the pump. The volumetric efficiency of a hydraulic machine is defined as the ratio between the real flow produced and the theoretical flow. Meanwhile, the mechanical efficiency is equal to the ratio between the real torque and the theoretical torque required to drive the pump. D is the hydraulic machine displacement and n is the shaft speed in RPM.

$$\eta_{\text{vol}} = \frac{Q}{Dn}, \qquad \eta_{\text{mec}_p} = \frac{D\Delta p}{T}. \quad (8)$$

Once the fluid is pressurized by the pump, it is transported by the hoses to the hydraulic motor where the power is transformed back into mechanical power and coupled to an electrical generator shaft.

$$P_{\text{GS}} = \eta_{\text{tot}_m} P_h. \quad (9)$$

The total efficiency in the case of the hydraulic motor is similar to the pump; the total efficiency is the same product between the volumetric and mechanical efficiency. The volumetric efficiency is expressed the same way; the difference is in the mechanical efficiency; the expressions are reversed.

$$\eta_{\text{mec}_m} = \frac{T}{D\Delta p}, \quad (10)$$

where T corresponds to the ideal hydraulic machine's shaft torque and ω is the rotational speed. Power on the generator shaft is expressed in the following equation.

$$P_{\text{GS}} = \omega_{\text{GS}} T_{\text{elec}}. \quad (11)$$

Finally, the energy is transformed from mechanical power to electrical power by means of the generator using the electromagnetic theory.

$$P_{\text{elec}} = \eta_{\text{gen}} P_{\text{GS}}. \quad (12)$$

3. Materials and Methods

3.1. Wind Speed. The rotational energy generated by a wind power turbine is proportional to the kinetic energy in the wind. The simulator replaces the wind turbine by a Brook Crompton Americas, 3-phase, 60 Hz, NEMA 56 electric motor (BF4N.33-2). The variation of the kinetic energy of the wind is simulated with the inclusion of a Teco Westinghouse electric inverter (JNEV-101-H1). The electric inverter varies the frequency at the motor power lines from 0 Hz

FIGURE 3: Test bench hardware setup.

to its nominal value of 60 Hz, proportionally affecting the rotational shaft speed of the motor between approximately 0 to 1800 rpm. These values of the frequency are assumed to be linearly proportional to the input wind speed, where 0 Hz corresponded to 0 wind speed and 60 Hz corresponded to the maximum permissible wind speed on a small wind power turbine (<50 kW) of approximately 15 m/s.

3.2. Hydraulic System. The electric motor is coupled to a 10.3 cc/rev variable displacement pump (Hydro-Gear, PG-1HCC-DL1X). This is the first component of a hydrostatic transmission, transducing the mechanical power from the electric motor into hydraulic power. The hydraulic power is then transmitted to a fixed displacement hydraulic gear motor (Double A Hydraulics, MFG 2), which converts the hydraulic energy back into mechanical energy. The hydraulic motor is coupled to a permanent magnet generator (GS Electric, B4CPM-102T) which converts the mechanical energy into electrical energy.

The variable displacement hydraulic pump regulates the amount of energy transferred to the hydraulic motor by adjusting the angle of the swashplate, hence adjusting the volumetric flow rate delivered by the pump. Regardless of the mechanical energy generated by the electric motor, the hydraulic power delivered to the hydraulic motor is adjusted to maintain an approximated constant power. The adjustment of the swashplate is accomplished by means of a linear actuator attached to the pump trunnion arm. The simulator regulates the angle according to the amount of the electrical energy required at the output of the DC generator. The rotation of the generator's shaft generates a current proportional to the rotational speed, which in turn is rectified into a DC signal (Figure 3).

3.3. Control. The wind speed is simulated by adjusting the driving frequency of the 3-phase electric motor. The frequency range can be adjusted from 0 to 60 Hz. The actual rotation speed depends on the mechanical load applied at the rotor. The wind profile tested is described in Section 5.

The driving frequency is controlled by using the EI. The frequency can be manually adjusted using a knob or via an analog signal at a predefined input port of the EI. The convenience of using an analog signal to control the frequency relies on the ability to implement an automatic system that programmatically changes the frequency following a given pattern or data set. This feature allows the system to simulate a wind profile.

Mechanical coupling between the electric motor and the hydraulic pump is achieved by the use of a jaw type couplings. These couplings include an elastomer insert between metal hubs. The insertion of the nonmetallic elastomer creates the opportunity to place a proximity sensor (Allen-Bradley, 872C-D3NN12-D4) to measure the rotational speed. A revolution of the rotor is measured by the number of teeth sensed. On a three teeth hub, a full revolution occurs when six teeth are sensed.

The rotational displacement of the rotor is transduced into hydraulic flow and pressure at the variable hydraulic pump. This variable capability is of crucial importance for the system. By adjusting the flow, the energy transduced by the electric generator can be controlled, hence maintaining a constant output voltage independent of the wind speed.

The flow rate at variable hydraulic pump is achieved by adjusting the angle of the swashplate. The swashplate is attached to a trunnion arm that can be externally manipulated to control the flow. A linear actuator is connected to the trunnion arm. The direction and rate of displacement of the actuator are performed by a power driver connected to a personal computer.

The hydraulic pump is connected to a hydraulic motor via hydraulic hoses. A flow meter and pressure sensor measure these two parameters at one of the hoses that connect the pump to the motor. Due to the nature of the system, the hydraulic flow is allowed to go in only one direction.

The hydraulic motor then transfers the energy to an electric generator. The mechanical coupling is accomplished through a jaw type coupling, where the rotational speed is measured using a hall effect sensor (Cherry, MP1005).

The permanent magnets of the electric generator create the excitation field necessary to induce current on the stationary coil. Additionally, a rectifier circuit converts the resulting AC voltage from the coil into a DC signal which is measured using the voltmeter (Tenma, 72-7735). Figure 4 shows a schematic representation of the hardware components.

3.4. Instrumentation and Control Algorithm. The instrumentation includes two hall effect sensors, a flow meter, and a pressure sensor. All these signals are sensed using a data acquisition card DAQ (NI USB-6221) connected to the personal computer. The multimeter is connected directly to the PC via its serial port.

Two output signals are controlled by the system, driving frequency at the EI, and flow rate at the hydraulic pump. The frequency is adjusted by an analog signal generated at the DAQ card, while the flow rate is controlled by adjusting the length of the linear actuator connected to the trunnion of the hydraulic pump.

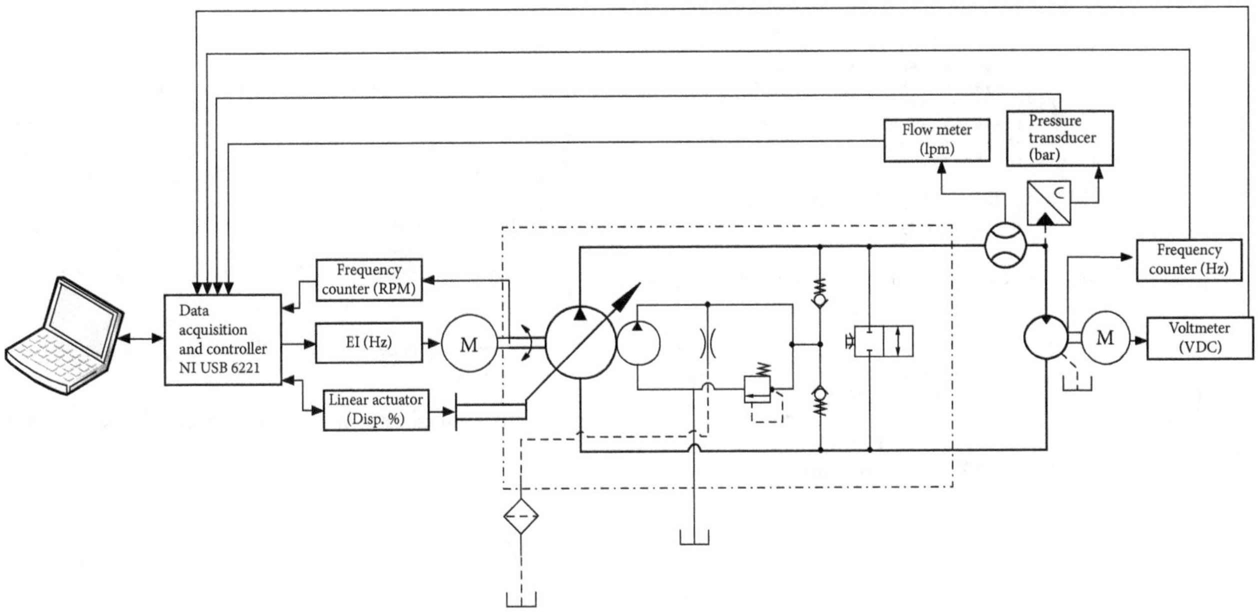

FIGURE 4: Schematic representation of test bench hardware setup.

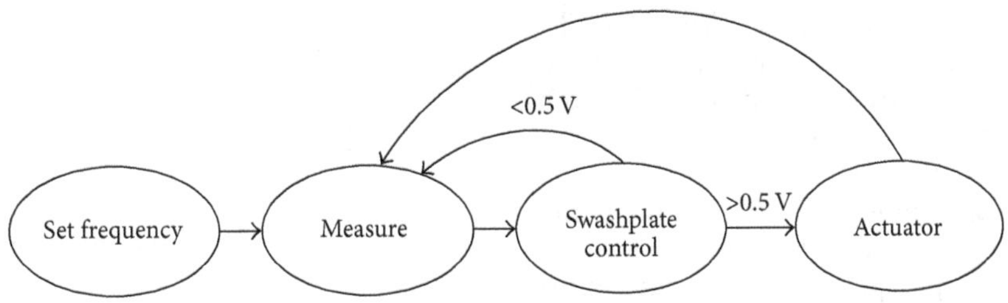

FIGURE 5: Control flow path.

The control algorithm follows a state machine architecture. Each state corresponds to a particular activity in the system as presented in Figure 5. The system starts by setting the desired frequency at the motor driver. This frequency corresponds to a particular simulated wind speed. Then, different variables are measured in the system: selected frequency, revolutions per minute (RPM) at the motor driver, RPM at the generator, hydraulic flow rate, pressure at outlet of the variable displacement pump, actuator position, and voltage at the electric motor.

The control of the variable displacement hydraulic pump is achieved by adjusting the swashplate using a linear actuator. A proportional control commands the adjustment of the actuator stroke length, either reducing or increasing it. The adjustment is performed until the signal error reaches ±0.5 V from the desired output voltage at the DC generator.

4. System Characterization

4.1. Hardware Characterization. The system was characterized to identify its operating range: frequency and DC output voltage. In the first series of tests, the angle of the variable swashplate was kept constant, starting at 25% of its maximum flow, increasing in steps of 25%, up to 100% displacement.

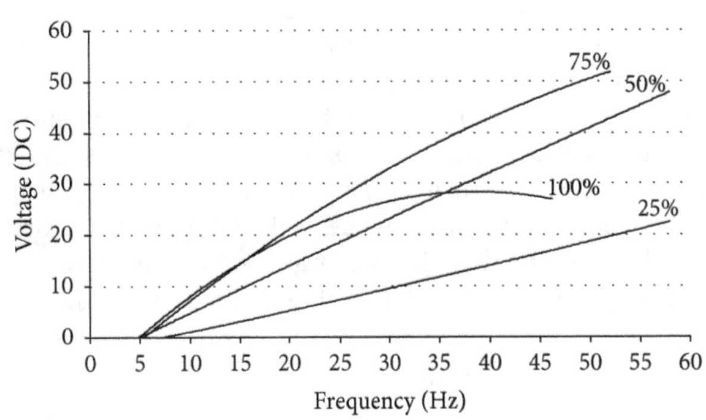

FIGURE 6: Voltage generation at different swashplate angles.

The control system was disabled and the electric load was the voltmeter. Figure 6 shows the polynomial fits of the collected voltage data; the fitted data is shown here to cleanly depict the frequency trends at various pump displacement values. It this figure, it can be seen that at 25% pump displacement the system can generate a maximum of 22 V. The maximum output voltage increases with the swash plate angle, up to 52 V. It is important to notice that increasing the swash plate angle does not necessarily implies reaching a higher output voltage.

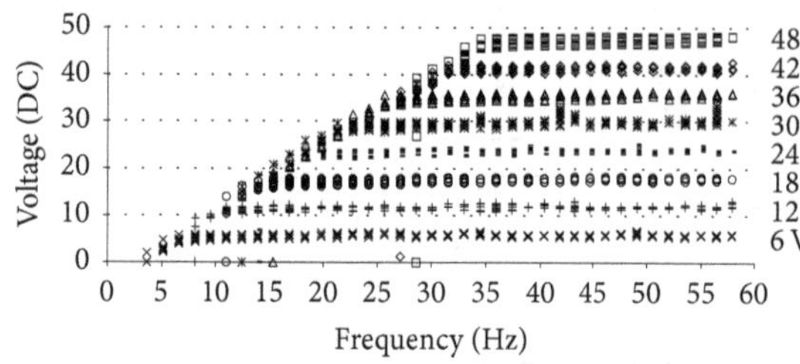

FIGURE 7: Voltage generation for different frequencies.

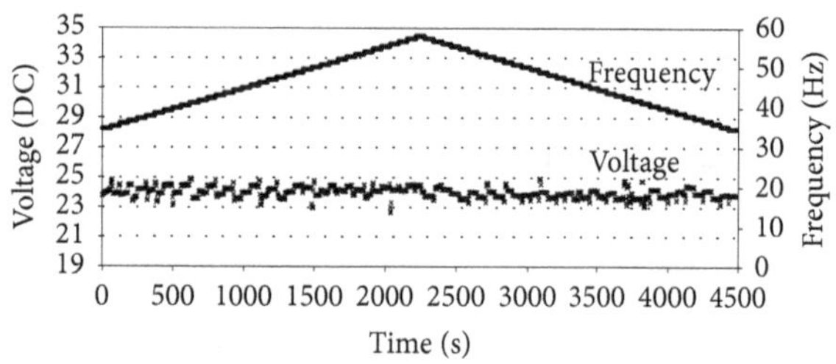

FIGURE 8: Voltage control for a constant ratio of frequency change.

This can be observed for the case when the angle is 100%. The voltage follows a similar path as per 75%; however at around 20 V the voltage decreases as the frequency increases. This can be explained because of the reduced capacity (displacement) of the hydraulic motor with respect to the pump. Because of its smaller size, the motor cannot admit 100% of the pump output; this creates an increase in the system pressure that in turn dampens the hydraulic pump, reducing the speed of the rotor. Hence, a hydraulic motor with a larger input capacity can produce a larger output voltage.

Due to the hardware constrains, it was decided to operate the system with a maximum output voltage of 48 V to avoid overheating of the input electric motor when the damping increases at the hydraulic pump. A second set of tests investigated the performance of the control at different output voltages. The frequency decreases on each case steadily by 1.5 Hz from 58 Hz down to the point where the rotor at the electric motor stops rotating due to damping. From Figure 7 it can be seen that there is a minimum voltage that any given frequency can generate, which follows the following equation.

$$V_{\min} = 1.5F - 4.5, \quad (13)$$

where $V_{\min}$ represents the minimum DC voltage and F is the generator's frequency in Hz.

To evaluate the performance of the controller, the output voltage set-point was fixed to 24 V, while the input frequency was varied between 35 and 58 Hz every 60 s, as shown in Figure 8. In all the cases the electric load at the DC generator was the voltmeter with a 10 MΩ input impedance.

Figures 9 and 10 present a detailed view of the test when the input frequency was increased and decreased, respectively. Right after the frequency changes, the output voltage increases its error relative to the set-point. Few seconds later, the system adjusts the flow to reduce this error to a maximum of ±0.5 V error band.

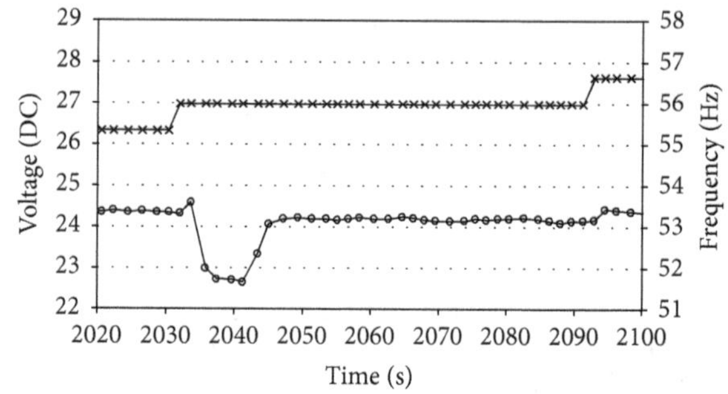

FIGURE 9: System adjustment as the frequency increases by 1.5 Hz.

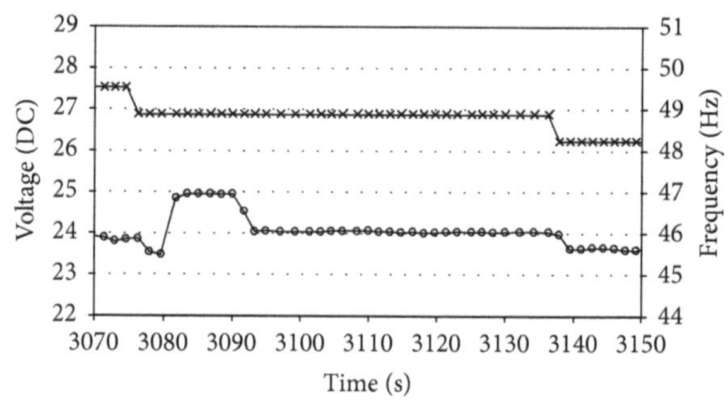

FIGURE 10: System adjustment as the frequency decreases by 1.5 Hz.

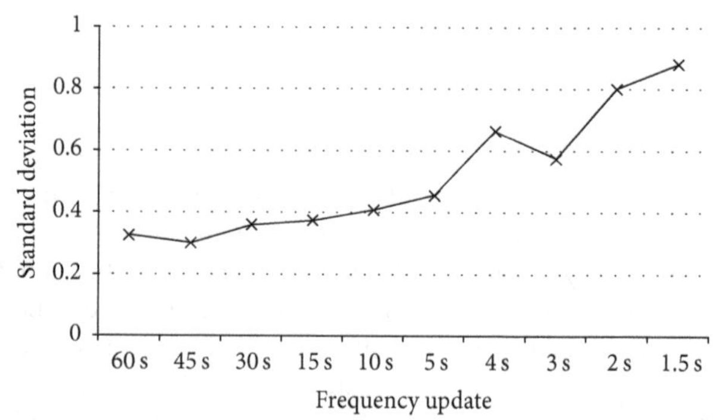

FIGURE 11: Standard deviation at various controller update rates.

The previous test was repeated reducing the time of the intervals between frequency changes, from 60 s to 1.5 s. Figure 11 shows the standard error deviation as the interval was reduced. The error increases as the time reduces; the system becomes less stable. At higher rates of change, the system cannot maintain the set voltage at the same pace as the frequency.

4.2. Computational Model of the System. A computational model of the hardware was built using Matlab Simulink. The system characterization presented in the previous section was used to model the hydraulic and electric systems present in the test bench. Figure 12 shows the full implementation

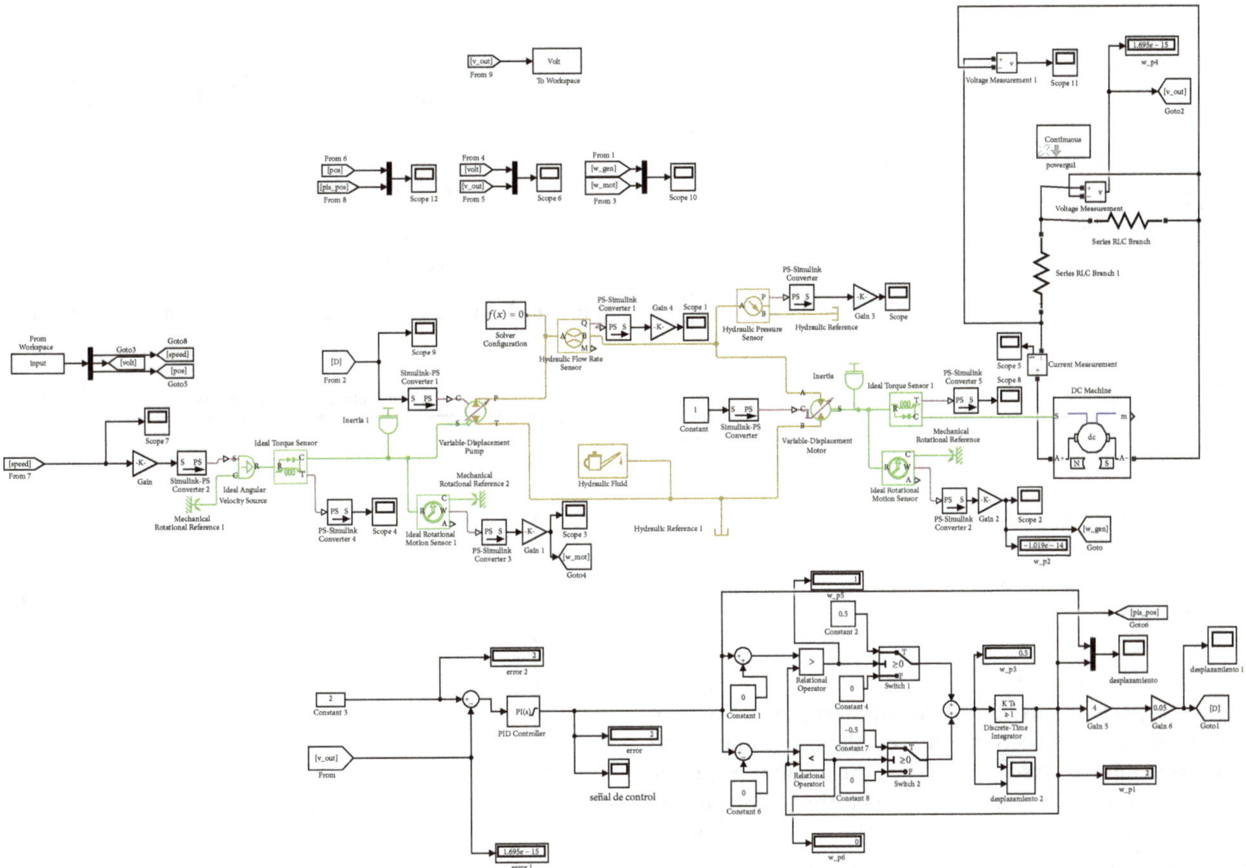

FIGURE 12: Computational model of the hydraulic wind power drive train.

TABLE 1: Model main parameters.

Parameter	Units	Value
Input shaft speed	RPM	Variable
Electric motor inertial mass	Kg·m^2	0.0008
Pump displacement	cm^3/rev	18.02
Fluid viscosity	cSt	55.4
Fluid density	Kg/m^3	890
Bulk modulus	bar	14560
Hydraulic motor displacement	cm^3/rev	0.29
Electric motor armature resistance	ohms	0.6
Torque constant	N·m/A	0.159

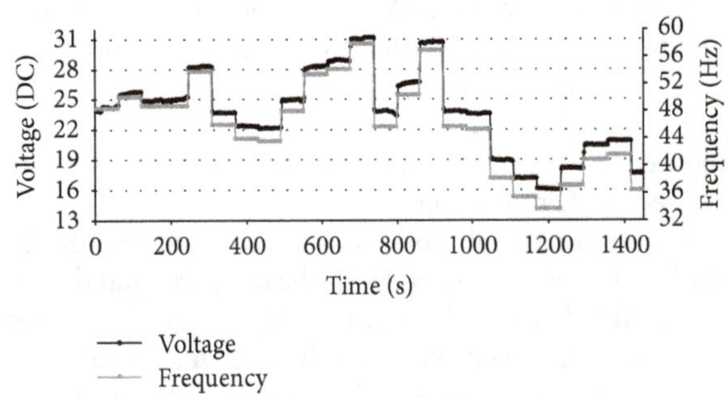

FIGURE 13: Voltage generation with control disabled.

of the model. The simulation included all of the hydraulic components, the DC electric generator, the controller, and the DC linear actuator. All of these models were validated using the experimental data and are summarized in Table 1.

5. Results and Discussion

A wind profile obtained from a real wind power turbine was used to evaluate the system. The data depicted in Figure 13 represents the wind measurements obtained over a period of 24 hours, one data point every hour. The data points were used in weighted estimates to span from 32 Hz to 58 Hz to represent wind speeds between 0.4 and 17.2 m/s. These frequencies were used to drive the electric motor effectively simulating a variable wind speed.

An initial test evaluated the response of the system as the simulated wind was being changed. In this first test no control was enabled. The electric generator was connected to a 24 V gear motor to simulate a constant load. The gear motor was estimated to consume an approximate 6 W of power. The results of this test are shown on Figure 13. The variable hydraulic pump was adjusted at the beginning of the test to generate 24 V at 47 Hz. The voltage at the generator follows the same path as the driving frequency of the electric

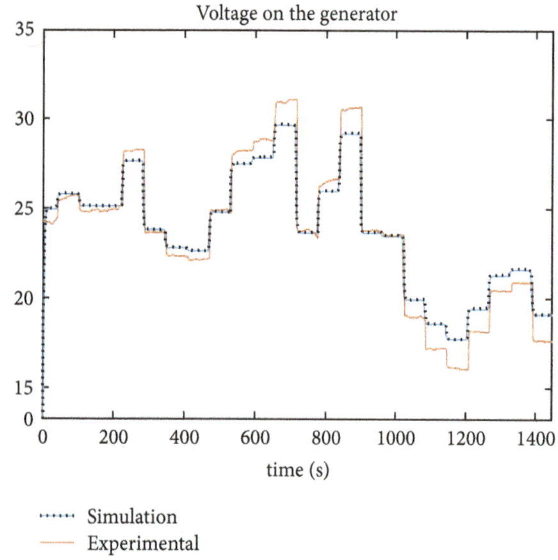

FIGURE 14: Voltage output for a varying wind speed input with control disabled.

motor, as expected. This case represents a fix gear ratio drive train, where the connection between the nacelle and the electric generator is fixed. This means the amount of power captured from the wind is reduced when the wind speed is lower than the nominal wind speed. However, if the wind speed is higher than the nominal value, the aerodynamic efficiency of the turbine is reduced to accommodate the optimal generator speed. At a certain high wind speed above the cut-off speed the nacelle is repositioned so that the turbine spins at a lower RPM. The generator output shown in Figure 13 demonstrates that undesirable variable voltage outputs are obtained at variable winds speeds; this system would be inefficient because the generated electric power would need to be rectified and regulated in order to be used by an electrical application.

The computational model was tested using the disabled controller. A comparison between the experimental data and the simulation is shown in Figure 14. The simulation followed the experimental results very closely with an accuracy as low as 12.5% at the worst case at 29 volts DC. The simulation was able to dynamically replicate the input signal without a significant time delay.

In the next experiment, the control was enabled and the test was performed under the same conditions described for the first experiment. It can be observed in Figure 15 that the system automatically adjusts itself to maintain a constant 24 V despite the input frequency. Overshoots can be observed when the frequency changes significantly; however, the voltage returns to the set-point within and average of 15 sec. These overshoots could be reduced if a hydraulic accumulator or another energy storage device was added to the system. A hydraulic accumulator is a device that captures hydraulic energy by storing pressurized fluid in a vessel; the pressurized vessel will increase its pressure as more fluid is pushed into the accumulator. Likewise the pressurized energy is released when the fluid is evacuated from the vessel. The

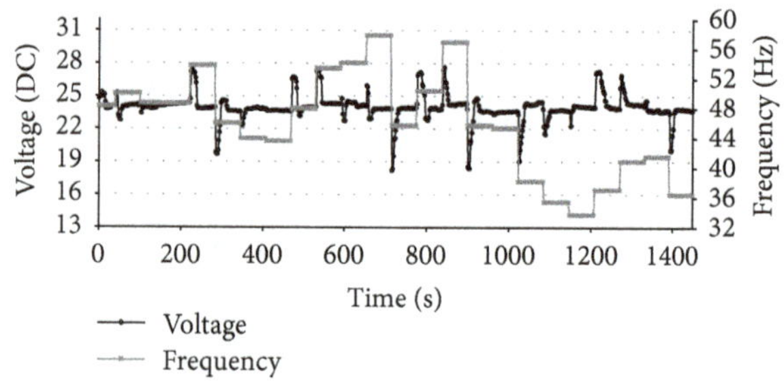

FIGURE 15: Hydraulic wind power drive train with enabled controller.

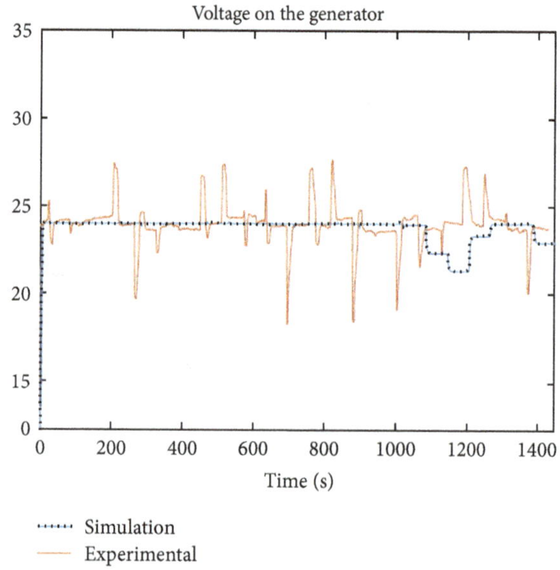

FIGURE 16: Voltage output for a varying wind speed input with control enabled.

accumulator contains a gas chamber that is allowed to change in volume such that when pressurized fluid rushes into the accumulator, the gas chamber is compressed, hence storing energy in the form of pressurized gas.

Results by Dutta et al. [17] show an estimated 4.1% percent increase in energy production on a 50 kW turbine by adding a 60 lt. accumulator. They also revealed how power generation can be regulated by absorbing voltage peaks and drops by adding such a hydraulic component.

The plot in Figure 16 shows a comparison of the simulated and the experimental behavior of the DC voltage generator when the controller is enabled. The simulation results show a lower magnitude of the overshoots when the frequency (wind speed) is changed, which means that the simulated system has not completely captured the dynamic effects of the physical system. There were instances where some drift from the desired voltage value was observed for the region of lower wind speeds recoded between 1100 s and 1200 s. This means the simulated controller was not able to adjust the angle of the swash plate in the pump to maintain a 24 VDC output at frequencies below 36 Hz. On the other hand, the experimental results also show that the experimental controller used was not able to effectively maintain the desired output at or below this simulated speeds corresponding to approximately 14.96 m/s.

6. Conclusions

A system capable of simulating a small wind power turbine drive train fitted with an infinitely variable transmission was built and tested in laboratory conditions. This system was demonstrated to effectively maintain the electric power output constant despite the input conditions at the nacelle, that is, a hydromechanical solution to maintain the RPM of the electric generator constant regardless of the simulated wind speed. The simple control strategy used to maintain the output voltage of the DC generator proved to be an acceptable method to improve wind power turbine generation at speeds higher than 15 m/s for a small wind power generator (10–100 kW). New control strategies will be implemented in the test setup for an improved wind speed range and dynamic behavior. Lastly, the experimental results show that adding an energy storage device could be an effective and inexpensive method to absorb voltage overshoots created by wind gusts and sudden changes in wind direction.

Conflicts of Interest

The authors declare that they have no conflicts of interest.

References

[1] P. J. Tavner, J. Xiang, and F. Spinato, "Reliability analysis for wind turbines," *Wind Energy*, vol. 10, no. 1, pp. 1–18, 2007.

[2] F. Spinato, P. J. Tavner, G. J. W. Van Bussel, and E. Koutoulakos, "Reliability of wind turbine subassemblies," *IET Renewable Power Generation*, vol. 3, no. 4, pp. 387–401, 2009.

[3] J. Carroll, A. McDonald, and D. McMillan, "Failure rate, repair time and unscheduled O&M cost analysis of offshore wind turbines," *Wind Energy*, vol. 19, no. 6, pp. 1107–1119, 2016.

[4] J. Ribrant and L. Bertling, "Survey of failures in wind power systems with focus on Swedish wind power plants during 1997-2005," in *Proceedings of the 2007 IEEE Power Engineering Society General Meeting, PES*, usa, June 2007.

[5] A. Froger, M. Gendreau, J. E. Mendoza, E. Pinson, and L.-M. Rousseau, "Solving a wind turbine maintenance scheduling problem," *Journal of Scheduling*, pp. 1–24, 2017.

[6] Y. Su, J. D. Kern, and G. W. Characklis, "The impact of wind power growth and hydrological uncertainty on financial losses from oversupply events in hydropower-dominated systems," *Applied Energy*, vol. 194, pp. 172–183, 2017.

[7] F. Ben Amar, M. Elamouri, and R. Dhifaoui, "Energy assessment of the first wind farm section of Sidi Daoud, Tunisia," *Renewable Energy*, vol. 33, no. 10, pp. 2311–2321, 2008.

[8] E. Muljadi and C. Butterfield, "Pitch-controlled variable-speed wind turbine generation," *IEEE Transactions on Industry Applications*, vol. 37, no. 1, Article ID 903156, pp. 240–246, 2001.

[9] B. Thul, R. Dutta, and K. Stelson, "Hydrostatic transmission for mid-size wind turbines," *Wind Energy Plants*, pp. 1017–1022, 2011.

[10] M. Deldar, A. Izadian, and S. Anwar, "Reconfiguration of a wind turbine with hydrostatic drivetrain to improve annual energy production," in *Proceedings of the 7th Annual IEEE Energy Conversion Congress and Exposition, ECCE 2015*, pp. 6660–6666, can, September 2015.

[11] J. Schmitz, N. Vatheuer, O. Reinertz, and H. Murrenhoff, "Hydrostatic drive train in wind energy plants," in *Proceedings of International Fluid Power Conference and Exposition*, pp. 1017–1022, Las Vegas, 2011.

[12] Z. Yang, M. Krishnamurthy, and J. M. Garcia, "Modeling and control of a continuously variable planetary transmission for a small wind turbine drivetrain," in *Proceedings of the ASME 2013 Conference on Smart Materials, Adaptive Structures and Intelligent Systems*, pp. 10–1115, Snowbird, UTAH, 2013.

[13] X. Zhao and P. Maißer, "A novel power splitting drive train for variable speed wind power generators," *Renewable Energy*, vol. 28, no. 13, 2003.

[14] S. Hamzehlouia, A. Izadian, A. Pusha, and S. Anwar, "Controls of hydraulic wind power transfer," in *Proceedings of the 37th Annual Conference of the IEEE Industrial Electronics Society, IECON 2011*, pp. 2475–2480, aus, November 2011.

[15] A. Pusha, M. Deldar, and A. Izadian, "Efficiency analysis of hydraulic wind power transfer system," in *Proceedings of the 2013 IEEE International Conference on Electro/Information Technology, EIT 2013*, usa, May 2013.

[16] Danish Wind Industry Association, "Wind Power Generators," Wind Turbine Generators, 2017.

[17] R. Dutta, F. Wang, B. F. Bohlmann, and K. A. Stelson, "Analysis of short-term energy storage for midsize hydrostatic wind turbinel," *Journal of Dynamic Systems, Measurement and Control, Transactions of the ASME*, vol. 136, no. 1, Article ID 011007, 2014.

10

Investigation of Performance and Rotor Tip Flow Field in a Low Speed Research Compressor with Circumferential Groove Casing Treatment at Varying Tip Clearance

Matthias Rolfes, Martin Lange, and Ronald Mailach

Technische Universität Dresden, Institute of Fluid Mechanics, Chair of Turbomachinery and Flight Propulsion, 01062 Dresden, Germany

Correspondence should be addressed to Matthias Rolfes; matthias.rolfes@tu-dresden.de

Academic Editor: Funazaki Ken-ichi

Experimental investigations in a single-stage low speed axial research compressor are presented. The influence of four different rotor tip clearances on the overall compressor performance and on the rotor tip flow field is investigated in configurations with and without circumferential groove casing treatments. Piezo-resistive pressure transducers are used to capture the unsteady flow field in the rotor tip region. The investigated casing groove is effectively working at the three largest investigated tip clearance sizes. The largest achieved operating range increase by the groove is 6.9%. The groove can delay the upstream movement of the flow interface between leakage flow and main flow and thus increase the stable operating range. Rotating instabilities are shown to exist at large tip clearance sizes in throttled operating conditions. Their amplitudes can be damped by the casing groove. No modal activities could be detected in the current single-stage compressor build.

1. Introduction

The relative motion between blades and casing dictates the necessity of radial clearances in an axial compressor. It is commonly known that an increase in tip clearance to chord ratios decreases pressure ratio, efficiency, and normally operating range of an axial compressor; see, for example, Wisler [1] and Cumpsty [2]. In service, the clearance ratios of a gas turbine's axial compressor are normally not constant but change due to wear, transient operations, and thermal expansions. Moreover, the minimum absolute clearance level can also be constrained by different thermal expansions of the components. As modern compressors tend to have very high pressure ratios, the last stages have very small blade heights which, together with the above-mentioned constraints, leads to larger clearance ratios here. The use of wide-chord blades is not able to totally compensate for this effect. These observations underline the fact that axial compressors at least partially run with tip clearances that are greater than favored by aerodynamic, which lead to the above-mentioned performance penalties.

One possible means to recover a loss in operating range due to a tip clearance increase is the use of circumferential groove casing treatments. Their simplicity and ease of installation put them in the focus of research for several decades. In the 1970s several experimental works [3–5] showed the general effectiveness of circumferential groove casing treatments. It is also in this early period that the potential of grooves to reduce the sensitivity towards large tip clearances has been proven. One example for a more recent investigation on circumferential groove CTs using CFD methods is the work by Shabbir and Adamczyk [6]. The authors showed that the flow of the compressor is stabilized by a momentum exchange between groove and main passage flow and furthermore that grooves reduce the aerodynamic blockage near the rotor tip. Concerning the working mechanism of casing groove some authors, for example, Lu et al. [7], Perrot et al. [8], and Rolfes et al. [9], showed that grooves work by delaying the upstream movement of the flow interface between tip leakage flow and passage flow and therefore shift the point where spill-over of leakage flow occurs towards lower mass flows. On the other hand, Houghton and Day [10] and Li et al. [11] showed

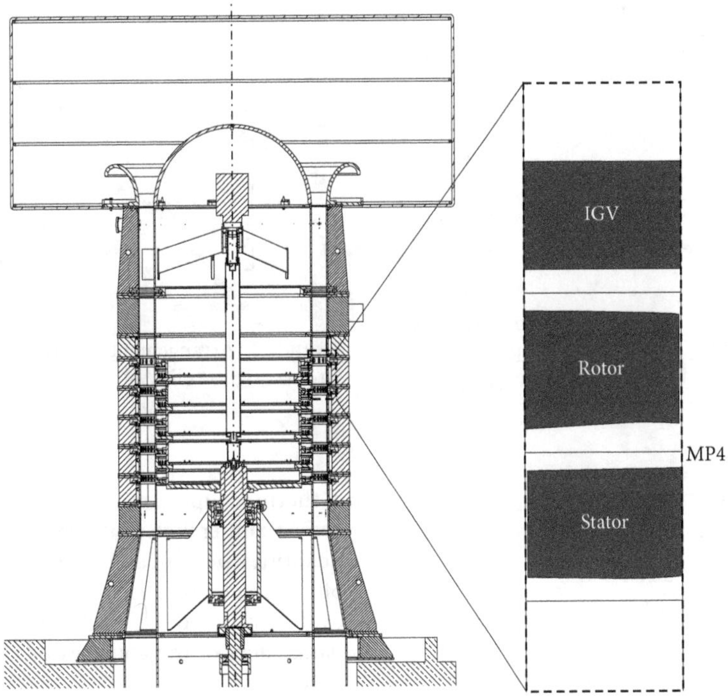

FIGURE 1: LSRC Dresden, cross section of single-stage build.

that the influence of the flow interface is not mandatory for a well working circumferential groove. This shows that the working mechanism of circumferential groove is still not universally understood and that different mechanisms might exist.

Compressor stability, the reason for surge or stall and the way towards it, is in the focus of interest since compressors in gas turbines became widely used. A very detailed overview of research in this field is given by Day [12]. Mailach et al. [13] and März et al. [14] investigated the highly controversial field of rotating instabilities (RI) within the topic of compressor stability. Mailach et al. found out that the observed disturbances are called forth by periodic fluctuations of the tip leakage vortex. Other authors, for example, Wisler et al. [15], connect the occurrence of RI in Mailach et al. to the fact that one stage of the four-stage compressor has been weakened by a tip clearance increase. Young et al. [16], however, found similar disturbances as Mailach et al. and März et al. that rotated relative to the rotor under near stall conditions in their investigated single-stage compressor with large tip clearance. According to Camp and Day [17] compressor stall can be caused by a local (mostly in the rotor tip) flow breakdown, the so called spikes, in only one or two blade passages that grow into stall cells. The authors also found modal oscillations that exist long before stall and are characterized by flow field oscillations of a wavelength of the compressor circumference or half the circumference. Modes are not directly linked with compressor stall, but can grow into a stall cell with further throttling. These authors concluded that modes appear when the compressor stalls with a positively (or zero) sloped characteristic and spikes when the characteristic is negatively sloped at stall.

In this paper a single-stage low speed axial compressor with and without casing grooves at four different rotor tip clearances is investigated. Besides overall performance, the influence of tip clearance increases and casing groove on the rotor tip flow field is shown. Their influence on rotating instabilities and modal oscillations will also be addressed to increase understanding of these phenomena.

2. Test Compressor and Measurement Techniques

The investigations have been performed on the Low Speed Research Compressor at Technische Universität Dresden. The reference configuration of the LSRC consists of 4 stages. One characteristic of the compressor is its vertical arrangement. The air passes 4 repeating stages from top to bottom. Different operating points are set by a throttle in the basement. The cross section of the compressor is shown in Figure 1 and reference build parameters can be found in Table 1. Boos et al. [18] and Künzelmann et al. [19] presented detailed information about the compressor installation and measurement results for the reference build. For the investigation at hand the compressor has been installed in a single-stage configuration. Different tip clearances were realized by milling the compressor blades to the desired height. Four rotor tip clearances were investigated. The smallest tip clearance (TC) is $s/c = 1.2\%$ and has been increased to $s/c = 2.5\%$, $s/c = 3.7\%$, and $s/c = 5.0\%$. Stator hub clearance has been kept constant at $s/c = 2.1\%$. The installed blading has been the same as in Rolfes et al. [9]. It has been designed by ALSTOM Power to model the rear stages of a high pressure compressor of a state of the art gas turbine.

TABLE 1: Data of the LSRC reference build.

Casing diameter	1500 mm
Hub-to-tip ratio	0.84
Blade count IGV	51
Blade count rotor	63
Blade count stator	83
Chord length rotor, MS	110 mm
Chord length stator, MS	89 mm
Solidity of rotor blades, tip	1.55
Mass flow, DP	$\dot{m} = 25.35$ kg/s
Reynolds number rotor inlet, MS	$\mathrm{Re} = 5.7 \cdot 10^5$
Mach number rotor inlet, MS	$\mathrm{Ma} = 0.22$

DP: design point; MS: midspan.

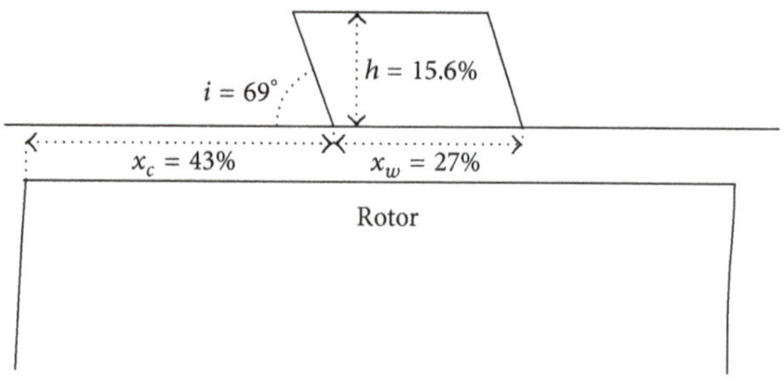

FIGURE 2: Geometry of circumferential casing groove.

The geometry of the circumferential casing groove is depicted in Figure 2. The performance of the compressor with $s/c = 5.0\%$ and this groove has already been described in Rolfes et al. [9].

This paper presents the characteristics of the compressor and results from unsteady wall pressure measurements with piezo-resistive pressure transducers which were installed flush in the rotor casing. All measurements were performed at design speed of 1000 rpm at standard conditions.

Compressor characteristics are composed of 11 throttle positions from open throttle to stall. At each of these positions the stator vanes are traversed in 57 uniformly distributed steps over two pitches. These data are arithmetically averaged to form one point on the speed line. The overall pressure rise is determined with an accuracy of ±0.02% by 12 casing static pressure taps in the exit plane and 6 casing static pressure taps in the inlet plane of the compressor. To calculate the total pressure ratio, the total pressures in the inlet and exit planes are derived by an iterative procedure from the static values by the assumption of a perfect gas. Thus, the curves of the total pressure ratio presented in this paper are derived from casing static pressure taps. To calculate the efficiency, data from a high-precision torquemeter (accuracy 0.01% full-scale at 5000 Nm) are used which results in a measuring accuracy of the efficiency of ±0.2 percentage points (representing the 95% confidence interval). Total temperature at the inlet of the compressor is measured via 8 resistance thermometers in the filter cage. In the exit plane 6 total temperature probes equipped with a thermocouple are equally spaced around the compressor circumference. Enthalpy is derived from the measured quantities of temperature and pressure by the assumption of a perfect gas. The stall point of the compressor has been explicitly tested for each configuration. The performance parameters of the compressor characteristics are calculated as follows:

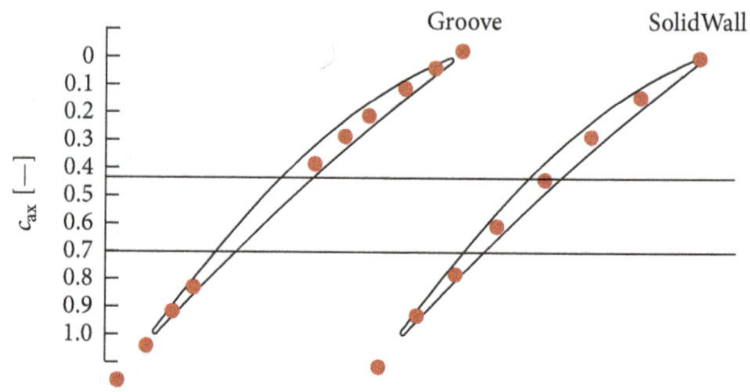

FIGURE 3: Arrangement of pressure transducers over rotor, left: Groove, right: SolidWall.

$$\eta = \frac{\dot{m}_{\text{in}} \cdot \left(h_{\text{total,is,out}} - h_{\text{total,in}}\right)}{M \cdot 2\pi n} \quad \text{isentropic efficiency}$$

$$\xi = \frac{\dot{m}_{\text{in}} \cdot \sqrt{T_{\text{total,in}}} / p_{\text{total}}}{\dot{m}_{\text{in,DP}} \cdot \sqrt{T_{\text{total,in,DP}}} / p_{\text{total,DP}}} \quad \text{flow rate coefficient}$$

$$\pi_t = \frac{p_{\text{total,out}}}{p_{\text{total,in}}} \quad \text{total pressure ratio} \tag{1}$$

$$\Psi_{t-s} = \frac{2 \cdot \left(p_{\text{out}} - p_{\text{total,in}}\right)}{u_{\text{MS}}^2 \cdot \rho} \quad \text{total-to-static pressure coefficient}$$

$$\phi = \frac{v_{\text{ax}}}{u} \quad \text{flow coefficient.}$$

Mounting piezo-resistive pressure sensors flush with the rotor casing allows the unsteady wall pressure over the rotor to be captured. For these measurements the stator vanes have been traversed in 21 steps over one pitch. The arrangement of the transducers along the blade chord can be found in Figure 3. The left hand side shows the sensors for the groove configuration and the right hand side that of the solid wall configuration. For clarity, the limits of the circumferential groove are plotted on both sides.

The standard deviation of the unsteady wall pressure, which is used in the presentation of the results, is defined as follows:

$$\sigma_p = \sqrt{\frac{1}{N} \sum_{i=0}^{N-1} \left(p_i(t) - \langle p(t) \rangle\right)^2}. \tag{2}$$

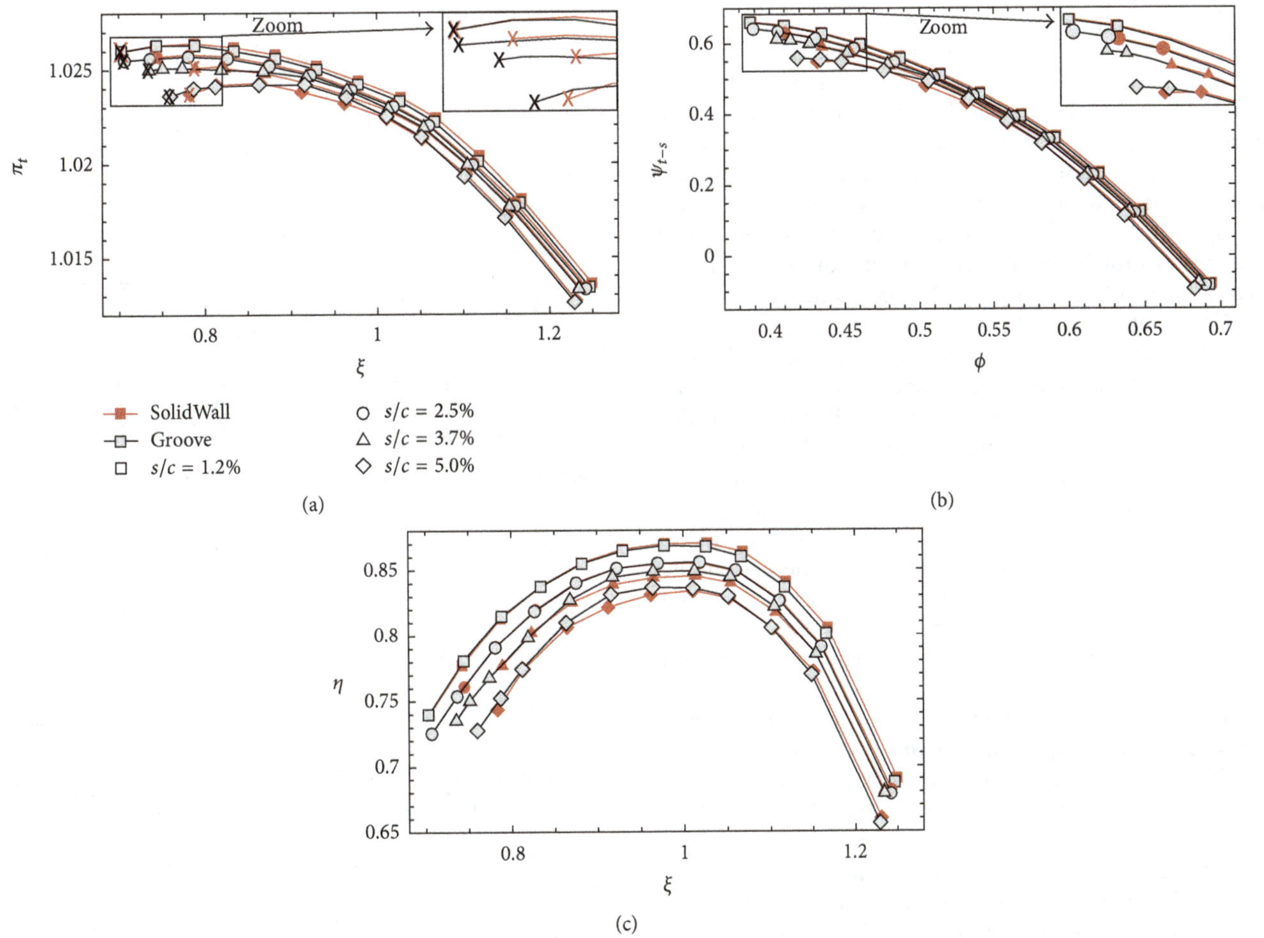

FIGURE 4: Compressor characteristics: (a) total pressure ratio, (b) total-to-static pressure coefficient, and (c) efficiency.

Here, p_i is the current pressure value, N is the number of revolutions, and $\langle p(t) \rangle$ is the ensemble-averaged pressure value.

3. Results

3.1. Compressor Characteristics. The compressor characteristics for all four tip clearance sizes are plotted in Figure 4. The throttle settings are identical between each configuration. Figure 4(a) shows the total pressure ratio over flow rate coefficient. In this part of the figure, the last stable operating points are marked by an X to better distinguish the stall points between SolidWall and Groove. The zoomed region has the same intention. Figure 4(c) shows isentropic efficiency over flow rate coefficient and the total-to-static pressure coefficient is plotted over flow coefficient in Figure 4(b). Here, the zoomed region only shows symbols for the last two measured points on the characteristic curve to simplify the identification of the slopes of the characteristics. The flow coefficients of the stall point and the percentage reduction in stall mass flow due to the Groove can be found in Table 2 for all tested configurations.

At the smallest tip clearance of $s/c = 1.2\%$ the compressor has the highest total pressure ratio and efficiency due to the lowest tip leakage flow loss. The Groove is not able to positively influence either total pressure ratio, efficiency, or stall mass flow (see Table 2) at this tip clearance size. At flow rate coefficients close to stall Groove and SolidWall show nearly identical total pressure ratios and efficiencies whereas at flow rates higher than design ($\xi \geq 1.0$) both are slightly reduced by Groove. One reason for the lack of operating range increase can be that the rotor is not tip-critical in this configuration.

TABLE 2: Stalling flow coefficients and percentage reduction of stall mass flow for each configuration.

s/c	Configuration	ϕ_{Stall}	$\Delta\dot{m}_{\text{Stall}}$
1.2%	SolidWall	0.388	0%
	Groove	0.387	
2.5%	SolidWall	0.410	5.1%
	Groove	0.389	
3.7%	SolidWall	0.435	6.9%
	Groove	0.405	
5.0%	SolidWall	0.432	3.0%
	Groove	0.419	

As expected, an increase of tip clearance to s/c = 2.5% leads to a loss in total pressure ratio and efficiency and an increase in stall mass flow. The casing treatment obviously starts working at this size of tip clearance. With the casing treatment the stall mass flow is reduced by 5.1% (Table 2). The compressor is now able to run at a flow rate coefficient as low as that of SolidWall of the smallest tested tip clearance size s/c = 1.2%. The Groove configuration still shows a small reduction in total pressure ratio which, however, is smaller than at s/c = 1.2%. Isentropic efficiency is not decreased by the Groove at this size of tip clearance.

With further increased tip clearance to s/c = 3.7% total pressure ratio, efficiency, and operating range decrease. The Groove remains effective and decreases the minimum flow rate coefficient by 6.9% compared to SolidWall at the same clearance level, which can be seen in Table 2. Groove at this tip clearance size now even reaches a slightly lower minimum mass flow than SolidWall of s/c = 2.5%. The highest decrease of flow rate coefficient leads to the assumption that the Groove is most effective at this clearance configuration. Another distinctive feature in this configuration is the increase in total pressure ratio and efficiency in regions $0.85 < \xi < 1.1$.

The last tip clearance increase to s/c = 5.0% leads to a stronger decreasing total pressure rise and efficiency of SolidWall near stall than in the steps of tip clearance increase before. Furthermore, the stall mass flow of SolidWall has not been increased in contrast to s/c = 3.7% but has been slightly reduced. This behavior is reproducible and has been measured several times. The Groove, however, is still able to decrease the minimum stall mass flow by 3% (Table 2). The Groove increases efficiency in the region from ξ = 1.1 to stall by a maximum of 1%.

According to Camp and Day [17] and Day [12] the total-to-static characteristics are helpful to identify the type of stall inception. With a negatively sloped total-to-static characteristic stall is likely to occur by spikes whereas a zero slope indicates modal stall inception. The total-to-static pressure coefficient clearly shows negative slopes at tip clearance sizes of s/c = 1.2%, s/c = 2.5%, and s/c = 3.7% for both SolidWall and Groove, which can be seen in the zoomed region in Figure 4(b). Here, stall is likely to occur via spikes. The largest tip clearance configuration however shows a zero slope near stall for SolidWall and thus, according to Camp and Day, possible modal activities. The Groove configuration of s/c = 5.0%, however, shows again a negative slope.

It is noteworthy that the zero slope of the total-to-static characteristics occurs at the tip clearance size that showed no increase in stall mass flow compared to the next smaller tip clearance.

The following is an addition to a comparison of different investigations in open literature that has been presented by Berdanier and Key [20]. They plotted the percentage increase in stall flow coefficient relative to the smallest tested tip clearance size $\Delta\phi_{\text{Stall}}$ over tip clearance size for results from Wisler [1], McDougall [21], and their own test results. In Figure 5, which is based on Berdanier and Key, the data from the present paper are added as well as data of both compressors A and B from Wisler et al. [15]. The clearance derivative δ (the increase in stall flow coefficient per 1% clearance increase) that has also been discussed by Berdanier and Key is shown in Table 3 for the same test cases as in Figure 5. Some of the investigations tested more than two tip clearance sizes, so the derivatives for each step of tip clearance increase are listed (TC1-2, e.g., is the first step of tip clearance increase of the particular investigation).

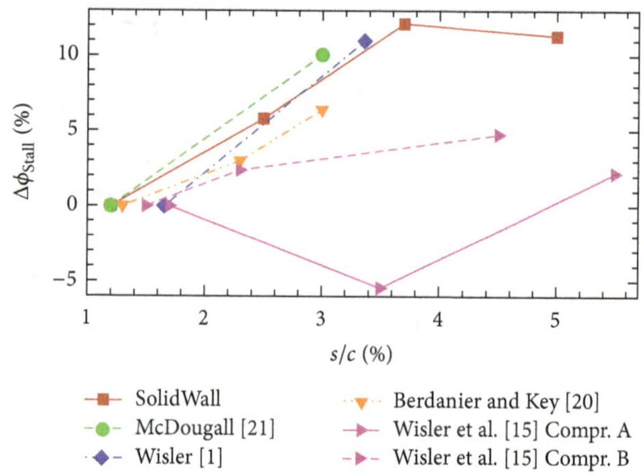

FIGURE 5: Stalling flow coefficient at different tip clearance sizes.

TABLE 3: Clearance derivatives of different investigations.

Investigation	Clearance derivative δ
Wisler [1]	6.5
McDougall [21]	5.6
Berdanier and Key [20] TC1-2	3.0
Berdanier and Key [20] TC2-3	4.8
Wisler et al. [15] Compr. A TC1-2	−3.0
Wisler et al. [15] Compr. A TC2-3	3.8
Wisler et al. [15] Compr. B TC1-2	3.0
Wisler et al. [15] Compr. B TC2-3	1.1
Present Paper TC1-2	4.5
Present Paper TC2-3	5.3
Present Paper TC3-4	−0.6

The increase in stall mass flow that resulted from a tip clearance increase from s/c = 1.2% to 2.5% (TC1-2) and from s/c = 2.5% to 3.7% (TC2-3) in the present paper fits well into the data presented by Berdanier and Key. The clearance derivatives of 4.5 and 5.3, respectively, are within the range of 3 to 6 that the other investigations show. This increase in clearance derivative from TC1-2 to TC2-3 has also been shown by Berdanier and Key for their compressor. The last tested tip clearance increase from s/c = 3.7% to 5.0% (TC3-4) in this paper, however, shows a remarkably different behavior because the stall mass flow is not increased but decreased and hence the derivative is −0.6. This behavior has not been detected by [1, 20, 21] because their largest tip clearance size is notably lower than the largest tested tip clearance size in this paper. Wisler et al. [15] tested two compressors A and B with largest tip clearance sizes of s/c = 5.5% and s/c = 4.5%,

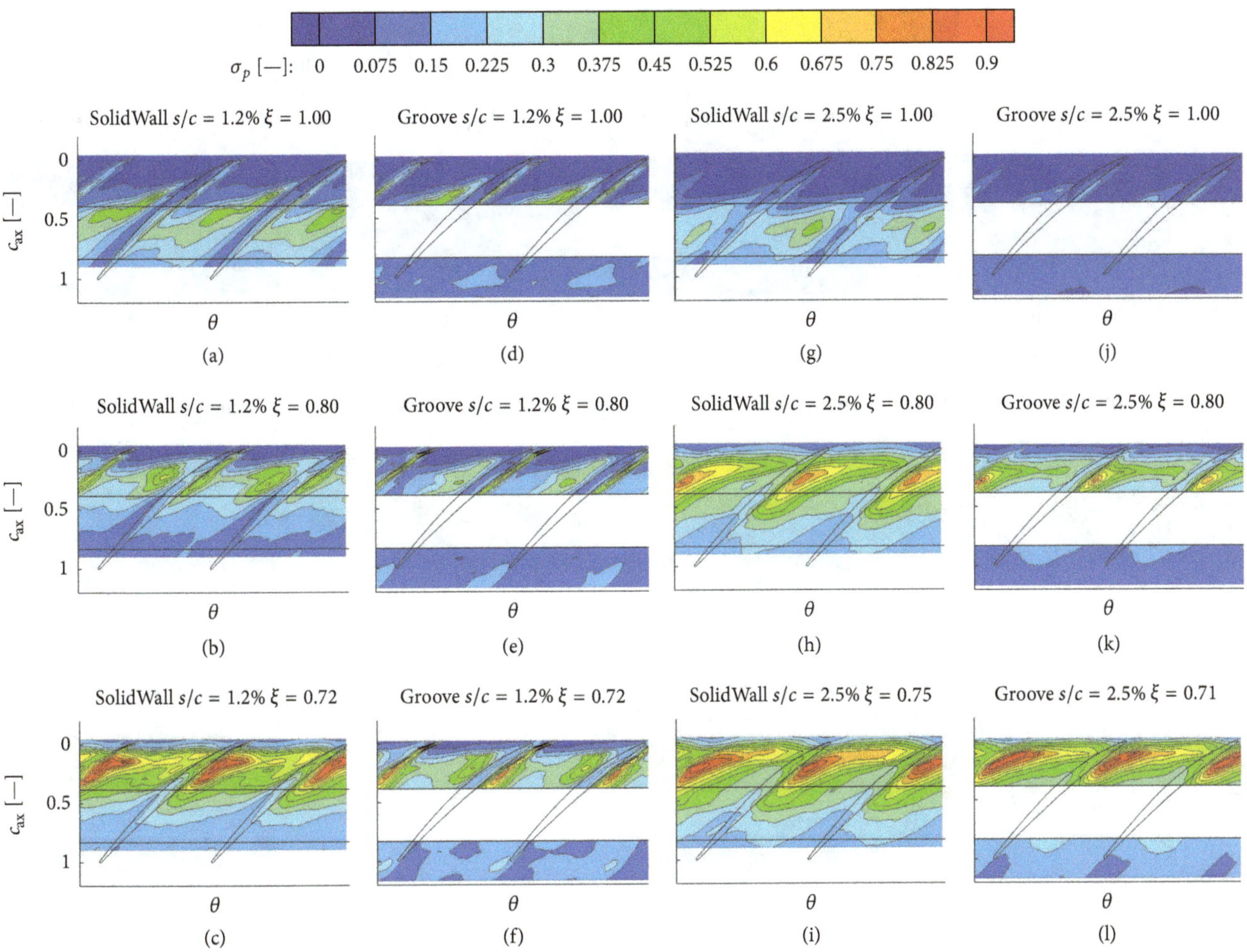

FIGURE 6: Standard deviation of casing static pressure over rotor for SolidWall and Groove at s/c = 1.2% and s/c = 2.5% at three different operating points ((a, d, g, j): design point, (b, e, h, k): throttled operating point, and (c, f, i, l): near stall of current configuration).

respectively. Both compressors show a different behavior than the results presented before. Compressor A shows at TC1-2 a decrease in stall mass flow which results in a clearance derivative of −3. According to Wisler et al. stability in compressor A is limited by tip and hub blockage whereas compressor B is more sensitive to tip blockage and therefore has a derivative of 3 at TC1-2. These results show that the clearance derivative is strongly dependent on the investigated compressor and that effects like a shift of the stall mechanism (either from hub to tip or possibly from rotor to stator) with tip clearance can also play a dominant role. The second tip clearance increase of compressor B in [15] shows a derivative of only 1.1. One could speculate that this reduced derivative of TC2-3 compared to TC1-2 hints to a similar behavior as in the present paper, a negative derivative at larger tip clearances. It is however not visible due to the limited number of investigated tip clearance sizes in [15] that result in an almost doubling of tip clearance size from TC1-2 to TC2-3. Thus, the fact that other investigations do not show a negative clearance derivative at larger tip clearances might be also due to a relatively low resolution between tightest and largest clearance.

The information from the present investigation is a valuable addition to data in open literature. It confirms the findings of Berdanier and Key [20] and add new information for tip clearances larger than $s/c = 4\%$.

3.2. Wall Pressure over Rotor. The following section deals with the results of the unsteady wall pressure measurements over the rotor.

The standard deviation of the static wall pressure (σ_p) is plotted over two rotor pitches in Figure 6 for s/c = 1.2% and s/c = 2.5% and in Figure 7 for s/c = 3.7% and s/c = 5.0%. Results for SolidWall and Groove are shown at three different operating points, the design point (ξ = 1.00), a throttled operating point (ξ = 0.80), and a near stall operating point of the respective configuration. Due to brevity, at s/c = 5.0% the operating points that are shown for the SolidWall configuration are only the design point and ξ = 0.80. Because it is not possible to place sensors in the Groove region on the casing radius, the Groove results show a blank space. This region is also indicated by horizontal black lines in the SolidWall results. In the presented plots, high values of

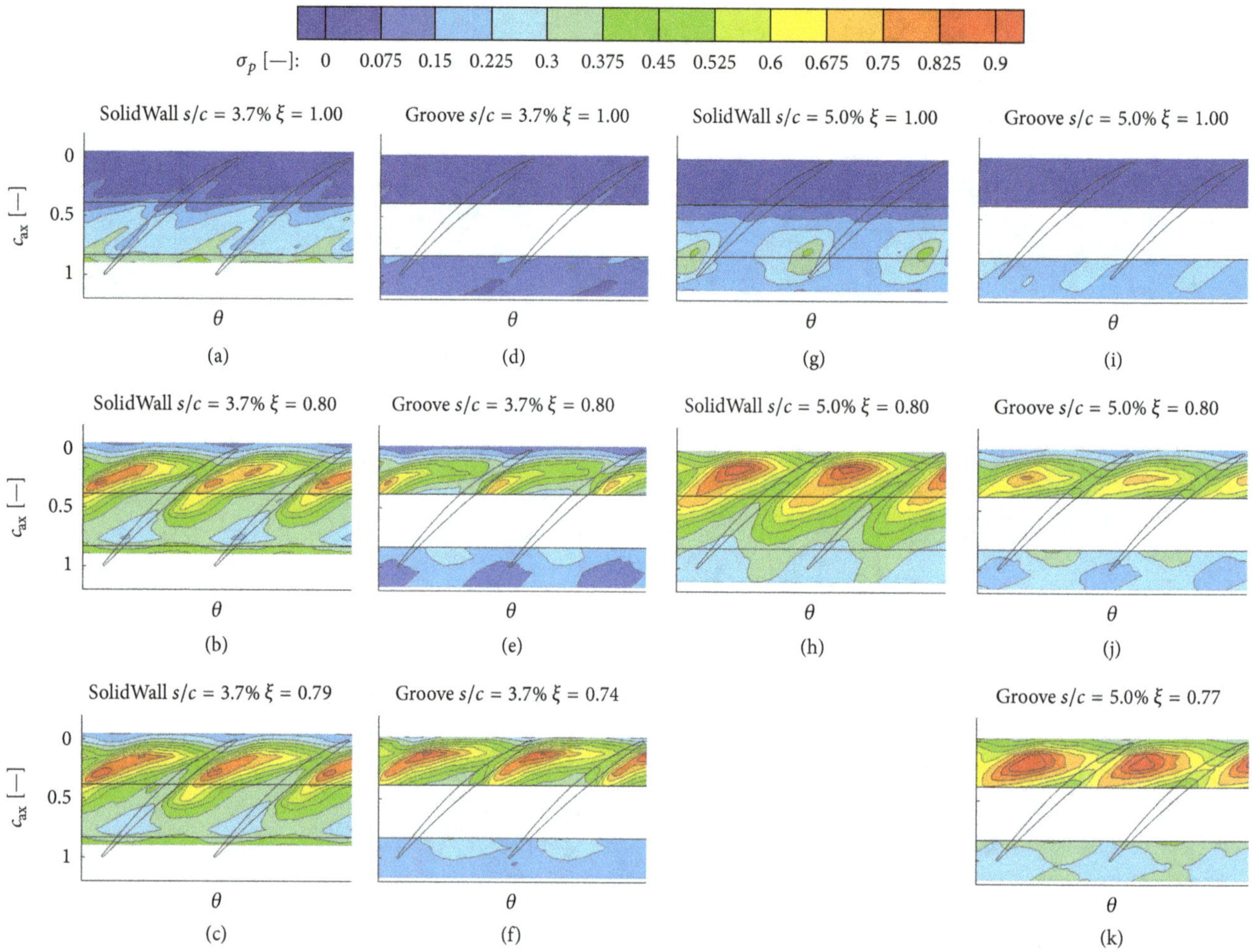

FIGURE 7: Standard deviation of casing static pressure over rotor for SolidWall and Groove at s/c = 3.7% and s/c = 5.0% at three different operating points ((a, d, g, i): design point, (b, e, h, j): throttled operating point, and (c, f, k): near stall of current configuration).

standard deviation can be found in regions where the tip leakage vortex interacts with the main flow.

3.2.1. s/c = 1.2%. At the design point of SolidWall (Figure 6(a)) high standard deviations (and thus the tip leakage vortex) start near the suction side of the blade around c_{ax} = 0.2. The vortex expands into the passage and in a region around c_{ax} = 0.6 it reaches the pressure side of the opposite blade. With increased throttling (Figure 6(b)) the region of high standard deviation moves upstream towards the leading edge. The vortex now travels towards the pressure side of the opposite blade in the front half of the blade and reaches it at approximately c_{ax} = 0.25. This means that also the vortex inclination against axial direction is increased. At the last stable operating point (Figure 6(c)) the upstream boundary of the tip leakage vortex is nearly vertical. The highest standard deviations can be found near the pressure side of the opposite blade, indicating a strong vortex fluctuation in this area. The increase in standard deviation from design point to near stall can be attributed to an increase in vortex strength with throttling.

At the design point of Groove (Figure 6(d)) vortex size and position are in the same region as in the SolidWall configuration. A manipulation of the vortex by the Groove cannot be detected under these operating conditions. At ξ = 0.80 (Figure 6(e)) differences between SolidWall and Groove occur. The values of high standard deviation near the blade pressure side are reduced in contrast to the SolidWall and the values of σ_p are lower in the passage. At the near stall operating point, the Groove's influence is enhanced. The standard deviations in the whole region upstream the Groove are lower than in the SolidWall configuration. The inclination of the upstream boundary of the tip leakage vortex has also been lowered. Despite the fact that the Groove influences the near casing flow field at the smallest tip clearance configuration, the Groove is not able to reduce the stall mass flow; see Figure 4. This underlines the previous assumption that stall at s/c = 1.2% is not triggered by the near casing flow in the rotor.

3.2.2. s/c = 2.5%. With increasing tip clearance to s/c = 2.5% the tip leakage vortex at the design point (Figure 6(g)) now starts more downstream around c_{ax} = 0.3. The region where the vortex reaches the pressure side is nearly unchanged in comparison with the next smaller tested tip clearance size. The Groove (Figure 6(j)) does not influence the starting point of the vortex at $\xi = 1.00$. At throttled operation in Figure 6(h) the vortex inclination has increased notably, as before. High values of σ_p can be found near the pressure side of the opposite blade around $c_{ax} = 0.3$. At the near stall operating point (Figure 6(i)) these values are further increased and the upper boundary of the tip leakage vortex is nearly vertical and blocks the whole near casing passage. This is in good agreement with one of Vo et al.'s [22] criteria for rotating stall onset, the spill-over of leakage fluid into the neighboring passage. The influence of the Groove can again be seen at the two last operating points (Figures 6(k) and 6(l)). The upper boundary of the tip leakage vortex at $\xi = 0.80$ is more downstream than in the SolidWall case and the standard deviations have been reduced. At the near stall operating point of Groove (which is 4 percentage points lower than that of SolidWall) a similar near casing flow pattern as in the SolidWall case (Figure 6(i)) exists. Thus, the Groove has been able to delay the upstream movement of the tip leakage vortex with throttling. The compressor can be throttled more intensely before the interface between the incoming flow and the tip leakage flow reaches the leading edge and causes a spill-over of leakage flow into the neighboring passage.

3.2.3. $s/c = 3.7\%$. The next tip clearance increase results at the design point (Figure 7(a)) in a further downstream movement of the point of vortex formation to approximately $c_{ax} = 0.4$. In contrast to the two smaller tested tip clearance sizes, high standard deviations can be found in the passage over the whole circumference and the exact point of the interaction of the vortex with the opposite blade cannot be detected. It can be seen that the Groove lies directly in the region of vortex formation and is expected to influence it. Directly downstream the Groove the standard deviation is notably smaller than in the SolidWall configuration. This underlines the above-stated assumption of a Groove's influence on the vortex formation. The lower values of standard deviation upstream the Groove in Figure 7(d) compared to the SolidWall are also attributed to the Groove's influence on the vortex. A reduction of vortex related blockage by the Groove is likely to have a positive upstream influence on the flow field. The throttled operating point of SolidWall in Figure 7(b) is at this clearance size already quite close to the near stall operating point. Thus the upper boundary of the tip leakage vortex is already placed near the blade leading edge. A similar pattern of standard deviation as in $s/c = 2.5\%$ (Figure 6(h)) is observed. A little further throttling to $\xi = 0.79$, shown in Figure 7(c), moves the upper boundary of the tip leakage vortex further upstream and is expected to cause spill-over of leakage fluid and thus to limit the stable operation. The Groove influence (Figures 7(e) and 7(f)) is also similar to the next smaller tested s/c-configuration. The delay of the upstream movement of the above-mentioned interface, which can be seen at $\xi = 0.8$ makes the compressor more stable in the rotor tip region and allows in this case an operation at flow rate coefficients that are lower by 5 percentage points in contrast to the SolidWall.

3.2.4. $s/c = 5.0\%$. At the largest tested tip clearance, the tip leakage vortex of SolidWall at the design point (Figure 7(g)) starts again more downstream than at $s/c = 3.7\%$. The starting point now approximately lies at $c_{ax} = 0.5$. This far downstream position of the starting point prevents an impingement of the vortex on the pressure side of the next blade. The Groove configuration at the design point (Figure 7(i)) shows a reduction of standard deviations downstream the Groove. The region of enhanced σ_p shows, in addition, a reduced spread in circumferential direction due to the Groove. This again shows the influence of the Groove on the near casing flow field. Due to the small difference between $\xi = 0.79$ and $\xi = 0.80$ (Figures 7(c) and 7(b)) at $s/c = 3.7\%$, here only the operating point $\xi = 0.80$ is shown in Figure 7(h). As in the configurations before, the upper boundary of tip leakage flow reaches the leading edge and causes stall at a slightly higher throttling. The mechanism of operating range improvement of Groove is also identical to the tip clearance sizes before. At $\xi = 0.80$, Figure 7(j), the vortex is less intensely inclined than at the SolidWall. At the near stall operating point, Figure 7(k), the upper vortex boundary reaches the leading edge, as in the near stall point of SolidWall.

3.3. Frequency Analysis. In this section a frequency analysis of wall pressure sensors is performed in order to find hints for rotating instabilities that have been found by Mailach et al. [13] in a different build of the same compressor that is investigated in this paper. Mailach et al. weakened the third stage of a four-stage build by locally increasing the tip clearance of the third rotor. At a tip clearance of $s/c = 4.3\%$ their frequency spectra showed typical humps at one-third of blade passing frequency that have been attributed to a fluctuating tip leakage vortex. It is now interesting to find out if these fluctuations also happen in a single-stage compressor with enlarged tip clearance or only in a weakened stage of a multistage compressor, where the weakened stage is stabilized by the surrounding stages and can therefore be operated at lower flow rate coefficients than the similar single-stage.

Frequency spectra of wall pressure sensors at $c_{ax} \approx 0.125$ are shown in Figure 8 for all tested tip clearance sizes at the throttled operating point. At $\xi = 1.00$ none of the four tested clearance sizes shows any activities at frequencies lower than BPF; therefore these results are omitted. For the smallest tested clearance this can also be observed at the throttled operating point. This is in good agreement with Mailach et al., who showed that RI only occur in the throttled operating region of a compressor with relatively large tip clearance size. At $s/c = 2.5\%$, however, a rise of amplitudes in the region 0.4–$0.5 \cdot$ BPF can be seen at $\xi = 0.80$ for the SolidWall. This is a first sign that rotating instabilities, as Mailach et al. called the fluctuations of the tip leakage vortex, occur in this compressor build. The rise in amplitudes is also visible for Groove but in a lower extent. It seems that the Groove is able to damp the observed fluctuations.

The throttled operating points of the two largest clearances again show increasing amplitudes below BPF. At $s/c = 3.7\%$ the SolidWall shows amplitudes below BPF that are nearly twice as high as in the next smaller tip clearance configuration. The frequencies, where the high amplitudes occur, are shifted towards lower values, compared with $s/c = 2.5\%$, and are now in the region of 0.35–$0.45 \cdot$ BPF. Here, it is now very obvious that the Groove is able to damp the high

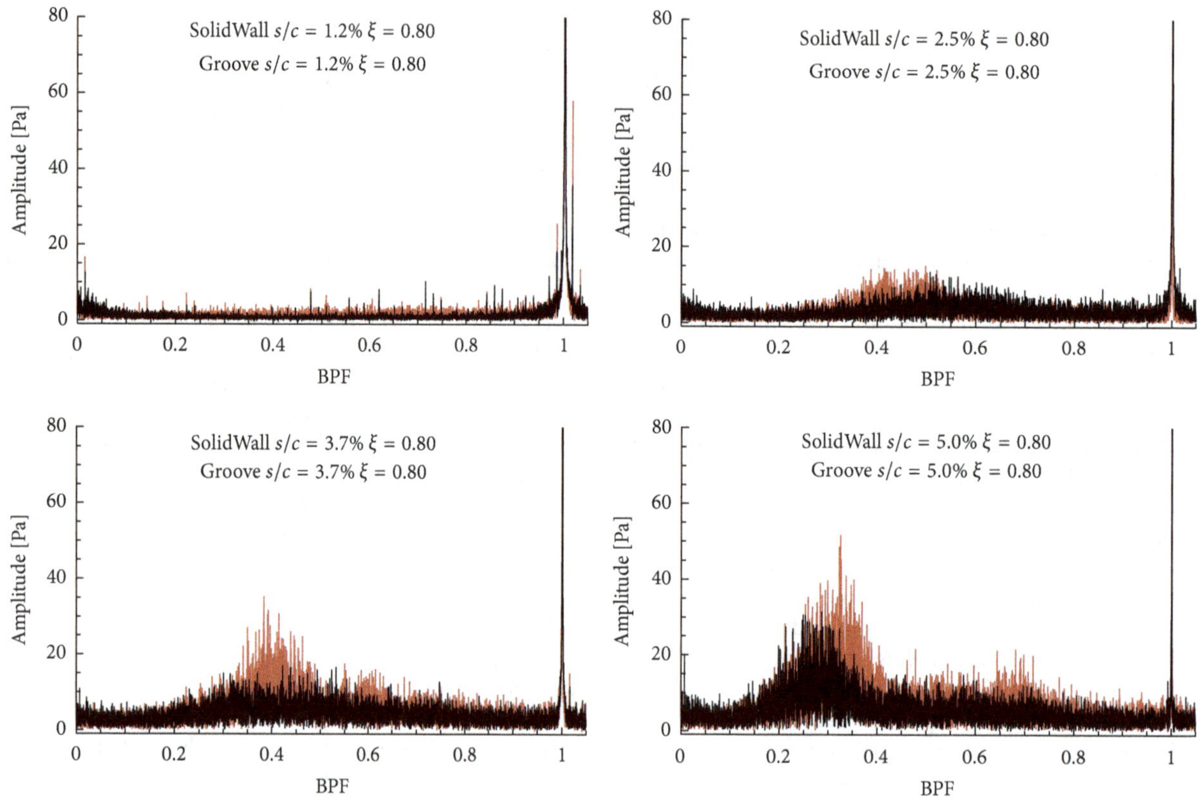

FIGURE 8: Frequency spectra of wall pressure sensor at $c_{ax} \approx 0.125$ for SolidWall and Groove at the throttled operating point for $s/c = 1.2\%$, 2.5%, 3.7%, and 5.0%.

amplitudes. The largest clearance shows a further increase in low frequency amplitudes and a further shift of these frequencies that now lie in a region of 0.3–0.4 · BPF. In this operating point a second hump is visible at around 0.7 · BPF which is probably a modulation of the hump around 0.3–0.4 · BPF and the BPF itself. Looking at the Groove configuration, the damping of the low frequency hump can be seen again.

In the frequency spectra, that are presented in Figure 8, the RI show amplitudes that are relatively small with up to 60 Pa. This can be attributed to the low periodicity of the disturbances and to the fact that the frequency spectra are derived from a dataset with approximately 5300 blade passings. In the raw data, the disturbance is much more pronounced and has amplitudes of more than 1000 Pa. This has also been addressed in [13].

The following conclusions can be drawn from this paragraph. No rotating instabilities occur at the tightest clearance. The other three investigated clearance sizes showed a rise in frequencies around one-half to one-third of BPF. Thus, the occurrence of RI is not limited to a weakened stage in a multistage compressor but is also detected in a single-stage compressor. The amplitudes of the above-mentioned frequencies rise with increased rotor tip clearance together with a shift towards lower frequencies. Furthermore, it has been shown that the investigated casing grooves are able to influence the rotating instabilities as they reduce their amplitudes in the frequency spectra.

3.4. Time-Resolved Casing Static Pressure. This section discusses time-resolved casing static pressures for all clearance configurations in order to detect a flow pattern that can be linked to the occurrence of rotating instabilities that have been identified in the previous section.

The time-resolved casing static pressures for all clearance configurations are shown in Figures 9 and 10. The data are derived from one sensor array (see Figure 3) at a fixed position on the compressor circumference. It is known from Mailach et al. [13] that RI rotate relative to the rotor with approximately 50% of rotor speed. The plotted results show snapshots of 10 rotor passages that have passed the measurement position. Snapshots at other instants of time show results that are analogous to the plotted results. Static pressure is depicted by the pressure coefficient which is defined as $c_p = (p - p_\infty)/(p_{in} - p_\infty)$ where p_{in} is the static pressure at rotor inlet and $p_\infty = 101325$ Pa. The operating point of the SolidWall configurations for the respective clearances is the near stall operating point; the operating point of Groove is identical to SolidWall. The dashed white lines connect the low pressure regions in each plot. These are traditionally linked with the trajectory of the tip leakage vortex. From other investigations (e.g., [23, 24]) it is known that this is a simplification and that, under certain operating conditions, the near casing flow field shows multiple vortical structures and not one clearly defined tip leakage vortex. This becomes obvious in our results at the largest tested tip clearance that is presented later.

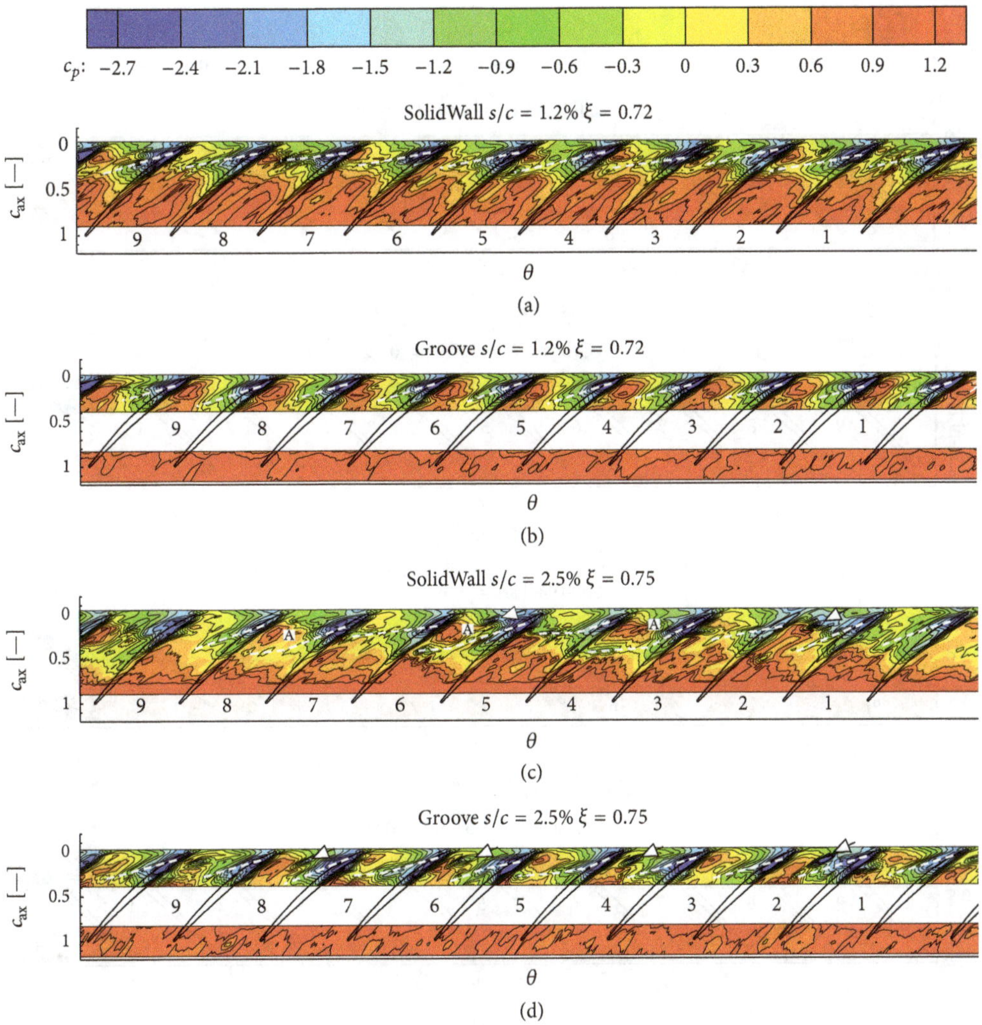

FIGURE 9: Time-resolved casing static pressure over rotor for SolidWall and Groove at s/c = 1.2% (a, b) and s/c = 2.5% (c, d). SolidWall operating points are near stall, Groove operating points identical to SolidWall.

At the smallest tested tip clearance size (Figures 9(a) and 9(b)) the tip leakage vortices start near the leading edge in the region $c_{ax} \approx 0.1$ at both SolidWall and Groove. The plotted tip leakage vortices of SolidWall slightly differ from passage to passage. In passages 5 and 7 the region of high c_p on the blade pressure side is interrupted by a low pressure spot which seems to stem from the tip leakage vortex. In passages 6 and 8, however, these low pressure spots do not appear. These differences are responsible for the increase of standard deviations in Figure 6(c). The Groove configuration of the same clearance size (Figure 9(b)) shows nearly the same flow pattern in each passage. A variation of tip leakage vortex trajectory cannot be detected. The above-mentioned low pressure spots on the blade pressure side vanished. Consequently the standard deviations (see Figure 6(f)) are lower than in SolidWall. It can be concluded that for s/c = 1.2% tip leakage vortex fluctuations are either small (SolidWall) or not present (Groove), which means that the Groove has a small influence on the near casing flow field that does not lead to an increase in operating range though.

With an increase of tip clearance to s/c = 2.5% the flow field in SolidWall (Figure 9(c)) is getting more irregular. The differences in vortex trajectory from passage to passage are now easy to detect. Moreover, the region of low c_p near the leading edge of the blade now differs strongly in every passage. Although only ten passages are shown in the figures, a repeating pattern can already be detected. Passages 4, 6, and 8 (and partially passage 2) all show a high pressure spot around c_{ax} = 0.25 on the blade pressure side (see region A) and have very similar vortex trajectories. Passages 3, 5, and 7 however show a higher inclination (against axial direction) of the tip leakage vortex trajectory. This change of flow pattern in every other passage that passes the measurement position leads to an increase of amplitudes in the frequency spectra around 0.5 · BPF (the frequency spectra for ξ = 0.75 are not shown here, for brevity, but show an amplitude increase at the same frequencies as the operating point ξ = 0.80 in Figure 8).

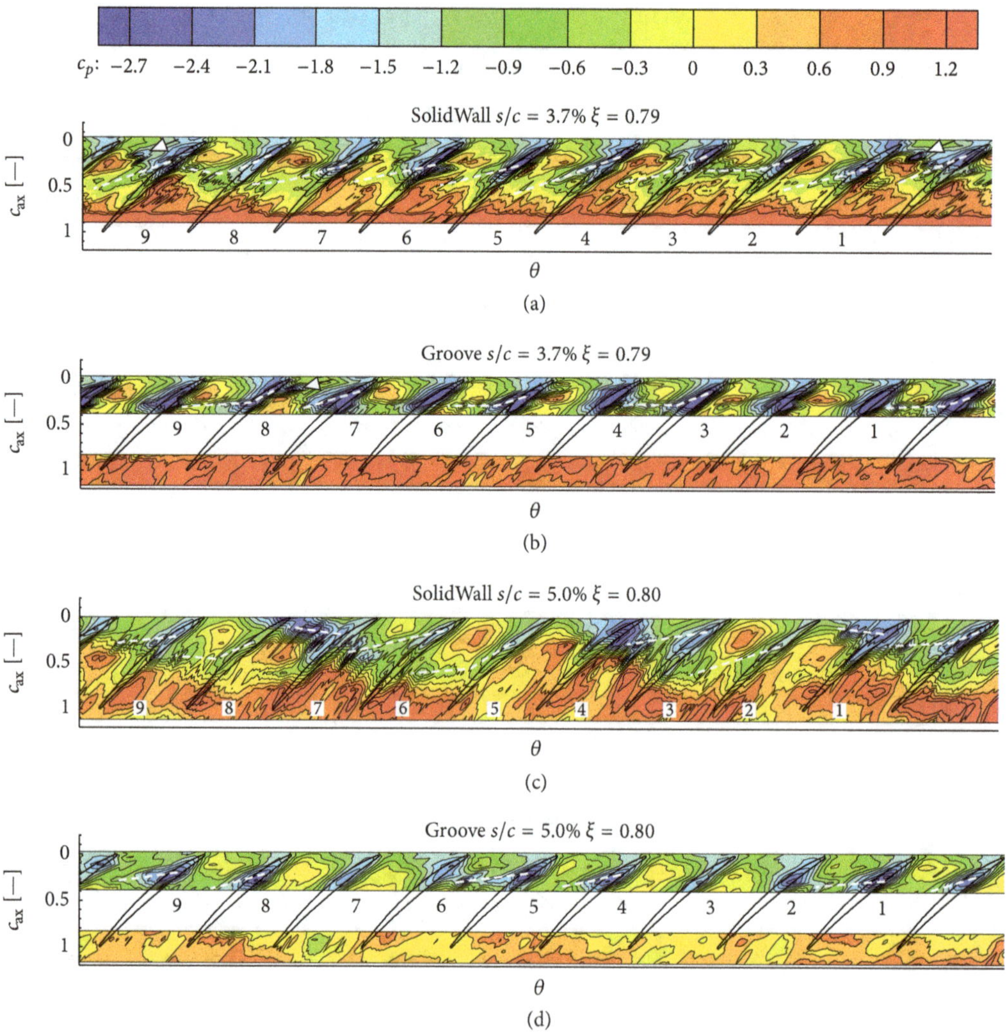

FIGURE 10: Time-resolved casing static pressure over rotor for SolidWall and Groove at s/c = 3.7% (a, b) and s/c = 5.0% (c, d). SolidWall operating points are near stall, Groove operating points identical to SolidWall.

The fact that the peak in the frequency spectra around $0.5 \cdot$ BPF is not sharp but blurred is due to the fact that the above-mentioned pattern is not repeating strictly every other passage but slightly changes.

The black-and-white arrows in passages 2 and 6 mark a low pressure spot which is positioned in the passage between suction side and pressure side of the blade. The source that causes this low pressure spot is not clear. It could be the radial vortex that Hah [25] identified in his investigation which is part of the tip leakage vortex. Inoue et al. [26] and Pullan et al. [27] also identified a radial vortex which stems from blade separation and also produces a low pressure spot in the instantaneous casing pressure. Weichert et al. [28] also linked their observed low pressure spot in the passage near the leading edge with a radial vortex as in [26, 27].

The flow field of Groove at s/c = 2.5% is shown in Figure 9(d). Here, different tip leakage vortex trajectories between the passages can also be observed. The pattern, however, is not as regular as in the SolidWall. In passages 2 and 8, for example, the trajectories are pretty similar and in both passages a distinct high pressure region near the pressure side exists. The vortex trajectories in passages 7 and 5 seem to reach the pressure side of the blade and thus the above-mentioned area of high pressure is interrupted. The low pressure spots that have been found in SolidWall are also present in Groove. They are again marked by black-and-white arrows. In this case, they appear in passages 2, 4, 6, and 8. This regular appearance, however, is not visible throughout the whole dataset. The authors in [26–28] all linked the appearance of their radial vortices to the formation of a spike that initiated compressor stall. The fact that Figure 9(d) shows the appearance of the low pressure spots at an operating point which is not near stall makes the same mechanism for its appearance as in [26, 27] less likely.

At the tip clearance of s/c = 3.7% the tip leakage vortex fluctuations increase; see Figure 10(a). In passages 3 and 9 the vortex stands nearly vertical in the passage whereas in the other passages it shows a notably lower inclination. Like in s/c = 2.5%, a strong variation of the minimum pressure spot over the blade is present. The blades on the right hand sides

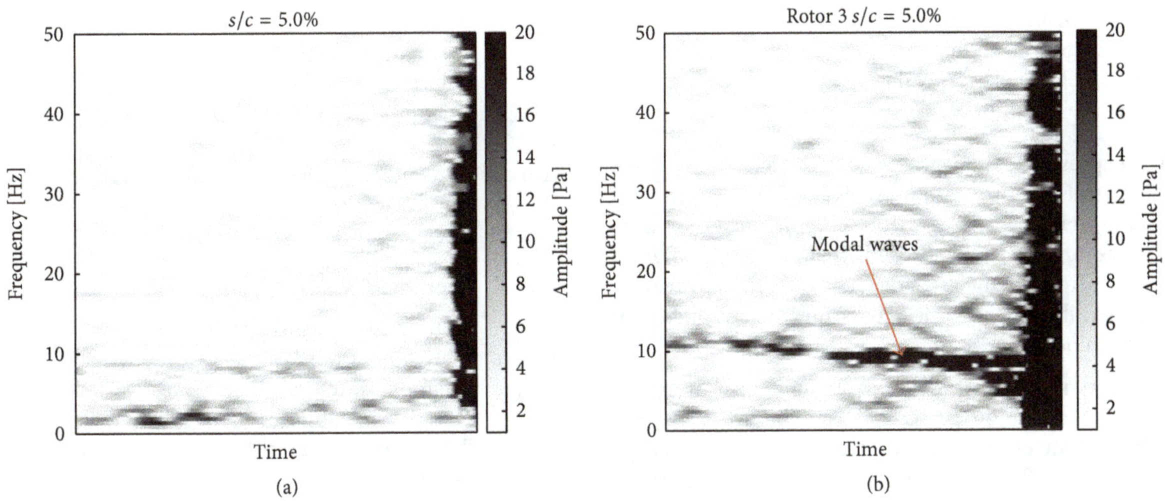

FIGURE 11: Spectrogram of microphone at $c_{ax} = -0.2$ (upstream rotor leading edge). (a) $s/c = 5.0\%$ of single-stage test. (b) $s/c = 5.0\%$ only in Rotor 3 of four-stage test.

of passages 2 and 8 both have a position of minimum over-blade pressure that is far more downstream than in the other passages. The low pressure spots marked by the arrow again show the potential existence of radial vortices near the blade leading edge. In the Groove configuration (Figure 10(b)) the position of minimum over-blade pressure is identical in each passage, which results in similar starting points of the tip leakage vortex. The position of the Groove makes it difficult to track the vortices in this configuration. It is, however, obvious that vortex fluctuations are still present. The spectral analysis in Figure 8 revealed that the dominant peak has been damped by the Groove. This means that the periodicity of appearance of the vortex fluctuations is unchanged compared to SolidWall, but its amplitudes have been damped.

The increase of rotor tip clearance to $s/c = 5.0\%$ drastically alters the near casing flow field. As mentioned earlier, especially at this tip clearance size, the dashed white lines are a strong simplification of the tip leakage vortex path. The figure suggests that multiple vortical structures are present. The SolidWall (Figure 10(c)) now exhibits very strong tip leakage vortex fluctuations. A periodic pattern can be identified. Starting in passage 3, the vortex starts around $c_{ax} = 0.3$ and proceeds within the passage. In passage 4, the starting point of the vortex moved upstream to $c_{ax} = 0.15$. The vortex trajectory is steeper than in passage 3 and the vortex interacts with the pressure side of the adjacent blade around $c_{ax} = 0.3$. Passage 5 then shows a very pronounced low pressure spot near the leading edge and from the measured pressure field the vortex trajectory cannot be estimated anymore. The following passage 6 is similar to passage 3 and the pattern of vortex fluctuation starts again. This pattern is responsible for the increase of amplitudes at frequencies around 0.3· BPF that were identified in Figure 8. In Figure 10(d), Groove configuration, the vortex fluctuations are not as pronounced as in SolidWall. Again, the Groove positions hinder the tracking of the vortex path. In some passages, for example, passages 2 and 6, the vortex is highly inclined against axial direction so that it is proceeding in the passage upstream the Groove. Passages 8 and 4 and passages 7 and 3 also show very similar patterns upstream the Groove which leads to the assumption that a similar pattern as in SolidWall is present. In both SolidWall and Groove of $s/c = 5.0\%$ the low pressure spots in the passage near the leading edge cannot be identified anymore, or are overlaid by the tip leakage vortex.

3.5. Modal Activities. The total-to-static characteristics in Figure 4 showed a zero slope near stall for SolidWall of $s/c = 5.0\%$ and according to general opinion hint to an existence of modal activities. Therefore in this section the occurrence of modal waves is investigated.

The spectrogram of a single microphone positioned 16 mm upstream the rotor leading edge is shown in Figure 11 during a slow closing of the throttle into stall. Figure 11(a) shows results of the SolidWall of $s/c = 5.0\%$ whereas Figure 11(b) shows results from third rotor of a four-stage test, where only rotor 3 runs with an increased clearance of $s/c = 5.0\%$. In the case of the weakened stage a distinct disturbance is visible throughout the whole time line which starts at frequencies slightly above 10 Hz that decrease with further throttling. Mailach [29] showed that these amplitudes are evoked by modal waves in the compressor. In the present single-stage test these disturbances cannot be detected neither at the largest tip clearance (see Figure 11(a)) nor at any other tip clearance size. Furthermore, no influence of the Groove on modal activities could be detected.

The results presented above reveal that the slope of the total-to-static characteristics does not seem to be a sufficient criterion to predict modal waves in an axial compressor.

4. Conclusions

In this paper, detailed measurements in a single-stage build of the Low Speed Research Compressor at Technische Universität Dresden have been presented. Four different tip clearance sizes of $s/c = 1.2\%$, 2.5%, 3.7%, and 5.0% have been investigated with and without a circumferential groove casing treatment. The overall compressor performance as well as the rotor tip flow field have been evaluated. The main conclusions of the paper are as follows:

(i) The present casing groove works well at the three largest tested tip clearance sizes ($s/c = 2.5\%$, 3.7%, and 5.0%). Here, it increases the compressor's operating range and partially the efficiency. At the smallest tested tip clearance size of $s/c = 1.2\%$ the grooves do not exhibit a negative influence on efficiency between design point and stall.

(ii) Results from unsteady wall pressure measurements show that the groove is able to delay the upstream movement of the flow interface between tip leakage and main flow and thereby increases compressor stability.

(iii) The peaks in the frequency spectra in regions around 0.3–0.5· BPF could clearly be linked to rotating instabilities.

(iv) The rotating instabilities increase in amplitude with tip clearance size. At the same time the frequency of appearance in the frequency spectra is decreased because the periodic pattern of RI increases from every other to every third blade passage.

(v) Low pressure spots near the leading edge in the passage have been observed that could be evoked by radial vortices either of the Inoue et al. [26] and Pullan et al. [27] type or by a mechanism described by Hah [25].

(vi) Although the total-to-static characteristics at SolidWall of $s/c = 5.0\%$ show a zero slope near stall, no modal activities could be detected.

Nomenclature

Greek Symbols

η: Isentropic efficiency
θ: Circumferential coordinate
ξ: Flow rate coefficient
π_t: Total pressure ratio
ϕ: Flow coefficient
$\Delta\phi_{\text{Stall}}$: Percentage reduction in stall flow coefficient relative to smallest tip clearance size
Ψ_{t-s}: Total-to-static pressure coefficient
ρ: Density [kg/m^3]
σ_p: Standard deviation of casing wall pressure
δ: Clearance derivative.

Latin Symbols and Abbreviations

BPF: Blade passing frequency
c: Chord length [m]
c_{ax}: Nondimensional axial chord length
DP: Design point
h: Depth of casing groove in % axial chord
h_{total}: Total enthalpy [m^2/s^2]
i: Inclination angle of casing groove [°]
is: Isentropic
IGV: Inlet guide vane
LSRC: Low Speed Research Compressor
$\Delta\dot{m}_{\text{Stall}}$: Percentage reduction of stall mass flow
$\dot{m}$: Mass flow [kg/s]
M: Torque [Nm]
MP: Measuring plane
MS: Midspan
n: Shaft speed [1/min]
p: Static pressure [Pa]
p_{total}: Total pressure [Pa]
RI: Rotating instabilities
s: Tip clearance size [m]
SW: SolidWall
T_{total}: Total temperature [K]
TC: Tip clearance
u: Blade speed [m/s]
v_{ax}: Axial component of absolute velocity [m/s]
x_w: Width of casing groove in % axial chord
x_c: Axial location of casing groove in % axial chord.

Disclosure

The responsibility for the content lies solely with its authors.

Conflicts of Interest

The authors declare that there are no conflicts of interest regarding the publication of this article.

Acknowledgments

The authors gratefully acknowledge the industrial partner ALSTOM Power for their permission to publish this paper.

References

[1] D. C. Wisler, "Loss reduction in axial-flow compressors through low-speed model testing," *Journal of Engineering for Gas Turbines and Power*, vol. 107, no. 2, pp. 354–363, 1985.

[2] N. A. Cumpsty, *Compressor Aerodynamics*, Krieger Publishing, 2004.

[3] W. M. Osborn, G. W. Lewis, and L. J. Heidelberg, "Effect of Several Porous Casing Treatments on Stall Limit and on Overall Performance of an Axial Flow Compressor Rotor, NASA Technical Note D-6537," Tech. Rep., 1971.

[4] D. C. Prince, D. C. Wisler, and D. E. Hilvers, "Study of Casing Treatment Stall Margin Improvement Phenomena, NASA Report CR 134552," Tech. Rep., 1974.

[5] H. Takata and Y. Tsukuda, "Stall margin improvement by casing treatment - its mechanism and effectiveness," *Journal of Engineering for Power*, vol. 99, no. 1, pp. 121–133, 1977.

[6] A. Shabbir and J. J. Adamczyk, "Flow mechanism for stall margin improvement due to circumferential casing grooves on axial compressors," *Journal of Turbomachinery*, vol. 127, no. 4, pp. 708–717, 2005.

[7] X. Lu, W. Chu, J. Zhu, and Y. Wu, "Mechanism of the interaction between casing treatment and tip leakage flow in a subsonic axial compressor," in *Proceedings of the 2006 ASME 51st Turbo Expo, GT2006-90077*, pp. 79–90, Barcelona, Spain, May 2006.

[8] V. Perrot, A. Touyeras, and G. Lucien, "Detailed CFD analysis of a grooved casing treatment on an axial subsonic compressor," in *Proceedings of the 7th European Conference on Turbomachinery: Fluid Dynamics and Thermodynamics, ETC7-016*, March 2007.

[9] M. Rolfes, M. Lange, and K. Vogeler, "Experimental investigation of circumferential groove casing treatments for large tip clearances in a low speed axial research compressor," in *Proceedings of the ASME Turbo Expo 2015: Turbine Technical Conference and Exposition, GT2015-42646*, Montreal, Quebec, Canada, June 2015.

[10] T. Houghton and I. Day, "Enhancing the stability of subsonic compressors using casing grooves," *Journal of Turbomachinery*, vol. 133, no. 2, Article ID 021007, 2011.

[11] J. Li, F. Lin, S. Wang, J. Du, C. Nie, and J. Chen, "Extensive experimental study of circumferential single groove in an axial flow compressor," in *Proceedings of the Proceeding of the ASME Turbo Expo 2014: Turbine Technical Conference and Exposition, GT2014-26859*, Düsseldorf, Germany, June 2014.

[12] I. J. Day, "Stall, surge, and 75 years of research," *Journal of Turbomachinery*, vol. 138, no. 1, Article ID 2478223, 2015.

[13] R. Mailach, I. Lehmann, and K. Vogeler, "Rotating instabilities in an axial compressor originating from the fluctuating blade tip vortex," *Journal of Turbomachinery*, vol. 123, no. 3, pp. 453–460, 2001.

[14] J. März, C. Hah, and W. Neise, "An experimental and numerical investigation into the mechanisms of rotating instability," *Journal of Turbomachinery*, vol. 124, no. 3, pp. 367–375, 2002.

[15] D. Wisler, B. F. Beacher, and H.-W. Shin, "Effects of loading and clearance variation on tip vortex and endwall blockage," in *Proceedings of the 9th International Symposium on Transport Phenomena and Dynamics of Rotating Machinery*, 10–14 February 2002.

[16] A. Young, I. Day, and G. Pullan, "Stall warning by blade pressure signature analysis," *Journal of Turbomachinery*, vol. 135, no. 1, Article ID 011033, 2012.

[17] T. R. Camp and I. J. Day, "A study of spike and modal stall phenomena in a low-speed axial compressor," *Journal of Turbomachinery*, vol. 120, no. 3, pp. 393–401, 1998.

[18] P. Boos, H. Möckel, J. M. Henne, and R. Seimeler, "Flow measurement in a multistage large scale low speed axial flow research compressor," in *Proceedings of the ASME 1998 International Gas Turbine and Aeroengine Congress and Exhibition, 98-GT-432*, Stockholm, Sweden.

[19] M. Künzelmann, R. Müller, R. Mailach, and K. Vogeler, "Steady and unsteady flow field in a multistage low-speed axial compressor - A test case," in *Proceedings of the 2008 ASME Turbo Expo, GT2008-50793*, pp. 447–455, Berlin, Germany, June 2008.

[20] R. A. Berdanier and N. L. Key, "The effects of tip leakage flow on the performance of multistage compressors used in small core engine applications," *Journal of Engineering for Gas Turbines and Power*, vol. 138, no. 5, Article ID 052605, 2016.

[21] N. M. McDougall, "A comparison between the design point and near stall performance of an axial compressor," *Journal of Turbomachinery*, vol. 112, no. 1, pp. 109–115, 1989.

[22] H. D. Vo, C. S. Tan, and E. M. Greitzer, "Criteria for spike initiated rotating stall," *Journal of Turbomachinery*, vol. 130, no. 1, Article ID 011023, 2008.

[23] D. Tan, Y. Li, H. Chen, I. Wilkes, and J. Katz, "The three dimensional flow structure and turbulence in the tip region of an axial flow compressor," in *Proceedings of the ASME Turbo Expo 2015: Turbine Technical Conference and Exposition, GT2015-43385*, Montreal, Quebec, Canada, June 2015.

[24] D. Tan, Y. Li, I. Wilkes, R. L. Miorini, and J. Katz, "Visualization and time-resolved particle image velocimetry measurements of the flow in the tip region of a subsonic compressor rotor," *Journal of Turbomachinery*, vol. 137, no. 4, Article ID 041007, 2015.

[25] C. Hah, "Effects of double-leakage tip clearance flow on the performance of a compressor stage with a large rotor tip gap," *Journal of Turbomachinery*, vol. 139, no. 6, Article ID 061006, 2017.

[26] M. Inoue, M. Kuroumaru, T. Tanino, and M. Furukawa, "Propagation of multiple short-length-scale stall cells in an axial compressor rotor," *Journal of Turbomachinery*, vol. 122, no. 1, pp. 45–54, 2000.

[27] G. Pullan, A. M. Young, I. J. Day, E. M. Greitzer, and Z. S. Spakovszky, "Origins and structure of spike-type rotating stall," *Journal of Turbomachinery*, vol. 137, no. 5, Article ID 051007, 2015.

[28] S. Weichert and I. Day, "Detailed measurements of spike formation in an axial compressor," *Journal of Turbomachinery*, vol. 136, no. 5, Article ID 051006, 2013.

[29] R. Mailach, *Experimentelle Untersuchung von Strömungsinstabilitäten im Betriebsbereich zwischen Auslegungspunkt und Stabilitätsgrenze eines vierstufigen Niedergeschwindigkeits-Axialverdichters [Doctoral thesis]*, Technische Universität Dresden, 2001.

11

Observations of the Growth and Decay of Stall Cells during Stall and Surge in an Axial Compressor

Adam R. Hickman and Scott C. Morris

University of Notre Dame, Notre Dame, IN 46556, USA

Correspondence should be addressed to Adam R. Hickman; adamrhickman@hotmail.com

Academic Editor: Pietro Zunino

This research investigated unsteady events such as stall inception, stall-cell development, and surge. Stall is characterized by a decrease in overall pressure rise and nonaxisymmetric throughflow. Compressor stall can lead to surge which is characterized by quasi-axisymmetric fluctuations in mass flow and pressure. Unsteady measurements of the flow field around the compressor rotor are examined. During the stall inception process, initial disturbances were found within the rotor passage near the tip region. As the stall cell develops, blade lift and pressure ratio decrease within the stall cell and increase ahead of the stall cell. The stall inception event, stall-cell development, and stall recovery event were found to be nearly identical for stable rotating stall and surge cases. As the stall cell grows, the leading edge of the cell will rotate at a higher rate than the trailing edge in the rotor frame. The opposite occurs during stall recovery. The trailing edge of the stall cell will rotate at the approximate speed as the fully developed stall cell, while the leading edge decreases in rotational speed in the rotor frame.

1. Introduction

The stable operation of an axial compressor at a given shaft speed is limited at low mass flows by compressor stall. Stall is characterized by a decrease in overall pressure rise and nonaxisymmetric throughflow. A region of low momentum fluid can develop in one or more rotor passages and rotate around the annulus. This region is often referred to as a stall cell. Rotating stall can lead to quasi-axisymmetric fluctuations in mass flow and pressure known as compressor surge. Axial compressors may enter periods of stable rotating stall and surge during operation at off-design conditions. This generally leads to decreased efficiency and increased blade vibration. This paper presents detailed unsteady experimental measurements from a transonic axial compressor in order to describe stall inception, stall cell development, and surging characteristics.

The inception of stall has been the topic of significant research for many years. Camp and Day [1] found that stall inception may occur locally from short length scale disturbances (spikes) or from long length scale disturbances (modes). The present work will focus on observations from a compressor that exhibits spike-type stall inception. Vo [2] found that a criterion for spike-type stall inception is that the interface created from the approach flow and the reverse tip-leakage flow becomes parallel with the rotor leading edge plane. Later Cameron et al. [3] obtained measurements supporting this concept using the compressor used in the present work. Tan et al. [4] provide a comprehensive review on the current understanding of stall inception. They found that the most common type of stall inception for modern compressor designs is a spike-type. This is when the initial stall cell develops on the order of a few rotor revolutions. Stall will typically develop in a single blade passage initially. There have been findings to support the existence of prestall spikes prior to stall inception. Initial disturbances have been found by Weichert and Day [5] found that initial disturbances can occur within the rotor passage at approximately 10–25% chord from the leading edge. Yamada et al. [6] also found that unsteady flow phenomena appear inside the rotor passage at near-stall.

After stall inception, a stall cell will increase in circumferential and radial extent. The stall cell will span multiple rotor passages as it rotates around the annulus. Cameron et al. [7] found that the leading edge of the stall cell will initially rotate at a slower rate in the laboratory frame. The trailing edge of the stall cell will initially rotate around the annulus at nearly the rotor speed. After a few revolutions the leading edge will increase in rotational speed and the trailing edge will decrease in rotational speed as the final size is reached. Day [8] gives a thorough review of the current understanding of rotating stall. Day found throughout many compressors described in the literature that smaller stall cells tend to travel faster in the laboratory frame than larger cells.

The system dynamics of stall and surge for a compression system has been studied extensively. Greitzer [9, 10] developed a lumped parameter model to predict the fluctuations in mass flow and plenum pressure during stall and surge. The Greitzer stall and surge model can be used to predict the point where, as a compressor reaches the stall limit, stall will transition to surge. In his work, Greitzer found that a single nondimensional parameter, termed the B parameter, was a good indicator to determine whether stall or surge would occur. Hickman and Morris [11] found that the poststall throttle point can also influence the onset of surge.

Surge can grow from rotating stall as the compressor interacts with compression system. Day [12] found that the flow breakdown around the rotor caused by a stall inception event is typically the reason for the onset of surge. Compressor surge is composed of multiple parts [13]. The developing stall cell produces an effective blockage of the compressor annulus. A sudden drop in pressure ratio and mass flow rate then follow [11]. Rotating stall may be present from stall inception until the surge recovery process, [14] depending on the compressor. The recovery process occurs where the compressor will recover from rotating stall briefly. During the brief stall recovery process, the pressure ratio returns to its prestall value. Another stall inception event and surge cycle repeat as the stall point is reached again [11].

Observations from measurements of stall inception, rotating stall, surge, and stall recovery are presented. This paper includes a description of the test facility and instrumentation used. Instantaneous shroud and rotor exit unsteady pressure contours are presented during transient events in order to better understand the process of stall inception and stall development. A comparison between stall and surge events is also given to describe how the onset of surge occurs [15].

2. Experimental Methods

2.1. Compressor Test Facility. Experiments were conducted using the Notre Dame Transonic Axial Compressor Facility (NDTAC). The NDTAC facility includes a butterfly valve and plenum upstream of the axial compressor and a second plenum and butterfly valve downstream. The upstream and downstream plenum volumes were approximately $1.52\,\text{m}^3$ and $0.70\,\text{m}^3$, respectively. The NDTAC facility employed a magnetic bearing system where the rotor shaft was levitated by a series of magnets. This enabled the facility to safely operate in rotating stall for extended periods of time and surge for short periods.

TABLE 1: ND Stage 04 design parameters.

Quantity	Value
Tip dia. [m]	0.457
Inlet hub dia. [m]	0.343
Exit hub dia. [m]	0.368
Rotor axial chord [cm]	3.56
Rotor blade count	20
Stator count	43
N_c [rpm]	14684
U_{tip} [m/s]	352

The compressor used in the experiments was ND Stage 04, a single stage transonic axial compressor with a mean hub-to-tip ratio of approximately 0.78. Figure 1 shows the schematic of the cross-section of ND Stage 04. Table 1 shows the design parameters for this stage. This compressor stage includes an annular contraction that incorporates a perforated plate which provided an approximate 3.5% turbulence intensity to the stage. The test article included a rotor and stator configuration.

2.2. Instrumentation. Steady and unsteady instrumentation were utilized for this work. Figure 1 shows the measurement planes. Standard Kiel probes to measure steady-state total temperature and total pressure were located at 3.2 axial rotor chords (c_{ax}) upstream of the rotor leading edge and $0.77c_{\text{ax}}$ downstream of the rotor trailing edge. Unsteady instrumentation was located upstream, downstream, and over the rotor.

Constant temperature hot wires were used to measure the unsteady stage inlet velocity and track the development of the stall cell during transient events. A circumferential array of 4 hot wires spaced every 90° were positioned approximately $8.9c_{\text{ax}}$ upstream of the rotor leading edge at midspan of the flow path. The hot wires were $5\,\mu$m tungsten wire probes and oriented normal to the flow and in the radial direction. Using the form of King's law presented by Bruun [16], a simple calibration was found to relate the voltage and flow velocity. The velocities from the four hot wires were averaged in order to estimate the time-resolved and the annulus average unsteady mass flow. The hot-wire frequency response was in excess of 100 kHz.

High-response pressure transducers were utilized for flow field measurements in the plenums and in the vicinity of the rotor. Figure 2 is a schematic of the locations of unsteady instrumentation around the rotor. The pressure sensors used in the experiments were silicon piezoresistive pressure sensors made by Kulite (model XTL-140-25A). These pressure transducers had a natural frequency of approximately 240 kHz. Kulite pressure transducers were located in the upstream and downstream plenums. A circumferential array of ten high-frequency pressure transducers, located approximately $0.18c_{\text{ax}}$ upstream of the rotor leading edge, were flush mounted to the flow surface of the shroud. This allowed tracking of the stall-cell development and calculation of the circumferential size of the stall cell. A line of thirteen

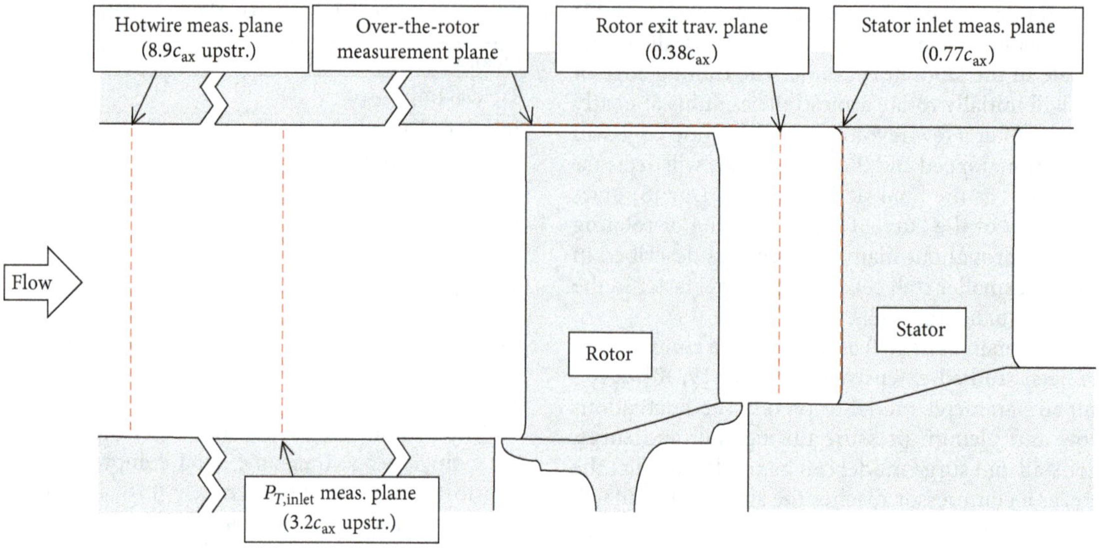

FIGURE 1: Cross-section of ND Stage 04.

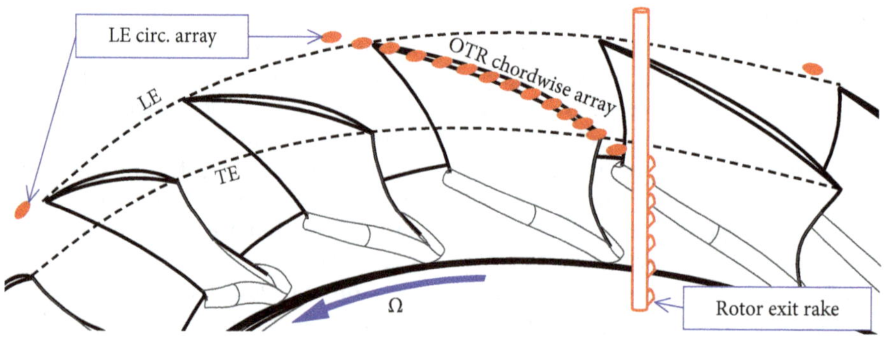

FIGURE 2: Schematic of near rotor unsteady instrumentation.

high-frequency pressure sensors was also located directly over-the-rotor (OTR) in a chord-wise array. The array of OTR transducers was flush mounted to the flow surface following the stagger angle of the rotor blade tip similar to Weichert and Day [5].

An unsteady total pressure rake was specifically designed for this work. The top of Figure 3(a) shows a photograph, and the bottom shows a schematic of the total pressure rake. The rake contained seven miniature Kulite transducers (XCQ-95-062) which were mounted into separate Kiel heads such that the transducer was directly exposed to the oncoming flow. Some have designed unsteady total pressure probes such that the transducer is in the body of probe [17]. Though this design can protect the sensor from dust and debris, it can limit the frequency response of the probe. The multiple sensors were positioned along the length of the probe and were concentrated towards the tip region of the rotor. It was installed at the rotor exit traverse plane shown in Figures 1 and 2. The unsteady total pressure rake was used to record unsteady total pressures simultaneously at multiple spanwise locations during stall and surge.

Because this was an unconventional design, a yaw sensitivity test was conducted for the developed unsteady total pressure rake. Figure 3(b) shows the total pressure normalized by the maximum as a function yaw angle for one of the transducers. All transducers contained nearly identical yaw sensitivity. The unsteady total pressure rake was found to have an acceptance angle limit of approximately ±20°. Beyond the acceptance range, the measured pressure is biased low and drops gradually compared to a standard Kiel probe at approximately 0.4% per degree.

Unsteady instrumentation was sampled at rate of 200 kHz. It was found that the sampling rate and frequency response of the sensors were sufficient to measure the frequencies associated with stall inception and rotating stall events at the speeds considered. The maximum frequencies of interest were found to be under 20 kHz. The Kulite sensors were calibrated before the measurement program and were found to have an uncertainty less than 1% of the full-scaled values. Hot-wire measurements were also calibrated immediately prior to the measurements and were found to have an uncertainty of approximately 3% of the measured values.

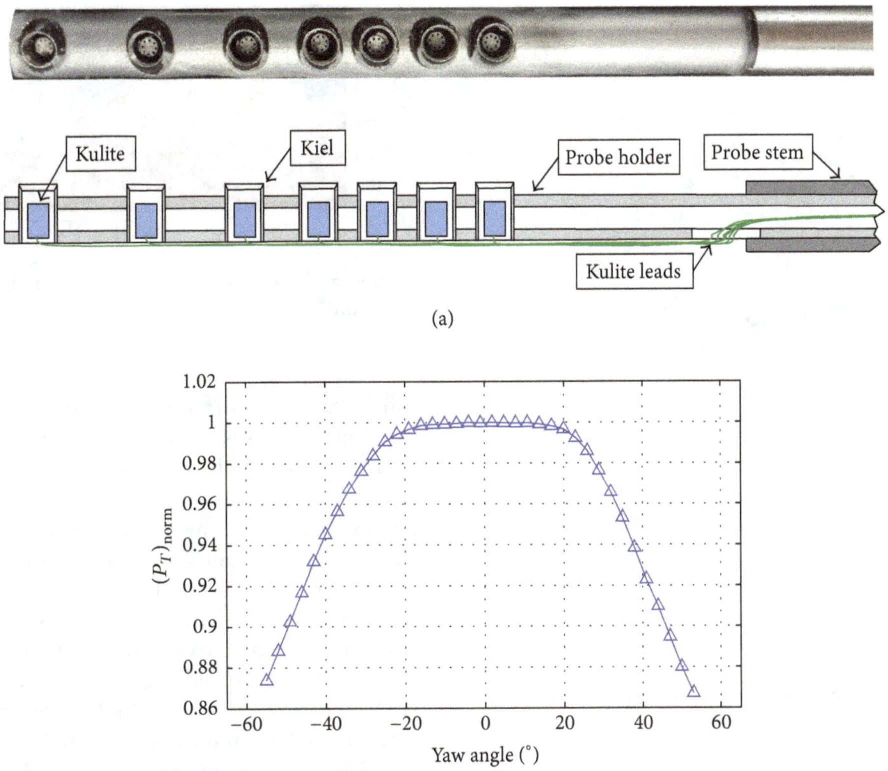

(b)

FIGURE 3: (a) Photograph and schematic of the unsteady total pressure rake and (b) the yaw response.

3. Steady and Unsteady Compressor Characteristics

This section presents the steady-state compressor characteristic and unsteady characteristics during two poststall events at 100% design speed. First, the steady-state normalized plenum pressure rise (ΔP) is shown as a function of flow coefficient (C_x/U) for ND Stage 04 at 100% design speed in Figure 4. The unstalled branch of the steady characteristic contains a steep negative slope. The stall point was found at approximately $C_x/U = 0.34$. Upon throttling the compressor beyond stall, the mass flow and pressure rise decrease abruptly. A black dashed line connects the last stable prestall point to the initial stall point. The rotating stall branch of the steady-state compressor characteristic below the value of $C_x/U = 0.29$ was approximated using data at lower speeds as discussed in Hickman and Morris [11]. This portion of the characteristic was used strictly for predictions of unsteady compressor characteristics and contains a nearly constant pressure rise as the mass flow decreases. The steady characteristic includes a stall/unstall hysteresis portion. Upon opening the valves, the characteristic deviates from the original path. The pressure rise remains relatively constant as the mass flow increases to approximately $C_x/U = 0.32$. This was found to be the critical recovery value for this compressor at 100% design speed. Upon opening the throttle valves beyond this point, the compressor recovered from rotating stall.

The measured unsteady compressor characteristics for a stall and surge case at 100% design speed are also shown in Figure 4. The markers along the measured curves indicate every 5 rotor revolutions. The unsteady events were recorded back-to-back with identical speeds and identical values of the upstream and downstream B Parameter ($B_U = 2.52$, $B_D = 1.71$). The single difference between the stall and surge case is the poststall throttle point. The ending throttle point for the stall case (blue) corresponded to the initial stall point. The compressor was "eased" into stall by closing the valves up to and not past the initial stall point. After the compressor stalled, the mass flow and pressure rise decreased past the initial stall point. The mass flow and pressure rise then increased, or rebounded, such that it did not surpass the critical recovery value of C_x/U. The compressor then entered stable stall.

For the surge case (red) in Figure 4, the poststall throttle point is denoted by the red dashed line. The compressor was throttled deeper into stall past the initial stall point. The decrease in pressure rise and mass flow are more dramatic. The rebounding motion of the compressor characteristic is such that it surpasses the critical recovery value of C_x/U and the compressor temporarily recovers from stall. The compressor then returns back to the initial stall point. Not shown in Figure 4 is that after the compressor is again brought to the stall point, another surge cycle begins similar to the first. A further analysis of the development of the stall cell and comparison between the stall and surge case will be discussed

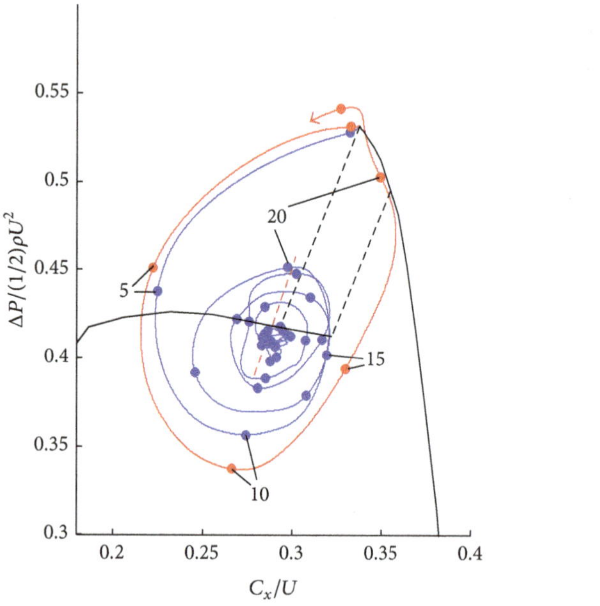

FIGURE 4: Steady and unsteady 100% corrected speed compressor characteristics.

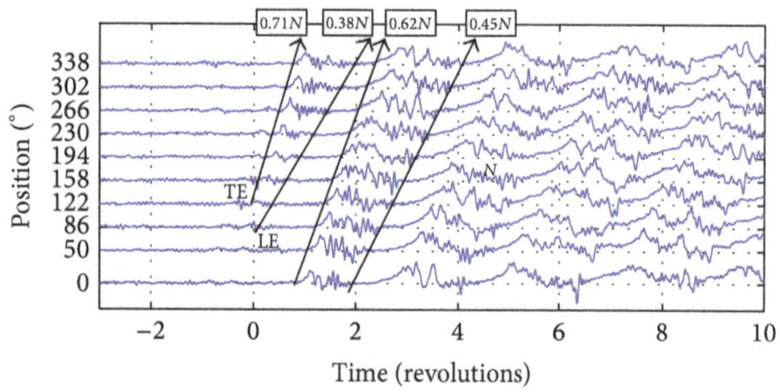

FIGURE 5: Time series from the circumferential array of static pressures upstream of the rotor leading edge.

in the following sections. All of the following data and results presented were acquired at the 100% design speed.

4. Stall Inception and Stall-Cell Development Measurements

4.1. Leading Edge Shroud Static Pressure Time Series. Stall inception events were recorded using the unsteady instrumentation described in Section 2.2. Figure 5 shows the pressure traces from the circumferential array of high-frequency pressure transducers during a typical stall inception event at 100% corrected speed. The abscissa is plotted in units of rotor revolutions and extends from −3 to 10 revolutions. The ordinate represents both the circumferential position and magnitude of the fluctuations at the position. Time series are low-pass filtered at 15 times the shaft rate (15N), just below the blade passing frequency such that the development of the rotating stall cell can be tracked. It was found that

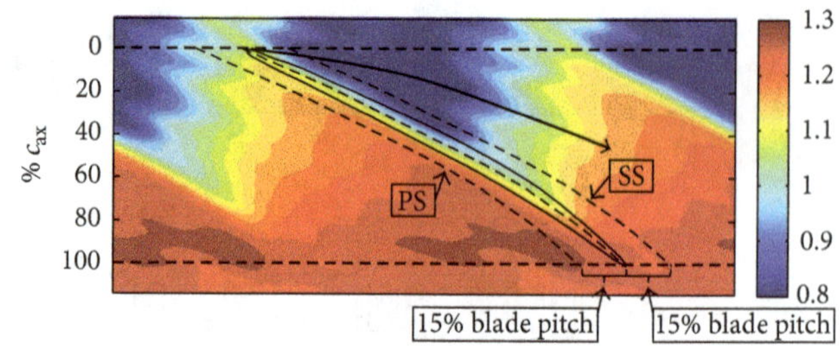

FIGURE 6: Ensemble average shroud static pressure at prestall conditions.

stall inception occurred in the same circumferential location repeatedly. This was most likely due to some slight asymmetry in the inlet flow path.

The origin of the time axis corresponds to the time when the fluctuations become greater than prestall fluctuations. At approximately revolution 0, a single stall cell or spike begins to form and rotate around the annulus. A static pressure rise occurs upstream of the blade row similar to Pullan et al. [18], which extends over the span of multiple blade passages. Initially, the trailing edge of the stall cell rotates at a rate of 0.71N and slows down in the laboratory frame as the stall cell increased in circumferential size. The leading edge rotates at 0.38N in the laboratory frame initially and increases rotational speed as the cell develops. These observations are consistent with short length scale, or "spike" type stall inception.

In the frame of the rotor the leading edge of the stall cell initially rotates at approximately 0.62N while the trailing edge rotates at only 0.29N. Thus, the stall cell grows on the leading edge side of the cell initially. Rotor passages at the leading edge of the stall cell stall at a faster rate relative to passages exiting the cell during the initial growth phase.

4.2. Prestall Shroud Static Pressure. To compare to stall inception measurements, the prestall phase-locked average shroud static pressure normalized by the inlet plenum pressure, P_{PU}, is presented in Figure 6. This contour corresponds to the shroud static pressure at the last stable operating point prior to stall inception. Evidence of the tip clearance vortex as discussed in Weichert and Day [5] can be seen by a low pressure core that extends from the suction side (SS) of the rotor leading edge towards the middle of the pressure side (PS) of the adjacent blade. The path is denoted by the black arrow in Figure 6.

4.3. Stall Inception. Prior to the earliest detectable signs by the circumferential array of high-frequency pressure transducers, evidence of stall inception can be seen within rotor passages. Figure 7 shows the shroud static (top) and rotor exit total (bottom) raw pressure time series. These data are measured using the OTR chord-wise array and unsteady total pressure rake described in Section 2.2. The centers of the contours in Figure 7 are at approximately revolution −0.8 at the 0° circumferential position. The raw data are displayed as a contour normalized by P_{PU}, where the ordinates are in space and the abscissa is in time. This gives an "instantaneous view" of the shroud static pressure and rotor exit total pressure,

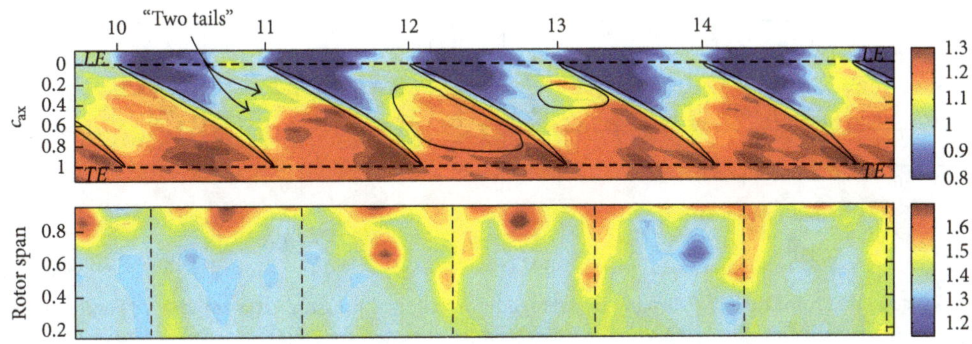

FIGURE 7: Instantaneous (space-time) shroud static and rotor exit pressure contour centered at revolution −0.8 of Figure 5 at the 0° circumferential position.

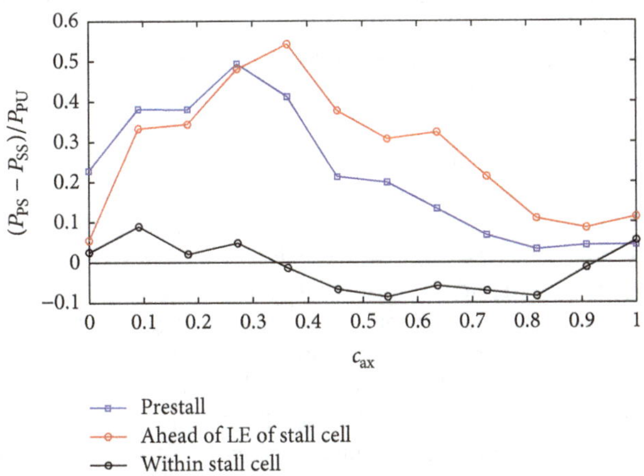

FIGURE 8: Rotor blade tip loading at prestall and poststall conditions.

assuming the flow field evolves slowly with respect to the blade passage time scale. Blade locations are overlaid for reference in the shroud contours. Approximate locations of the rotor wake are denoted with vertical dashed lines in the rotor exit contours. All instantaneous contours in this section are captured during the unsteady event shown in Figure 5, but the characteristics described in this section were found to be similar throughout the majority of stall inception events for this compressor.

Many repeatable features were found similar to Weichert and Day [5] such as the existence of the "two-tail" phenomenon. This feature can be seen in the shroud static pressure contour between blades 10 and 11 of Figure 7. This is an indication that the tip clearance vortex is disturbed. The two passages to the right show regions of low pressure. These regions are circled and indicate signs of prestall spikes. The blade loading for blades 12 and 13 decreases as a result of the low pressure regions. The rotor passage between blades 13 and 14 returns close to prestall values as compared to Figure 6.

The rotor exit pressure contours in Figure 7 show a large degree of unsteadiness from one passage to another. For this compressor, the vertical streaks of higher total pressure are typically associated with the blade wakes as these measurements are in the absolute frame. This contour shows no obvious signs of disturbed flow, and it will be shown that the values closely resemble prestall values for this moment in time.

4.4. Blade Lift and Spanwise Rotor Exit Total Pressure. To better quantify instantaneous values, the blade loading was approximated and compared to prestall values. As there was no unsteady instrumentation on the rotor to calculate loading on the blade, the blade loading was approximated from the shroud static pressure measurements from the OTR chord-wise array. The dashed line to the left of the blade in Figure 6 was defined as the sample location for the pressure on the PS of the rotor blade. The sample location was approximately 15% of a blade pitch from the mean camber line of the blade. A similar method was used to define the pressure on the SS of the rotor blade. The difference between the PS and SS at each axial location was assumed to be the blade loading at the tip of the rotor.

This method was used to determine the blade loading during stable prestall operation and poststall events. Figure 8 shows the average blade loading for prestall conditions and two poststall conditions. Curves in this figure are plotted as blade loading normalized by P_{PU} as a function of axial location. The blue curve in Figure 8 is the average prestall blade loading prior to stall inception. During prestall conditions, the rotor blade is more highly loaded towards the leading edge of the blade and exhibits the highest loading at approximately $0.27c_{ax}$.

The blade lift at the rotor tip was approximated during the steady prestall operation and poststall events. The blade lift was found by integrating the blade loading curve along the axial direction. For poststall events, blade lift was calculated as individual blades passed by the over-the-rotor chord-wise array. Figure 9(a) shows the blade lift at the 0° circumferential position during the stall inception event shown in Figure 5. Values are normalized by the mean of the prestall lift. The prestall lift can fluctuate from unsteadiness in the flow through the rotor. Statistics were performed on the lift in the prestall region immediately prior to stall inception. The grey band in Figure 9(a) represents values bounded by plus and minus one standard deviation from the mean of the prestall region. The spanwise average total pressure was also calculated and is shown in Figure 9(b). Values are normalized by the mean of the prestall rotor exit total pressure. The

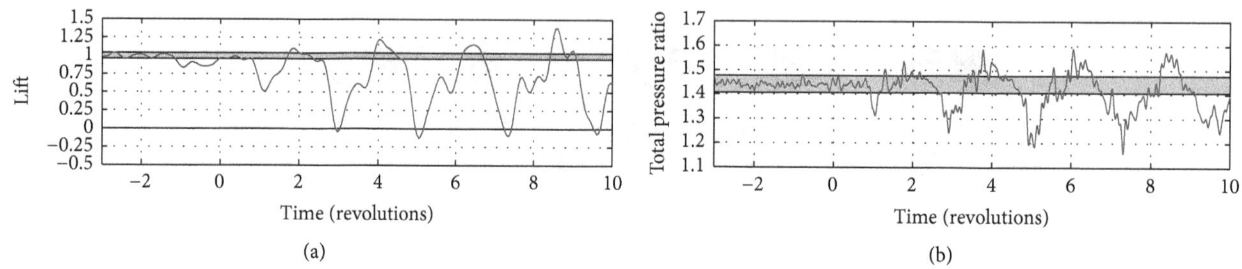

FIGURE 9: (a) Blade lift at rotor tip and (b) spanwise average rotor exit total pressure ratio at the 0° circumferential location.

spanwise average total pressure was found by averaging multiple pressures along the span measured by the unsteady total pressure rake at the 0° circumferential position. The grey band in Figure 9(b) also represents values bounded by plus and minus one standard deviation from the mean of the prestall region.

At approximately −0.8 revolutions, corresponding to the instantaneous contours in Figure 7, there is a statistically significant decrease in lift at the tip. At the same moment in time, the spanwise average total pressure ratio shows no deviation from the prestall. The first moment that the compressor experiences a significant decrease in total pressure is at approximately revolution 1.1, which corresponds to the first stall cell passing. By this time, the stall cell has grown in circumferential size and is clearly noticeable in Figure 5. This result agrees well with the work of Weichert and Day [5] that the first sign of stall inception spikes occurs within a rotor passage.

4.5. Stall-Cell Development. As the stall cell develops, blade lift and rotor total pressure fluctuate dramatically. The first instance the stall cell passes the OTR chord-wise array and unsteady total pressure rake is between revolutions 0.8 and 1.7. The trailing edge of the stall cell is at approximately 0.8 revolutions and the leading edge is at approximately 1.7 revolutions. Moving forward in time, as the trailing edge of the stall cell propagates to the 0° circumferential location at 0.8 revolutions the lift and total pressure ratio begin to decrease. By revolution 1, within the stall cell, the lift decreases to approximately 50% of the prestall value. The pressure ratio decreases to nearly 1.3. The stall cell has not yet fully developed, and the decrease in lift and pressure ratio are not as drastic as future stall cell passing. At revolution 1.7, ahead of the leading edge of the stall cell, the lift and pressure ratio are both slightly above prestall values. This suggests increased incidence and turning ahead of the stall cell.

It is useful to analyze the raw unsteady pressure time series to better visualize the growth of the stall cell. Figure 10 shows the instantaneous shroud static pressure and rotor exit total pressure contours during the first stall cell passing. Contours in Figure 10(a) are centered at revolution 0.9 and show the trailing edge of the stall cell. In this figure, passages to the left of blade 7 are exiting the stall cell and are in the process of recovering. The passages increase in blade loading and pressure ratio to the left of blade 7. Passages to the right of blade 7 are stalled with little blade loading and pressure ratio.

Figure 10(b) shows the leading edge of the stall cell. This contour is centered at revolution 1.1. Blade 12 exhibits a relatively high pressure difference across the blade in the shroud static pressure contour. The rotor exit contour shows pressures near prestall values. To the left of blade 12, the blade loading and rotor exit total pressure progressively decrease. Passages between blades 10 and 12 are entering the leading edge of the stall cell and are in the process of stalling. Passages to the left of blade 10 are stalled and show little pressure rise across the rotor.

Passages ahead of leading edge of the stall cell are shown in Figure 10(c), centered at revolution 1.7. These blades show an increase in blade loading and pressure ratio above the prestall values as discussed previously. These passages are overpressured and are approaching the stall cell.

After rotor revolution 2, the stall cell has become fully developed. Figure 9 shows the blade lift and pressure ratio for three more stall cell revolutions. Within the fully developed stall cell, the blade lift decreases to slightly below zero at revolutions 3, 5.1, and 7.3. The blade loading at approximately revolution 3 is shown in Figure 8 as the black curve. The blade loading between $0.35c_{ax}$ and $0.9c_{ax}$ is less than zero for this moment in time. This is evidence of localized back-flow through the rotor passage as the stall cell passes. The rotor total pressure ratio decreased as low as 1.2 within the stall cell (Figure 9(b)).

The blade lift increased significantly ahead of the stall cell at revolutions 4, 6.4, and 8.6. At revolution 8.6, blade lift increased approximately 40% above prestall values. The blade loading at approximately revolution 4.1 is shown in Figure 8 as the red curve. The loading from the rotor leading edge to $0.25c_{ax}$ is slightly below the average prestall loading. From $0.25c_{ax}$ to the trailing edge of the rotor, the blade shows increased loading compared to the prestall. The rotor exit total pressure also increased significantly ahead of the developed stall cell as high as approximately 1.59 as shown in Figure 9(b).

5. Stall and Surge Time Series

The development of the stall cell was remarkably similar between cases when the compressor entered stable rotating stall and when the compressor surged. This section will provide a detailed comparison between the two unsteady events shown on the left of Figure 4.

Both the stall and surge cases exhibited a sharp decrease in C_x/U and ΔP during the stall inception event. Figures 11(c) and 11(d) show a comparison of C_x/U and ΔP between the stall and surge cases. In these figures, the blue curve is

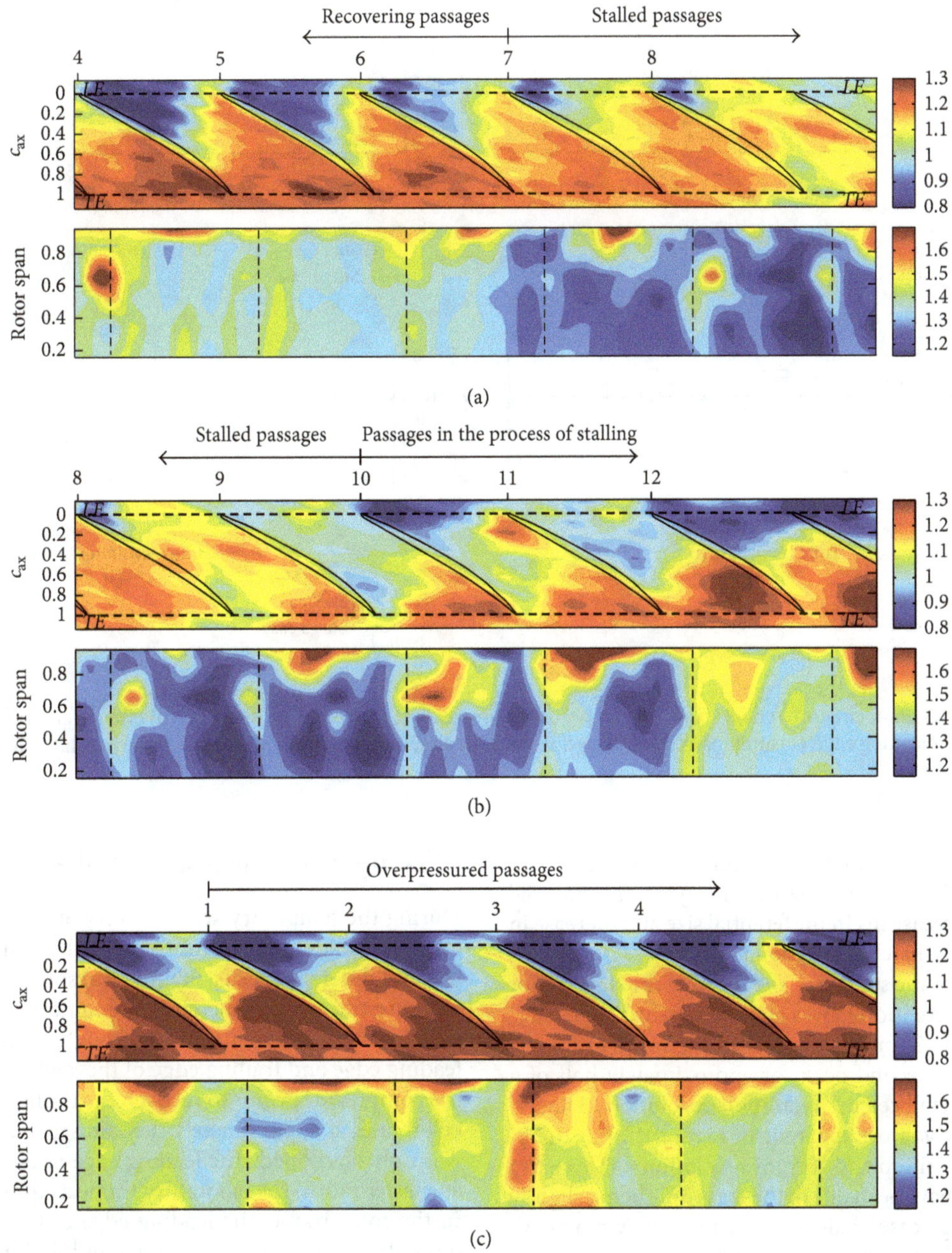

FIGURE 10: Instantaneous (space-time) shroud static and rotor exit pressure contours centered at revolution (a) 0.9, (b) 1.1, and (c) 1.8 of Figure 5 at the 0° circumferential position.

the 100% corrected speed stall case and the red curve is the 100% corrected speed surge case. The surge case shows a larger decrease in plenum pressure rise and mass flow compared to the stall case. The larger decrease produces a large enough return swing during the recovery to surpass the critical recovery value of C_x/U [11]. The critical recovery value of C_x/U is shown as the black line in Figure 11(c). The fluctuations in the stall case are just small enough such that the trajectory of the unsteady characteristic never surpasses the critical recovery value of C_x/U. This can be seen more clearly in Figure 4. The trajectory of the blue experimental curve approaches the critical recovery value of C_x/U on three occasions, but does not go above. It is the existence of a relatively large stall/unstall hysteresis branch in the steady-state characteristic that allows this to occur.

Figures 11(a) and 11(b) show the low-pass filtered time traces of shroud static pressure and upstream hot wires for both the stall and surge cases. These figures illustrate the similarities between the two stall inception events. From rotor revolutions 0 to 14 the traces are similar. Stall inception occurs at approximately the same circumferential location. The initial growth and rotation rates of the stall cell during the first few revolutions between the two cases are nearly identical. At approximately revolution 5, the speed of the stall cell was $0.44N$ in the laboratory reference frame. At this point, the stall cell was at its largest size and rotating the

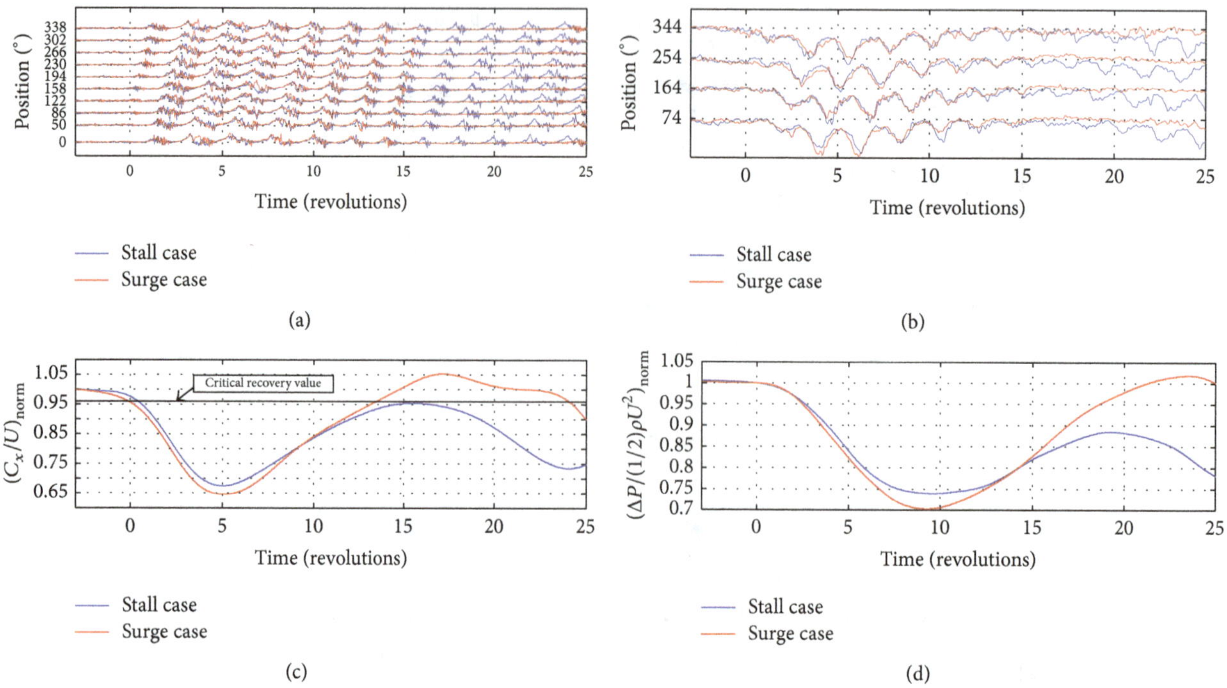

FIGURE 11: Time series of (a) circumferential array of static pressures upstream of the rotor leading edge, (b) circumferential array of hot wires upstream of the rotor leading edge, (c) annulus average mass flow, and (d) plenum-to-plenum pressure rise during stall (blue) and surge (red).

slowest. This coincides with the local minimum on the C_x/U trace in Figure 11(c). As the mass flow recovers, the stall cell then begins to decrease in circumferential size and increase in speed until approximately revolution 15 where C_x/U was at its local maximum for the stall case. The speed of the stall cell for the stall case at revolution 15 was approximately $0.52N$.

The shroud static pressure traces in Figure 11(a) for the surge case show a second peak for approximately half of a revolution of the stall cell between rotor revolutions 3 and 4. This second peak is most apparent in Figure 11(a) at the 0° circumferential position and at rotor revolution 4. This extra disturbance may have been due to the mass flow being slightly lower for the surge case. A decrease in mass flow may have increased the incidence angle possibly allowing for a second spike to briefly form at the leading edge of the main stall cell.

In the surge case, the critical recovery value of C_x/U was surpassed at approximately rotor revolution 13.5. It was at this point that the stall recovery process began for the compressor in the surge case. This is when the shroud static pressure traces and upstream hot-wire traces for the surge case begin to deviate from the stall case. With the compressor recovered from rotating stall, the plenum pressure rise recovered until the compressor was brought to the stall point again in the surge case. At revolution 24 of the surge case, a second stall inception event occurred and another surge cycle began. Figures 11(a) and 11(b) show how similar the stall inception events are for the stall and surge case. These results show it is purely the poststall throttling point in conjunction with the system dynamics that determine whether surge will grow from rotating stall. For a further analysis of the rotating stall case see Hickman and Morris [19].

6. Temporary Surge and Stall Recovery

During the temporary stall recovery at the end of the surge cycle, the leading edge and trailing edge of the stall cell will travel at different speeds. Figure 12(a) shows the shroud static pressure traces from the circumferential array for the surge case from rotor revolutions 10 through 18. The speeds of the leading edge and trailing edge of the stall cell are labeled. As the compressor begins to recover, the trailing edge of the stall cell rotates at approximately the same speed as when the cell was fully developed. The leading edge of the stall cell rotates at a faster rate in the laboratory frame as the cell is shrinking. In the rotor frame, the leading edge of the stall cell rotates more slowly until the leading edge and trailing edge meet and the stall cell disappears. This is very similar to a compressor recovering from stable rotating stall.

Data were acquired during the recovery from stable rotating stall. Figure 13 shows the unsteady characteristic for a rotating stall and stall recovery event. The markers along the solid blue curve represent every 2 rotor revolutions. The zero location was chosen as the approximate point where stall cell dies out. The unsteady characteristic during the stall event is shown as the blue dashed line for reference in Figure 13. In this case the compressor entered stable rotating stall by closing the valves. The valves were then opened and the compressor recovered from rotating stall. The solid blue line is the unsteady characteristic during the stall recovery event. The characteristic starts at the initial stall point and at approximately rotor revolution −2, the unsteady characteristic surpasses the critical recovery value of C_x/U, and the stall recovery process begins. The unsteady characteristic

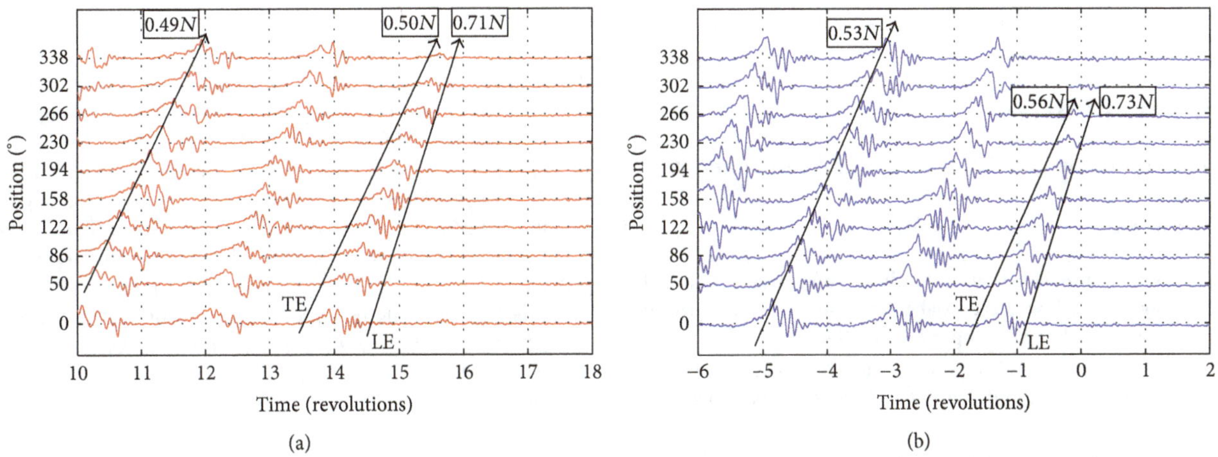

FIGURE 12: Time series of circumferential array of static pressures upstream of the rotor leading edge during (a) temporary recovery at the end of a surge cycle and (b) stall recovery with an opened outlet valve.

transiently attaches to the unstalled branch of the steady-state compressor characteristic.

In Figure 12(b) are the leading edge traces for the stall recovery event shown in Figure 13. The time axis corresponds to the labeled markers in Figure 13. The speeds of the trailing edge and leading edge of the stall cell during stall recovery are similar to the temporary stall recovery at the end of a surge cycle. The trailing edge rotates at approximately the same speed as the developed stall cell, while the stall cell decays at the leading edge. Figure 12 illustrates that stall recovery and temporary surge recovery are similar events.

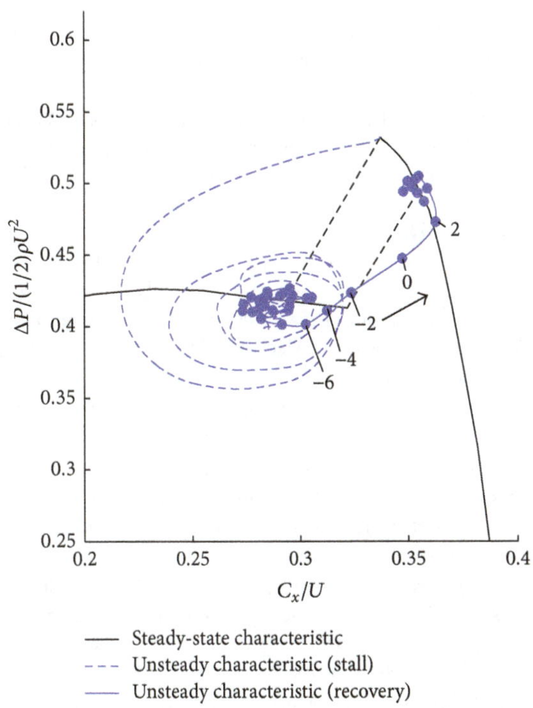

FIGURE 13: Unsteady compressor characteristic during stall and stall recovery.

7. Conclusions

Observations from unsteady measurements during stall inception, stall-cell development, rotating stall, and surge were presented. The first signs of stall inception spikes were found within rotor passages. As the stall cell grows after stall inception, the leading edge of the cell will rotate at a higher rate than the trailing edge in the rotor frame suggesting that the stall cell initially increases in size at the leading edge.

Blade loading at the tip and rotor exit total pressure ratio decrease dramatically within the stall cell. Blade lift can decrease to negative values within the stall cell suggesting back-flow. Ahead of the leading edge of the stall cell, the blade lift and rotor exit total pressure ratio surpass prestall values. This is a sign of higher incidence and turning ahead of the stall cell.

The compressor investigated stalls with a single full span stall cell. A single stall cell is also present during surge. There is a striking resemblance between the development of the stall cell for rotating stall and surge cases. The initiation location, initial speed, and initial growth rate of the stall cell are nearly indistinguishable between stall and surge cases. The unsteady event becomes a surge event when the critical recovery value of C_x/U is surpassed.

Temporary stall recovery at the end of a surge cycle and recovery from rotating stall are similar events. The trailing edge of the stall cell will rotate at the approximate speed as the fully developed stall cell, while the leading edge decreases in rotational speed in the rotor frame. During recovery from compressor stall, the stall cell decays at the leading edge.

Compressor surge was long thought to be an axisymmetric phenomenon. This research has shown that while global fluctuations in mass flow and pressure are present during surge events, local asymmetric instabilities similar to those seen during rotating stall may also take place. Fully understanding these instabilities may lead to new methods for delaying or mitigating their effects.

Conflicts of Interest

The authors declare that they have no conflicts of interest.

References

[1] T. R. Camp and I. J. Day, "A study of spike and modal stall phenomena in a low-speed axial compressor," *Journal of Turbomachinery*, vol. 120, no. 3, pp. 393–401, 1998.

[2] H. D. Vo, "Rotating stall suppression in axial compressors with casing plasma actuation," *Journal of Propulsion and Power*, vol. 26, no. 4, pp. 808–818, 2010.

[3] J. D. Cameron, M. A. Bennington, M. H. Ross et al., "The influence of tip clearance momentum flux on stall inception in a high-speed axial compressor," *Journal of Turbomachinery*, vol. 135, no. 3, Article ID 051005, 2013.

[4] C. S. Tan, I. Day, S. Morris, and A. Wadia, "Spike-type compressor stall inception, detection, and control," *Annual Review of Fluid Mechanics*, vol. 42, pp. 275–300, 2010.

[5] S. Weichert and I. Day, "Detailed measurements of spike formation in an axial compressor," *Journal of Turbomachinery*, vol. 136, no. 5, Article ID 051006, 2013.

[6] K. Yamada, H. Kikuta, M. Furukawa, S. Gunjishima, and Y. Hara, "Effects of tip clearance on the stall inception process in an axial compressor rotor," in *Proceedings of the ASME Turbo Expo 2013: Turbine Technical Conference and Exposition (GT '13)*, June 2013.

[7] J. D. Cameron, S. C. Morris, S. T. Barrows, and J.-P. Chen, "On the interpretation of casing measurements in axial compressors," in *Proceedings of the 2008 ASME Turbo Expo*, pp. 653–665, American Society of Mechanical Engineers, New York, NY, USA, June 2008.

[8] I. J. Day, "Stall, surge, and 75 years of research," *Journal of Turbomachinery*, vol. 138, no. 1, Article ID 2478223, 2015.

[9] E. M. Greitzer, "Surge and rotating stall in axial flow compressors—part i: theoretical compression system model," *Journal of Engineering for Gas Turbines and Power*, vol. 98, no. 2, pp. 190–198, 1976.

[10] E. M. Greitzer, "Surge and rotating stall in axial flow compressors—part ii: experimental results and comparison with theory," *Journal of Engineering for Gas Turbines and Power*, vol. 98, no. 2, pp. 199–211, 1976.

[11] A. R. Hickman and S. C. Morris, "Stall and Surge Characteristics of a Two-Volume Compression System," in *Proceedings of the 52nd AIAA/SAE/ASEE Joint Propulsion Conference*, Salt Lake City, UT, USA, 2016.

[12] I. J. Day, "Axial compressor performance during surge," *Journal of Propulsion and Power*, vol. 10, no. 3, pp. 329–336, 1994.

[13] H. Schoenenborn and M. de Vries, "Aeroelasticity at reversed flow conditions-part III: reduction of surge loads by means of intentional mistuning," *Journal of Turbomachinery*, vol. 135, no. 3, Article ID 041009, 2013.

[14] T. Abe, H. Mitsui, and Y. Ohta, "Coexisting phenomena of surge and rotating stall in an axial flow compressor," *Journal of Thermal Science*, vol. 22, no. 6, pp. 547–554, 2013.

[15] A. R. Hickman, *The Transition From Rotating Stall to Surge in an Axial Compressor [Ph.D. thesis]*, University of Notre Dame-Turbomachinery Laboratory, 2017.

[16] H. H. Bruun, *Hot-wire Anemometry: Principles and Signal Analysis*, Oxford University Press, 1995.

[17] J. Lepicovsky and E. P. Braunscheidel, "Measurement of flow pattern within a rotating stall cell in an axial compressor," NASA/TM—2006-214270, NASA, 2006.

[18] G. Pullan, A. M. Young, I. J. Day, E. M. Greitzer, and Z. S. Spakovszky, "Origins and structure of spike-type rotating stall," in *Proceedings of the ASME Turbo Expo 2012: Turbine Technical Conference and Exposition (GT '12)*, pp. 2567–2579, June 2012.

[19] A. R. Hickman and S. C. Morris, "Characteristics of stable rotating stall cells in an axial compressor," in *Proceedings of ASME Turbo Expo 2017: Turbomachinery Technical Conference and Exposition*, pp. GT2017–GT64901, 2017.

12

The Impact of Convective Fluid Inertia Forces on Operation of Tilting-Pad Journal Bearings

Thomas Hagemann, Christopher Zeh, Maximilian Prölß, and Hubert Schwarze

Institute of Tribology and Energy Conversion Machinery, Clausthal University of Technology, Leibnizstraße 32, 38678 Clausthal-Zellerfeld, Germany

Correspondence should be addressed to Thomas Hagemann; hagemann@itr.tu-clausthal.de

Academic Editor: Fadi Dohnal

This paper presents a combination of experimental data, CFD analyses, and bearing code predictions on emergence of convective inertia fluid forces within the lube oil flow of tilting-pad journal bearings. Concordantly, experimental data and CFD analyses show a significant rise of local pressure at the transition between inlet and leading edge of tilting-pad, especially for high-speed applications with surface speeds up to 100 m/s. This effect can be related to convective inertia forces within fluid flow as cross-sections and flow character rapidly change at the pad entrance. An energy-based approach is implemented in the bearing code in order to provide enhanced boundary conditions for Reynolds equation considering this effect. As a result, predictions of bearing code achieved significant improved correlation with measured pressure distributions and CFD-data. Further, beside the local influence, a nonnegligible impact on characteristic parameters of bearing operation such as maximum temperature and stiffness and damping coefficients is observed. Finally, the results are critically analyzed and requirements to gain more distinct and reliable data are specified.

1. Introduction

The theoretical analysis of the operating characteristic of journal bearings commonly focuses on modelling fluid flow within the lubricant gap by using Reynolds equation. Here, fluid inertia forces are negligible for many practical applications as viscous forces are much higher than inertia ones. However, Reynolds equation is not valid within the whole bearing. For example, within lubricant pockets, boundary layer flow occurs [1–3] and faces combined Poiseuille and Couette flow on the leading edge of the downstream pad [3]. The significant local changes of flow cross-section and flow character potentially generate convective inertia forces as local fluid velocity gradients exist. Nevertheless, most analyses neglect this effect, though it should be of importance especially for tilting-pad bearings due to their tilting degree of freedom and comparably low angular pad span.

This paper presents experimental evidence and theoretical investigations on inertia effects occurring in transition region between pad surfaces and inlet area and concentrates on their influence on local pressure distribution. Further, the impact on selected characteristic parameters of bearing operation is studied.

2. Materials and Methods

2.1. Bearing Code Model. The theoretical bearing model is based on a two-dimensional (2D) extended and generalized Reynolds equation and a full three-dimensional (3D) energy equation for bearing, oil film, and journal. Due to the high rotor speed, the temperature distributions in the journal are assumed to be 2D as the temperature does not vary in circumferential direction. The foundation of the entire bearing model used for the upcoming analyses is comprehensively described in [4–6]. Therefore, the consecutive explanations are reduced due to their relevance for the particular investigations performed here.

Within the lubricant film following extended and generalized Reynolds equation is applied at stationary operation:

$$\frac{\partial}{\partial x}\left(\frac{\rho F_2}{K_x}\frac{\partial p}{\partial x}\right)+\frac{\partial}{\partial z}\left(\frac{\rho F_2}{K_z}\frac{\partial p}{\partial z}\right) = U\frac{\partial}{\partial x}\left[\rho\left(h-\frac{F_1}{F_0}\right)\right]. \tag{1}$$

Equation (1) considers the 3D viscosity distribution due to variable temperature in all three space directions of the fluid film by the following factors [7]:

$$F_0 = \int_0^h \frac{dy}{\eta},$$
$$F_1 = \int_0^h \frac{y}{\eta}dy, \tag{2}$$
$$F_2 = \int_0^h \frac{y}{\eta}\left(y-\frac{F_1}{F_0}\right)dy.$$

Additionally, the effects of local turbulent flow or Taylor vortices are included using the coefficients K_x and K_z [3]. Constantinescu and Galetuse provide an appropriate model to consider inertia effects in Reynolds equation [8]. It assumes that the shape of flow profiles and, therefore, wall shear stresses remain unchanged and do not alter by inertia forces. Consequently, (2) can be extended by inertia terms according to [8]

$$\begin{aligned}\frac{\partial}{\partial x}\left(\frac{\rho F_2}{K_x}\frac{\partial p}{\partial x}\right)+\frac{\partial}{\partial z}\left(\frac{\rho F_2}{K_z}\frac{\partial p}{\partial z}\right) &= \frac{\partial}{\partial x}\left(\rho U\left(h-\frac{F_1}{F_0}\right)\right)\\ &-\frac{\partial}{\partial x}\left(\frac{\rho F_2}{K_x h}\left(\frac{\partial I_{xx}}{\partial x}+\frac{\partial I_{xz}}{\partial z}\right)\right)\\ &-\frac{\partial}{\partial z}\left(\frac{\rho F_2}{K_z h}\left(\frac{\partial I_{xz}}{\partial x}+\frac{\partial I_{zz}}{\partial z}\right)\right).\end{aligned} \tag{3}$$

Herein, the integrals of inertia forces are

$$I_{xx} = \int_0^h u^2 dy = \alpha_1 U_m^2 h - \alpha_2 U_m U h + \alpha_3 U^2 h,$$
$$I_{xz} = \int_0^h uw\,dy = \alpha_1 U_m W_m h - \frac{\alpha_2}{2} W_m U h, \tag{4}$$
$$I_{zz} = \int_0^h w^2 dy = \alpha_1 W_m^2 h.$$

The parameters α_1, α_2, and α_3 are specific for the flow regime [8]. In addition to laminar and turbulent flow, the bearing model considers the transition range between laminar and turbulent flow, where Taylor vortices appear and no complete turbulent flow is present. Here, a linear interpolation between the parameters of laminar and turbulent flow is used for the turbulence model as well as for the inertia forces in order to stabilize the iterative calculation procedure by preventing step phenomena:

Laminar flow ($\mathrm{Re}_{\mathrm{eff}} \leq \mathrm{Re}_{\mathrm{cr}}$) is

$$\alpha_1 = \frac{6}{5},\quad \alpha_2 = \frac{1}{5},\quad \alpha_3 = \frac{2}{15}. \tag{5}$$

Transition ($\mathrm{Re}_{\mathrm{cr}} < \mathrm{Re}_{\mathrm{eff}} \leq 2\cdot\mathrm{Re}_{\mathrm{cr}}$) is

$$\begin{aligned}\alpha_1 &= \frac{7}{5}-\frac{1}{5}\cdot\frac{\mathrm{Re}_{\mathrm{eff}}}{\mathrm{Re}_{\mathrm{cr}}},\\ \alpha_2 &= \frac{2}{5}-\frac{1}{5}\cdot\frac{\mathrm{Re}_{\mathrm{eff}}}{\mathrm{Re}_{\mathrm{cr}}},\\ \alpha_3 &= \frac{4}{15}-\frac{0.885}{(2\cdot\mathrm{Re}_{\mathrm{cr}})^{0.367}}-\left(\frac{2}{15}-\frac{0.885}{(2\cdot\mathrm{Re}_{\mathrm{cr}})^{0.367}}\right)\cdot\frac{\mathrm{Re}_{\mathrm{eff}}}{\mathrm{Re}_{\mathrm{cr}}}.\end{aligned} \tag{6}$$

Turbulent flow ($\mathrm{Re}_{\mathrm{cr}} < \mathrm{Re}_{\mathrm{eff}} \leq 2\cdot\mathrm{Re}_{\mathrm{cr}}$) is

$$\alpha_1 = 1,\quad \alpha_2 = 0,\quad \alpha_3 = \min\left(\frac{0.885}{\mathrm{Re}_{\mathrm{eff}}^{0.367}},\frac{2}{15}\right). \tag{7}$$

The modified Reynolds equation (3) is solved using Elrod's algorithm [9] neglecting compressibility of the lubricant.

Assuming a boundary layer character of fluid flow in the inlet region the radial distribution of circumferential velocities can be approximated by third-order polynomial approach based on the local layer thickness. Mittwollen [1] proposes a model to predict local boundary layer thickness using the isoviscous boundary layer theory [10]. It is based on Blasius' solutions of Prandtl's boundary layer equations. The one-percent boundary layer thickness of a lengthwise flow of a plane plate can be determined by [10]

$$\delta = 5\cdot\sqrt{\frac{\nu\cdot x}{U_\infty}}. \tag{8}$$

Considering the pad trailing edge oil flow as a boundary condition at the beginning of the layer, the local laminar boundary layer thickness can be predicted by [1, 10]

$$h_\delta = 5\cdot\sqrt{\frac{{C_R}^2\cdot\varphi_\delta\cdot\eta_m^*}{\mathrm{Re}^*}+\left(\frac{h_E\cdot\Theta_E}{5}\right)^2}. \tag{9}$$

Here φ_δ is the angular span of the layer. The transition between laminar and turbulent flow in the inlet region occurs at a Reynolds number of $\mathrm{Re}_{\delta,\mathrm{eff,cr}} = 1511$. Flow transition is a local phenomenon, and, consequently, either the whole inlet

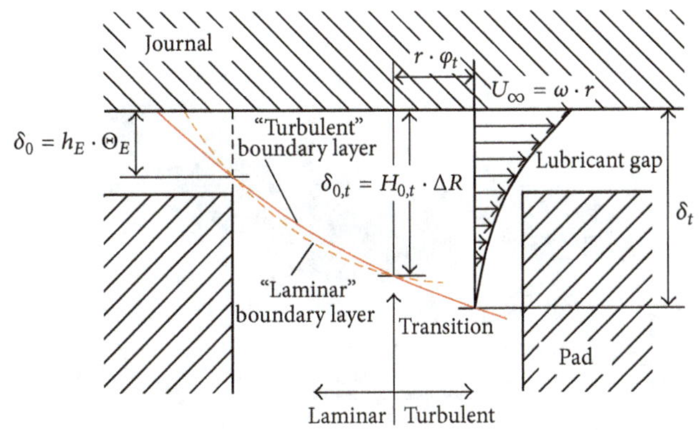

FIGURE 1: Boundary layer in the inlet region at laminar and turbulent flow according to [1].

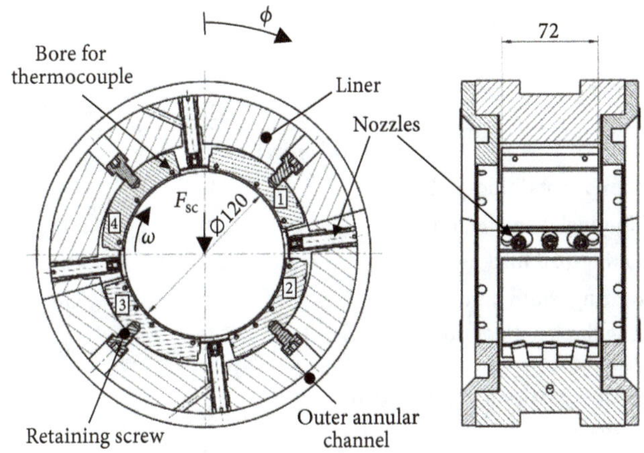

FIGURE 2: Tilting-pad bearing (TPB120).

region or a certain part of it can be turbulent if the criterion $Re_{\delta,eff,cr} > 1511$ is fulfilled. Thus, the initial thickness of the turbulent boundary layer can be equal to the effective filled gap of the upstream pad or it matches the laminar boundary layer thickness in case of transition within the inlet region. A power ansatz is used to approximate the velocity profile of the turbulent boundary layer. The 1/7-power ansatz using additional empirical data leads to [10]

$$\delta_t = 0{,}37 \cdot x \cdot Re_x^{-1/5}. \quad (10)$$

Here, Re_x is the Reynolds number calculated using the virtual length of the turbulent boundary layer. Introducing the nondimensional thickness $H = h/C_R$ provides a consistent expression for the turbulent boundary layer thickness based on all previously mentioned boundary conditions [1]:

$$h_{\delta_t} = 0{,}37 \cdot \left[\frac{\varphi_t}{\psi} + \left(\frac{H_E \cdot \Theta_E}{0{,}37} \cdot \left(\frac{Re}{\eta_m^*}\right)^{1/5}\right)^{5/4}\right]^{4/5} \left(\frac{Re}{\eta_m^*}\right)^{-1/5} \cdot C_R. \quad (11)$$

Figure 1 presents the basic formation of laminar and turbulent boundary layer. Dissipation in the inlet region can be determined by the wall shear stress according to [1, 10].

2.2. Investigated Bearings. Investigations are conducted for two tilting-pad journal bearings. The first one depicted in Figure 2 is a four-pad bearing with an eccentrical rocker-pivot typically applied in smaller turbines or compressors. Oil is supplied via three nozzles between each pad. The clearance between journal and baffle is about 3.5 times the minimum radial clearance of the lubricant gap to provide a nearly unrestricted discharge oil flow. Moreover, the baffles contain several bores in order to reduce the side flow resistance. References [11–13] provide a description of the test rig that the bearing is investigated with.

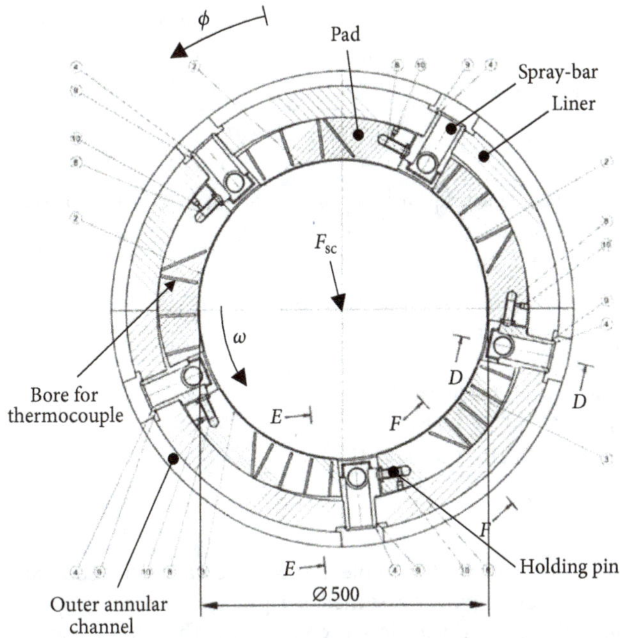

FIGURE 3: Tilting-pad bearing (TPB500).

The second bearing is a flooded lubricated five-pad bearing for large turbine applications already investigated in [4–6] and shown in Figure 3. The bearing features an elliptical contact between pads and liner with a pivot offset of 0.6. Further, sealing baffles with a radial clearance of $C_{R,baffle}$ = 1.0 mm are arranged on both bearing sides.

Figures 2 and 3 include a description of the load situation, the origin of the angular coordinate, and conventions for pad declarations that the subsequent results are related to.

The key parameters of the analysis and the geometry of the bearings are listed in Table 1. Measurement data for the two bearings is provided by previous investigations [5, 6, 12].

2.3. CFD Analyses. CFD analyses are conducted for TPB120 bearing. The model consists of a three-dimensional fluid and

TABLE 1: Key parameters of analysis.

Parameter	Value TPB120	Value TPB500
Geometrical properties		
Number of tilting-pads	4	5
Nominal diameter, mm	120	500
Tilting-pad thickness, mm	20	72.5
Bearing width, mm	72	350
Angular length of tilting-pads, degrees	70	56
Relative position of tilting-pad support	0.6	0.6
Radial clearance, μm	108	300
Preload	0.50	0.23
Static analysis parameters		
Bearing load, kN	175–438	175–438
Rotational speed, rpm	6000–15000	1500–3000
Lubricant supply temperature, °C	50	50
Lubricant flow rate, l/s	1.5	7.0
Lubricant properties		
Lubricant	ISO VG 32	ISO VG 32
Lubricant density kg/m^3	865 @ 40°C	844 @ 40°C
Lubricant specific heat capacity kJ/(kg K)	2.0 @ 20°C	2.17 @ 70°C
Thermal boundaries and material properties		
Lubricant thermal conductivity, W/(m K)	0.13	0.13
Pad and journal thermal conductivity, W/(m K)	45	45
Pad ambient temperature, °C	60	65

structure part and is implemented in commercial software. The structure part is assumed as rigid, and accordingly, only its influence on heat flow is considered. Thermomechanical deformation and radial pad deflection due to pivot elasticity $s(\phi, z)$ are imported from the bearing code and determined by approximate formulas [4]. The local film thickness of pad number i is described by

$$h(\phi, z)_i = C_R \cdot \left(\frac{1}{1-m_i} - \frac{m_i}{1-m_i} \cdot \cos(\phi - \varphi_{F,i}) - \frac{\delta_i}{\psi} \cdot \left(1 + \psi_{\text{pad},i}\right) \cdot \sin(\phi - \varphi_{F,i}) + \varepsilon \cdot \cos(\phi - \beta) \right) + s(\phi, z). \quad (12)$$

Herein the preload of the pad m and the relative pad thickness ψ_{pad} are defined as

$$m = 1 - \frac{C_R}{R_B - r}, \qquad \psi_{\text{pad}} = \frac{t_B}{r}. \quad (13)$$

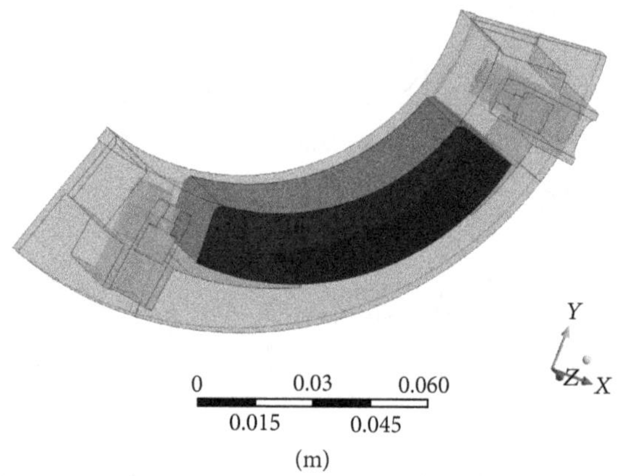

FIGURE 4: Fluid-model (light grey) and structure-model (dark grey) for CFD analyses.

Equation (12) equals the expression applied in the bearing code. Due to limited computational capacities, many investigations are performed with a simplified model consisting of one pad and an upstream and downstream inlet region. This model shown in Figure 4 consists of approximately 1.4 million cells and assumes axial symmetry to the middle plane. At the selected fluid-structure interfaces the grid is refined in order to model heat transfer within the small temperature boundary layer due to high Prandtl numbers reliably. As investigations actually focused also on heat transfer among surfaces directly facing the inlet region in circumferential direction conjugate heat transfer is defined here in addition to the sliding surface. To reduce deviations between models of bearing code and CFD analyses heat transfer coefficients are set on the backside and the lateral boundary surfaces of the pad. Further, geometry of the bearing with its thin lubricant film region and comparably high circumferential and axial length scales provides negative effects for the grid quality that could be measured by comparably high aspect ratios. Therefore, a comprehensive verification procedure is conducted to determine a robust grid resolution. The origin of axial orientated z-axis is located on the lateral edge of the pad.

A key objective of CFD-analysis is to investigate special phenomena that are not measurable reliably in order to develop models that are suitable to be integrated into the time efficient bearing code. A major benefit of the CFD-model approach described above is that the model can easily be adapted to the bearing model by boundary conditions. Hence, deviations between predicted results of CFD analyses and bearing code reduce to a limited number of physical phenomena as several other effects that are present in a complete bearing model are suppressed. These physical phenomena are expected to be captured more precisely by CFD analyses or not included in the bearing code.

The inlet boundary conditions for CFD analyses visible by black arrows in Figure 5 are imported from the bearing code. These include temperature and flow velocity distributions transferred to the CFD grid by interpolation at the trailing edge of the upstream pad as well as in the inlet of the nozzles.

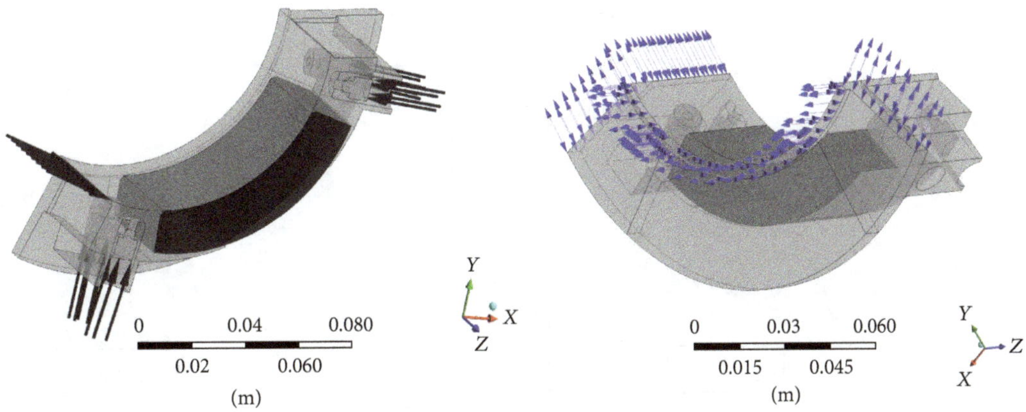

(a)

(b)

FIGURE 5: Inlet (a) and outlet (b) boundary conditions for CFD analyses.

At the axial outlet between baffle and journal and the remaining flow cross-sections in circumferential direction opening boundary conditions are defined which are noticeable by blue arrows in Figure 5(b).

The remaining surfaces represent walls with defined heat transfer coefficients. Furthermore, journal speed and temperature match the ones of the bearing code.

3. Results and Discussion

In a first step, a verification procedure between bearing code and CFD-analysis is conducted. For this purpose, two approaches are applied. In the first one, all parameters of the local film thickness function (12) are set equal to produce equal film thickness in the pad area in both software packages. Secondary, an iterative procedure is implemented in CFD-analysis modifying journal eccentricity and tilting angle until mechanical equilibrium for the same load on pad as determined by the bearing code is reached. The investigations indicated that the latter one produces better comparable results. The constraints of the former one prove too restrictive and generate different results, though differences concerning tilting angle and eccentricity are comparably small. Figure 6 contains a comparison of central oil film pressure distributions for pad number 3. A significant difference can be observed at the trailing edge of the pad, where CFD analyses show a step-shaped increase of pressure from the level of the inlet region where pressure equals the ambient one.

The interfaces between the inlet region and the pad represent nearly stepwise changes of flow cross-section, which are only reduced by bevels or curvatures applied at the leading and trailing edge of the pad. Therefore, influence of fluid inertia forces at the entrance and exit of the lubricant gap in circumferential direction have to be expected with high probability. According to the previously described model for the fluid flow in the inlet region, boundary layer thickness corresponds with the local film thickness at the trailing edge of the pad. Consequently, velocity gradients are small and concordantly inertia forces are comparably low. This assumption seems reasonable looking at trailing edge conditions in Figure 7 as thin boundary layer overflows a section of low fluid velocities without significant disturbances. However, at the leading edge of downstream pad significant velocity gradients occur. Therefore, maximum pressure within this region is located in the upper right corner comparable to lid driven cavity flow. Black arrows indicate the mayor flow directions in Figure 7(b).

Beside the complex flow in transition area between pad and inlet boundary layer flow is present in the inlet and combined Poiseuille and Couette flow on the pad as described earlier and illustrated in Figure 8.

A simple approach predicting entrance pressure at the leading edge and appropriate to implement in a bearing code based on Reynolds equation can be derived from the energy theorem first proposed by Pan [14] and comprehensively discussed by Constantinescu [15]. The radial distribution of local circumferential velocity $u_{\mathrm{bl},\phi}$ within boundary layer is well approximated by a third-order polynomial approach, as local boundary layer thickness is known from (9) or (11). Using the subsequent boundary conditions

$$\begin{aligned} u_{\mathrm{bl},\phi}\left(y'=0\right) &= U, \\ u_{\mathrm{bl},\phi}\left(y'=\delta\right) &= 0, \\ \frac{\partial u_{\mathrm{bl},\phi}}{\partial y'}\left(y'=\delta\right) &= 0, \\ \frac{\partial^2 u_{\mathrm{bl},\phi}}{\partial y'^2}\left(y'=\delta\right) &= 0, \end{aligned} \tag{14}$$

provides the polynomial coefficients in order to determine local fluid velocity distribution in radial direction:

$$u_{\mathrm{bl},\phi}\left(\phi, y', z\right) = U \cdot \left(1 - 3 \cdot y' + 3 \cdot y'^2 - y'^3\right). \tag{15}$$

Consequently, local mean velocity of the boundary layer facing the lubricant film results from

$$\overline{u}_{\mathrm{bl},\phi}\left(\phi, z\right) = \frac{1}{h} \cdot \int_0^h u_{\mathrm{bl},\phi}\left(\phi, y', z\right) dy'. \tag{16}$$

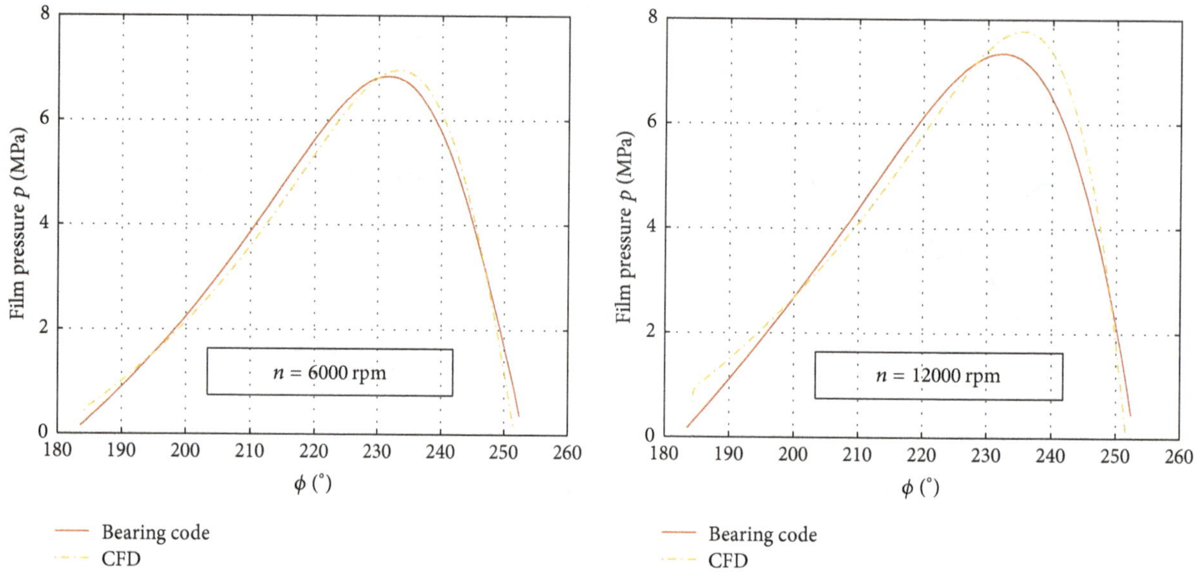

FIGURE 6: Comparison of predicted pressure distributions on pad 3 (TPB120, p_q = 2.0 MPa, Q = 90 l/min, T_{sup} = 50°C, z = 36 mm).

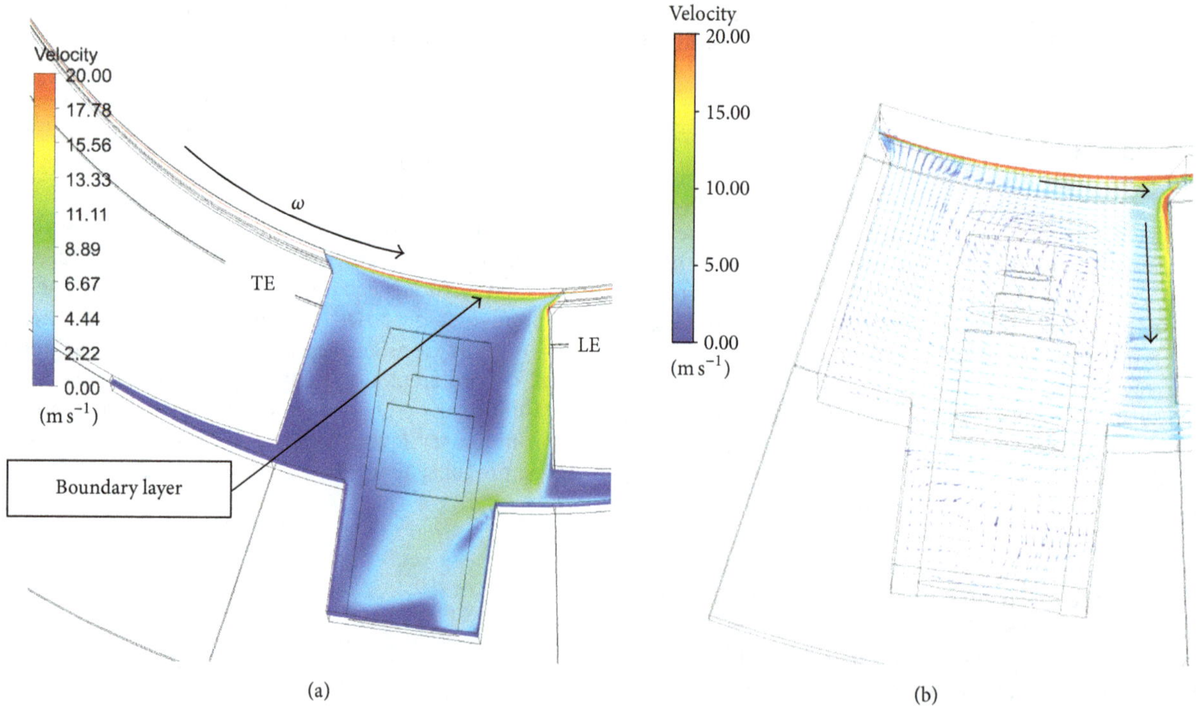

FIGURE 7: Absolute fluid velocities in the inlet region between pads 2 and 3 (TPB120, n = 12000 min^{-1}, p_q = 2,0 MPa, Q = 90 l/min, T_{sup} = 50°C, and z = 30 mm).

Considering the mean velocity at the leading edge of pad region $\overline{u}_{\text{pad,le},\phi}$ the pressure increase can be derived from energy approach:

$$\Delta p(z) = \frac{\rho}{2}\left(\overline{u}_{\text{bl},\phi}^2 - \overline{u}_{\text{pad,le},\phi}^2\right). \tag{17}$$

The lower limit of $\overline{u}_{\text{bl},\phi}$ is $U/2$ if the boundary layer thickness matches the film thickness. Boundary layer thickness increases with a reduction of speed. Consequently, mean relative velocity in the boundary layer facing the leading edge gap $\overline{u}_{\text{bl},\phi}/U$ increases. However, rising wall velocity U implies an increasing difference of the quadratic velocities in (17), and therefore, it results in higher values of Δp. The previously described assumptions neglect that oil which is not transported into the lubricant film meets inertia forces as wall boundary condition provokes a change of the flow direction.

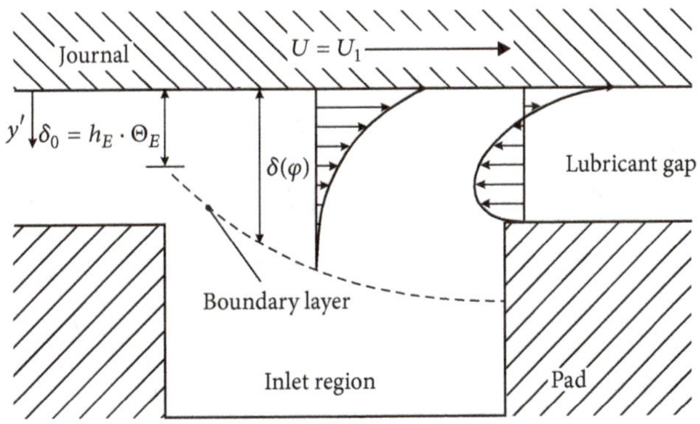

FIGURE 8: Flow profile in inlet and leading edge pad region.

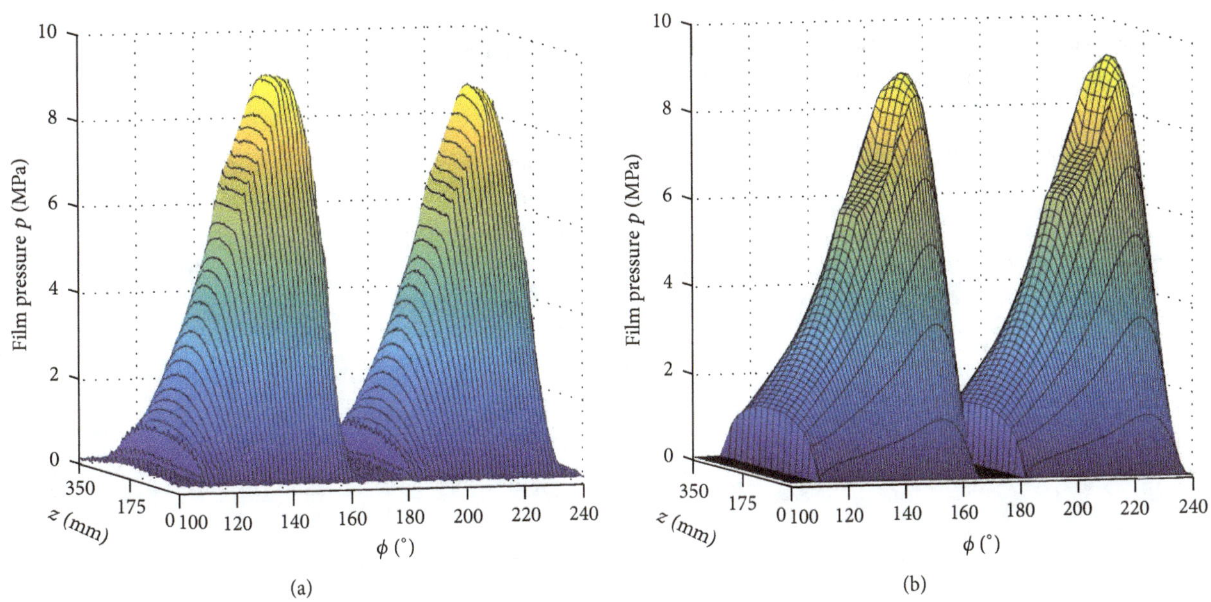

FIGURE 9: Measured [5, 6] and predicted pressure distribution on the loaded pads 3 and 4 (TPB500, $n = 3000\,\text{min}^{-1}$, $p_q = 2.0\,\text{MPa}$, $Q = 420\,\text{l/min}$, and $T_{\text{sup}} = 50^{\circ}\text{C}$).

This relation tends to increase the pressure difference while the existent lateral flow slightly decreases it. However, the pressure increase is no exact step phenomena, and hence, (17) seems to be a good compromise in order to provide a suitable approach to model this effect. Gauge pressure from (17) is defined as an offset to the pressure in the inlet region located on its last nodes in circumferential direction. This approach is chosen with respect to numerical robustness and not to a certain physical argument. The boundary condition remains unchanged during the iterative procedure to find the mechanical equilibrium of the tilting-pad for a certain journal position and is numerically damped in order to provide robust convergence of the complete calculation.

Experimental evidence for the presence of the pressure rise at the leading edge exists within measurement results for both bearings introduced in Table 1. The test rig introduced in [5, 6, 16] is able to measure a pressure and film thickness distribution within the whole film by a lateral movement of the shaft. Two capacitive proximity probes and two piezoelectric pressure sensors are mounted in the shaft rotated 90° to each other. In circumferential direction the resolution is fixed at 240 values and in axial direction it increases with rising speed up to about 4000 values per angular measurement point. Figure 9 shows a comparison of measured and predicted pressure distribution on the two maximum loaded pads of the TPB500 at a rotor speed of 3000 rpm and a specific load of 2.0 MPa. The pressure increase is clearly observable within measurement data at both pads in Figure 9(a). The lateral resolution accounts for 2400 values. The shape of the pressure rise determined by simulation applying the above-described model matches the experimental results; however, the absolute values are overpredicted. According to Figure 10, this characteristic could be observed within the whole operating range. Consequently, measurement results disagree with the hypothesis that the theoretical model should provide smaller values.

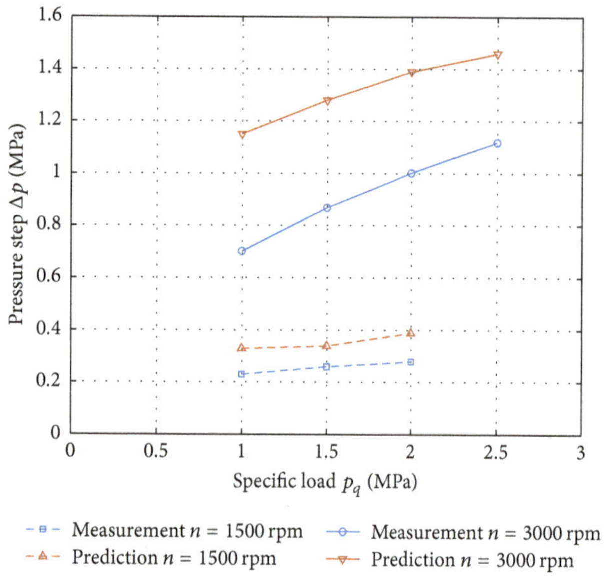

FIGURE 10: Measured [5, 6] and predicted pressure increase at leading edge of pad 4 (TPB500, $Q = 420$ l/min, and $T_{sup} = 50^{\circ}$C).

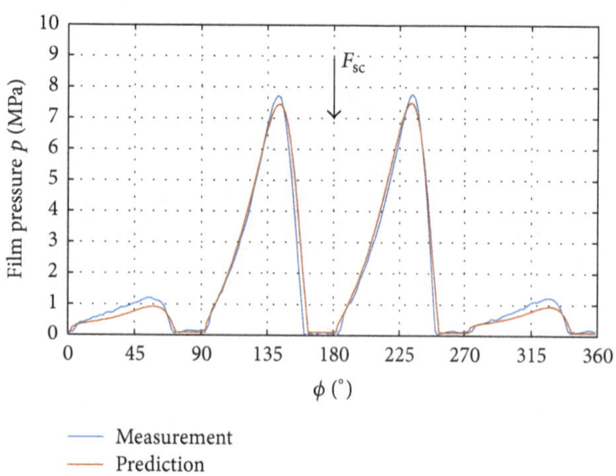

FIGURE 11: Measured and predicted pressure distribution (TPB120, $n = 6000\,\text{min}^{-1}$, $p_q = 2.0$ MPa, $Q = 90$ l/min, $T_{sup} = 50^{\circ}$C, and $z = 36$ mm).

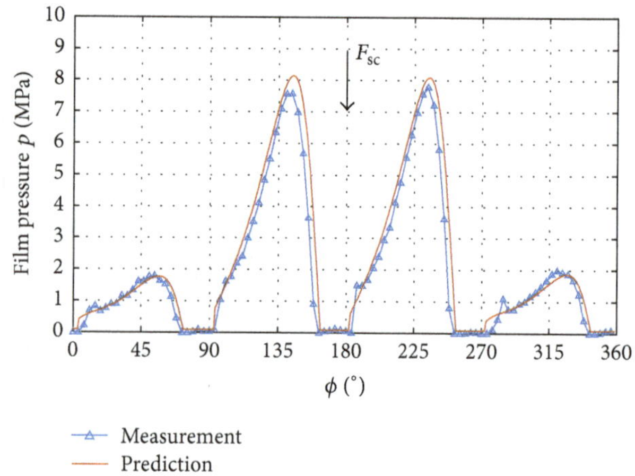

FIGURE 12: Measured and predicted pressure distribution (TPB120, $n = 12000\,\text{min}^{-1}$, $p_q = 2.0$ MPa, $Q = 90$ l/min, $T_{sup} = 50^{\circ}$C, and $z = 36$ mm).

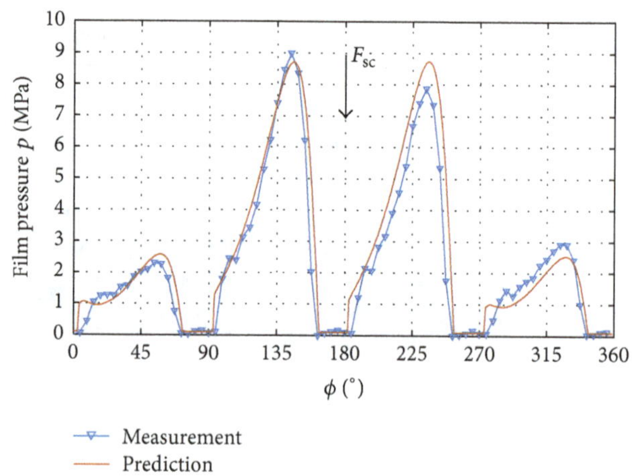

FIGURE 13: Measured and predicted pressure distribution (TPB120, $n = 15000\,\text{min}^{-1}$, $p_q = 2.0$ MPa, $Q = 90$ l/min, $T_{sup} = 50^{\circ}$C, and $z = 36$ mm).

The TPB120 was investigated experimentally in [12] and theoretically by the bearing code and CFD analyses. Figures 11–13 show a comparison of predictions performed by the bearing code and measurement data for different rotor speeds at constant bearing load. In general, good agreement is reached. Concordantly, inertia effects at leading edge are negligible at 6000 rpm, while they are clearly observable at 12000 rpm and 15000 rpm. At 15000 rpm a local pressure maximum on the upper pads is predicted shortly behind the leading edge. Measurement data at 12000 rpm and 15000 rpm tends to prove this phenomenon.

Further, the results can be compared to CFD-data in Figure 14. The correlation between CFD-data and bearing code prediction in Figure 14 is good for 6000 rpm and excellent for 12000 rpm. Results of CFD-simulation for the leading edge pressure step prove to be very robust and insensitive concerning the grid within the inlet region. The comparison between prediction and measurement data in Figures 11–14 shows generally good agreement. The impact of inertia forces at the leading edge of pad is observable distinctly. However, discrepancies can also be detected. The markers in Figures 12 and 13 indicate the limited resolution of the measurement which is reduced with increasing speed as the sampling frequency remains at 40 kHz. Hence, at 15000 rpm 80 values are detected in circumferential direction and the distance between two points accounts for $\Delta\phi = 4.5^{\circ}$. Moreover, the piezoelectric pressure sensors and their adapters mounted in two lateral planes in the journal according to Figure 15 have a diameter of approximately 12 mm and they are not smoothing completely to the sliding surface of the journal. Consequently, the detected pressure values represent a mean

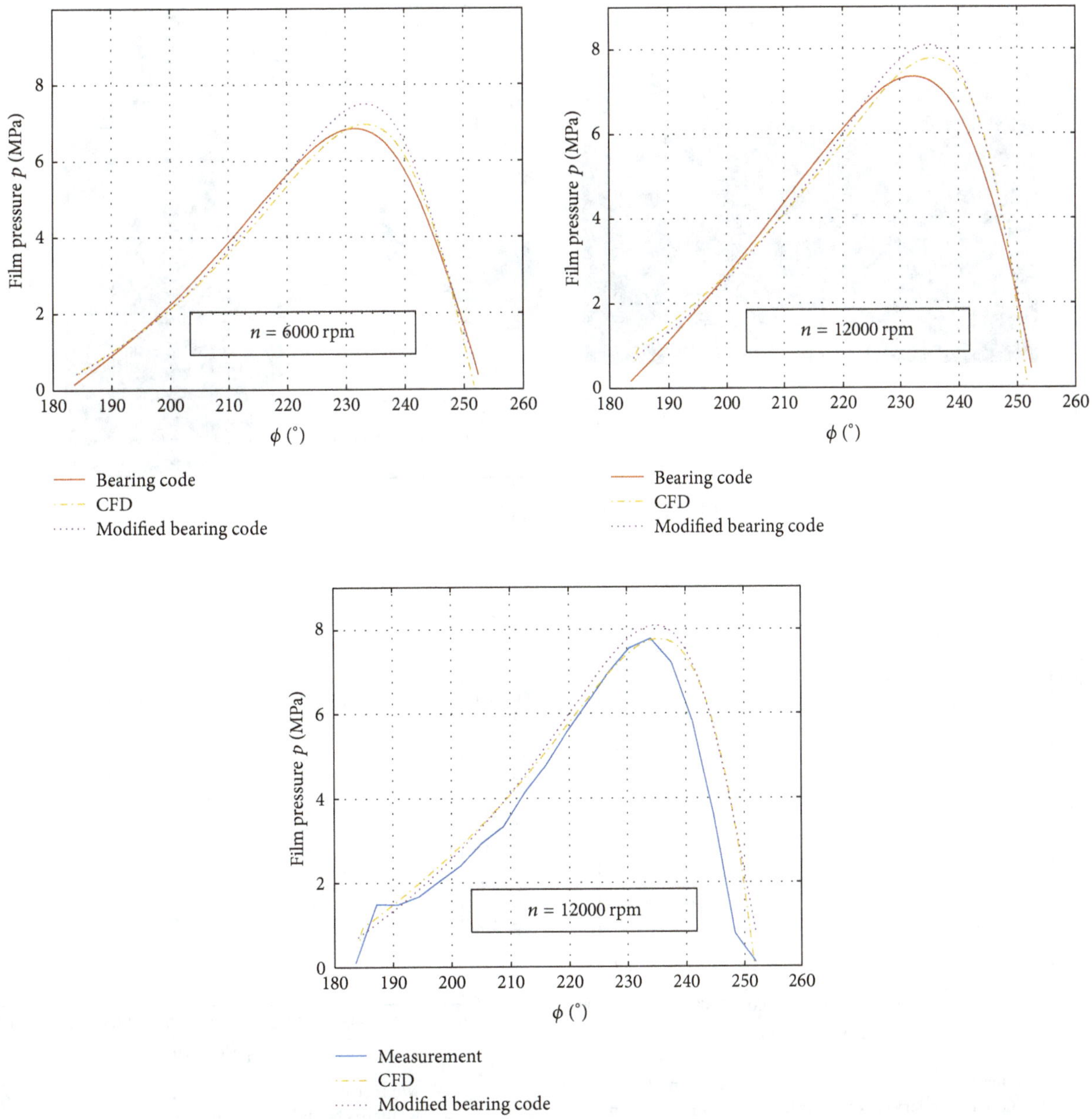

FIGURE 14: Comparison of measured and predicted pressure distributions on pad number 3 considering enhanced boundary conditions (TPB120, $p_q = 2.0$ MPa, $Q = 90$ l/min, $T_{sup} = 50$°C, and $z = 36$ mm).

up to 11°. Therefore, high pressure gradients are measured with higher uncertainties.

The subsequently presented results refer to sensitivity of selected characteristic bearing parameters on modelling of convective inertia forces. Here, predictions first neglecting inertia forces, secondly using only the boundary condition from (17), and finally considering the boundary condition from (17) and inertia forces in the lubricant gap are compared. As expected, the maximum sensor temperature depicted in Figure 16 referring to the thermocouple locations shown in Figure 2 is predicted nearly equally for lower speeds. However, with increasing speed deviations occur and rise up to $\Delta T = 5.4$ K at 15000 rpm. As a reference, Figure 16 includes measurement results.

The main stiffness coefficients derived from the stationary results by perturbation method and presented in Figure 17 indicate more sensitivity related to modelling than the main damping coefficients in Figure 18. This property might be related to the impact of pressure gradients on stiffness coefficients. Figure 19 includes a comparison of predicted local pressure distributions in order to illustrate local deviations due to modelling which are present on all four pads.

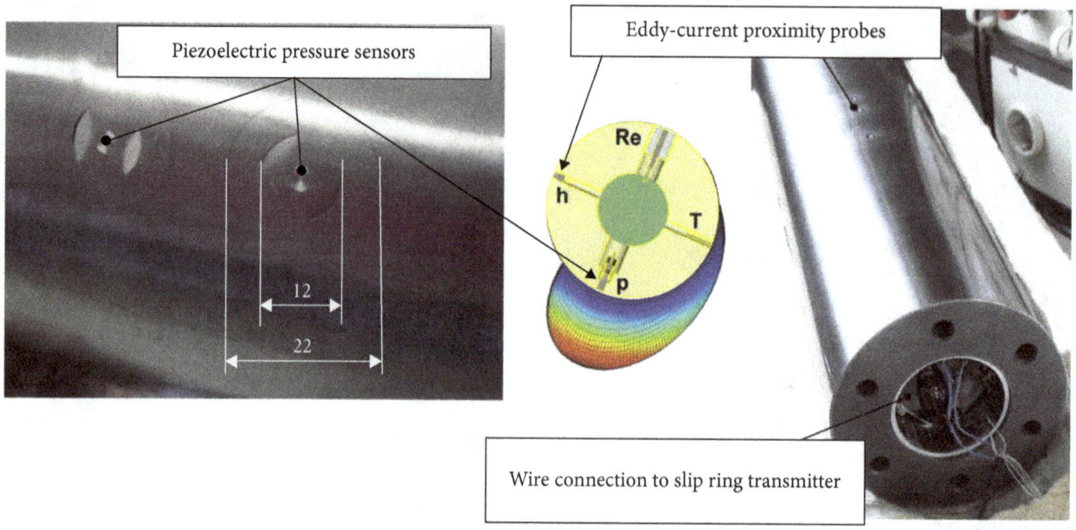

FIGURE 15: Rotor-sided applied sensors [11].

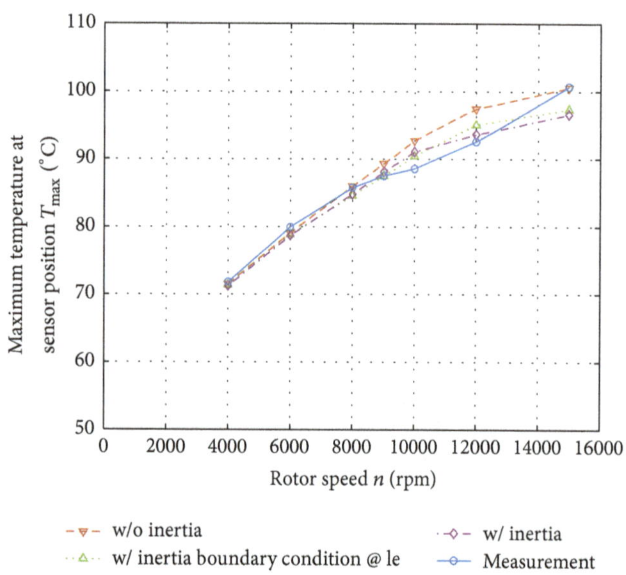

FIGURE 16: Comparison of measured and predicted maximum sensor temperatures for different simulation boundary conditions (TPB120, $p_q = 2.0$ MPa, $Q = 90$ l/min, $T_{sup} = 50$°C, and $z = 36$ mm).

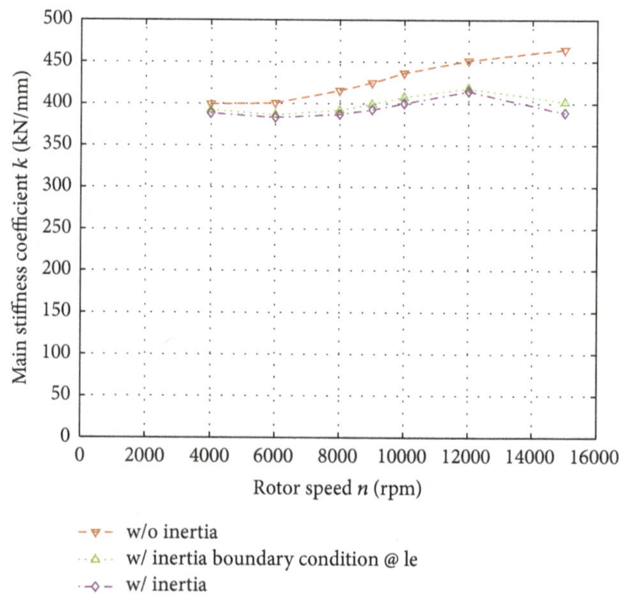

FIGURE 17: Comparison of predicted synchronous main bearing stiffness for different simulation boundary conditions (TPB120, $p_q =$ 2.0 MPa, Q = 90 l/min, T_{sup} = 50°C, z = 36 mm, and k_{pivot} = 0.8 kN/mm).

4. Conclusions

The investigations show the impact of fluid convective inertia forces at the leading edge of tilting-pad journal bearings. Concordantly, measurement data and CFD analyses indicate that the pressure characteristic is significantly influenced by this phenomenon, especially at elevated rotor speeds. A simple energy approach is proposed and implemented in the bearing code to improve the quality of boundary conditions for Reynolds equation. The enhanced bearing model shows good agreement with experimental and CFD-data. Here, results provide evidence that inertia effects within the pad area become more important due the impact of leading edge boundary condition.

However, the results presented here were identified within investigations with a different focus. Therefore, the quality of the CFD grid is not sufficient enough to give a detailed answer on the transition of the flow profile at leading edge where it changes from boundary layer to combined Poiseuille and Couette flow. Further, a higher sampling rate and a modified pressure measurement should be aspired for a more detailed resolution of the leading edge pressure

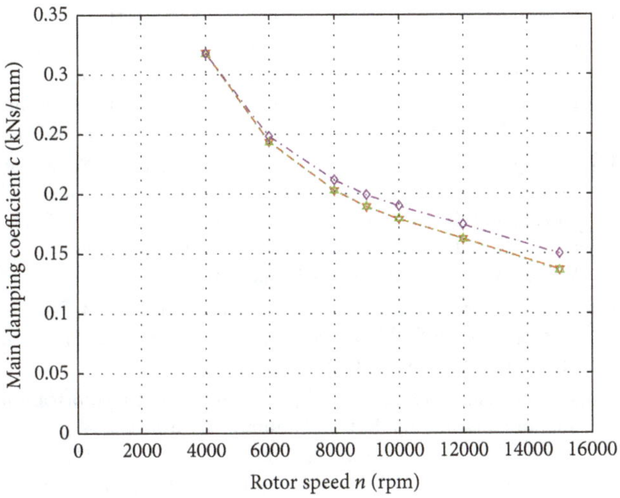

FIGURE 18: Comparison of predicted synchronous main bearing damping for different simulation boundary conditions (TPB120, p_q = 2.0 MPa, Q = 90 l/min, T_{sup} = 50°C, z = 36 mm, and k_{pivot} = 0.8 kN/mm).

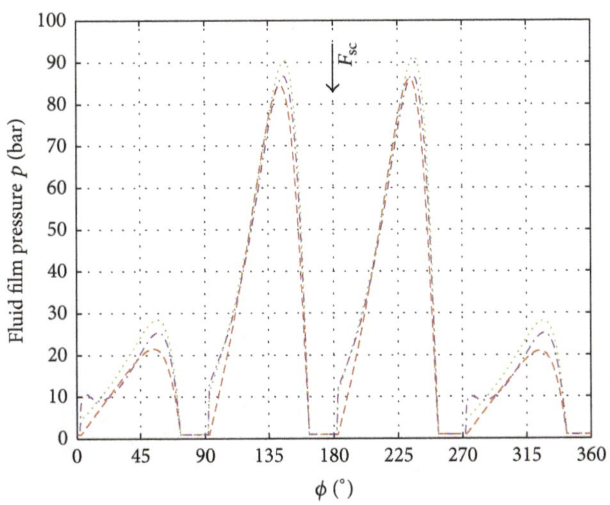

FIGURE 19: Comparison of local pressure predicted for different simulation boundary conditions (TPB120, p_q = 2.0 MPa, Q = 90 l/min, T_{sup} = 50°C, and z = 36 mm).

distribution. The available measurement data depicted in Figure 9 and Figures 11–14 includes high uncertainties in quantifying the actual pressure characteristic at leading edge. These analytical steps would provide the ability to improve modelling and to judge about special questions of implementation since predictions indicate a nonnegligible influence on characteristic parameters of bearing operation.

Nomenclature

Symbols

B: Axial bearing length
C_R: Bearing radial clearance, $C_R = R - r$
c_p: Lubricant specific heat
D: Bearing diameter
F: Bearing force
h: Film thickness
H: Relative lubricant film thickness, $H = h/C_R$
I_{xx}, I_{xz}, I_{zz}: Integrals of inertia forces
K_x, K_z: Turbulence factors
m: Pad preload
n: Rotational speed
p: Pressure
p_q: Specific bearing load, $p_q = F/(BD)$
Q: Lubricant flow rate
R: Journal bearing inside radius
R_B: Pad bore radius
r: Journal radius
Re: Reynolds number, $\text{Re} = \rho\omega r C_R/\eta$
Re_δ: Reynolds number related to boundary layer thickness, $\text{Re} = \rho\omega r h_\delta/\eta$
Re_{eff}: Effective Reynolds number, $\text{Re} = \rho\omega r C_{R,\text{eff}}/\eta_{\text{eff}}$
Re^*: $\text{Re}^* = \text{Re}\cdot\psi$
s: Radial displacement
T: Temperature
t_B: Pad thickness
U: Surface speed
u: Flow velocities
x, y, z: Cartesian coordinates
$\alpha_1, \alpha_2, \alpha_3$: Inertia flow parameters
β: Attitude angle of the journal
δ: Tilting angle, boundary layer thickness
ε: Relative eccentricity of the journal
η: Lubricant dynamic viscosity
η_m^*: Ratio of mean and related lubricant dynamic viscosity
Θ: Local fill factor
ν: Lubricant kinematic viscosity
ρ: Density
φ: Angular coordinate in circumferential direction
φ_F: Angular pivot position
ψ: Relative bearing clearance.

Subscripts

bl: Boundary layer
cr: Critical
eff: Effective
le: Leading edge
sup: Supply.

Conflicts of Interest

The authors declare that there are no conflicts of interest regarding the publication of this article.

Acknowledgments

The authors gratefully thank the German Federation of Industrial Research Associations (AiF e. V.) for the financial support of these investigations. The authors also thank the expert committees of the German Research Communities FVA e. V. and FVV e. V. for the technical and scientific steering of this research project.

References

[1] N. Mittwollen, *Betriebsverhalten von Radialgleitlagern bei Hohen Umfangsgeschwindigkeiten und Hohen Thermischen Belastungen Theoretische Untersuchungen*, vol. 187 of *VDI Series 1*, VDI-Verlag, Dusseldorf, Germany, 1990.

[2] C. M. Ettles, "Discussion: "Mixing inlet temperatures in hydrodynamic bearings" (Heshmat, H., and Pinkus, O., 1986, ASME J. Tribol., 108, pp. 231–244)," *Journal of Tribology*, vol. 108, no. 2, pp. 245–247, 1986.

[3] V. N. Constantinescu, S. Galetuse, and F. Kennedy, "On the comparison between lubrication theory, including turbulence and inertia forces, and some existing experimental data," *Journal of Lubrication Technology*, vol. 97, no. 3, pp. 439–449, 1975.

[4] T. Hagemann, *Ölzuführungseinfluss bei schnell laufenden und hoch belasteten radialgleitlagern unter berücksichtigung des lagerdeformationsverhaltens [Ph.D. Thesis]*, TU Clausthal, Germany, 2011.

[5] T. Hagemann, S. Kukla, and H. Schwarze, "Measurement and prediction of the static operating conditions of a large turbine tilting-pad bearing under high circumferential speeds and heavy loads," in *Proceedings of the ASME Turbo Expo 2013: Turbine Technical Conference and Exposition, GT2013-95004*, American Society of Mechanical Engineers, San Antonio, Tex, USA, June 2013.

[6] T. Hagemann, S. Kukla, and H. Schwarze, "Measurement and prediction of the dynamic operating conditions of a large turbine tilting-pad bearing under high circumferential speeds and heavy loads," in *Proceedings of the ASME Turbo Expo 2013: Turbine Technical Conference and Exposition, GT2013-95074*, American Society of Mechanical Engineers, San Antonio, Tex, USA, June 2013.

[7] D. Dowson, "A generalized Reynolds equation for fluid-film lubrication," *International Journal of Mechanical Sciences*, vol. 4, no. 2, pp. 159–170, 1962.

[8] V. N. Constantinescu and S. Galetuse, "Operating characteristics of journal bearings in turbulent inertial flow," *Journal of Lubrication Technology*, vol. 104, no. 2, pp. 173–179, 1982.

[9] H. G. Elrod, "A cavitation algorithm," *Journal of Lubrication Technology*, vol. 103, no. 3, pp. 350–354, 1981.

[10] H. Schlichting and E. Gersten, *Boundary Layer Theory*, Springer-Verlag, Berlin, Germany, 10th edition, 2006.

[11] P. Pfeiffer, H. Blumenthal, and H. Schwarze, "Hochleistungs-Gleitlagerprüfstand zur Ermittlung von tribologischen Kengrößen hydrodynamischer Gleitlager für Turbomaschinenanwendungen," *Tribologie und Schmierungstechnik*, vol. 3, no. 62, p. 31ff, 2015.

[12] T. Hagemann, P. Pfeiffer, X. Si, C. Zeh, and H. Schwarze, *Einfluss der Ölzuführung auf die Hydraulischen, Energetischen und Mechanischen Vorgänge in Schnell Laufenden und Hoch Belasteten Radialkippsegmentlagern*, 1184, Forschungsvereinigung Antriebstechnik e.V., Frankfurt am Main, Germany, 2016.

[13] T. Hagemann, H. Blumenthal, C. Kraft, and H. Schwarze, "A study on energetic and hydraulic interaction of combined journal and thrust bearings," in *Proceedings of the ASME Turbo Expo 2015: Turbine Technical Conference and Exposition*, p. V07AT31A019, Montreal, Quebec, Canada, 2015.

[14] C. H. T. Pan, "Calculation of pressure, shear, and flow in lubricating films for high speed bearings," *Journal of Lubrication Technology*, vol. 96, no. 1, pp. 80–94, 1974.

[15] V. N. Constantinescu, "Special problems concerning lubrication phenomenas at large Reynolds numbers," in *Proceedings of the 2nd Leeds-Lyon Symposium of Tribology*, pp. 77–84, 1975.

[16] G. Hopf and D. Schüler, "Investigations on large turbine bearings working under transitional conditions between laminar and turbulent flow," *Journal of Tribology*, vol. 111, no. 4, pp. 628–634, 1989.

13

Observations on Rotating Instabilities and Spike Type Stall Inception in a High-Speed Multistage Compressor

Johannes Schreiber, Benoit Paoletti, and Xavier Ottavy

Laboratoire de Mécanique des Fluides et d'Acoustique, UMR CNRS 5509, Université de Lyon, Ecole Centrale de Lyon, UCB Lyon I, INSA, 36 Av. Guy de Collongue, 69134 Écully Cedex, France

Correspondence should be addressed to Johannes Schreiber; johannes.schreiber@ec-lyon.fr

Academic Editor: Feng Lin

This work investigates high-frequency measurements for the contribution to the understanding of different stall inception types in high-speed multistage compressors. A 3.5-stage high-speed axial multistage compressor is investigated with a 2 MW test rig in the Laboratoire de Mécanique des Fluides et d'Acoustique (LMFA) at Ecole Centrale de Lyon, France. Two different types of instabilities arise in this compressor as a function of shaft speed. At part speed, a controversy called "rotating instability" type flow field modulation is identified with the measurements. New results are the demonstration of the periodic behavior of this instability and the analogy to classical frequency modulation, periodic to one revolution of the instability. Furthermore, the amplitude of the instability is modulated by the time period of a rotor revolution. At nominal speed, the abrupt spike type stall inception is detected, taking usually less than five rotor revolutions.

1. Introduction

At the low mass flow rate end of the operating range, every axial compressor will ultimately experience rotating stall, one or several local rotating detached flow cell(s). This can be followed by surge, which is a system instability with a possible global flow reversal. There are two widely recognized rotating stall inception types for low- and high-speed axial compressors: modal and spike [1], though weak rotating perturbations can be present before the rotating stall inception, allowing a stable compression system operation. They are often reported as rotating instabilities in literature [2].

The need for detailed measurements in high-speed axial compressors and their analysis was clearly stated by Day [1] in 2015 reviewing 75 years of stall research and derived from flow features that can only be found in high-speed compressors. That environment poses a challenge in terms of measurement cost and technique but here the measurements are necessary to understand the instabilities arising in a real aircraft engine compressor. Numerical studies can be used for complementary flow field understanding but their limitations (particularly induced by the use of RANS modeling) need to be considered.

The spike stall inception describes an onset mechanism starting from a local strong perturbation of the flow field in the tip region of a rotor. Recent numerical and experimental works converge towards explaining the typical increase and decrease in wall pressure by a vortex structure forming at the leading edge of the rotor blade and spanning to the casing ("tornado vortex"). Inoue et al. [3] identified this numerically at first, and recent numerical and experimental works by Pullan et al. [4] support those results. In high-speed compressors, the spike type structure evolves very quickly (a few rotor revolutions) into a rotating stall cell.

Rotating disturbances can be observed already at a mass flow rate greater than the stall inception point and appear as a regular modulation of the flow field prior to surge. This disturbance does not induce a significant flow blockage in the concerned rotor contrary to the rotating stall. They have been sometimes reported as high-frequency rotating stall [5] or pressure waves [6], with cells numbers and rotation speed higher than the ones observed with rotating stall.

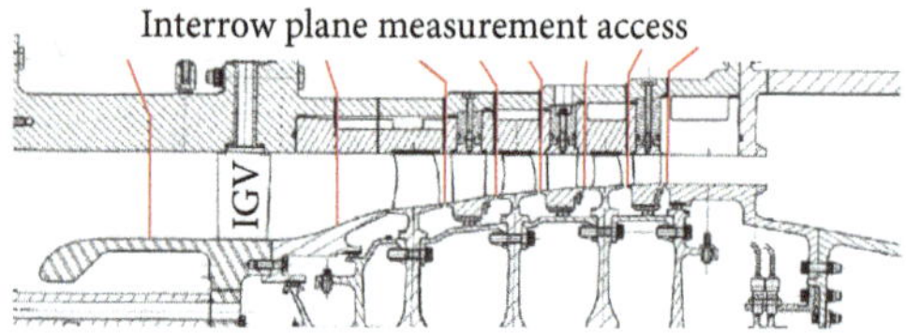

FIGURE 1: Meridian view of CREATE and measurement planes.

A certain controversy called rotating instability (RI) could be understood as a subtype of the rotating disturbances. The RI induces a characteristic frequency bump (or rake) in a pressure spectrum compared to a single peak for the rotating disturbances. The source mechanism of the RI appears to be closely related to the spike formation as demonstrated by Inoue et al. [7] and Yamada et al. [8]. The RI occurs mostly in low-speed compressors and with large rotor tip gaps. A flow separation has to occur at the leading edge of usually a rotor blade, causing a probable leading edge vortex. The flow separation can also be provoked by a corner separation in a case without tip gaps as identified by Beselt et al. [9]. What eventually induces the "unstable" behavior, causing the multiple frequencies, has to be identified yet.

Firstly, the high-speed multistage compressor test case and the experimental methods will be presented. Then, two types of instabilities will be discussed: spike and rotating instability. The high-frequency measurements allow a characterization of the abrupt spike type stall inception at nominal shaft speed. Furthermore, this work will present original results about the periodic behavior of a rotating instability at part shaft speed, contrary to what would be expected from the naming in literature.

2. Test Case (CREATE)

For this work, the research compressor *CREATE* (Compresseur de Recherche pour l'Etude des effets Aérodynamiques et TEchnologiques) is investigated in the laboratory of fluid mechanics and acoustics *(LMFA)* at Ecole Centrale de Lyon, France. The geometry and rotation speed of the 3.5-stage high-speed axial research compressor are representative of high-pressure compressor median rear blocks of a modern turbojet engine, designed by Safran Aircraft Engines.

The compressor and the measurement planes are presented in Figure 1. The rotor shaft is driven at a design speed of 11543 rpm, whereby the flow is slightly transonic in the first stage and fully subsonic in the last two stages. More characteristics of *CREATE* are given in Table 1. The spatial circumferential periodicity of the 3.5 stages is reduced to $2\pi/16$ by choosing all numbers of blades as a multiple of 16, as stated in Table 2. Consequently one temporal period corresponds to 1/16th of a rotor revolution. A profound knowledge-base has been acquired with CREATE for numerical and experimental methods aiming at flow field investigations in the high-speed multistage compressor environment [10–13]. Standard tip gap sizes are implemented compared to increased tip gaps in rotors 2 and 3 in the previous version of CREATE.

TABLE 1: Characteristics of CREATE at design point.

Casing diameter	0.52 m
Core speed	11543 rpm
Rotor 1 tip speed	313 m/s
Rotor 1 tip inlet relative Mach number	0.92

TABLE 2: Number of blades of CREATE.

Row	IGV	R1	S1	R2	S2	R3	S3
Number of blades per row ($2\pi/16$)	2	4	7	5	7	5	7

TABLE 3: Relative measurement uncertainties for high frequency static pressure measurements estimated for CREATE by [14].

	Weak gradient zones	Strong gradient zones
Rotor 1	0.47%	3.3%
Rotor 2	0.51%	3.0%
Rotor 3	0.60%	3.0%

3. Experimental Methods

The main characteristics of the high-frequency casing pressure measurements are summarized hereinafter. Furthermore, global performance measurements are conducted but not described in more detail here. Ottavy et al. [10] state more details about this test rig and give an exhaustive introduction of the experimental methods, whereas only a brief summary is given hereinafter.

3.1. High-Frequency Wall Static Pressure Measurements. High-frequency wall static pressure measurements have been carried out for a general description of the flow field and the study of the onset of instabilities near surge and the surge inception. For capturing all the unsteady features of the flow, the wall static pressure measurements have been acquired simultaneously with a frequency of 500 kHz using 48 sensors (Kulite: XTE-190). Six probes can be installed per interrow plane and 12 probes per rotor tip. The sensors, used without their protective grid, have a correct frequency-response up to about 160 kHz, where the BPFs of rotors 2 and 3 correspond to 15 kHz. More details about the high-frequency-response acquisition system are stated in the works of Gourdain et al. [11] and Courtiade [14], and the relative measurement uncertainty is given in Table 3.

3.2. Data Treatment Methods. The data treatment is done consistently in this work and defined hereinafter. The mass flow rate, isentropic efficiency, and pressure ratio are normalized by their respective maximum values or given as difference. Entropy, temperature, and pressure are normalized by their overall time-mean rise in the compressor according to (1). The time axis might be normalized by the rotating frequency of the shaft. The nondimensional data is labeled "normalized" (or "norm.")

$$X_{\text{norm}} = X_{\text{mes}} \times \frac{1}{\overline{X_{\text{outlet}}} - \overline{X_{\text{inlet}}}}. \quad (1)$$

Static variables are arithmetically averaged in space, whereas convected variables (X_c) are weight-averaged by the axial velocity (V_{x_i}) according to

$$\overline{X_c} = \frac{1}{N \cdot \overline{V_x}} \cdot \sum_{i=0}^{N-1} X_i \cdot V_{x_i}. \quad (2)$$

Temporal averages are performed as arithmetic averages.

In this work, rotating frequencies of flow structures are obtained with the help of cross-correlations between the signals of probes (e.g., signal 1 and signal 2) located at identical axial positions but different circumferential positions. The cross-correlation allows the estimation of the time lag between the passing of a flow structure in front of probe 1 and that in front of probe 2. The time lag is estimated between the maximum of the autocorrelation of signal 1 and the maximum of the cross-correlation between signals 1 and 2.

Knowing the circumferential distance between the probes ($\Delta\theta_{12}$ in radians) and the time lag (Δt in seconds), a rotating frequency ($\omega_{\text{flow structure}}$) can be calculated according to

$$\omega_{\text{flow structure}} = \frac{1}{2\pi} \cdot \frac{\Delta\theta_{\text{probes}}}{\Delta t}. \quad (3)$$

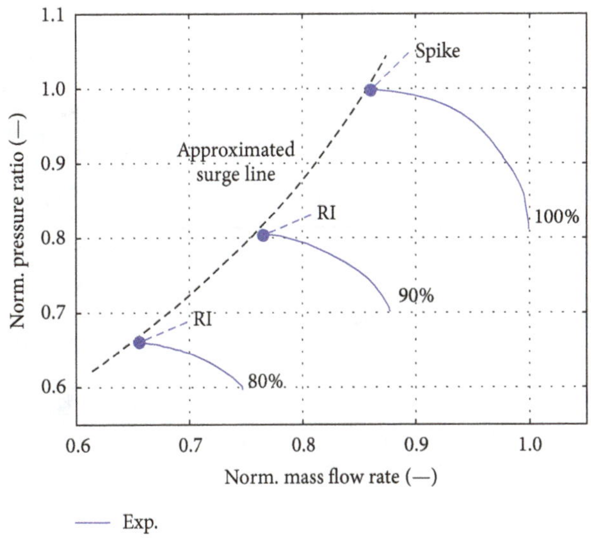

FIGURE 2: Experimental results for total pressure ratio at part and nominal rotation speeds.

4. Stall Inception in the Compressor CREATE

The stall inception type is a function of the rotor speed in this high-speed compressor. In this work it is therefore possible to present observations on rotating instabilities and spikes.

4.1. Global Overview. The global total pressure ratios at 80%, 90%, and 100% shaft rotation speed are presented in Figure 2 and give an overview of identified instabilities in this compressor. The pressure rise distributions reach a rather flat part towards the low mass flow rate before the system falls into surge (see "approximated surge line"). The often called "peak" of the pressure rise distribution is not exceeded here at any rotation speed. The measurements allow the detection of a spike type stall inception at 100% shaft speed and a rotating instability (RI) type stall inception at part shaft speed. This work will present observations concerning these two instabilities with detailed measurement data at 80% (representative of the part speed flow field) and 100% shaft rotation speed.

In this compressor and as usual in a well matched multistage compressor [1], the instability and stall inception occurs in rotor 1. The general characteristics of the instability onset are described with help of Figure 3 for 80% shaft speed on the left and 100% shaft speed on the right. Six signals are presented per speed. They are obtained from the high-frequency casing pressure probes that are positioned around the circumference at a plane just upstream of rotor 1. For these surge inception measurements, the outlet throttle valve is closed continuously and very slowly from a stable operating point, giving quasi-steady operating conditions and avoiding the generation of "steps" in the pressure signals when the throttle is closed step by step. The signals are low-pass-filtered below the BPF and they are shown for 23 rotor revolutions.

At both shaft speeds, a single rotating stall cell forms between rotations 14 and 18. The circumferential extension of the cell is growing rapidly during two rotations while the cell's LE is rotating at 0.6 Ω (estimated with cross-correlations according to (3)). The compressor falls abruptly into surge between revolutions 18 and 20.

At 80% shaft speed, the instability is present in hundreds of rotations prior to the rotating stall cell (quantified later in Figure 7). It is seen as somewhat of noise in the signals here. The rotating stall cell forms (indicated with the black dashed circle) without any detectable trace of a spikelike flow perturbation in this compressor. The difference between the rotating stall cell and RI signature will be discussed with a wavelet analysis later on.

At 100% shaft speed, the typical increase and decrease related to a spike type perturbation are found [4]. The increase and decrease of the static pressure extend each over four rotor blade passing instances (not detailed here). At the first detection, the rotating speed of the spike is estimated to be 0.9086 Ω with the help of cross-correlations between the six signals, applying (3). The spike structure evolves within less than 2 rotor revolutions into a single rotating stall cell, which is forming at revolution 16. The whole stall inception process is abrupt at 100% shaft speed and takes less than 20 rotor revolutions in any measured case and a stall warning system could not react in this short span of time.

At all investigated shaft speeds, temporal mode 1 (normalized by the shaft rotating frequency) is present in the flow field in rotor 1. At 80%, this mode causes a modulation of the rotating instability and can be found as a wavelike modulation of the signals in Figure 3(a). At 100% shaft speed, mode 1

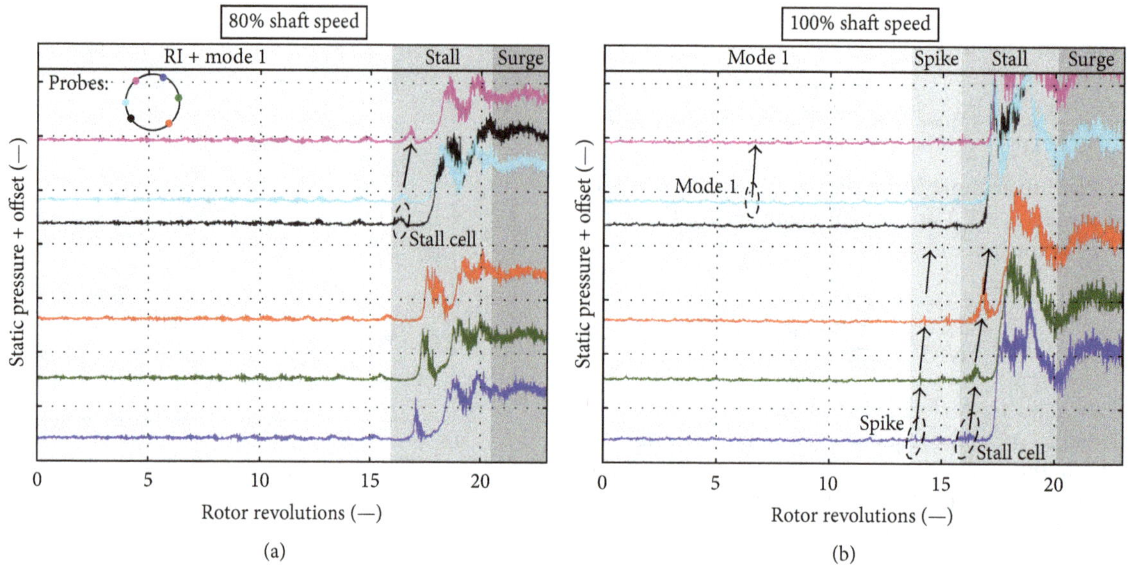

FIGURE 3: Static pressure signals (low-pass filtered below BPF) of six probes around the circumference upstream of the rotor 1 LE during stall inception at 80% (a) and 100% (b) shaft speed with an identical scale.

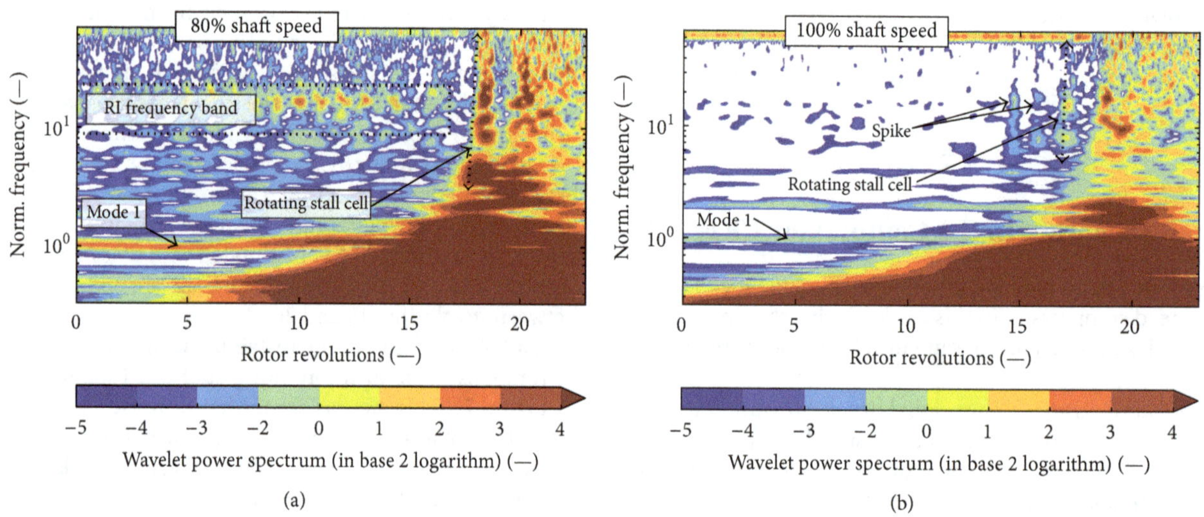

FIGURE 4: Wavelet power spectra for the green signals (2nd circ. pos. from the bottom) in Figure 3 for 80% (a) and 100% (b).

is visibly better because no instability is present in the flow field. The rotation speed of this mode corresponds to the rotation speed of the shaft. It might be therefore the signature of a geometrical default in the rotor. This default might be found in about 4–8 blades (temporal extension of pressure peaks) with a different tip gap than the surrounding blades but further tip gap measurements are necessary to support this hypothesis. It might be the geometrical trigger for the onset of the spike inception. Weichert and Day [15] found that, in 90% of the investigated stall inception events, the spike perturbation would form in the same rotor passage due to a geometrical default.

Wavelet spectra shall clarify the timeline of the formation of the instability structures during the stall inception. Figure 4 shows the wavelet power spectra computed on the green signal (2nd signal from the bottom) in Figure 3. The wavelet spectra are obtained using the tools of Torrence and Compo [16] with a Morlet wavelet, which has been proven useful for turbomachinery applications, for example, by Lin et al. [17]. The wavelet analysis is conducted in each case with an input signal normalized by the standard deviation computed on the first half of the signal. The vertical axis traces the normalized frequency (by the shaft rotating frequency) and the horizontal axis the rotor revolutions, which correspond to the ones presented in Figure 3.

At 80% speed, the RI is seen as a wide frequency band with elevated power. The origin of these frequencies will be discussed later on in this work. The BPF is located at the upper limit of the spectrum. It corresponds to the normalized frequency 64 whereas the RI frequency band can be seen

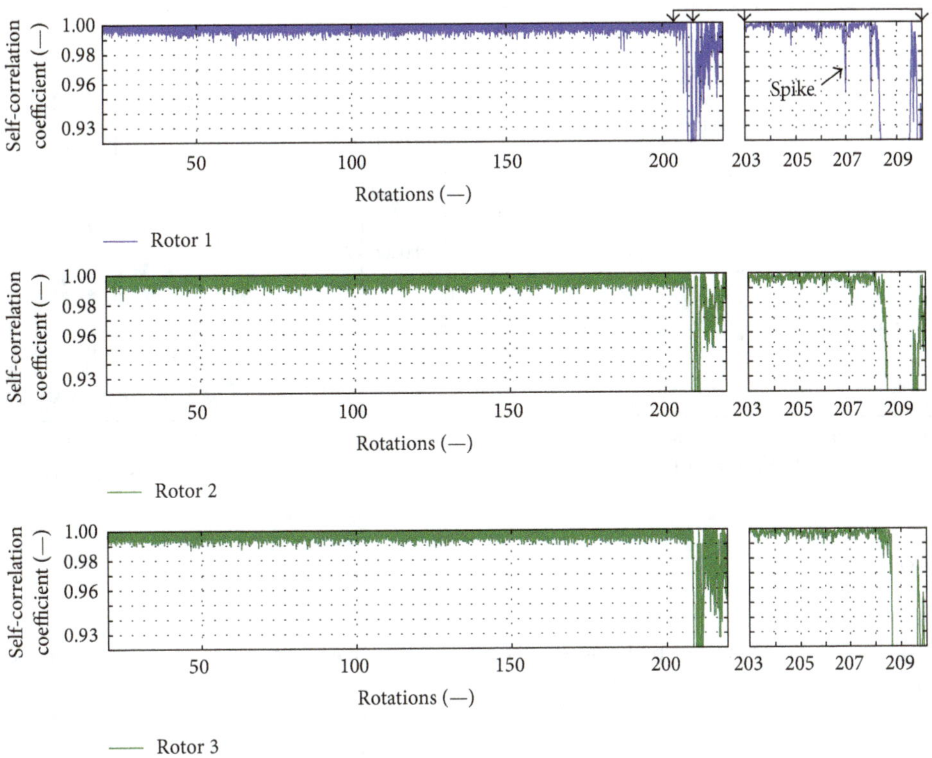

FIGURE 5: Disturbance detection by "self-correlation" of casing pressure signals at LE of rotors 1 to 3 at 100% Ω.

around 18. The BPF is not disturbed by the rotating instability. The rotating stall cell appears as a vertical trace in the wavelet spectrum. Low frequency content related to the rotating stall and surge causes the zone of high power (red). The absence of the interblade pressure gradient due to the detached flow (rotating stall cell) is found as an interruption of the BPF trace in the wavelet spectrum. Surge is eventually seen as a chaotic spectrum without a clear frequency content starting from about rotor revolution 19.

At 100% shaft speed, mode 1 is present as well (Figure 4(b)) up to the rotating stall inception. Tracing of mode 1 is no longer possible once the low frequency content arises (large white zone) and a clear link between mode 1 and the spike is therefore not possible. The spike is seen at normalized frequency 16. That corresponds to the inverse of the temporal extension. With 64 rotor blades, the temporal extension corresponds thus to 25% of the rotor blades passing in time. Recent research work demonstrates a vortexlike structure as the initial form of the spike, for example, the work of Pullan et al. [4]. This present arrangement of measurement probes does not detect this vortexlike structure with less than one pitch extension. The spike structure might grow too rapidly circumferentially in between the measurement probe locations.

The flow perturbation is understood as a rotating stall and no longer spike or RI, when the BPF is disturbed because of the possible absence of the interblade pressure gradient. This frontier is indicated with a dashed line (with arrows at both ends) for both shaft rotation speeds.

A fine spike detection method and the origin of the RI frequency band are discussed in the following parts of this paper.

4.2. Spike Detection Method. The temporal and spatial locations of the stall inception are identified with an analysis of the casing pressure signals above the rotors (close to the LE) during more than 200 rotor revolutions prior to any surge inception. One example is presented for the spike type stall inception at 100% shaft speed in Figure 5. For this detection method, the signals are cut in segments representing each temporal period ($2\pi/16\,\Omega$) and are low-pass-filtered below the blade passing frequency at 7 kHz, allowing the detection of small (compared to BPF) stall inception related disturbances. A cross-correlation is calculated between each segment and the corresponding one a revolution earlier ("self-correlation"). Normalized by the autocorrelation, a correlation coefficient of 1 represents a perfect match between two signals. The maxima of the cross-correlations per temporal period are eventually plotted and give a fine measure for meaningful changes in the flow field. This stall precursor detection method resembles the one proposed by Dhingra et al. [18]. A spike would decrease the maximum correlation coefficient. Please note that the rotor revolution numbers in Figure 5 are not related to the one seen in Figure 5 earlier.

Globally, the unsteadiness increases in flow direction. This is seen as increasing background noise or band of fluctuations between 0.99 and 1. The surge inception occurs between revolutions 208 and 209 and is found as drop of the coefficient below values of 0.925. In this example, a

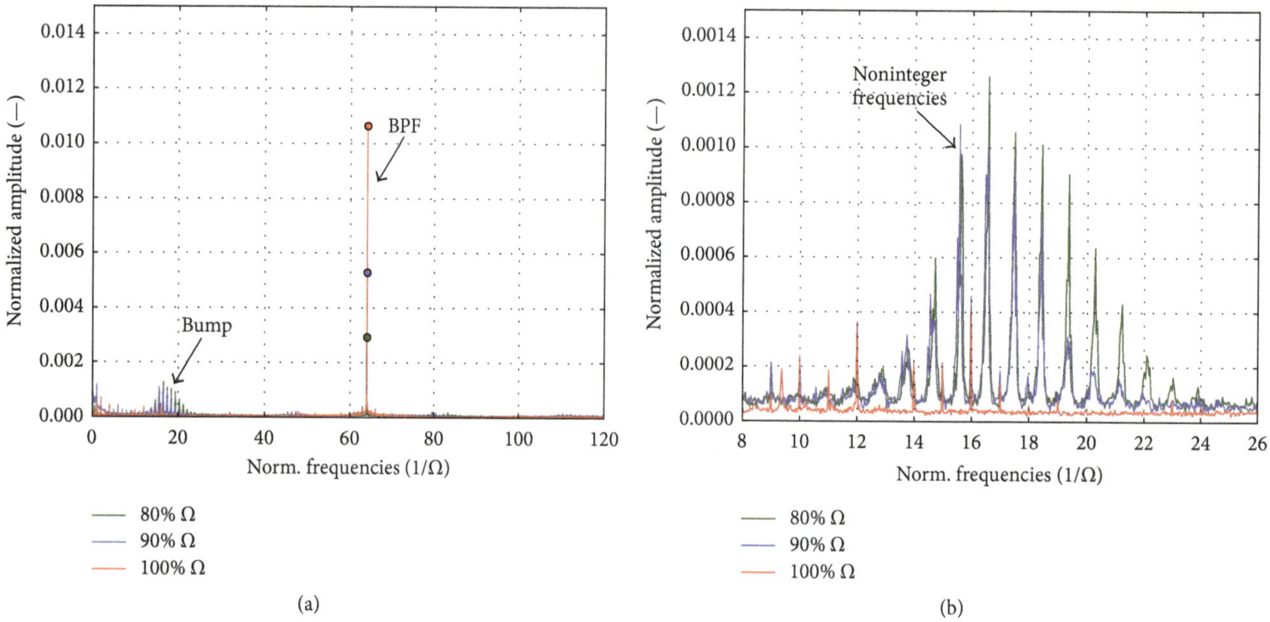

FIGURE 6: Frequency spectra of a casing pressure probe close to the rotor ~ 1 LE in comparison of three shaft rotation speeds (80% Ω, 90% Ω, and 100% Ω), "full" spectrum (a), and zoom on frequency bump (b).

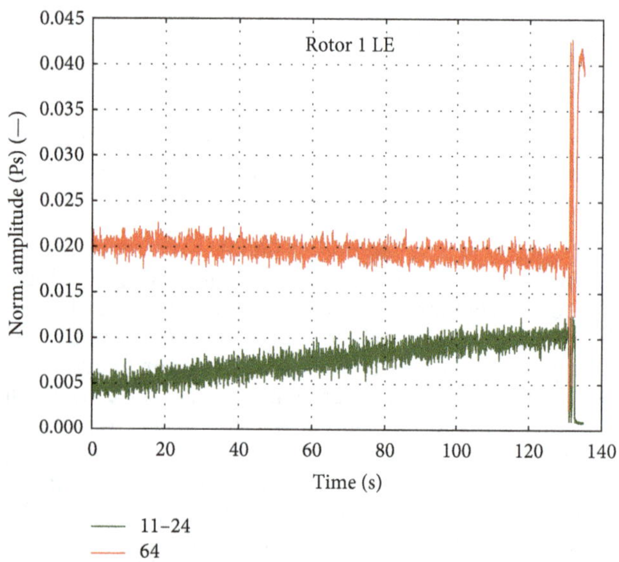

FIGURE 7: Temporal evolution of rotating instability modes (11 to 24) and rotor blade passing frequency during surge inception measurement.

disturbance appears about 20 revolutions prior to the surge inception in rotor 1 (understood as "predisturbance"). It causes the coefficient to drop to a value outside of the noise band. With the instrumentation setup, this is not always observed before surge inception in this compressor. In all surge inception measurements though, the "last" disturbance prior to surge occurs first in rotor 1. This can be seen in the zoom in Figure 5 on the right at revolution 207. During the same revolution, a disturbance is also found in rotor 2. In none of the measurements, rotor 2 shows the disturbance prior to the one in rotor 1. Furthermore, the predisturbances occur only in rotor 1 and never in rotor 2 or 3. A disturbance arises always last in rotor 3, only as a consequence of the disturbances upstream. In conclusion, disturbances leading to the stall inception of the compressor occur in rotor 1.

This method was also applied to measurements at part shaft speed but a unique perturbation such as the spike was not detected (not shown here). The observations on the RI and the stall inception with the RI present in the flow field will be presented hereinafter.

4.3. Rotating Instability (RI). For an operating point close to the stall limit, frequency spectra are obtained from signals of a high-frequency casing pressure probe axially located close to rotor 1 leading edge and presented for three shaft speeds (80% Ω, 90% Ω, and 100% Ω) in Figure 6. The input signals have a length corresponding to about 400 rotor revolutions, and the spectra are integrated over bands of 6 Hz/Ω to reduce the noise. Always when approaching the stall limit at part speeds (80% Ω and 90% Ω) with the CREATE compressor, an increase of several amplitudes (forming a "bump") is observed in these frequency spectra at about 30% of the dimensionless blade passing frequency (12 to 24 compared to 64). The maximum amplitude of the bump reaches about 12% of the BPF of the rotor 1. That is also the case in the measurements for a previous version of the compressor at part speed but has not been investigated before. These frequencies do not arise at 95% Ω in the previous compressor version. Thus the switch from spike type stall inception to a type with multiple frequencies happens probably between 90% Ω and 95% Ω. The available measurement data does not allow linking this switch to a change in the flow field.

The multiple peaks are characteristic of rotating instabilities as commonly reported in literature. The discrete peaks are found at about the same normalized frequencies comparing multiple measurements for the two part shaft speeds presented in this paper. The RI peaks arise at nonintegers of the shaft speed. This is different from the 100% shaft speed, where peaks are found in the same frequency range but are harmonics of the shaft rotation speed, or correspond to harmonic 16 which is linked to the compressor periodicity. At 100% shaft speed, a noninteger exception is the normalized frequency of about 9.3. It is not related to the RI but probably related to an acoustic phenomenon in the downstream rotors, as referred to in the work of Courtiade and Ottavy [6].

The frequency gap between the RI peaks is constant and is a characteristic of RI. This link between the peaks and the rotating frequency was also identified by Kameier and Neise [19]. It corresponds to the rotating frequency of the RI. In the present case, the rotating speed (ω_{RI}) has been calculated from the cross-correlations between signals obtained at different circumferential positions just upstream of rotor 1, applying (3). It leads to $\omega_{RI} = 0.925\,\Omega$, which perfectly fits the frequency gaps between the peaks of the RI bump.

Rotating frequencies of $0.5\,\Omega$ to $0.65\,\Omega$ have been reported in several works on RI ([2, 20–22]) and $0.7\,\Omega$ was reported by Inoue et al. [7]. Note that among these references the works of Baumgartner et al. [20] and Vignau-Tuquet and Girardeau [22] were conducted on high-pressure compressors. Also a rotation speed of $0.9\,\Omega$ has been reported by Zhang et al. [23] in a numerical work on a turbine working locally as a compressor at low mass flow rate. In the present case, $\omega_{RI} = 0.925\,\Omega$ appears high but not out of range of what has been reported in literature. The maximum amplitude is found close to the rotor 1 LE and there is no trace of the instability in rotors 2 and 3 (not detailed here). Likewise, Mailach et al. [2] reported the maximum RI amplitude in the first 30% of rotor blade chord. Furthermore we know from their work (and others) that the RI trace is usually limited to 10% of span towards the casing.

Figure 7 presents the RI bump's amplitude evolution over hundreds of rotations towards surge inception. The temporal evolutions of the RI (integrated over the normalized frequencies from 11 to 24) and the rotor 1 BPF are presented based on the casing pressure signal from a probe positioned close to the rotor 1 LE. The temporal evolution is obtained from windowed Fourier transforms using signals with a length of three rotor revolutions per window and an overlap of 1.5 revolutions. The measurements start at a loaded stable operating point and the outlet throttle valve is closed very slowly giving quasi-steady operating conditions. The compressor falls into stall and surge after about 131 seconds from the beginning of the presented acquisition. During the valve closing, the BPF decreases in amplitude by less than 10% but the integrated amplitude of the RI doubles. The RI amplitude reaches a critical level (close to stall) when it surpasses half of the BPF amplitude in this compressor configuration. This information could be fed into a stall warning system. Eventually, the RI is evolving into rotating stall without a detectable spike type stall inception as shown in Figure 3.

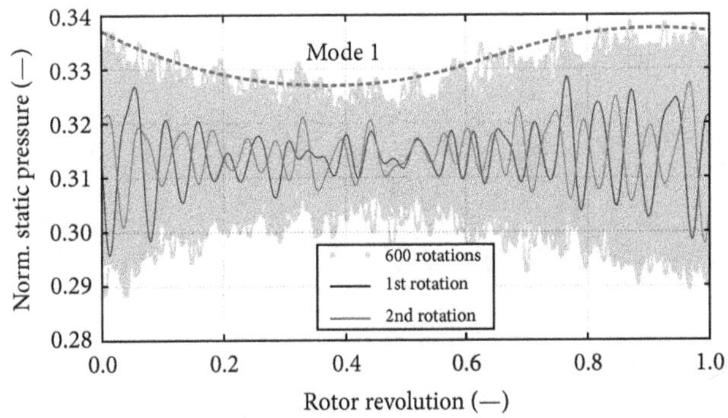

FIGURE 8: Casing pressure signal of 600 rotor revolutions stacked and superposed on two example revolutions (low-pass filtered at 5 kHz, below BPF) at a location above rotor 1 close to the LE.

The main focus in the next section is put on the periodic rotating behavior of the RI in this compressor, contrary to the unstable behavior suggested by the naming "rotating instability."

4.3.1. Variation of Temporal Extension of RI Disturbances. The observations on the temporal extension variations of the RI disturbances will be presented from raw signal plots to a detailed frequency analysis in this section.

Figure 8 shows unsteady casing pressure signals acquired with a measurement probe located close to the LE of rotor 1 at the casing. The signals are low-pass-filtered below the BPF at 5 kHz. An example of two successive rotor revolutions is plotted in red and black. The oscillations in those signals correspond to RI passing in front of the probe. The grey envelope is obtained in superimposed time trace of the pressure signals over 600 rotor revolutions. It reveals temporal mode 1 over one rotor revolution. The minimum modulation is found at about 0.4 rotor revolutions and the maximum modulation at the beginning or ending of the rotor revolutions. The amplitude modulation is in the order of magnitude of the RI induced fluctuations. With the RI present, the amplitude modulation might present the influence of the compressor on the RI. One could imagine that, once per rotor revolution, rotor blades with different tip gaps than the average pass by the measurement probe and modulate the RI amplitude. This is a hypothesis but could not be verified because time-resolved tip gap measurements were not available.

Figure 9 presents also the two previous successive pressure signals, but with a time axis now equal to the RI revolution time period ($t_{RI} = 1/\omega_R = 1/0.925$ rotor rev.). Those two pressure signals are well in phase. This could be shown for any two arbitrary successive signals. It reflects the temporal periodicity of the RIs (related to the rotating frequency ω_{RI}) over the RI revolution time period. Furthermore, a regular spaced cosine is also plotted in the figure as a reference (black dashed line). The peaks of the two pressure signals are aligned only very few times with the reference cosine. That implies a temporal extension variation of the RI disturbances out of the perspective of the fixed probe in the casing. This temporal extension variation appears periodic because of the

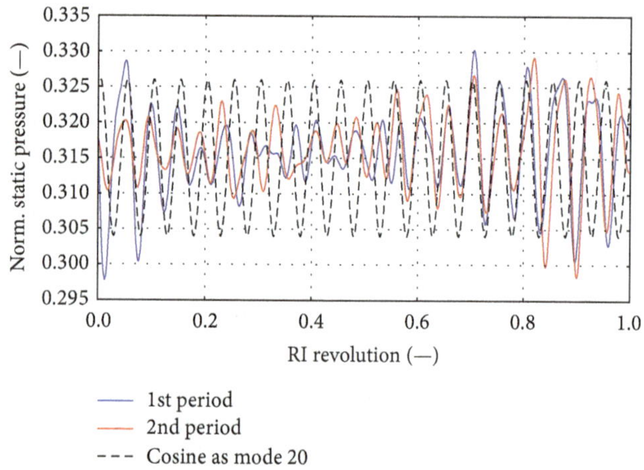

FIGURE 9: Two successive RI revolutions of the rotating instability in the casing pressure signal from above the rotor 1 LE (low-pass filtered at 5 kHz, below BPF, colored lines), and cosine as reference (black dashed line).

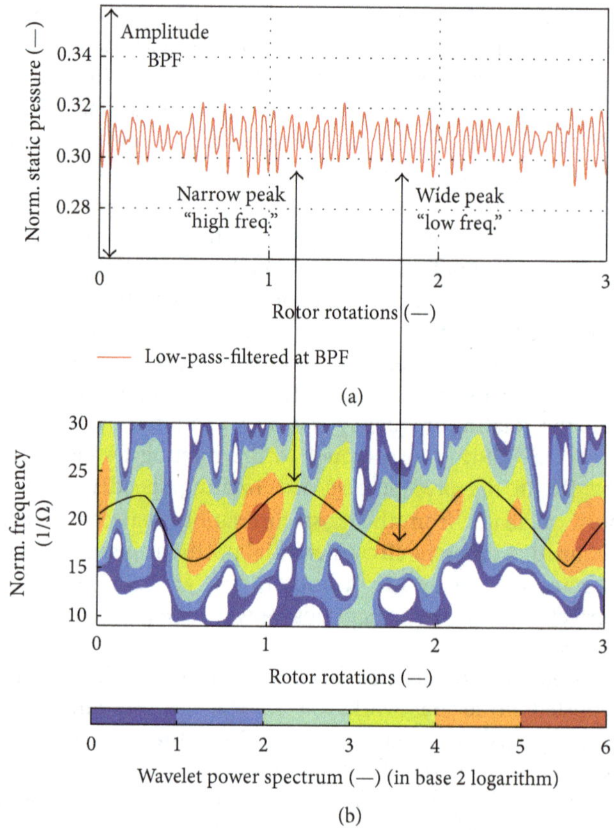

FIGURE 10: Rotating instability characterization with casing pressure signals over three rotor revolutions (a) and wavelet power spectrum (b).

fairly good match between the two periods. Additionally, the effect of the previously demonstrated amplitude modulation (AM), associated with mode 1 in Figure 8, is visible here. The AM causes peaks of different amplitudes between two RI revolutions ("red" versus "blue") at a given time point of a RI revolution. That can be explained by the fact that the AM is not periodic to a revolution of the RI but to a rotor revolution.

In conclusion, the pressure signals present fluctuations with nonconstant periods which are in phase over a time corresponding to one RI revolution, but an amplitude modulation of the RI is imposed by the rotor periodicity.

Hereafter, the variation of the temporal extension of the RI induced fluctuations is evaluated with a wavelet analysis.

Figure 10 shows the casing pressure signal (a) low-pass-filtered at 7 kHz (below BPF). Counting the RI induced peaks in the signal does not yield a constant number per rotor revolution. The short signal length of three rotor revolutions might give another impression because of the temporal periodicity being very close to one shaft rotation ($1/\omega_R$ = 1/0.925 rev. = 1.081 rev.).

At the bottom of the same figure, the wavelet power spectrum is shown with a zoom on the rotating instability frequencies. The variation of the temporal extension of one RI disturbance is seen as a modulation in the wavelet spectrum. A high frequency is related to a narrow peak (short temporal extension) in the low-pass-filtered signal, and vice versa for a low frequency. The temporal evolution of the "frequency with the maximum wavelet power" can be extracted as a signal, which is sketched with the black solid line.

This "frequency with the maximum wavelet power" signal is presented for 10 rotations in Figure 11(a). Faulty peaks can be explained by the limited temporal resolution of the wavelet spectrum. With Fourier transforms of this extracted signal, the frequency spectrum is computed and shown in Figure 11(b). The main frequencies constituting this signal are the rotating frequency of the rotating instability (ω_{RI} = 0.925 Ω) and its harmonics.

This result allows the interpretation that the rotation of the disturbance relative to the blade row seems to cause the periodic modulation of the disturbance over the circumference. Consequently, the temporal period of the modulation corresponds to $1/\omega_{RI} = 1.081$ rotor revolutions. In other words, one RI revolution corresponds to 1.081 rotor revolutions. Counting the peaks passing over 600 RI revolutions leads to a mean number of 20.097 peaks per RI revolution, with a standard deviation of 1.02 peaks. Note that RI revolutions with 21 peaks are followed immediately by RI revolutions with 19 peaks, which explains the standard deviation. This shows that there is no change in the number of peaks or rotation speed of the instability over hundreds of RI revolutions up to rotating stall inception.

4.3.2. Discussion. The result of a classical RSI would be a regular spaced (in time and space) lobed structure according to Tyler and Sofrin [24] and induce a single peak in the frequency spectra of a pressure signal. For a multistage compressor, Courtiade et al. [25] demonstrated that the amplitude of the mode could change with the time, but not its frequency. The RI is not a classical RSI because it is found as an irregular spaced lobed structure in time. That means its structure is not imposed by a blade row with constant blade pitches. The periodic temporal extension variations resemble the result of a classical frequency modulation (FM), which consists of a

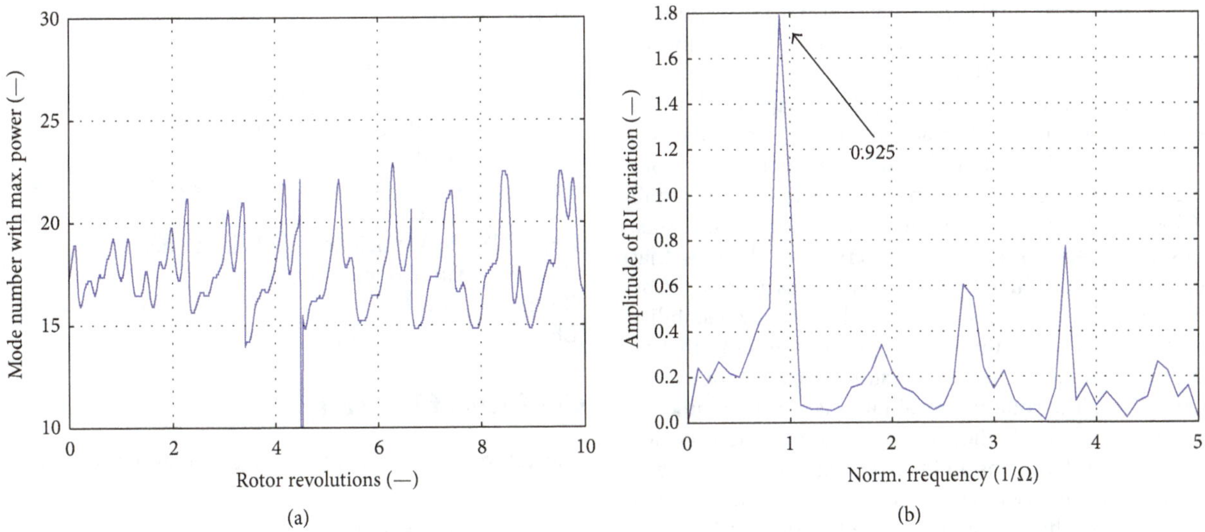

FIGURE 11: Frequency variation of rotating instability over time: temporal evolution (a); frequency spectrum (b).

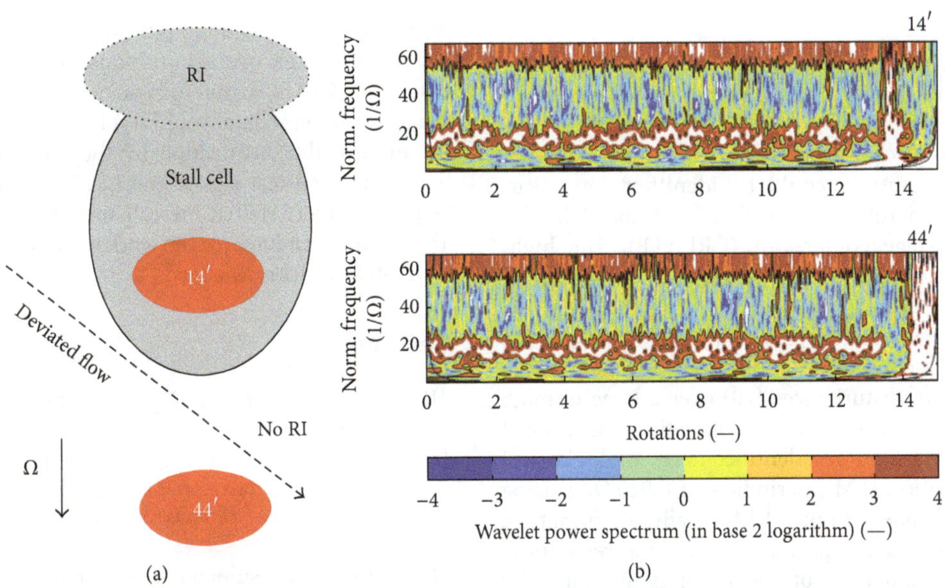

FIGURE 12: Rotating instability trace in wavelet analysis (b) and sketch (a) based on two signals positioned upstream of rotor 1 at 14′ and 44′ (red ellipse).

high frequency (the carrier signal) and a modulation signal. Furthermore, a FM would also be seen as multiple peaks in a frequency spectrum. In the case of the RI, the carrier frequency is a theoretical unmodulated rotating disturbance with a frequency of $f_c = 20 \times 0.92\,\Omega = 18.5\,\Omega$. The rotating disturbance might be caused by a separation vortex on the rotor blade suction surface and its interaction with the neighboring blade passages [3] but cannot be resolved with the fairly coarse measured grid (12 synchronized wall pressure probes). The modulation signal is equal to the rotating frequency of the RI with $f_m = \omega_{\mathrm{RI}} = 0.925\,\Omega$ (as found with the wavelet analysis). A relative motion of the disturbance to the blades would be also observed in a compressor cascade, where RI can also occur as shown by Beselt et al. [9].

4.3.3. RI at Stall Inception. We observed in this compressor that once a rotating stall cell is detected (e.g., local vanishing of interblade pressure gradient), the frequency trace of the rotating instability disappears as shown in Figure 12. The signals of two probes (positioned at 14′ and 44′ upstream of rotor 1) are analyzed with wavelet transforms, and the amplitude variations of the normalized frequencies up to the rotor 1 BPF are given in the two power maps. The signals are zoomed in about 13 rotations prior to rotating stall inception. The wavelet transforms are conducted with a temporal resolution which allows the detailed analysis of the amplitude evolution. A wavelet analysis is always a trade-off between frequency and temporal resolution. Thus the frequency resolution of the BPF is not ideal here and is found as a large horizontal band above normalized frequency 60. Again, the rotating instability is found in the band of normalized frequency of 11 to 24.

The stall inception is seen as vertical stripe of elevated power (white) at rotation 13.5 at the top (14′) and rotations 14 at the bottom (44′). Note that the rotor is turning from probe 1 (14′) to probe 2 (44′), and so does the rotating stall cell. Close to the probe at 14′ the rotating stall cell forms, induces a blockage, and causes a diversion of the flow that decreases the loading in this part of the rotor, as introduced by Emmons et al. [26]. As a consequence, the rotating instability is no longer seen in the unloaded sector of the rotor, that is, by the probe at 44′. This behavior is summarized in the sketch in Figure 12(a). The rotating instability disappears thanks to the reorganization of the flow field. This shows the very fast response of the rotating instability to changes in the flow field. The rotating instability does not reappear once the rotating stall cell is established as can be seen after rotation 14 of the signal at 14′.

5. Conclusion

This work made use of high-frequency measurements for the study of rotating instability and spike type stall inception in a high-speed compressor.

Widely reported characteristics of rotating instabilities in low-speed compressors could be identified here (e.g., frequency rake and rotation speed) at part speed in this high-speed multistage compressor (CREATE). The high-speed environment and standard tip gap sizes make this an unprecedented test case.

An original new result for the rotating instability is the demonstration of a constant rotation speed ($0.925\,\Omega$) with a constant number of disturbances (20) over a large number of rotating instability revolutions. The variation over time of the disturbance extension was identified to resemble classical frequency modulation (FM), periodic to $1/0.925\,\Omega$, causing the well-known frequency rake. Additionally, an amplitude modulation (AM) occurs, periodic to a rotor revolution, and marks the modulation of the rotating instability by the compressor system (e.g., geometry defaults). Finally the rotating instability characterized by a bump (frequency rake) in the spectra appears in the present case to be very stable, contrary to the controversy chosen naming.

The modulation of possible rotating separation vortices (forming the rotating instability) needs to be proven with higher spatially resolved measurements. In future collaborative works, the newly identified characteristics of the rotating instability could be verified on existing measurement databases of cases presented in literature.

For nominal compressor speed, a fine spike detection method was applied but the first spikelike perturbations appear usually only up to 5 rotor revolutions prior to surge in this compressor. That demonstrates the abrupt stall onset in the high-speed axial compressor environment. This makes a useful stall warning system impossible. Pressure measurements on the rotor blade surface close to the tip might be necessary for an earlier and finer detection of the spike inception.

Nomenclature

BPF:	Blade passing frequency
RI:	Rotating instability
Ω:	Rotating speed of the compressor shaft
ω_{RI}:	Rotating frequency of rotating instability
PE, L, NS:	Peak efficiency, loaded, and near surge operating points
Rev.:	Revolutions
LE:	Leading edge.

Conflicts of Interest

The authors declare that they have no conflicts of interest.

Acknowledgments

The authors are grateful to Safran Aircraft Engines for supporting the CREATE compressor research program and for granting permission to publish these results. The funding provided via the Consortium Industrie-Recherche en Turbomachines (CIRT) is gratefully acknowledged. The analysis of the measurement data would not have been possible without the high-quality acquisition by the measurement team for the high-speed test rigs at the LMFA, formed of Sébastien Goguey, Gilbert Halter, Pierre Laucher, and Lionel Pierrard. Their help and advice for the work with the experimental data are gratefully acknowledged.

References

[1] I. J. Day, "Stall, surge, and 75 years of research," *Journal of Turbomachinery*, vol. 138, no. 1, Article ID 011001, 16 pages, 2015.

[2] R. Mailach, I. Lehmann, and K. Vogeler, "Rotating instabilities in an axial compressor originating from the fluctuating blade tip vortex," *Journal of Turbomachinery*, vol. 123, no. 3, pp. 453–460, 2000.

[3] M. Inoue, M. Kuroumaru, T. Tanino, S. Yoshida, and M. Furukawa, "Comparative studies on short and long length-scale stall cell propagating in an axial compressor rotor," *Journal of Turbomachinery*, vol. 123, no. 1, pp. 24–32, 2001.

[4] G. Pullan, A. M. Young, I. J. Day, E. M. Greitzer, and Z. S. Spakovszky, "Origins and structure of spike-type rotating stall," *Journal of Turbomachinery*, vol. 137, no. 5, Article ID 051007, 11 pages, 2015.

[5] I. J. Day, T. Breuer, J. Escuret, M. Cherrett, and A. Wilson, "Stall inception and the prospects for active control in four high-speed compressors," *Journal of Turbomachinery*, vol. 121, no. 1, pp. 18–27, 1999.

[6] N. Courtiade and X. Ottavy, "Experimental study of surge precursors in a high-speed multistage compressor," *Journal of Turbomachinery*, vol. 135, no. 3, Article ID 061018, 9 pages, 2013.

[7] M. Inoue, M. Kuroumaru, S. Yoshida, T. Minami, K. Yamada, and M. Furukawa, "Effect of tip clearance on stall evolution process in a low-speed axial compressor stage," in *Proceedings of the ASME Turbo Expo*, pp. 385–394, Vienna, Austria, June 2004.

[8] K. Yamada, H. Kikuta, M. Furukawa, S. Gunjishima, and Y. Hara, "Effects of tip clearance on the stall inception process in

an axial compressor rotor," in *Proceedings of the ASME Turbo Expo 2013*, San Antonio, Tex, USA, June 2013.

[9] C. Beselt, B. Pardowitz, R. Van Rennings, and P. U. Thamsen, "Influence of the clearance size on rotating instability in an axial compressor stator," in *Proceedings of the 10th European Conference on Turbomachinery Fluid dynamics & Thermodynamicsics*, Lappeenranta, Finland, 2013.

[10] X. Ottavy, N. Courtiade, and N. Gourdain, "Experimental and computational methods for flow investigation in high-speed multistage compressor," *Journal of Propulsion and Power*, vol. 28, no. 6, pp. 1141–1155, 2012.

[11] N. Gourdain, F. Wlassow, and X. Ottavy, "Effect of Tip clearance dimensions and control of unsteady flows in a multi-stage high-pressure compressor," *Journal of Turbomachinery*, vol. 134, no. 5, Article ID 051005, 13 pages, 2012.

[12] J. Schreiber, X. Ottavy, G. N. Boum, S. Aubert, and F. Sicot, "Numerical simulation of the flow field in a high speed multistage compressor-study of the time discretization sensitivity," in *Proceedings of the ASME Turbo Expo 2015*, Montreal, Canada, June 2015.

[13] J. Schreiber, X. Ottavy, G. N. Boum, and N. Gourdain, "Influence of rotor tip flow field mis-prediction on rotating disturbance near surge in a high speed multistage compressor," in *Proceedings of the ASME Turbo Expo 2016*, Seoul, South Korea, June 2016.

[14] N. Courtiade, *Experimental analysis of the unsteady flow and instabilities in a high-speed multistage compressor [Ph.D. thesis]*, Ecole centrale de Lyon, Ecully, France, 2012.

[15] S. Weichert and I. Day, "Detailed measurements of spike formation in an axial compressor," *Journal of Turbomachinery*, vol. 136, no. 5, Article ID 051006, 9 pages, 2013.

[16] C. Torrence and G. P. Compo, "A practical guide to wavelet analysis," *Bulletin of the American Meteorological Society*, vol. 79, no. 1, pp. 61–78, 1998.

[17] F. Lin, J. Chen, and L. Meilin, "Wavelet analysis of rotor-tip disturbances in an axial-flow compressor," *Journal of Propulsion and Power*, vol. 20, no. 2, pp. 319–334, 2004.

[18] M. Dhingra, Y. Neumeier, J. V. R. Prasad, and H.-W. Shin, "Stall and surge precursors in axial compressors," in *Proceedings of the 39th AIAA/ASME/SAE/ASEE Joint Propulsion Conference and Exhibit*, Huntsville, Ala, USA, July 2003.

[19] F. Kameier and W. Neise, "Experimental study of tip clearance losses and noise in axial turbomachines and their reduction," *Journal of Turbomachinery*, vol. 119, no. 3, pp. 460–471, 1997.

[20] M. Baumgartner, F. Kameier, and J. Hourmouziadis, "Non-engine order blade vibration in a high pressure compressor," in *Proceedings of the in 12th International Symposium on Airbreathing Engines*, Melbourne, Australia, 1995.

[21] J. März, C. Hah, and W. Neise, "An experimental and numerical investigation into the mechanisms of rotating instability," *Journal of Turbomachinery*, vol. 124, no. 3, pp. 367–374, 2002.

[22] F. Vignau-Tuquet and D. Girardeau, "Aerodynamic rotating vortex instability in a multi-stage axial compressor," in *Proceedings of the in 17th International Symposium on Airbreathing Engines*, Munich, Germany, 2005.

[23] L. Y. Zhang, L. He, and H. Stüer, "A Numerical investigation of rotating instability in steam turbine last stage," *Journal of Turbomachinery*, vol. 135, no. 1, Article ID 011009, 2012.

[24] J. M. Tyler and T. G. Sofrin, "Axial flow compressor noise studies," *SAE Transactions*, vol. 70, pp. 309–332, 1962.

[25] N. Courtiade, X. Ottavy, and N. Gourdain, "Modal decomposition for the analysis of the rotor-stator interactions in multistage compressors," *Journal of Thermal Science*, vol. 21, no. 3, pp. 276–285, 2012.

[26] H. W. Emmons, C. E. Pearson, and H. P. Grant, "Compressor Surge and Stall Propagation," *Transaction of the ASME*, vol. 77, pp. 455–469, 1955.

14

Effects of Geometry Design Parameters on the Static Strength and Dynamics for Spiral Bevel Gear

Zhiheng Feng[1] and Chaosheng Song[2]

[1]*Guangxi Liugong Machinery Co., Ltd, Liuzhou, Guangxi 545007, China*
[2]*The State Key Laboratory of Mechanical Transmissions, Chongqing University, Chongqing 400030, China*

Correspondence should be addressed to Chaosheng Song; chaoshengsong@hotmail.com

Academic Editor: Paolo Pennacchi

Considering the geometry design parameters, a quasi-static mesh model of spiral bevel gears was established and the mesh characteristics were computed. Considering the time-varying effects of mesh points, mesh force, line-of-action vector, mesh stiffness, transmission error, friction force direction, and friction coefficient, a nonlinear lumped parameter dynamic model was developed for the spiral bevel gear pair. Based on the mesh model and the nonlinear dynamic model, the effects of main geometry parameters on the contact and bending strength were analyzed. Also, the effects on the dynamic mesh force and dynamic transmission error were investigated. Results show that higher value for the pressure angle, root fillet radius, and the ratio of tooth thickness tend to improve the contact and bending strength and to reduce the risk of tooth fracture. Improved gears have a better vibration performance in the targeted frequency range. Finally, bench tests for both types of spiral bevel gears were performed. Results show that the main failure mode is the tooth fracture and the life was increased a lot for the spiral bevel gears with improved geometry parameters compared to the original design.

1. Introduction

As a key component of drive axles, the failure of spiral bevel gears is detrimental to the performance and quality of the driveline system. Therefore, understanding the influence of design parameters, such as pressure angle, helix angle, root fillet radius, and tooth thickness ratio, on gear performance, is essential and can be applied to improve the dynamic performance of spiral bevel gears. This investigation is undertaken because prior studies on effects of spiral bevel gear design parameters on driveline dynamics are not widely seen in the open literature. Zhao and Dong [1] proposed that gears having 28° pressure angle have better contact and bending strength than those with 20° pressure angle. Huang et al. [2] determined that gears of bigger pressure angle have better lubrication performance than those with nominal 20° pressure angle based on elastohydrodynamic lubrication theory.

Previously, there have been extensive studies performed on static strength and dynamics for spiral bevel and hypoid gears. Fan [3, 4] described a method of determining an optimized face cone element for spiral bevel and hypoid gears based on the incorporation of calculation of tooth surface and root geometries, the conjugate relationship of the mating gear members, the ease-off topography, and the tooth contact analysis. Also, he proposed a method of correcting tooth surface errors for spiral bevel and hypoid gears generated by the face-hobbing process using computer numerically controlled hypoid gear generators. Kahraman [5, 6] proposed an accurate and practical methodology based on ease-off topography to perform loaded tooth contact analysis of spiral bevel and hypoid gears having both types of local and global deviations. Also, he investigated the impact of misalignments on root stresses of hypoid gear sets experimentally and theoretically. Yang et al. [7] proposed a pinion development approach in order to obtain excellent transmission performance of the face-milled spiral bevel and hypoid gears. Kawasaki et al. [8] proposed a method for remanufacturing pinion member of large-sized skew bevel gears using a CNC machining center and respecting an existing gear member.

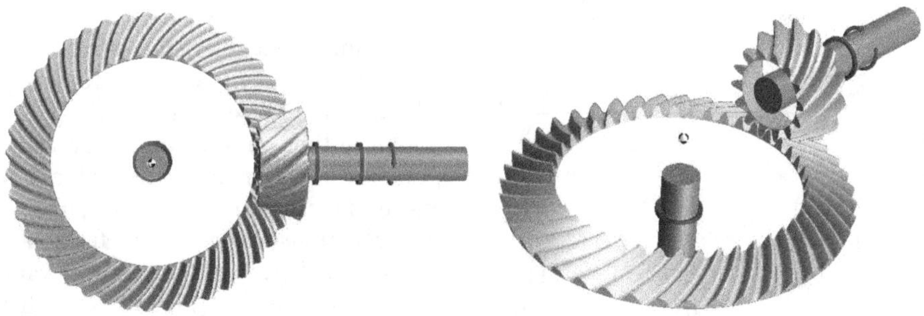

FIGURE 1: Mesh model of the spiral bevel gear pair.

The gear geometry and kinematics of space curve meshing were investigated and the mathematical model of conical-helix bevel gears from the idea of logarithmic spiral bevel gears was proposed [9]. Simon [10, 11] presented a method for the determination of optimal tooth modifications in spiral bevel gears based on improved load distribution and reduced maximum tooth contact pressure and transmission errors. Also, he proposed an optimization methodology to systematically define the optimal head-cutter geometry and machine-tool settings to simultaneously minimize the tooth contact pressure and angular displacement error of the driven gear and to reduce the sensitivity of face-hobbed spiral bevel gears to the misalignments. Yang and Lim [12] developed a multicoordinate dynamic model of a hypoid geared rotor system considering the propeller shaft bending effect. Peng [13] developed the multibody dynamic model for hypoid gear pair and investigated the effects of misalignments, torque loads, and eccentricity on the dynamic responses. Most of the research efforts on the optimization and comparison of the design parameters of spiral bevel gears are mainly based on the static strength analysis. Those studies ignored the influences on the dynamic characteristics that directly control the vibration, shock, and noise responses of running gears. Furthermore, there are even fewer studies considering the influence of the lubricating oil film. When the lubricating oil film effect is examined, most previous models assumed full film lubrication, which is not always true in practice because there is often not enough oil to create a full film contact condition. In most situations, spiral bevel gears operate in a condition of mixed lubrication with both full film and asperity contacts. This, of course, affects the static mesh and dynamics. Hence, a more realistic analysis of the friction characteristics that accurately describe the influence of friction force and friction coefficient on the dynamics of spiral bevel gears is critically needed.

Accordingly, the goals of this study are given as follows: (1) study the effects of the design parameters of spiral bevel gears on tooth bending strength and dynamics; (2) develop a nonlinear lumped parameter dynamic model considering the effect of the friction; (3) build a tooth mesh friction coefficient model for the mixed lubrication condition; (4) compare and analyze the dynamic responses of improved gears to original gears; (5) verify the computational results by comparing to the bench test results.

TABLE 1: Gearing design parameters.

	Pressure angle (°)	Tooth thickness ratio	Root fillet radius (mm)	
			Pinion	Wheel
Original	20	2.09	1.5	2.5
Optimized	22.5	2.21	1.9	3.9

2. Geometry Design and Strength Analysis

A set of spiral bevel gears with a high rate of failure were considered in this study. The weakness of these gears was found to be the poor tooth bending strength. An improved set of design parameters with optimized pressure angle, tooth thickness ratio, and tooth root fillet were determined. First, instead of 20° pressure angle, 22.5° pressure angle was applied to increase the bending strength. In fact, bevel gears with larger pressure angle are commonly used in heavy duty applications. Then, the tooth root fillet radius that is known to play an important role in gear performance was increased to reduce the tooth root stress concentration. The value of tooth root fillet radius depends on the fillet radius of the blade tip. There are many considerations for choosing the maximum blade tip fillet radius: avoidance of gear interference while running, manufacturability of the blade tip fillet, and scuffing prevention of the tooth flank due to the nonworking face of the blade using a commercial gear cutter; the cutter parameters can be chosen to achieve the desired gear parameters. Also, the cutter offset can be adjusted to obtain the specified tooth thickness ratio. The original and improved design parameters are listed in Table 1.

Based on the geometry design parameters, a commercial software Romax was used to create the mesh model and perform the loaded tooth contact analysis for the spiral bevel gear pair with the original and optimized parameters. The mesh model is shown in Figure 1. And the load condition with 4895 Nm as the input torque and 200 Rpm as the input speed was used for the subsequent calculations. Then the mesh stiffness, transmission error, line-of-action, equivalent mesh radius, and other mesh parameters can be calculated [14] as the input for the subsequent dynamic model.

The calculated results for the contact strength and bending strength are shown in Table 2 for the original spiral

TABLE 2: Contact and bending stress and safety factor of original spiral bevel gears.

	Pinion	Wheel
Number of gear meshes	1	1
Life factor at contact stress	1.432	1.432
Working stress s_w (contact) (MPa)	2628	2882
Contact stress s_c (contact) (MPa)	2734.53	
Safety factor (contact)	0.96	1.05
Life factor at bending stress	0.91	1.21
Working stress s_w (bending) (MPa)	535.8	610.9
Bending stress s_c (bending) (MPa)	470.7	628.1
Safety factor (bending)	1.14	0.97

TABLE 3: Contact and bending stress and safety factor of optimized spiral bevel gears.

	Pinion	Wheel
Number of gear meshes	1	1
Life factor at contact stress	1.364	1.364
Working stress s_w (contact) (MPa)	2674	2912.
Contact stress s_c (contact) (MPa)	2604.47	
Safety factor (contact)	1.027	1.118
Life factor at bending stress	0.81	0.95
Working stress s_w (bending) (MPa)	538.54	623.1
Bending stress s_c (bending) (MPa)	420.34	494.28
Safety factor (bending)	1.281	1.261

bevel gears. It can be observed that the contact stress of original spiral bevel gears is 2734.53 MPa. The bending stress is 470.67 MPa for pinion and 628.10 MPa for wheel. The contact safety factor of pinion is 0.96, which indicates high risk of surface pitting. This can be avoided by using shot peening in tooth surface to intensify the contact strength. Also, the bending safety factor of the wheel is relatively low, which may lead to tooth fracture.

The calculated results for the improved spiral bevel gears are shown in Table 3. The contact stress is 2604.47 MPa, which has been reduced by 4.76% as compared to the performance of the original gears. Bending stress is 420.34 MPa for pinion and 494.28 MPa for wheel, which represent a decrease of 10.69% and 21.31%, respectively. Contact and bending safety factors also increased. From the results shown in the tables, it can be concluded that the gears with 22.5° pressure angle have a relatively lower contact and bending stress level compared to the 20° pressure angle gears. The bending stress of wheel shows a significant reduction. The lower bending stress will improve the fatigue life of the spiral bevel gear.

3. Dynamic Modelling Considering Mixed Elastohydrodynamic Lubrication

3.1. Dynamic Model. Spiral bevel gear has very complex tooth geometries and kinematics. Unlike spur and helical gears, the mesh point, mesh force, and line-of-action vector for spiral bevel gear are time-varying in the 3-dimensional space. These spatial mesh characteristics along with time-varying transmission error and mesh stiffness are the primary cause of gear vibration and noise. To tackle the gear vibration problem, a nonlinear time-varying lumped parameter dynamic model is proposed as shown in Figure 2, which considers time-varying effects of mesh point, mesh force, line-of-action vector, mesh stiffness, transmission error, friction force direction, and friction coefficient as the result of varying contact areas at different roll angles [15, 16]. Note that the time-variation behavior also occurs even under constant gear speed.

In this dynamic model, the pinion and wheel are both defined as rigid bodies with complete degrees of translation and rotation freedom. The driver and load have one degree of rotational freedom each and are connected to the pinion and wheel through a set of torsional stiffness element. Gears are supported by a set of infinitesimal lumped stiffness and damping elements as the equivalent structure of shafts, bearings, and housing. In addition to the direction of the friction force as the parametric excitation parameter, the kinematic transmission error excitation and backlash nonlinearity are also included as described earlier.

The resultant 14 degrees of freedom dynamic equation can be written as

$$[M]\{\ddot{q}\} + [C]\{\dot{q}\} + [K]\{q\} = [F], \tag{1}$$

where $[M]$ is the lumped mass matrix, as

$$[M] = \text{diag}\left[I_D, M_{px}, M_{py}, M_{pz}, I_{px}, I_{py}, I_{pz}, M_{gx}, M_{gy}, M_{gz}, I_{gx}, I_{gy}, I_{gz}, I_L\right], \tag{2}$$

where $[q]$ is the generalized coordinates matrix, as

$$[q] = \text{diag}\left[\theta_D, x_p, y_p, z_p, \theta_{px}, \theta_{py}, \theta_{pz}, x_g, y_g, z_g, \theta_{gx}, \theta_{gy}, \theta_{gz}, \theta_L\right]. \tag{3}$$

And where $[F]$ is force and friction force and it can be represented by

$$[F] = \left\{T_D, h_p F_m - g_p F_{mf}, -h_g F_m + g_g F_{mf} - T_L\right\}^T, \tag{4}$$

where I_D and I_L are mass moments of inertias of driver and load and I_{li} is the mass moment of inertia about the axis li in the coordinate system S_l ($i = x, y, z$; $l = p, g$ for pinion and wheel, resp.). Furthermore, $[K]$ and $[C]$ are the stiffness and damping matrices of shafts and bearings. The damping matrix $[C]$ is assumed to be proportional viscous type representing the dissipation of energy due at the gear mesh and shaft-bearing components. Here, T_D and T_L are the torques applied to the engine and load, respectively. h_l and g_l ($l = p, g$) are the directional transformation vectors.

The dynamic transmission error can be given by

$$\delta = \{h_p\}\{q_p\} - \{h_g\}\{q_g\}. \tag{5}$$

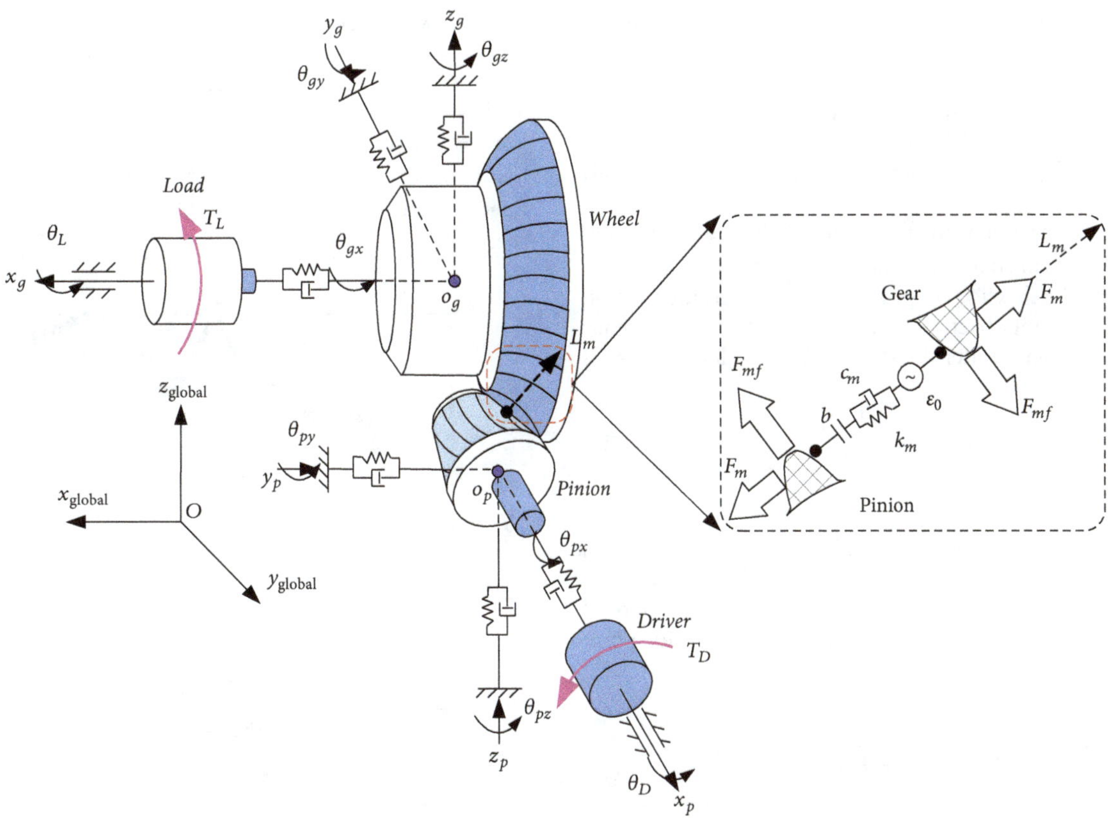

FIGURE 2: The dynamic model for spiral bevel gears.

The dynamic mesh force can be given by

$$F_m = \begin{cases} k_m(\delta - e - b) + c_m(\dot{\delta} - \dot{e}), & \delta - e > b \\ 0, & -b < \delta - e < b \\ k_m(\delta - e + b) + c_m(\dot{\delta} - \dot{e}), & \delta - e < b, \end{cases} \quad (6)$$

where e is the static transmission error and b is the gear backlash. k_m and c_m are the mesh stiffness and damping element at the mesh point with the direction along the line-of-action. The mesh damping is shown as

$$c_m = 2\xi_g \sqrt{\frac{k_{\mathrm{am}} I_{pz} I_{gz} \lambda_{gz}^2 \lambda_{pz}^2}{\lambda_{gz}^2 I_{gz} + \lambda_{pz}^2 I_{pz}}}, \quad (7)$$

where ξ_g is the mesh damping ratio, k_{am} is the average mesh stiffness, and λ_{lz} $(l = p, g)$ is the projection of line-of-action along the tangential direction of rotational motion.

The dynamic friction force is written as

$$F_{fm} = \mu F_m, \quad (8)$$

where μ is the friction coefficient.

3.2. Friction Model. Spiral bevel gears are typically operating in a condition consisting of full film and asperity contacts. In this state, the tooth contact area is divided into two situations. The first is the full film lubrication zone where two contact surfaces are completely separated by a lubricant film and the other one is the asperity contact zone with direct metal-to-metal contact. Therefore, in order to obtain more reasonable results, the time-varying friction coefficient in the mixed lubrication condition is analyzed. The tooth meshing friction coefficient can be represented by

$$\mu^{\mathrm{ML}} = \frac{F_{\mathrm{total}}}{P_{\mathrm{total}}}. \quad (9)$$

The total tangential load F_{total} is consisted of full lubrication (F_{EL}) and asperity contact (F_{BL}), as

$$F_{\mathrm{total}} = F_{\mathrm{EL}} + F_{\mathrm{BL}}. \quad (10)$$

Likewise, the total normal contact load P_{total} consists of full lubrication (P_{EL}) and asperity contact (P_{BL}), as

$$P_{\mathrm{total}} = P_{\mathrm{EL}} + P_{\mathrm{BL}}. \quad (11)$$

It may be noted that the load share coefficient of full lubrication is f_λ $(0 \leqslant f_\lambda \leqslant 1)$. Thus, the load share coefficient of asperity contact is $1 - f_\lambda$. Then, it can be shown as

$$\begin{aligned} P_{\mathrm{EL}} &= P_{\mathrm{total}} f_\lambda \\ P_{\mathrm{BL}} &= P_{\mathrm{total}} (1 - f_\lambda). \end{aligned} \quad (12)$$

Likewise, similar to the normal contact load component, the tangential load can be written as

$$F_{\mathrm{EL}} = \mu^{\mathrm{EL}} P_{\mathrm{EL}} = \mu^{\mathrm{EL}} P_{\mathrm{total}} f_{\lambda} \tag{13}$$

$$F_{\mathrm{BL}} = \mu^{\mathrm{BL}} P_{\mathrm{BL}} = \mu^{\mathrm{BL}} P_{\mathrm{total}} \left(1 - f_{\lambda}\right), \tag{14}$$

where μ^{EL} is the friction coefficient of full film lubrication and μ^{BL} is the friction coefficient of asperity contact. In addition, Winter and Michaelis [17] proposed an important relation between the average friction coefficient and the instantaneous friction coefficient under the state of full film lubrication, as shown below:

$$\mu^{\mathrm{EL}} = \mu_i^{\mathrm{EL}} \left(f_{\lambda}\right)^{0.2}. \tag{15}$$

Thus, (7) can be written as

$$\mu_i^{\mathrm{ML}} = \mu_i^{\mathrm{EL}} \left(f_{\lambda}\right)^{1.2} + \mu_i^{\mathrm{BL}} \left(1 - f_{\lambda}\right). \tag{16}$$

Zhu and Hu [18] and Castro and Seabra [19] proposed the following definition of the load share function for point contacts given by

$$f_{\lambda,\mathrm{zhu}} = \frac{1.21\lambda^{0.64}}{1 + 0.37\lambda^{1.26}}$$
$$f_{\lambda,\mathrm{zhu}} = 0.82\lambda^{0.28}, \tag{17}$$

where λ is defined as the ratio of the average oil film thickness lubricant to the root-mean square of surface roughness.

Also, λ is an important indicator to distinguish the lubrication state. Based on the research of Castro and Seabra [14], the film thickness ratio between $0.36 \leqq \lambda \leqq 1.8$ is thought to be a state of mixed film lubrication. In the mixed lubrication zone, the load share coefficient of full lubrication f_{λ} increases with the increase of λ, until it reaches a state of full film lubrication. For our calculation, a simplified value for the film thickness ratio 1.08 was used.

The instantaneous friction coefficient in mixed lubrication condition is determined by equation (14). The dynamic response of spiral bevel gear system considering the mixed lubrication time-varying friction coefficient can be calculated by substituting (14) into the dynamic model.

4. Dynamic Response Analysis

In this section, numerical integration applying the explicit 4/5th order Runge-Kutta formula is used to solve the nonlinear dynamic model. And the gear dynamic responses including dynamic mesh force and dynamic transmission error were computed and analyzed. The analysis compares the dynamic responses of the original and improved spiral bevel gears. The main geometry parameters are shown in Table 4. The dynamic mesh force and dynamic transmission error are shown in Figures 3 and 4.

From Figure 3, it can be seen that little differences of the dynamic mesh force between original spiral gears and improved spiral bevel gears can be observed for the frequency response up to 2000 Hz. However, in this area the dynamic mesh force has a relatively large amplitude and frequency fluctuations. The main reason is in the lower frequency zone; the variation range of friction coefficient is larger than that in the higher frequency range. In the frequency range from 2000 to 3500 Hz, the improved gears have a lower maximum value and smaller fluctuation than the original design. On the other hand, for the dynamic transmission error as shown in Figure 4, the improved gears are obviously better in dynamic performance than the response of the original gears in the

TABLE 4: Design parameters of spiral bevel gear.

<table>
<tr><th></th><th>Pinion</th><th>Wheel</th></tr>
<tr><td>Module</td><td colspan="2">11</td></tr>
<tr><td>Number of teeth</td><td>8</td><td>37</td></tr>
<tr><td>Helical angle</td><td colspan="2">35</td></tr>
<tr><td>Hand of spiral</td><td>LH</td><td>RH</td></tr>
<tr><td>Speed ratio</td><td colspan="2">4.625</td></tr>
</table>

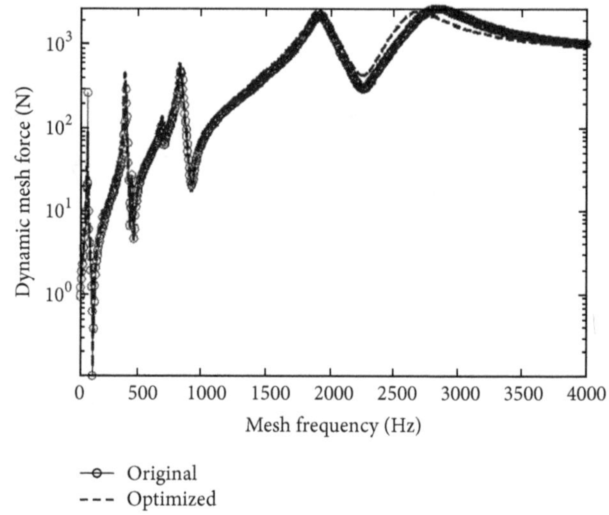

FIGURE 3: Dynamic mesh force response.

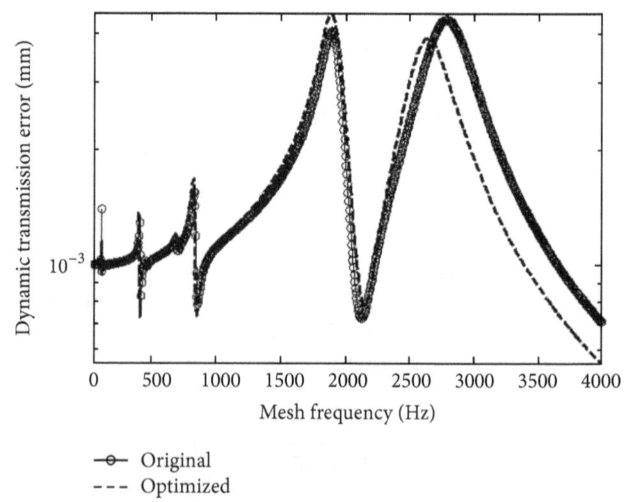

FIGURE 4: Dynamic transmission error.

FIGURE 5: Bench testing setup.

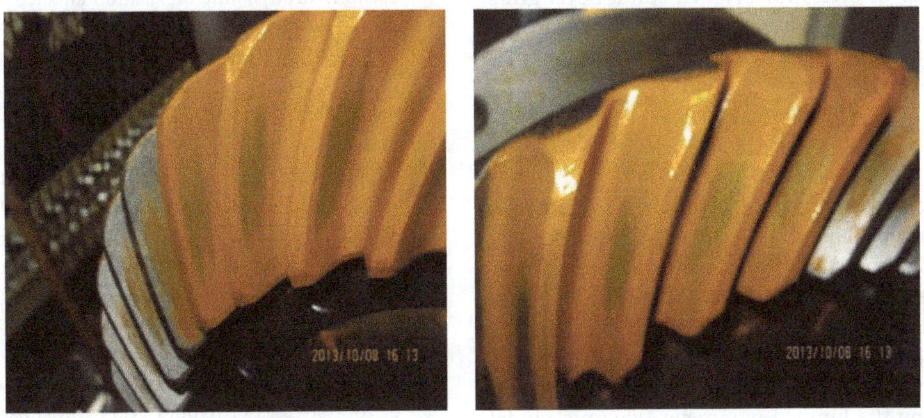

(a) Convex surface (b) Concave surface

FIGURE 6: Tested contact pattern for original design parameters.

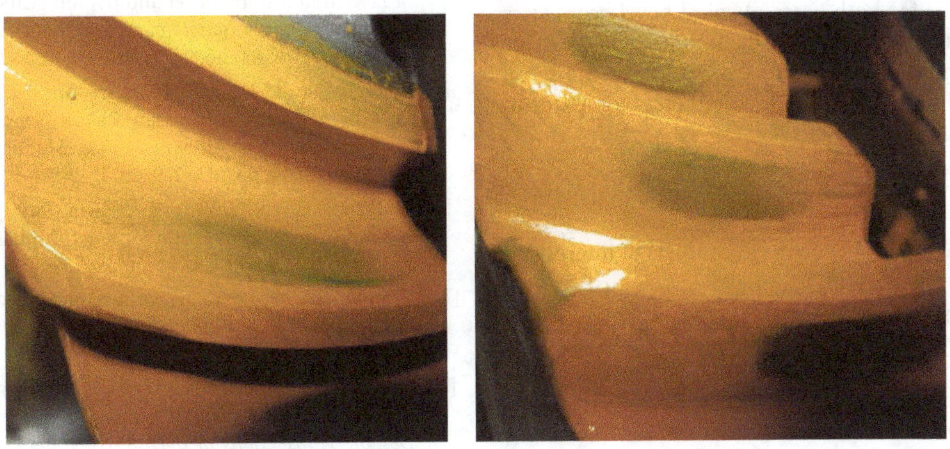

(a) Convex surface (b) Concave surface

FIGURE 7: Tested contact pattern for optimized design parameters.

frequency range from 2000 to 4000 Hz. In practical application, the improvement of the dynamic responses indicates the reduction of the vibration and noise of gears.

5. Bench Test

Bench tests for both types of spiral bevel gears were conducted to verify the improvement of gears. The test rig is shown in Figure 5. The tested contact pattern for original and optimized design parameters underrated load condition are shown in Figures 6 and 7. The contact pattern is approximately located in the middle of the convex and concave tooth surface for the original geometry design parameters. But the location moved to the toe slightly for the optimized geometry design parameters which is better from the engineering view considering the actual load condition.

The bench test ran until the tooth fracture or serious pitting happened. Record running cycle times after the test

TABLE 5: Results of the bench test.

	Torque/N·m	Cycles/10^4	Failure mode
Original	110367	42.2	Pinion teeth fracture
		31.5	Pinion teeth fracture
		26.38	Wheel teeth fracture
Optimized	110367	53.5	Pinion pitting
		40.8	Wheel teeth fracture
		49.34	Wheel teeth fracture

finish. The test results are shown in Table 5. It can be seen that the average number of cycles for original spiral bevel gears is 333,600. And the average number of cycles for improved spiral bevel gear is 478,800, which is an increase by 44% compared to the original gears.

6. Conclusions

(1) A quasi-static mesh model of spiral bevel gears was established and the effects of main geometry parameters on the contact and bending strength were investigated. Higher value for the pressure angle, root fillet radius, and the ratio of tooth thickness tend to improve the contact and bending strength and to reduce the risk of tooth fracture.

(2) The dynamic analysis of original and improved spiral bevel gears applying the lumped parameter dynamic model and the theory of mixed elastohydrodynamic lubrication was performed. The results indicate that improved gears have a better vibration performance in the targeted frequency range.

(3) Bench tests for both types of spiral bevel gears were performed. Results show that the main failure mode is the tooth fracture and the life was increased a lot for the spiral bevel gears with improved geometry parameters compared to the original design.

Conflicts of Interest

The authors declare that they have no conflicts of interest.

Acknowledgments

The authors would like to thank the National Natural Science Foundation of China (Grant no. 51405043), China Postdoctoral Science Foundation (Grant no. 2015M582517), Chongqing Research Program of Basic Research and Frontier Technology (no. cstc2016jcyjA0415), Postdoctoral Special Projects Funded of Chongqing (no. Xm2016004), Fundamental Research Funds for Central Universities (106112017CDJPT280002), and the support by Driveline R&D of Liugong. Also, thanks are due to Professor Teik C. Lim (College of Engineering and Applied Science at the University of Cincinnati).

References

[1] Y. J. Zhao and K. Dong, "Mechanical performance of involute spur gear with 28 degrees pressure angle," *Journal of Mechanical Transmission*, vol. 28, no. 1, pp. 52-53, 2004.

[2] J. L. Huang, X. X. Miu, and S. M. Luo, "Influence analysis on the elasto hydrodynamic lubrication affected by pressure angle of gear," *Journal of Machine Design*, vol. 23, no. 12, pp. 14-15, 2006.

[3] Q. Fan, "Tooth surface error correction for face-hobbed hypoid gears," *Journal of Mechanical Design, Transactions of the ASME*, vol. 132, no. 1, pp. 0110041–0110048, 2010.

[4] Q. Fan, "Optimization of face cone element for spiral bevel and hypoid gears," *Journal of Mechanical Design, Transactions of the ASME*, vol. 133, no. 9, Article ID 091002, 2011.

[5] M. Kolivand and A. Kahraman, "An ease-off based method for loaded tooth contact analysis of hypoid gears having local and global surface deviations," *Journal of Mechanical Design, Transactions of the ASME*, vol. 132, no. 7, pp. 0710041–0710048, 2010.

[6] M. A. Hotait, A. Kahraman, and T. Nishino, "An investigation of root stresses of hypoid gears with misalignments," *Journal of Mechanical Design, Transactions of the ASME*, vol. 133, no. 7, Article ID 071006, 2011.

[7] Y. Yang, S. Mao, W. Guo, and Y. Kuang, "Pinion development of face-milled spiral bevel and hypoid gears based on contact attributes," *International Journal of Advanced Manufacturing Technology*, vol. 84, no. 9–12, pp. 2347–2356, 2016.

[8] K. Kawasaki, I. Tsuji, H. Gunbara, and H. Houjoh, "Method for remanufacturing large-sized skew bevel gears using CNC machining center," *Mechanism and Machine Theory*, vol. 92, article no. 2504, pp. 213–229, 2015.

[9] R. Tan, B. Chen, C. Peng, and X. Li, "Study on spatial curve meshing and its application for logarithmic spiral bevel gears," *Mechanism and Machine Theory*, vol. 86, pp. 172–190, 2015.

[10] V. Simon, "Head-cutter for optimal tooth modifications in spiral bevel gears," *Mechanism and Machine Theory*, vol. 44, no. 7, pp. 1420–1435, 2009.

[11] V. V. Simon, "Optimal machine-tool settings for the manufacture of face-hobbed spiral bevel gears," *Journal of Mechanical Design, Transactions of the ASME*, vol. 136, no. 8, Article ID 081004, 2014.

[12] J. Yang and T. C. Lim, "Influence of propeller shaft bending vibration on drivetrain gear dynamics," *International Journal of Automotive Technology*, vol. 16, no. 1, pp. 57–65, 2015.

[13] T. Peng, *Coupled multi-body dynamic and vibration analysis of hypoid and bevel geared rotor system [Ph.D. thesis]*, University of Cincinnati, Ph.D. Dissertation, Cincinnati, OH, USA, 2010.

[14] Z. Feng, *Research on multi-body muti-DOF nonlinear dynamic of helical bevel gear [Ph.D. thesis]*, Chongqing University, Ph.D. Dissertation, Chongqing, China, 2010.

[15] Z. Feng, S. Wang, T. C. Lim, and T. Peng, "Enhanced friction model for high-speed right-angle gear dynamics," *Journal of Mechanical Science and Technology*, vol. 25, no. 11, pp. 2741–2753, 2011.

[16] S. Wang, Z. Feng, S. Lei, and H. Xiao, "Effect of time-varying friction coefficient on dynamics of spiral bevel gears," *China Mechanical Engineering*, vol. 22, no. 2, pp. 148–152, 2011.

[17] H. Winter and K. Michaelis, "Scoring load capacity of gears lubricated with EP-oils," Tech. Rep., AGMA Fall Technical Meeting, Montreal, Canada, 1983.

[18] D. Zhu and Y.-Z. Hu, "A computer program package for the prediction of EHL and mixed lubrication characteristics, friction, subsurface stresses and flash temperatures based on measured 3-D surface roughness," *Tribology Transactions*, vol. 44, no. 3, pp. 383–390, 2001.

[19] J. Castro and J. Seabra, "Coefficient of friction in mixed film lubrication: gears versus twin-discs," *Proceedings of the Institution of Mechanical Engineers, Part J: Journal of Engineering Tribology*, vol. 221, no. 3, pp. 399–411, 2007.

15

Bubble Flow Analysis of High Speed Cylindrical Roller Bearing under Fluid-Solid Thermal Coupling

Li Cui, Wenxia Wang, and Zhang Yanlei

College of Engineering, Shanghai Polytechnic University, Shanghai, China

Correspondence should be addressed to Li Cui; cuili@sspu.edu.cn

Academic Editor: Luis A. San Andrés

Heat generation model of high speed cylindrical roller bearing is constructed by calculating the local friction in the bearing. Bubble flow calculation model of roller bearing considering fluid-solid thermal coupling is constructed based on two-body fluid model and k-ε turbulent model, in which diameter and size of bubbles, breakup, and coalescence model of bubbles are considered. Using dynamic mesh method, a new method for evaluating bearing temperature is set up treating the rolling elements as moving heat sources. Based on these models and finite element method, bubble flow of a high speed roller bearing is studied based on FLUENT software. The numerical study reveals the relationship between velocity of bearing, air volume fraction, and velocity and pressure of oil-air flow. An increase of air content in the oil produces a lower pressure at the bearing outlet while the exit fluid velocity increases. When fluid-solid thermal coupling effect is considered, velocity and pressure at outlet of the bearing both become larger, while temperature of bearing is lower than that without coupling. In comparison, the coupling effects on flow pressure and temperature are obvious. For a given rotating speed, there is an optimal value for air volume fraction, such that temperature rise of the bearing reaches the lowest value. Experiments verify the outcomes of the method presented in this paper.

1. Introduction

High speed rolling bearing in the spindle of high speed machine tool and aviation engine parts often use oil-air lubrication. Oil-air lubrication system has advantages of high cooling efficiency, high lubricating efficiency, and environmental benefits because of two-phase flow [1]. Hydrodynamic properties of two-phase flow have an important influence on the capacity and thermal characteristics of the bearing [2, 3]. Therefore, it is of important theoretical significance and practical value to analyze bubble flow of rolling bearing under oil-air lubrication.

Heat analysis of high speed rolling bearing has attracted substantial attention because of heat stability effects due to friction of bearing. Jeffrey and Nelson [4] investigated operating temperatures of a rolling bearing which combined dynamics calculation with finite element thermal analysis method, which can predict transient temperature responses of the bearing but the calculation will cost too much time. Yan et al. [5] developed the network approach to analyze transient thermal properties of spindle bearing system, in which thermal-deformation coupling is considered. Experiment results indicate that temperature prediction by the transient model is much more accurate than the steady model. Ma et al. [6] constructed a three-dimensional finite element analysis model to conduct transient thermal-structure analysis of a high speed spindle bearing system. Results showed that the 3D FEA model was more accurate than the traditional model, and the thermal equilibrium experiments were conducted to validate the effectiveness of the model. These studies focus on the temperature prediction method for the rolling bearing; however, the friction model inside the bearing is also very important to the temperature results. In order to obtain more accurate temperature results, this paper will present a local friction model to construct the heat generation model of the roller bearing.

In recent years, some researchers have studied two-phase flow in rolling element bearing. Glahn and Wittig [7] gave a theoretical calculation method on liquid film flows in aeroengine bearing considering interfacial shear and gravity forces, which provided a simple analytical approach using empirical equations. In order to obtain more accurate results,

CFD analysis should be done. Aidarinis et al. [8] constructed a CFD modeling for air flow in aeroengine bearing, in which the air flow parameters were developed by the experiments. The efficiency of the lubrication and cooling system was optimized based on the CFD calculation results. Chen et al. [9] developed a numerical model for predicting the lubricant flow inside an aeroengine bearing chamber, in which film/droplet/air interactions are considered. CFX software was used to simulate the pressure and velocity distributions of lubricant flows; it was proved that the results are more accurate when film/droplet/air interactions were considered. Jiang and Mao [10] developed an experiment setup to investigate the oil-air lubrication for the high speed ball bearing; performance tests of ball bearings had bend done. In order to obtain the lowest temperature rise, oil-air supply pipe length, bearing preload, oil type, oil viscosity, and nozzle should be well designed.

These studies have a positive effect on revealing that the influence of oil and air flow in the bearing on the working performance. However, most of the existing research did not explore the effect of two-phase flow state under the working state of bearing on its working performance. Bubble flow is a kind of oil and air two-phase flow in the bearing cavity under the condition of high speed [11]. Under the condition of bubble flow, air in the bearing is no longer a whole to participate in the lubrication but is divided into different diameter, which is mixed with the lubricating oil to form a bubble air-liquid two-phase medium. The characteristics of bubble size, breakup, and coalescence effect will influence flow velocity and pressure in the bearing [12]. Bubble flow in the air-oil lubricated rolling bearing should be considered. Furthermore, fluid-solid thermal coupling existed in rolling bearing under oil-air lubrication. With consideration of the coupling, the turbulence intensity and the turbulent kinetic energy are predicted to be lower because part of the heat is carried out by the flow due to the heat transfer [13].

Bubble flow of high speed cylindrical roller bearing under fluid-solid thermal coupling has not been performed. An in-depth research should be done for a comprehensive understanding about the fluid-solid thermal coupling analysis of cylindrical roller bearing.

This paper presents a new model of two-phase bubble flow in rolling bearing under fluid-solid thermal coupling, in which the rolling elements are considered as moving heat sources. Velocity and pressure field distribution in rolling bearing are analyzed. The effects of bubble flow and fluid-solid thermal coupling on temperature rising are given.

2. Heat Generation Model

A new heat generation model of cylindrical roller bearing is proposed by analyzing local friction in the bearing. The friction power of high speed cylindrical roller bearings includes sliding friction power between roller and raceway, sliding friction power between cage and guide surface of rings, and flow resistance power caused by rollers move in oil-air mixture [14].

For the high speed cylindrical roller bearing, the contact between roller and raceway is designed as line contact, and the contact area is designed as rectangle considering contact deformation. According to Hertz contact theory, the actual contact surface always shrinks into an oval at the end because edge of roller is designed as crown, as shown in Figure 1. Main sliding contact friction between roller and raceway is parallel to the rolling direction of Y.

Because of the influence of machining precision and crown roller, the contact load between roller and raceway is distributed nonuniformly. Therefore, contact surface between roller and raceway is divided into many slides as shown in Figure 1; w is the width of the narrow area, λ is the slide number, and l is roller length.

The normal contact stress with y away from central position is given as follows [15]:

$$\sigma_{\lambda 1j} = \frac{2Q_{1j}}{\pi l b_{1j}} \left\{1 - \left(\frac{y}{b_{1j}}\right)^2\right\}^{0.5}, \qquad \sigma_{\lambda 2j} = \frac{2Q_{2j}}{\pi l b_{2j}} \left\{1 - \left(\frac{y}{b_{2j}}\right)^2\right\}^{0.5}, \tag{1}$$

where $\sigma_{\lambda 1j}$ is contact stress between the λth slide of jth roller and inner ring, Q_{1j} is contact load between the jth roller and inner ring, b_{1j} is half contact width between the jth roller and inner ring, $\sigma_{\lambda 2j}$ is contact stress between the λth slide of jth roller and outer ring, Q_{2j} is contact load between the jth roller and outer ring, and b_{2j} is half contact width between the jth roller and outer ring.

Surface shear stress with y away from central position is given as follows [15]:

$$\tau_{\lambda 1j} = \frac{2\mu Q_{1j}}{\pi l b_{1j}} \left\{1 - \left(\frac{y}{b_{1j}}\right)^2\right\}^{0.5}, \qquad \tau_{\lambda 2j} = \frac{2\mu Q_{2j}}{\pi l b_{2j}} \left\{1 - \left(\frac{y}{b_{2j}}\right)^2\right\}^{0.5}, \tag{2}$$

where $\tau_{\lambda 1j}$ is shear stress between the λth slide of the jth roller and inner ring, $\tau_{\lambda 2j}$ is shear stress between the λth slide of the jth roller and outer ring, and μ is drag coefficient of the lubricant.

It is assumed that 1/3 of the elastic deformation occurs on the roller, and 2/3 of the elastic deformation occurs on the raceway. Sliding velocity between the λth slide of jth roller and inner ring and outer ring can be written as

$$\begin{aligned} V_{\lambda 1j} &= 0.5\left\{\left[d_m + \left(D_\lambda + \frac{2}{3}\delta_{\lambda 1j}\right)\right]\omega_1 - \left(D_\lambda - \frac{1}{3}\delta_{\lambda 1j}\right)\omega_j\right\}, \\ V_{\lambda 2j} &= 0.5\left\{\left[d_m + \left(D_\lambda + \frac{2}{3}\delta_{\lambda 2j}\right)\right]\omega_2 - \left(D_\lambda - \frac{1}{3}\delta_{\lambda nj}\right)\omega_j\right\}, \end{aligned} \tag{3}$$

where $\delta_{\lambda 1j}$, $\delta_{\lambda 2j}$ are contact deformation between the λth slide of jth roller, inner ring, and outer ring. d_m is pitch

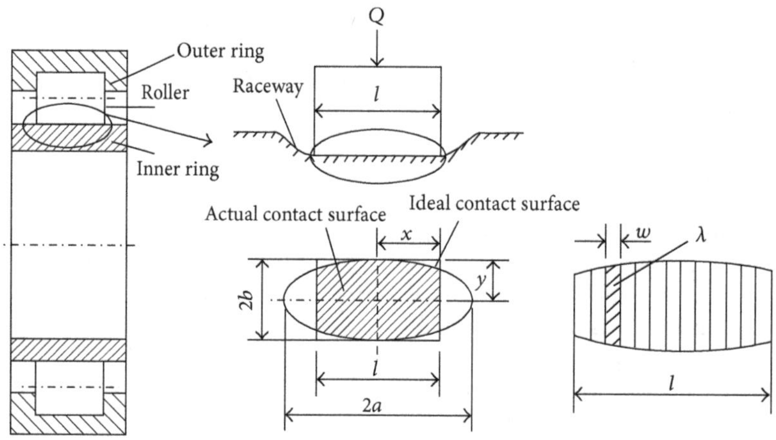

FIGURE 1: Schematic view of the roller bearing and elliptic contact surface.

diameter of roller bearing, D_λ is the equivalent roller diameter of λth slide, ω_1, ω_2 are rotating speed of inner ring and outer ring, and ω_j is the rotating speed of the jth roller.

According to the relationship of motion and frictions of the cylindrical roller bearing, friction power in bearing can be determined as follows.

Sliding friction power consumption between the jth roller and raceways of inner ring and outer ring can be given as [16]

$$H_{1j} = \sum_{\lambda=1}^{N} \tau_{\lambda 1j} V_{\lambda 1j} dA, \qquad H_{2j} = \sum_{\lambda=1}^{N} \tau_{\lambda 2j} V_{\lambda 2j} dA, \tag{4}$$

where N is the total number of the slides; dA is area of the λth slide.

Flow resistance power caused by the jth roller move in oil-air mixture can be written as [17]

$$H_{dj} = 0.5 d_m F_{dj} \left(\omega_j - \omega_o\right)^{0.81}, \tag{5}$$

where F_{dj} is drag friction between the jth roller and lubricant; ω_o is angular velocity of rolling elements.

Total power consumption of the jth roller in the bearing is

$$H_{jB} = \sum_{\lambda=1}^{N} \left(H_{1j} + H_{2j} + H_{dj}\right). \tag{6}$$

The heat generation rate of the contact ellipse region on the surface between the jth roller and rings can be calculated [18]:

$$q_{jB} = \frac{I \times H_{jB}}{A}, \tag{7}$$

where A is contact area between rollers and rings; I is heat partition coefficient and can be determined as [14]

$$I = \frac{\left(\sqrt{a_{e1}/V_1}/k_{f1}\right) / \left(\sqrt{a_{e1}/V_1}/k_{f1} + \sqrt{a_{ej}/V_j}/k_{fj}\right) + \left(\sqrt{a_{e2}/V_2}/k_{f2}\right) / \left(\sqrt{a_{e2}/V_2}/k_{f2} + \sqrt{a_{ej}/V_j}/k_{fj}\right)}{2}, \tag{8}$$

where a_{e1}, a_{e2}, a_{ej} are heat dissipation coefficient of the surface of inner ring, outer ring, and jth roller. V_1, V_2, V_j are velocities of the surface of inner ring, outer ring, and the jth roller. k_{f1}, k_{f2}, k_{fj} are thermal conductivity of inner ring, outer ring, and the jth roller.

The forced convective heat transfer coefficient between inner surface of the bearing and lubricant is given as [19]

$$\alpha_l = 0.0986 \left(\frac{\omega_1 \left(1 + D_b \cos\alpha / d_m\right)}{\left(\omega_o d_m / 2\right)}\right)^{0.5} k P_r^{0.33}, \tag{9}$$

where k is the average thermal conductivity, $k = (k_{f1} + k_{f2} + k_{fj})/3$. P_r is Prandtl constant, which changes with temperature; $P_r = 0.703$ when temperature is 20°C. α is contact angle in rolling bearing, $\alpha = 0$°C for cylindrical roller bearing.

3. Oil-Air Two-Phase Flow and Bubbles Model

3.1. Oil-Air Two-Phase Flow Model. Oil-air flow follows the N-S equation [20], in which continuity equation of oil phase is expressed as

$$\frac{\partial \left(\rho_l \alpha_l\right)}{\partial t} + \nabla \cdot \left(\mathbf{u}_l \rho_l \alpha_l\right) = 0. \tag{10}$$

The momentum equation of air phase is expressed as

$$\frac{\partial(\rho_g\alpha_g)}{\partial t}+\nabla\cdot(\alpha_g\rho_g\mathbf{u}_g\otimes\mathbf{u}_g)-\nabla\alpha_g \cdot[\mu_g\nabla(\mathbf{u}_g+\mathbf{u}_g{}^T)]=-\alpha_g\nabla p-\mathbf{F}_{gl}. \quad (11)$$

The momentum equation of the oil phase can be written as

$$\frac{\partial(\rho_l\alpha_l)}{\partial t}+\nabla\cdot(\alpha_l\rho_l\mathbf{u}_l\otimes\mathbf{u}_l)-\nabla\alpha_l\cdot[\mu_l\nabla(\mathbf{u}_l+\mathbf{u}_l{}^T)] =-\alpha_l\nabla p-\mathbf{F}_{gl}, \quad (12)$$

where α is air volume fraction, ρ is density of the phase, p is pressure, $\mathbf{u}$ is velocity vector, and μ is dynamic viscosity. F_{gl} stands for transient exchange force between air phase and oil phase. Symbol ∇ stands for gradient, $\otimes$ stands for dyadic operation, subscript "l" is liquid phase, "g" is gas phase, and superscript "T" is transpose of a matrix.

Under the condition of bubble flow, turbulent flow is formed in the liquid phase due to the mixing of the bearing element. Turbulent model is used to describe the oil flow. Turbulent flow energy equation and dissipation rate equation of the oil phase are given [20]:

$$\frac{\partial(\rho_l\alpha_lk_l)}{\partial t}+\nabla\cdot\alpha_l(\rho_l\mathbf{u}_lk_l-\mu_e\nabla k_l)=\alpha_l(\mathbf{G}_l-\rho_l\varepsilon_l),$$
$$\frac{\partial(\rho_l\alpha_l\varepsilon_l)}{\partial t}+\nabla\cdot\alpha_l(\rho_l\mathbf{u}_l\varepsilon_l-\mu_e\nabla\varepsilon_l) =\alpha_l\left(C_1\mathbf{G}_l\frac{\varepsilon_l}{k_l}-C_2\rho_l\frac{\varepsilon_l{}^2}{k_l}\right), \quad (13)$$

where $\mathbf{G}_l=\mathbf{F}_{gl}\cdot(\mathbf{u}_l{}^T-\mathbf{u}_l)/\rho_l$, μ_e is effective dynamic viscosity, μ_l is laminar dynamic viscosity, k_l is turbulent kinetic energy of liquid phase, ε_l is dissipation rate of liquid phase, $C_1=1.44$, and $C_2=1.92$ [21].

The turbulence effect of the continuous liquid phase on dispersed air phase is imposed by transient exchange force between air phase and oil phase.

3.2. Breakup and Coalescence Model of Bubbles. Fluid domain model of cylindrical roller bearing is shown in Figure 2. The bearing is lubricated in air/oil. The air bubbles can be broken and aggregated by the action of the bearing under high speed rotation when the bearing is in air/oil lubrication [19]. Because of the large number of air bubbles, the bubble diameter can be calculated according to the size of the population, and the bubbles can be divided into different populations. By considering breakup and coalescence phenomena of the populations, the model of breakup and coalescence of bubbles can be constructed.

Population balance equation for the ith population of bubbles is written as [20]

$$\frac{\partial(\rho_g\alpha_gf_i)}{\partial t}+\nabla\cdot(\mathbf{u}_g\rho_g\alpha_gf_i)=S_i, \quad (14)$$

where α_g is cavitation ratio of bubbles, f_i is volume fraction of ith population, and S_i stands for breakup and coalescence of bubbles.

The breakup model of bubbles can be constructed by [22], which is presented by Luo and Svendsen. The coalescence model of bubbles can be constructed by [23], which is presented by Prince and Blanch.

Table 1: Bearing parameters.

Parameters	Value
Inner diameter	40 (mm)
Outer diameter	75 (mm)
Roller diameter	7 (mm)
Roller length	7 (mm)
Roller number	13
Density of bearing material	7800 kg/m^3
Thermal conductivity	48 W/(m·k)

4. Modeling and Parameter Setting

4.1. Calculation Model and Grid. Taking cylindrical roller bearing NU308 as an example to build the model, the bearing parameters are shown in Table 1. In order to improve the calculation efficiency, self-rotation moving of rollers and cage is ignored. The calculation area is the fluid domain in the bearing cavity. The rotation speed of bearing is set as 5000 r/min. The bearing is loaded with 3000 N in rotation direction.

3D model of the cylindrical roller bearing is established in UG software. Fluid regions of the bearing are meshed in ICEM software. The block meshing is used for the flow field, and the tetrahedral element is used for other regions. Global element scale factor is set as 0.5; max element of global element seed size is set as 16. Refined mesh grids are applied in inlet and outlet area, which should be studied sophisticatedly. The whole bearing is divided into 185231 elements. Meshing model of the flow field in roller bearing is shown in Figure 3.

The meshed model is imported to FLUENT software. Inner ring, outer ring, and rollers are set as the static physical parameters of the 45 steel. The material of nozzle is rubber. The material properties parameters of each component are shown in Table 2, in which kinematic viscosity is defined in 40°C. HM68 lubricant is adopted and physical parameter of air is set as ideal in-compressible air. Physical properties parameters include density, thermal conductivity, specific heat capacity of the fluid and the solid is set in FLUENT software.

4.2. Setting of Dynamic Grid and Fluid-Solid Thermal Coupling. Heat generation rate of rollers is calculated by formula (7), and rollers are set as heat sources in the software. Considering moving of heat sources, dynamic grid is set in FLUENT by importing the UDF file to activate the dynamic mesh model. In the model, spring constant factor is set as 0.8, boundary point relaxation factor is set as 0.8, and grid reconstruction method is chosen as 2.5D. The rollers are set

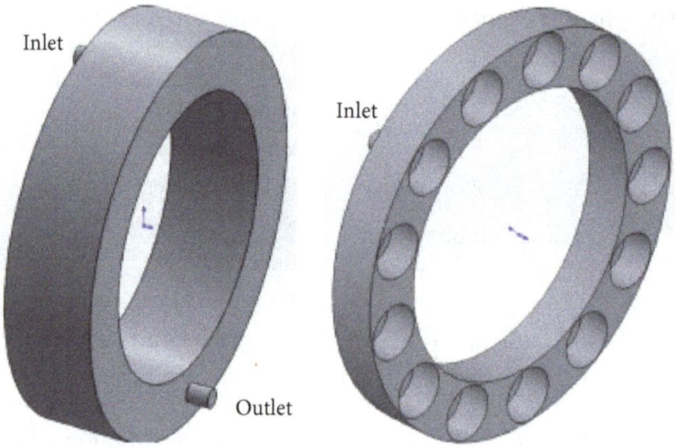

(a) Fluid cavity model (b) Section of cylindrical roller bearing

FIGURE 2: The fluid domain in the cylindrical roller bearing.

FIGURE 3: Meshing model of bearing cavity.

as rotating wall, and rotation speed of the rollers is set as follows:

$$\omega_R = \frac{(d+D)}{4D_w}\omega_1\left(1-\left(\frac{2D_w}{(d+D)}\right)^2\right), \tag{15}$$

where ω_R is rotating speed of the rollers; d, D, D_w are diameter of inner ring, outer ring, and roller, respectively.

Oil is set as the first phase; air is the second phase. In order to consider fluid-solid thermal coupling effect, contacting surface between the internal fluid and solid parts is set as default coupling-surface, and convection heat transfer is calculated automatically. Velocity boundary is set at the inlet of the bearing, which is set as 60 m/s. Pressure boundary is set at the outlet of the bearing. Finite difference method is used for discrete continuous equation and RNG k-ε two-equation

TABLE 2: Material properties of air-oil.

Parameters	Value
Initial velocity of oil	60 m/s
Density of oil	800 kg/m^3
Thermal conductivity of oil	0.148 W/(m·k)
Kinematic viscosity of oil	0.048 kg/(m·s)
Initial velocity of air	60 m/s
Density of air	1 kg/m^3
Thermal conductivity of air	0.0271 W/(m·k)
Kinematic viscosity of air	1.7×10^{-5} kg/(m·s)

model is used in the calculation. The wall of the roller bearing is set as no-slip boundary. SIMPLEC algorithm is used to simulate the coupling effect of pressure and velocity.

4.3. Simulation of Bubbles and Calculation Setting. Population balance module in FLUENT is used to simulate breakup and coalescence of bubbles and analyze multiphase flows considering population balance of bubbles. The population balance module is loaded into ANSYS FLUENT through the text user interface. The module can only be loaded when a valid ANSYS FLUENT case file has been set. The text command to load the module is "define/models/addon-module."

Multiphase flow model is set as Eulerian model with an Eulerian phase number of two, and the turbulent multiphase flow model is set as mixture. Then the population balance module can be loaded. Four bubble populations are considered, the minimum diameter of bubbles is set as 2 mm, 4 mm, 6 mm, and 8 mm, respectively, and the maximum diameter of bubbles can be calculated by the particle volume coefficient, which is set as default, 0.523. The aggregation kernel and breakage kernel are specified as Luo-model, and the surface tension for population balance is set as 0.07.

The residual error is set as $1e^{-6}$, and the iteration step number is 600; then the iteration can be solved. The results of velocity, pressure, and temperature distribution in the bearing can be obtained when the calculation is converged.

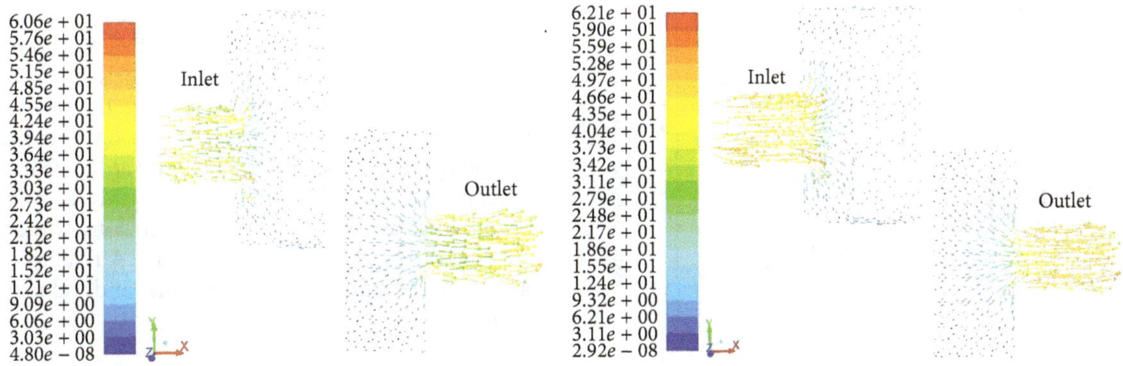

(a) Without coupling (b) With coupling

FIGURE 4: Velocity vector of oil-air flow at inlet and outlet.

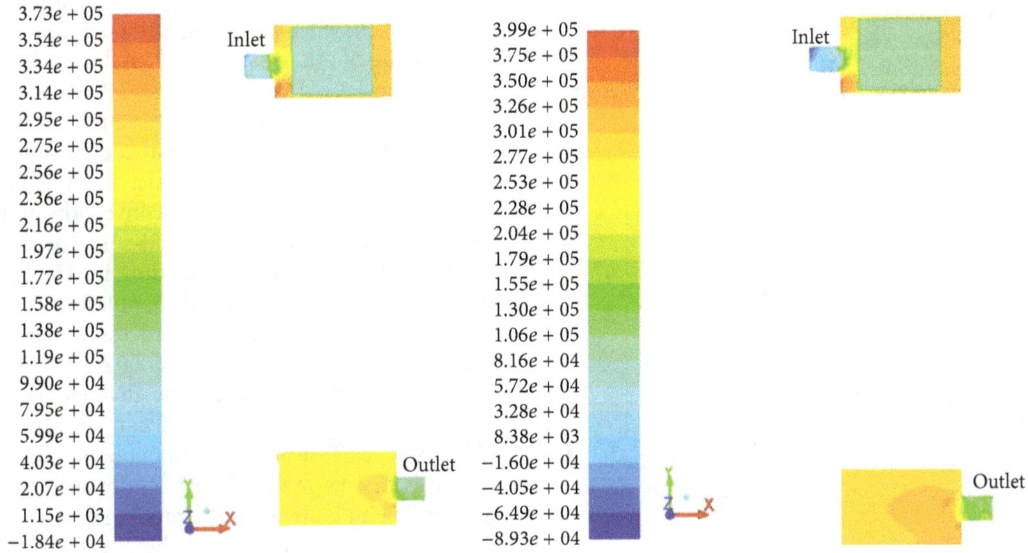

(a) Without coupling (b) With coupling

FIGURE 5: Pressure contour of oil-air flow at inlet and outlet.

5. Results and Discussions

5.1. Comparison of Velocity and Pressure in Bubble Flow. Assuming air volume fraction is 0.5, Figure 4 gives velocity vectors at inlet and outlet of roller bearing NU308. Figure 4(a) shows the velocity results without fluid-solid thermal coupling effect, while Figure 4(b) shows that the results of coupling effect are considered. Color of the arrows stands for speed value, and direction of arrows stands for flow directions. As seen in Figure 4, most of the oil-air mixture flows consistently with rotation direction of bearing, while flow directions change suddenly close to inlet because of resistance of rollers. Velocity value with coupling is larger than that without coupling. The maximum velocities of the flow at outlet are 54.5 m/s and 56.3 m/s, respectively, as shown in Figures 4(a) and 4(b).

The pressure contours at inlet and outlet of the bearing are shown in Figures 5(a) and 5(b). It can be seen that pressure reduces from inlet to outlet. It is similar for overall pressure contours whether the influence of fluid-solid thermal coupling is considered or not. The maximum pressure with coupling is higher than that without coupling effects, which are 3.99×10^5 Pa and 3.73×10^5 Pa, respectively.

Figure 6 shows the effect of air volume fraction on velocity and pressure at the outlet of the bearing cavity. An increase of air volume fraction produces the fact that pressure at outlet decreases obviously. The reason is that the increase of air components will lead to the decline of density of lubricating medium, which leads to the decrease of pressure. The decrease of pressure at the outlet will cause decrease of back flow of the lubricant. When air volume fraction increases from 0.1 to 0.9, pressure at outlet reduces by 26.1% with coupling and 22.6% without coupling, respectively. Figure 6 also shows that velocity at outlet increases slowly with increase of air volume fraction; it is because the velocity at the outlet is mainly affected by the motion of lubricant and rotor speed. When air volume fraction increases from 0.1 to 0.9, velocity at outlet increases by 7.2% with coupling and 5.7%

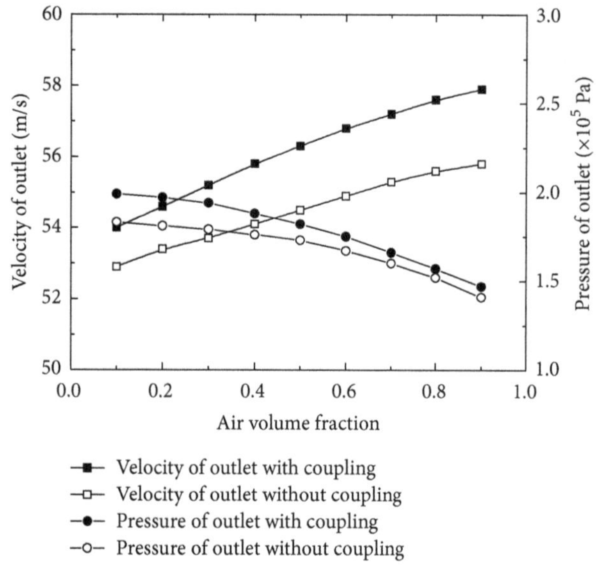

FIGURE 6: Effect of air volume fraction on velocity and pressure.

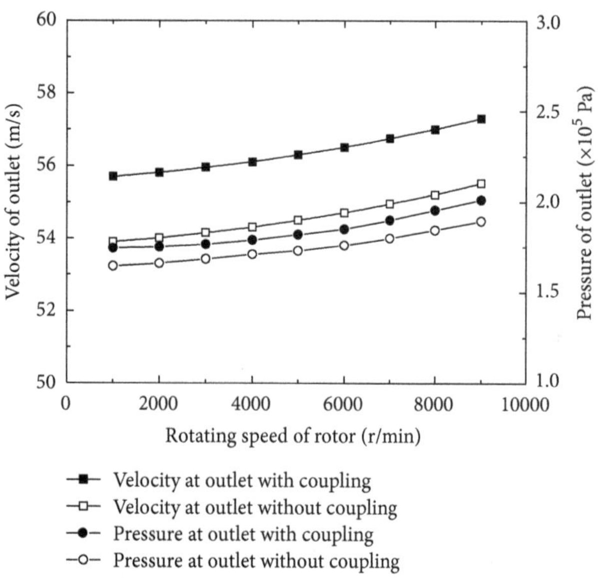

FIGURE 7: Effect of rotating speed on velocity and pressure.

without coupling, respectively. When fluid-solid thermal coupling effect is considered, velocity value and pressure value both become larger than those without coupling.

Figure 7 shows the influence of rotating speed on velocity and pressure at outlet of the bearing. It can be seen that velocity and pressure at outlet both increase with the increase of rotating speed. This is because the increase of rotation speed can accelerate the movement of oil-air in the bearing, and it also can converse the kinetic energy into potential energy, which causes the increase of pressure and velocity at the outlet. When rotating speed increases from 1000 rpm to 9000 rpm, velocity at outlet increases by 2.9% with coupling and 3.1% without coupling; pressure at outlet increases by 14% with coupling and 17.5% without coupling.

When fluid-solid thermal coupling effect is considered, velocity and pressure at outlet of the bearing both become larger than those without coupling. An increase of air content in the oil produces a lower pressure at the bearing outlet while the exit fluid velocity increases. However, an increase of rotating speed causes a higher pressure at the bearing outlet while the exit fluid velocity increases slightly. In comparison, the coupling effect on flow pressure is obvious, while coupling effect has little influence on the velocity. Air volume fraction has important influence on pressure and velocity in the bearing.

5.2. Analysis of Bubble Breakup and Temperature. Considering the influence of fluid-solid thermal coupling, temperature distribution at end face of the rollers is given in Figure 8(a). It can be seen that the maximum value occurs opposite the nozzle. Similar results can be obtained in another case shown in Figure 8(b), where coupling is not considered. The highest temperature of rollers is 347 K when fluid-solid thermal coupling is considered, and the value is 353 K when coupling is not considered. It is found that the temperature of the rollers is lower when fluid-solid thermal coupling is considered. It is because convective heat transfer in the bearing cavity is accelerated by the increase of convection and conduction coefficient under the effect of fluid-solid thermal coupling.

Assuming air volume fraction is 0.5, bubble flow is studied by the calculation with population equation considering breakup and coalescence process of bubbles. Assuming rotating speed increases from 1000 rpm to 9000 rpm, the effects of rotating speed on the diameter and number of bubbles are shown in Figure 9. It can be seen that diameters of bubbles decrease and the number of bubbles increases; the reason is that large bubbles broke down into small bubbles gradually with increase of the rotating speed.

Figure 10 gives relationship between average diameter of initial bubbles and the maximum temperature of rollers. It shows that roller temperature increases gradually with the increase of rotating speed, and the temperature also increases with diameter of initial bubbles. When rotating speed increases from 1000 rpm to 9000 rpm, the maximum temperature difference between diameters of bubbles is 4 mm and no bubble increases from 5 K to 10 K, and that between diameters of bubbles is 8 mm and no bubble increases from 11 K to 15 K. The temperature difference becomes more and more obvious with the increase of rotating speed. The reason is that pressure drops at outlet when bubbles exist, which will result in the decline of backflow ability of oil-air. Therefore, an increase of bubble diameter will cause the increase of roller temperature.

Assuming rotating speed is 5000 rpm, Figure 11 studies the relation between maximum temperature of rollers and air volume fraction. Results show that the temperature of rollers decreases first and then increases with the rise of air volume fraction. The reason is that oil resistance is higher when air volume fraction is smaller, which will cause temperature rise. Meanwhile, pressure at outlet decreases and oil supply nearby the outlet is insufficient when air volume fraction is too large, which will also cause temperature rise. When the air volume fraction is 0.65, it is found that the bearing temperature comes to the lowest value at 5000 rpm.

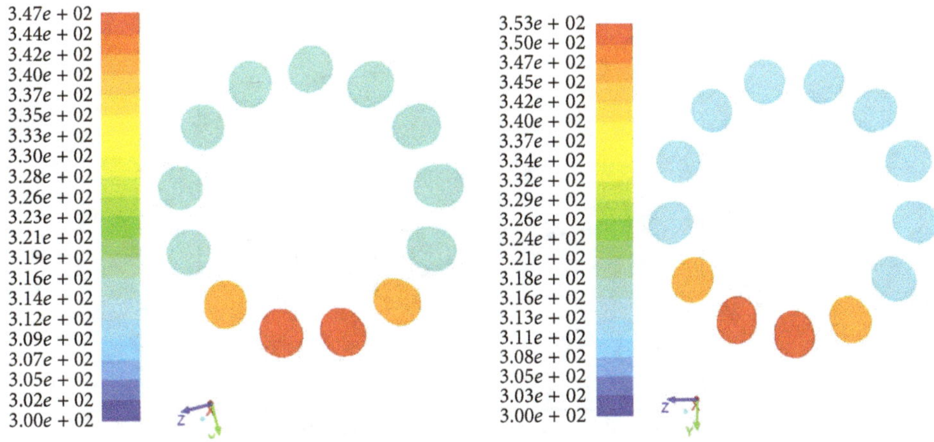

(a) With coupling (K) (b) Without coupling (K)

FIGURE 8: Temperature of the end face of rollers.

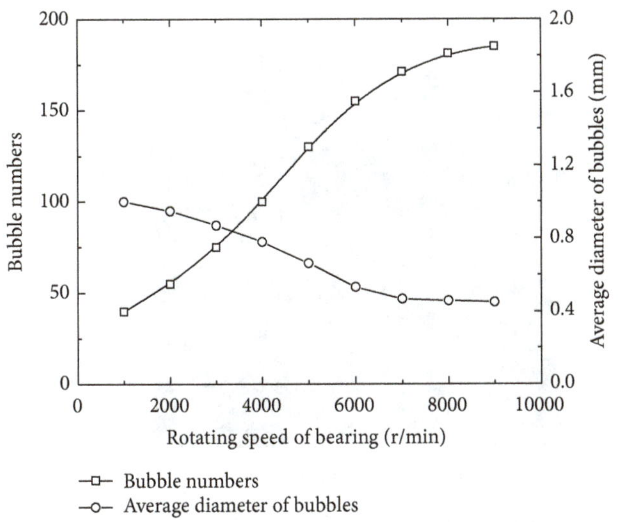

FIGURE 9: Bubble number and diameter varying with rotating speed.

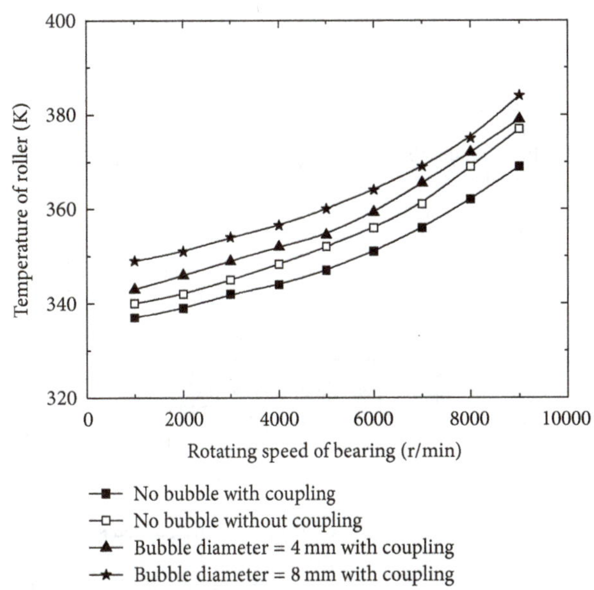

FIGURE 10: Effect of diameter of bubbles on temperature of roller.

5.3. Experimental Verification. Figure 12(a) shows the schematic view of high speed spindle bearing equipment. The equipment includes water cooling device, oil/air lubrication generator, air compressor, high speed spindle bearing system, sensors, and data sample system. The compressed air is sent into an air reservoir by a filter and then supplied to an oil-air distributor. A pump sends oil to the oil-air mixer, and oil-air volume fraction can be adjusted by a valve. Cooling water is pumped into the sleeve of the spindle and flux of the water can be adjusted. A pair of test bearings is back-to-back mounted on a shaft in the spindle bearing system. Continuous oil film is formed gradually inside the oil-air pipes, and the oil film moves along the inner wall of the oil-air pipe to the test bearings. The bearings are lubricated by oil-air mixture and cooled by the cycling water.

Figure 12(b) gives the view of the experiment setup. The rotating speed of bearing can reach 18000 rpm. The temperatures at outer rings of the bearings are measured by two thermocouples with an accuracy of 0.1°C. The vibration signal of the bearings is measured by two vibration acceleration sensors. The sampled data are sent to the computer.

Ambient temperature is kept as 25°C by an air-conditioner during the test. The oil-air lubrication system and cooling water are started fifteen minutes prior to the beginning of the experiment. The air volume fraction can be adjusted; flux of cooling water is set as 3.5 L/min. It takes about ten minutes for the bearing to reach the thermal equilibrium at a specific rotating speed. Temperature rising tests of roller bearing NU308 under different rotating speed have been done using this equipment. During the test, rotating speed is increased from 1000 rpm to 9000 rpm; the temperature data of bearing is recorded when thermal equilibrium occurs.

When air volume fraction is set as 0.5, Figure 13(a) gives contrast between experimental results and calculation

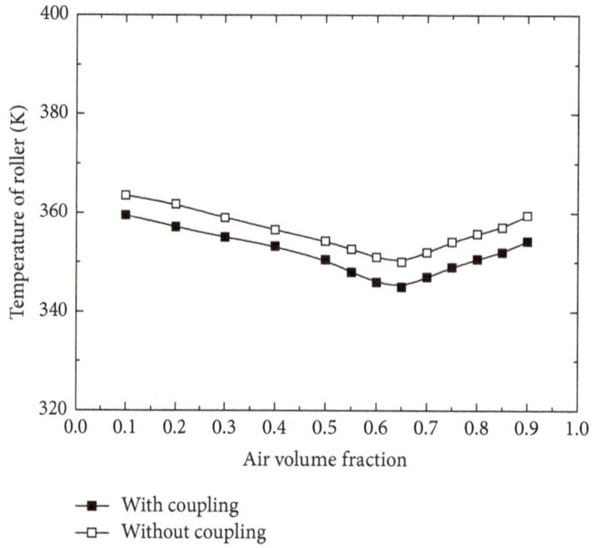

FIGURE 11: Relation of maximum roller temperature and air volume fraction at 5000 rpm.

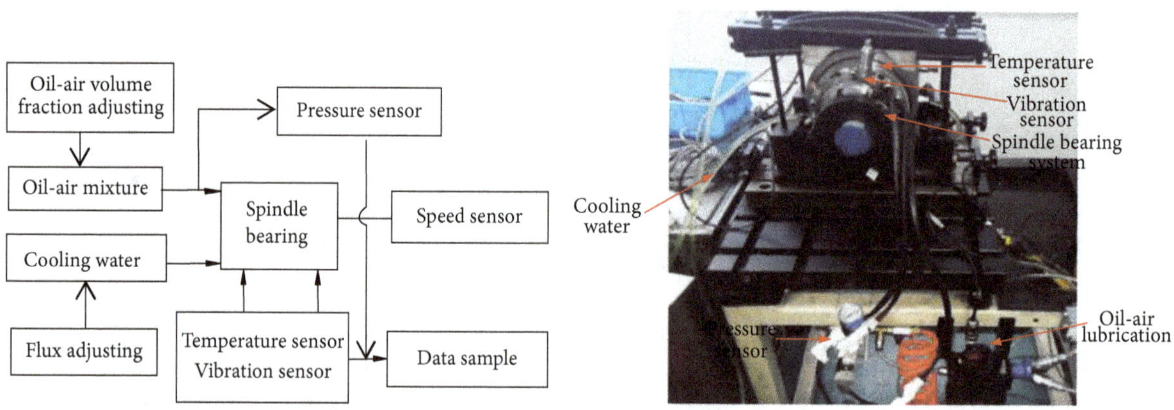

(a) Schematic view (b) The experiment setup

FIGURE 12: High speed spindle bearing equipment under oil-air lubrication.

results of temperature rising at the outer ring of the bearing. It can be seen that temperature rising increases gradually with the increase of rotating speed. When rotating speed increases from 1000 rpm to 9000 rpm, the difference between calculation results without coupling and test results increases from 6.9% to 12.5%. While the difference between calculation results with coupling and test results increases from 4.3% to 5.7%. Deviation between calculation results and test results increases with rotating speed; the reason is that vibration of the bearing increases due to large unbalance force of the spindle system at high speed, which causes the fact that actual temperature is larger than the calculation result. When fluid-solid thermal coupling effect is considered, temperature value is lower than that without coupling. The results demonstrate that calculation results with fluid-solid thermal coupling are more close to the experimental results, which verify the outcomes of the method presented in this paper.

When air volume fraction is adjusted from 0.1 to 0.9, temperature rising at outer ring of the bearing is measured by the test equipment. Figure 13(b) gives test results under different air volume fraction at 3000 rpm, 5000 rpm, and 7000 rpm, respectively. It can be seen that temperature rising decreases first then increases with the rise of air volume fraction. When rotating speed is 3000 rpm, 5000 rpm, and 7000 rpm, the temperature rising has the lowest value when air volume fraction is 0.5, 0.65, and 0.75. The conclusions verify the calculation result in Figure 11. In order to obtain the lowest temperature rising of bearings, the air volume fraction should be adjusted according to the rotating speed during the operation.

6. Conclusions

(1) When fluid-solid thermal coupling effect is considered, velocity and pressure at outlet of the cylindrical roller bearing both become larger than those without coupling. An increase of air content in the oil produces a lower pressure at the bearing outlet while the exit fluid velocity increases slightly. However, an increase of rotating speed produces a higher

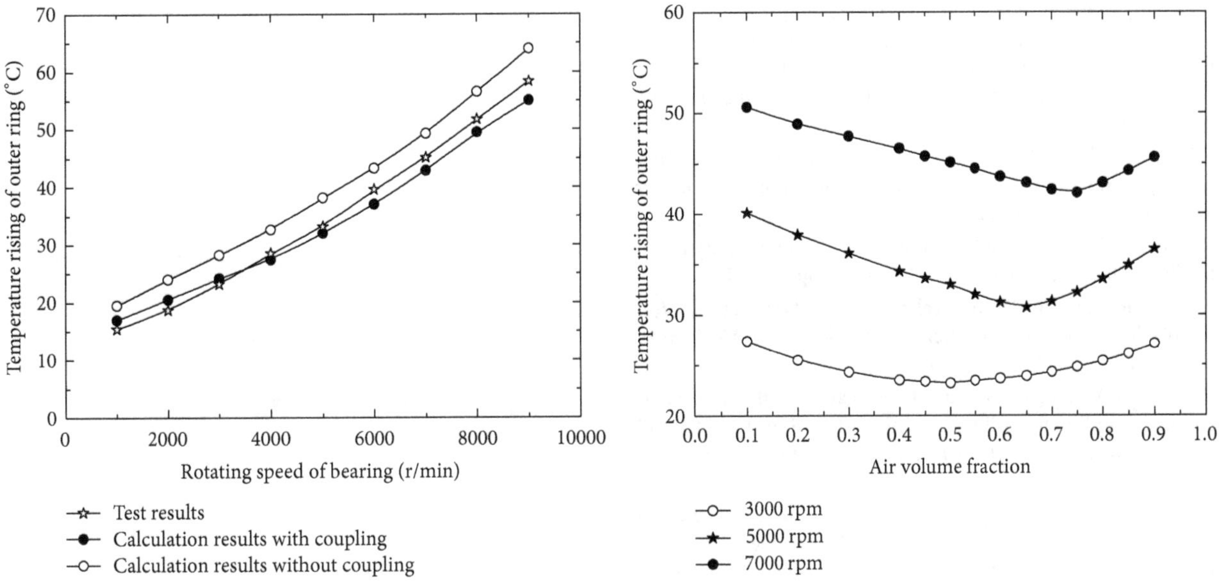

(a) Contrast when air volume fraction is 0.5

(b) Test results when air volume fraction changes

FIGURE 13: Contrast between experimental and calculation results.

pressure at the bearing outlet while the exit fluid velocity increases slightly. In comparison, the coupling effect on flow pressure is obvious, while coupling effect has little influence on the velocity. Air volume fraction has important influence on pressure and velocity in the bearing.

(2) Temperature value of roller with fluid-solid thermal coupling is lower than that without coupling, and the difference of temperature values becomes more and more obvious with the increase of rotating speed. For a given rotating speed, there is an optimal value for air volume fraction, such that temperature rise of the bearing reaches the lowest value. In order to obtain the lowest temperature rising of bearings, the air volume fraction should be adjusted according to the rotating speed during the operation.

(3) Pressure, velocity at the outlet, and temperature of bearing all increase with the increase of rotating speed. Large bubbles break down into small bubbles gradually with increase of speed, and the diameter of bubbles decreases. When bubbles exist, an increase of diameter of initial bubble in oil-air flow causes a higher temperature of roller in bearing.

(4) Experiment proves the calculation accuracy considering fluid-solid thermal coupling effect. Fluid-solid thermal coupling should be considered in the calculation for predicting the properties of the flowing in rolling bearing under oil-air lubrication.

Conflicts of Interest

The authors declare that there are no conflicts of interest regarding the publication of this paper.

Acknowledgments

This work was financially supported by the National Natural Science Foundation of China (no. 51675323), Shanghai Natural Science Foundation (no. 14ZR1416800), and the Key Subject of Shanghai Polytechnic University (Material Science and Engineering, XXKZD1601).

References

[1] S. Y. Jiang and H. B. Mao, "Investigation of the high speed rolling bearing temperature rise with oil-air lubrication," *ASME Journal of Tribology*, vol. 133, Article ID 021101, 9 pages, 2011.

[2] G. B. Daniel and K. L. Cavalca, "Evaluation of the thermal effects in tilting pad bearing," *International Journal of Rotating Machinery*, vol. 2013, Article ID 725268, 17 pages, 2013.

[3] S.-R. Chen, G.-C. Xu, and H.-G. Ruan, "Applications of computational fluid dynamics in design of oil-mist collection," *Design*, vol. 11, pp. 910–913, 2009.

[4] R. B. Jeffrey and H. F. Nelson, "Development of numerical models for heat generation and temperature prediction in lightly lubricated bearings," *ASME Journal of Tribology*, vol. 5, pp. 23–28, 2003.

[5] K. Yan, J. Hong, J. Zhang, W. Mi, and W. Wu, "Thermal-deformation coupling in thermal network for transient analysis of spindle-bearing system," *International Journal of Thermal Sciences*, vol. 104, pp. 1–12, 2016.

[6] C. Ma, X. Mei, J. Yang, L. Zhao, and H. Shi, "Thermal characteristics analysis and experimental study on the high-speed spindle system," *International Journal of Advanced Manufacturing Technology*, vol. 79, no. 1-4, pp. 469–489, 2015.

[7] A. Glahn and S. Wittig, "Two-phase air/oil flow in aero-engine bearing chambers—Assessment of an analytical prediction method for the internal wall heat transfer," *International Journal of Rotating Machinery*, vol. 5, no. 3, pp. 155–165, 1999.

[8] J. Aidarinis, D. Missirlis, K. Yakinthos, and A. Goulas, "CFD modeling and LDA measurements for the air-flow in an aero engine front bearing chamber," *Journal of Engineering for Gas Turbines and Power*, vol. 133, no. 8, Article ID 082504, 2011.

[9] G. Chen, B. Chen, Y. Liu, and C. Chen, "Exploring through numerical simulation lubrication flow inside an aeroengine bearing chamber with film/droplet/air interactions taken into consideration," *Journal of Northwestern Polytechnical University*, vol. 29, no. 1, pp. 62–67, 2011.

[10] S. Y. Jiang and H. B. Mao, "Investigation of the high speed rolling bearing temperature rise with oil-air lubrication," *Journal of Tribology*, vol. 133, no. 2, pp. 655–664, 2011.

[11] X. Yin and S. Kumar, "Lubrication flow between a cavity and a flexible wall," *Physics of Fluids*, vol. 17, no. 6, pp. 1791–1804, 2005.

[12] H. Wu and G. Chen, "Numerical research on two-phase gas/oil bubble flow in bearing chamber," *Chinese Journal of Mechanical Engineering*, vol. 44, no. 9, pp. 70–78, 2008.

[13] B. Tong, G. Wang, and X. Sun, "Investigation of the fluid-solid thermal coupling for rolling bearing under oil-air lubrication," *Advances in Mechanical Engineering*, vol. 7, no. 2, Article ID 835036, 2015.

[14] L. Q. Wang, G. C. Chen, L. Gu, and D. Z. Zheng, "Operating temperature in high-speed ball bearing," *Proceedings of the Institution of Mechanical Engineers, Part C: Journal of Mechanical Engineering Science*, vol. 221, no. 3, pp. 353–359, 2007.

[15] T. A. Harris, *Rolling Bearing Analysis*, John Wiley and Sons, New York, NY, USA, 4th edition, 2001.

[16] L. Cui and J. Zheng, "Nonlinear vibration and stability analysis of a flexible rotor supported on angular contact ball bearings," *Journal of Vibration and Control*, vol. 20, no. 12, pp. 1767–1782, 2014.

[17] J. H. Rumbarger, E. G. Filetti, and D. Gubernick, "Gas turbine engine mainshaft roller bearing system analysis," *Journal of Lubrication Technology*, vol. 95, no. 4, pp. 401–416, 1973.

[18] A. Cameron, *Basic Lubrication Theory*, Ellis Horwood, Herts, UK, 1981.

[19] L. Q. Wang, G. C. Chen, L. Gu, and D. Z. Zheng, "Study on operating temperature of high-speed cylindrical roller bearings," *Journal of Aerospace Power*, vol. 23, no. 1, pp. 179–183, 2008.

[20] H. Wu and G. Chen, "Numerical research on two-phase gas/oil bubble flow in bearing chamber," *Chinese Journal of Mechanical Engineering*, vol. 44, no. 9, pp. 70–75, 2008.

[21] H. Gu and L. Guo, "Numerical simulation of turbulent bubble flow in vertical pipe," *Journal of Chemical Industry and Engineering*, vol. 55, no. 4, pp. 563–568, 2004.

[22] H. Luo and H. Svendsen, "Theoretical model for drop and bubble break-up in turbulent dispersions," *Chemical Engineering Science*, vol. 66, no. 5, pp. 766–776, 1996.

[23] M. J. Prince and H. W. Blanch, "Bubble coalescence and break-up in air-sparged bubble columns," *AIChE Journal*, vol. 36, no. 10, pp. 1485–1499, 1990.

16

Study on Frequency Characteristics of Rotor Systems for Fault Detection using Variational Mode Decomposition

Kai Chen,[1,2] Xin-Cong Zhou,[1,2] Jun-Qiang Fang,[1,2] and Li Qin[3]

[1]*Key Laboratory of High Performance Ship Technology (Wuhan University of Technology), Ministry of Education, Wuhan 430063, China*
[2]*Reliability Engineering Institute, School of Energy and Power Engineering, Wuhan University of Technology, Wuhan 430063, China*
[3]*School of Machinery and Automation, Wuhan University of Science and Technology, Wuhan 430081, China*

Correspondence should be addressed to Kai Chen; kaichen@whut.edu.cn

Academic Editor: Adam Glowacz

Due to the complicated structure, vibration signal of rotating machinery is multicomponent with nonstationary and nonlinear features, so it is difficult to diagnose faults effectively. Therefore, effective extraction of vibration signal characteristics is the key to diagnose the faults of rotating machinery. Mode mixing and illusive components existed in some conventional methods, such as EMD and EEMD, which leads to misdiagnosis in extracting signals. Given these reasons, a new fault diagnosis method, namely, variation mode decomposition (VMD), was proposed in this paper. VMD is a newly developed technique for adaptive signal decomposition, which can decompose a multicomponent signal into a series of quasi-orthogonal intrinsic mode functions (IMFs) simultaneously, corresponding to the components of signal clearly. To further research on VMD method, the advantages and characteristics of VMD are investigated via numerical simulations. VMD is then applied to detect oil whirl and oil whip for rotor systems fault diagnosis via practical vibration signal. The experimental results demonstrate the effectiveness of VMD method.

1. Introduction

For a long time, faults of rotating machinery were mainly diagnosed by spectrum analysis method of vibration signal, which determine the failure by analyzing the frequency spectrum or frequency characteristics of vibration signals [1]. Practically, the vibration signal is generally nonstationary and nonperiodic when the fault occurs, so the traditional FFT cannot meet the needs of the time-varying and nonstationary signals detection and time-frequency analysis [2].

Like many fault signals, the vibration signal of rotor system is a typical nonlinear and nonstationary time-varying signal whose frequency components change over time [3]. The feature extracting of fault signal not only affects the accuracy of fault diagnosis, but also may lead to misdiagnosis. Therefore, effective extraction of vibration signal characteristics is the key to diagnose the faults of rotating machinery [4].

Analysis, processing, and feature extraction of nonlinear and nonstationary signal always are one of the hot topics concerning engineers and researchers [5–7]. EMD is an effective signal analysis method which is suitable for dealing with nonlinear and nonstationary signals [8]. It consists in a local and fully data-driven separation of a signal with fast and slow oscillations. However, EMD experiences some problems, such as end effect and mode mixing [9]. To overcome shortcomings, the ensemble empirical mode decomposition (EEMD) was proposed [10], which can suppress the appearance of modal mixing. But both of them lack theoretical support. Besides being recursive decomposition, error of envelope line will be spread; thus they cannot eliminate the problem of modal mixing completely.

Dragomiretskiy and Zosso [11] have proposed a new adaptive decomposition method called variational mode decomposition (VMD) in 2014, which can nonrecursively decompose a signal into a number of intrinsic mode functions (IMFs). In the process of decomposition, by using alternate direction multipliers method (ADMM), the center frequency and bandwidth of each modal were updated, to search the optimal solution of the signal decomposition. The updating process of each mode is carried out synchronously and makes the component of each modal in the corresponding baseband, finally, to achieve the effective decomposition of the signal. Compared with EMD and EEMD, VMD does not strip the signal step by step, but decomposes the signal synchronously. It is nonrecursive and reduces the spread of error. This method also has a good theoretical basis which is easy to understand. Besides, it has stronger robustness to noise [12, 13]. With its advantages, VMD is a wide application in the research field of signal denoising [14, 15], feature extraction [16, 17], image processing [18], energy and economics price forecasting [19, 20], speech signals detection [21], and especially in fault diagnosis [22–25]. However, these literatures mainly focus on the advantages of VMD comparing with the EMD and EEMD, rather than investigating the own capabilities of the VMD in signal analysis. The research of VMD needs to be conducted in depth. For this reason, a further investigation of the VMD is conducted in this paper via both numerical simulation and experimental approaches.

In this paper, the characteristics of VMD are investigated based on the theoretical of VMD, as well as the advantages of VMD. And then, VMD is applied to detect the fault of oil whirl and oil whip in rotor systems. This provides an effective solution for fault diagnosis of rotor system.

2. Variational Mode Decomposition

2.1. EMD and EEMD Method. EMD is a time-frequency signal analysis method for nonlinear signals, which can decompose the data adaptively and obtain a series of IMFs [9]. These IMFs reflect the characteristics of the signal itself. EMD algorithm is very suitable for analyzing the nonstationary signal.

Each signal could be decomposed into a number of intrinsic mode functions (IMFs), each of which must satisfy the following definition [9]:

(1) In the whole data set, the number of extrema and the number of zero-crossings must either be equal or differ at most by one.

(2) At any point, the mean value of the envelope defined by local maxima and the envelope defined by the local minima is zero.

With the definition, any signal $x(t)$ can be decomposed as follows [9]:

(1) Identify all local maxima and all local minima in the signal $x(t)$.

(2) Connect all local maxima and all local minima by using a cubic spline line as the upper envelope $e_u(t)$ and the lower envelope $e_l(t)$, respectively. The mean $m_1(t)$ of upper and low envelope is calculated by using the equation

$$m_1(t) = \frac{(e_u(t) + e_l(t))}{2}. \quad (1)$$

(3) Calculate the equation

$$h_1(t) = x(t) - m_1(t) \quad (2)$$

and examine whether the resultant $h_1(t)$ is an IMF satisfying the two aforementioned conditions. If $h_1(t)$ is not an IMF, regard $h_1(t)$ as original signal and repeat steps (1)–(3) until $h_1(t)$ is an IMF.

(4) Separate the first IMF from $x(t)$ and then get the residual component $r_1(t)$:

$$r_1(t) = x(t) - h_1(t). \quad (3)$$

Treat $r_1(t)$ as original data and repeat the above processes; therefore the second IMF $h_2(t)$ is obtained. Iterate the previous calculations n times and n IMFs of the signal could be got. Then

$$r_n(t) = r_{n-1}(t) - h_n(t). \quad (4)$$

The decomposition process can be stopped when the final residual component $r_1(t)$ is a monotonic function or a constant from which no more IMF can be extracted.

(5) The original signal $x(t)$ finally can be expressed as

$$x(t) = \sum_{i=1}^{n} h_i(t) + r_n(t). \quad (5)$$

Thus, one can achieve a decomposition of the signal into n-empirical modes, and a residue $r_n(t)$, which is the mean trend of $x(t)$. The IMFs $h_1(t), h_2(t), \ldots, h_n(t)$ include different frequency bands ranging from high to low.

EMD is a self-adaptive signal decomposition method, and the obtained IMFs have the advantages of mutual orthogonality. However, practice has disclosed that the EMD also shows the following defects in signal processing and feature extraction.

(1) No Mathematical Foundation. Without strict mathematical proof and mathematical model, EMD lacks theoretical support. And there is no valid way to stop the decomposition process in the sifting process of EMD.

(2) End Effect. In the decomposition process, endpoint will be used as an extreme point for fitting the upper and lower envelopes. But if the endpoint is not an extreme point, the fitted curve using cubic spline line will have some error. This error will impact on each subsequent decomposition step; the error will accumulate gradually and then is likely to pollute the internal decomposed results from both ends.

(3) Mode Mixing. Mode mixing occurs in EMD; when steps change, pulse interference and outside noise will show in the signal. There are two forms of mode mixing; one form is a single IMF including oscillations of dramatically disparate scales and the other is a component of a similar scale residing in different IMFs. Once the mode mixing is generated, subsequent decomposition step will be influenced, even making IMFs meaningless in physical interpretation.

To alleviate the mode mixing problem occurring in EMD, an ensemble empirical mode decomposition (EEMD) is presented [26]. The essential essence of EEMD algorithm is decomposing the original signal added Gaussian noise repeatedly by EMD method, and the original vibration data is decomposed into a series of IMFs with different scales and continuous characteristics because of the characteristic of frequency uniform distribution of Gaussian white noise, which can suppress the appearance of modal mixing.

2.2. VMD Method. VMD is a new signal analysis method for nonlinear and nonstationary signal, which aims to decompose the signal into different discrete modes [11]. The VMD has a solid mathematical foundation, which is able to decompose a signal into an ensemble of band-limited intrinsic mode functions (IMFs) simultaneously. IMFs are redefined as amplitude modulated frequency modulated (AM-FM) signals, written as

$$u_k(t) = A_k(t)\cos(\varphi_k(t)), \tag{6}$$

where $u_k(t)$ is IMF, $A_k(t)$ is nonnegative envelope, and $\varphi_k(t)$ is the phase and a nondecreasing function.

The core of VMD is to construct and solve the variational problem, the decomposition process of VMD algorithm is the solution of the variational problem. There are three important concepts in the signal processing, Wiener Filtering, Hilbert transform, and frequency mixing, which constitute the building blocks of VMD model.

Assume each mode u_k is band-limited and compactly distributed around with a center frequency w_k, so the solution of variational problem turns to seeking for k modes u_k to make the sum of the bandwidth minimized, while the constraint condition is the sum of modes equal to the input signal f. The principle of the VMD algorithm is the following:

(1) The analytic signal of mode u_k was computed by means of Hilbert transform, so obtain the unilateral spectrum of the analytic signal.

$$\delta(t) + \frac{j}{\pi(t)} * u_k(t). \tag{7}$$

(2) Shift the frequency spectrum of each mode to the respective estimated central frequency.

$$\left[\delta(t) + \frac{j}{\pi(t)} * u_k(t)\right]\cdot e^{-jw_k t}. \tag{8}$$

(3) The bandwidth is estimated through the Gaussian smoothness of the demodulated signal, that is, the squared L^2-norm of the gradient. The resulting constrained variational problem is given by the following:

$$\min_{\{u_k,w_k\}} \left\{\sum_k \left\|\partial_t\left[\delta(t) + \frac{j}{\pi(t)} * u_k(t)\right]\cdot e^{-jw_k t}\right\|_2^2\right\}$$
$$\text{s.t} \quad \sum_k u_k = f, \tag{9}$$

where $\{u_k\} := \{u_1,\ldots,u_k\}$ and $\{w_k\} := \{w_1,\ldots,w_k\}$ are the identified set of modes and their central frequencies.

(4) A quadratic penalty term α and a Lagrangian multiplier factor $\lambda(t)$ are introduced in order to render the problem unconstrained. The quadratic penalty is a classic way to encourage reconstruction fidelity, and the Lagrangian multipliers are a common way of enforcing constraints strictly. The augmented Lagrangian is shown as follows:

$$\begin{aligned} L(\{u_k\},\{w_k\},\lambda) &= \alpha\sum_k \left\|\left[\delta(t) + \frac{j}{\pi(t)} * u_k(t)\right]\cdot e^{-jw_k t}\right\|_2^2 \\ &+ \left\|f(t) - \sum_k u_k(t)\right\|_2^2 \\ &+ \left\langle \lambda(t), f(t) - \sum_k u_k(t)\right\rangle. \end{aligned} \tag{10}$$

(5) An alternate direction multipliers method (ADMM) is applied to solve the original minimization problem, to find the saddle point of the augmented Lagrange expression via updating u_k^{n+1}, w_k^{n+1}, and λ_k^{n+1} alternately. So the signal was decomposed into different discrete modes adaptively.

The flowchart of VMD methods is shown in Figure 1, and the detailed steps are as follows.

(1) Initialize $\{\hat{u}_k^1\}$, $\{w_k^1\}$, $\hat{\lambda}^1$ and $n = 0$.

(2) $n = n+1$; update $\hat{u}_k$ and w_k according to (11) and (12).

$$\hat{u}_k^{n+1}(w) = \frac{\hat{f}(w) - \sum_{i\neq k}\hat{u}_i^n(w) + \hat{\lambda}^n(w)/2}{1 + 2\alpha(w - w_k^n)^2}, \tag{11}$$

$$w_k^{n+1} = \frac{\int_0^\infty w\left|\hat{u}_k^{n+1}(w)\right|^2 dw}{\int_0^\infty \left|\hat{u}_k^{n+1}(w)\right|^2 dw}, \tag{12}$$

where $k = \{1, 2, \ldots, K\}$.

(3) Update λ.

$$\hat{\lambda}^{n+1}(w) = \hat{\lambda}^n(w) + \tau\left(\hat{f}(w) - \sum_k \hat{u}_k^{n+1}(w)\right). \tag{13}$$

(4) Repeat steps (2) and (3), until the following convergence condition is satisfied:

$$\frac{\sum_k \left\|\hat{u}_k^{n+1} - \hat{u}_k^n\right\|_2^2}{\left\|\hat{u}_k^n\right\|_2^2} < \varepsilon, \tag{14}$$

where $\varepsilon \geq 0$ is the convergence tolerance.

To show the advantage of VMD, the simulated signal in (15) is decomposed by using EMD, EEMD, and VMD, respectively; the results are shown in Figure 2.

$$\begin{aligned} y_1 &= \sin(2\pi 30t), \\ y_2 &= \sin(2\pi 10t) * (t > 0.2), \\ y_3 &= \sin(2\pi 50t) * (t < 0.2), \\ y &= y_1 + y_2 + y_3. \end{aligned} \tag{15}$$

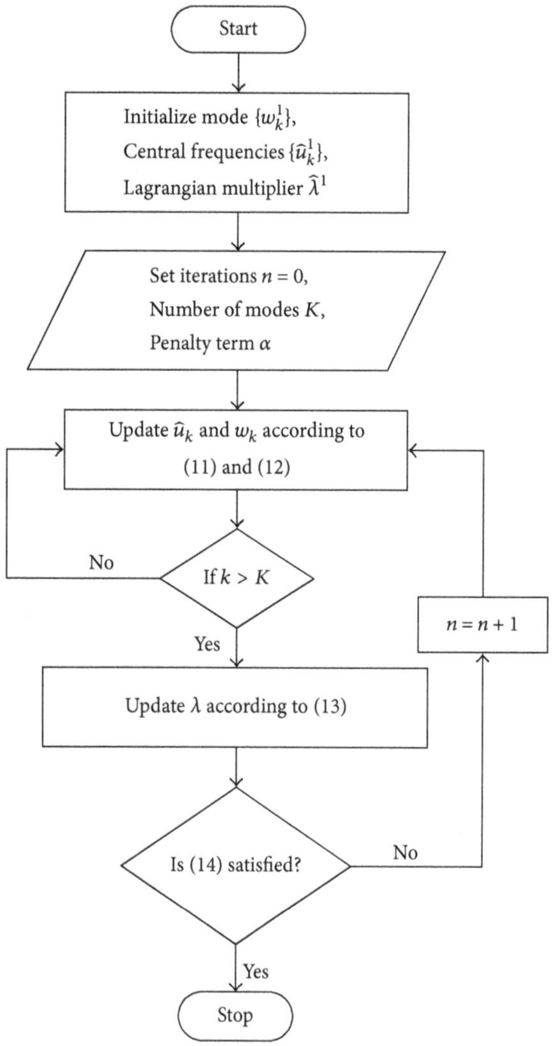

FIGURE 1: Flowchart of VMD.

It can be seen from Figure 2 that the three components of original signal were decomposed by VMD without mode mixing, while it is decomposed by VMD completely. The IMF2, IMF3, and IMF4 decomposed by EEMD are basically the same as the three components in the original signal, but mode mixing was not completely eliminated, and end effect still exists; furthermore, there is a high-frequency component IMF1. And the IMFs decomposed by EMD are hard to distinguish while the mode mixing still exists. In summary, VMD is better than EEMD and EMD in feature extraction of vibration signal.

3. Characteristics of VMD

Compared to EMD and EEMD, VMD has advantages in solving the problem of mode mixing and end effect. What is more, according to the principle and decomposition steps of VMD, the following characteristic of VMD can be obtained.

(1) Adaptivity. Unlike wavelet analysis which needs to select the best wavelet base, the signal was decomposed by VMD without prior knowledge. Various signals can be decomposed into different IMFs according to information about the signal itself, and the IMFs correspond to the components of different frequency range and center frequency in the signal. This reflects adaptability of the VMD method.

(2) Reconfigurability. As mentioned before, the core of VMD is to solve the variational constraint problem, while the constraint condition is the sum of modes equal to the input signal. So the original signal can be reconstructed by the IMFs decomposed by the VMD, which was proved by formula (9) theoretically. The simulation signal of (15) was taken as an example to show that VMD is reconfigurable. The original signal, the reconstructed signal, and error curve are shown in Figure 3. It can be seen from the figure that the error is very small to explain the reconfigurability of VMD.

(3) Smoothing Filtering. VMD also has certain filtering characteristics. The simulation signal of (15) added a random noise which was decomposed by the VMD and reconstructed. The time and frequency domain of the original simulation signal, the simulation signal added noise, and the reconstructed signal were illustrated in Figure 4.

By comparing the original signal and the reconstructed signal, the added noise was eliminated, and the characteristic peaks of spectrum were more prominent and obvious after reconstitution. And, in the time domain, fluctuating character of the reconstructed signal is more clear than the signal added noise, which is hard to identify. VMD does have the smoothing filter characteristic.

This can also be deduced from the VMD algorithm in theory. It is observed from formula (11) that the quadratic penalty term α is in the denominator. When α increases, the bandwidth of each mode will be decreased, the spectrum becomes smooth, and the filtered and smoothed modal components are discarded as noise.

(4) Orthogonality. A series of modal components (IMFs) were decomposed by VMD; each modal component represents a feature of the original signal theoretically, so the modal components are orthogonal to each other. However, it is impossible to achieve absolute orthogonality between any two IMFs in practical applications; this is because of end effect and mode mixing. Firstly, the end effect of the VMD may result in IMF data diverge at both ends, thus creating a false component. The endpoint effect is more serious in the shorter length signal. In addition, there is no mode mixing in and between the IMFs theoretically, but there still exists mode mixing practically.

4. Examples of Application

4.1. Experiment Setup and Signal Acquisition. To evaluate the performance of VMD method, experimental analysis of a rotor system has been carried out in a rotor test rig. As shown in Figure 5, the rotor test rig is composed of electrical motor, shaft, rotor, bearing seat, coupling, sensor support, and foundation. An electrical motor with frequency converter, whose speed is up to 10,000 r/min, is the drive. The

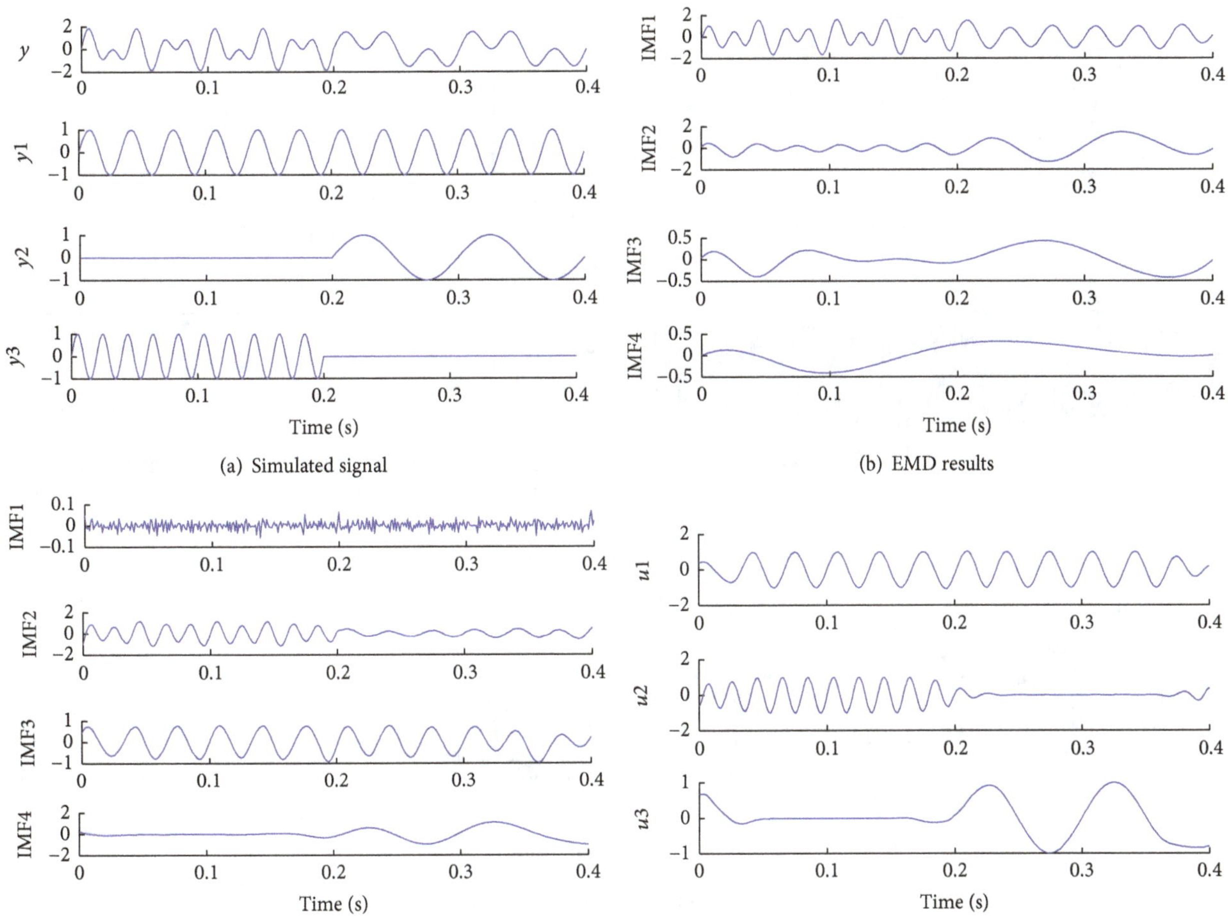

(a) Simulated signal

(b) EMD results

(c) EEMD results

(d) VMD results

FIGURE 2: Comparison results of EMD, EEMD, and VMD.

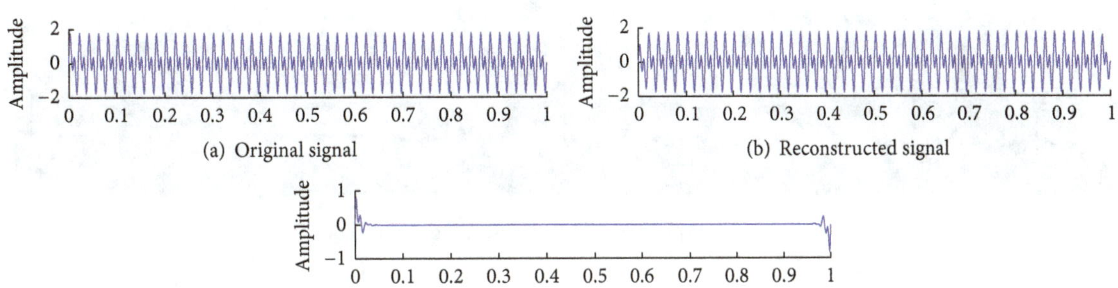

(a) Original signal

(b) Reconstructed signal

(c) Error

FIGURE 3: Reconfigurability of VMD.

oil film bearing and oil cup were designed to simulate the phenomenon of oil whip.

A phase sensor closed to the motor is fixed to measure the rotating speed. A velocity sensor (ZA-HV-2-5) is fixed on bearing seat in the middle of test rig to measure the vibration of bearing seat. Two eddy current sensors, fixed on sensor support closed to oil film bearing, are used to pick up horizontal and vertical directions displacement of rotor.

4.2. Results and Discussion. The oil film instability of the rotor system usually consists of two stages, oil whirl and oil whip. The whirl is a form of motion in which the rotor rotates around its own axis, while the axis rotates around the center of the bearing. When the rotational speed of the rotor reaches a certain speed, the oil whirl is generated, and the whirling frequency is about half of that speed. When the rotational speed is up to 2 times of the first-order critical speed of rotor system, the oil whirl turns into the oil whip, the amplitude

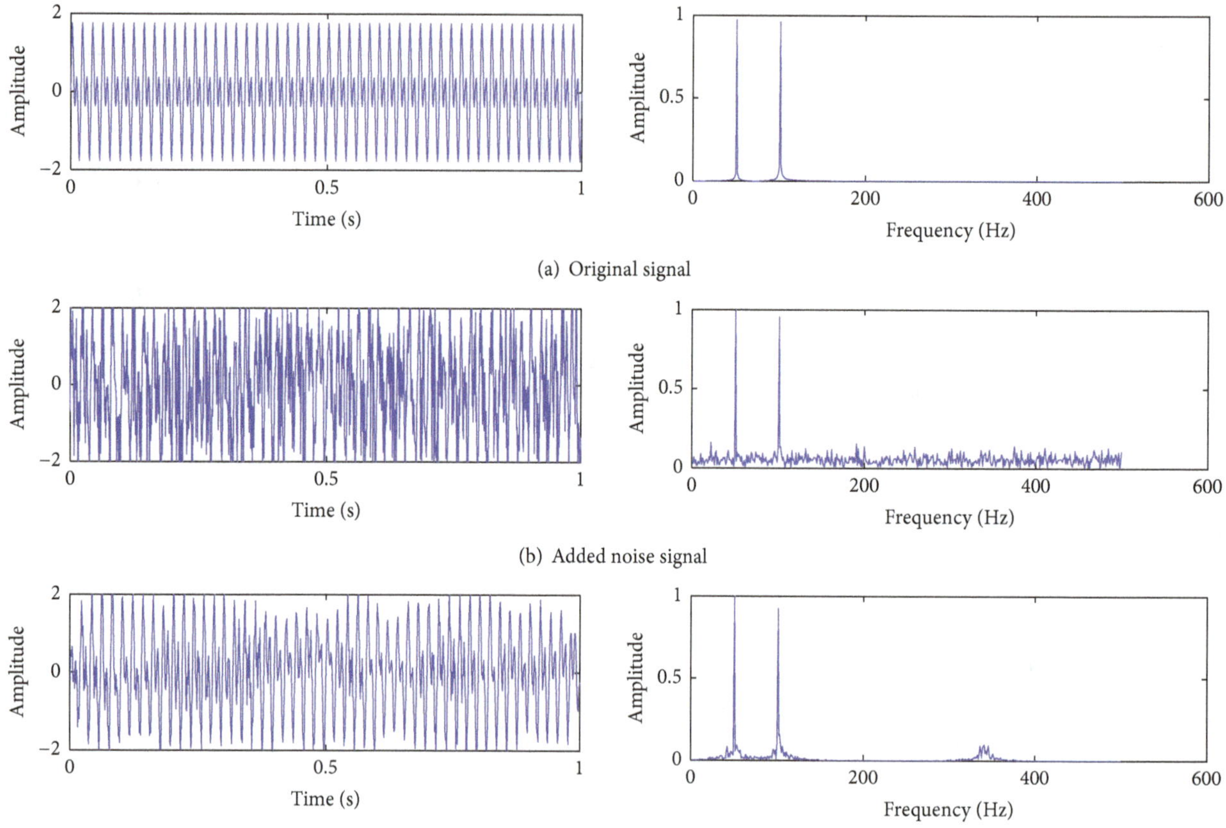

(a) Original signal

(b) Added noise signal

(c) Reconstructed signal

FIGURE 4: The filtering characteristics of VMD.

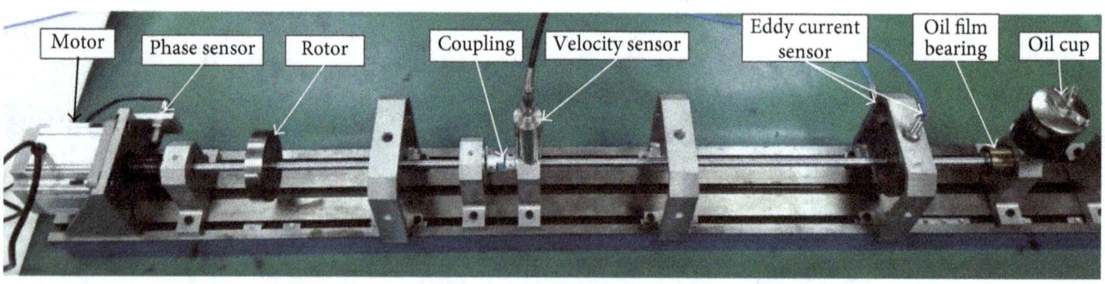

FIGURE 5: Experimental setup and sensor arranged.

of the rotor increases, and the whip frequency stabilizes in a certain value (first-order critical speed of rotor system). The fault feature extraction of oil whirl and oil whip is the main basis to diagnose these type faults.

In the process of oil film instability, the oil whirl occurs near the speed of 3100 r/min. When the rotational speed is up to 4500 r/min, the rotor system test rig shows bigger vibration because of the oil whip.

(1) Oil Whirl. The vibration signal and its spectrum of the oil whirl are shown in Figure 6. In the spectrum, the oil whirl frequency 1/2x (≈26 Hz) and rotating frequency 1x (≈52 Hz) can be seen. The value of the oil whirl frequency is about half value of rotating frequency and the amplitude is higher than rotating frequency.

Five IMFs decomposed by VMD for oil whirl signal are shown in Figure 7(a), and it is decomposed completely. IMF1 is corresponding to the oil whirl frequency 1/2x and IMF2 is corresponding to rotating frequency 1x. IMF3 is corresponding to 2x. IMF5 is corresponding to 4x. The VMD results not only show the fault feature of oil whirl, but also present more periodicity components.

For comparison, IMFs analyzed by EEMD and EMD are shown in Figures 7(b) and 7(c). Although IMF2 and

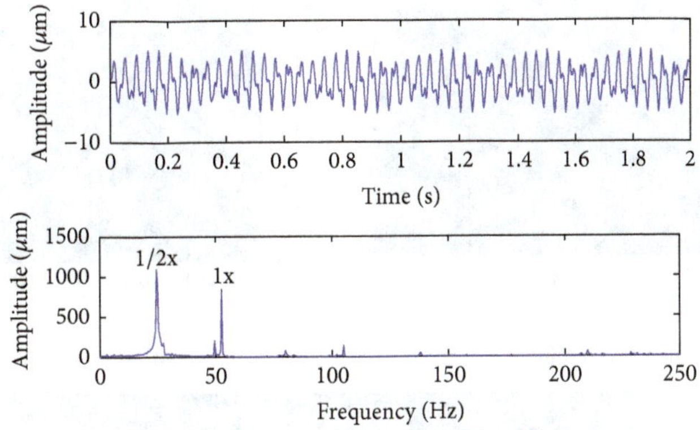

FIGURE 6: Time Domain waveform and spectrum of oil whirl signal.

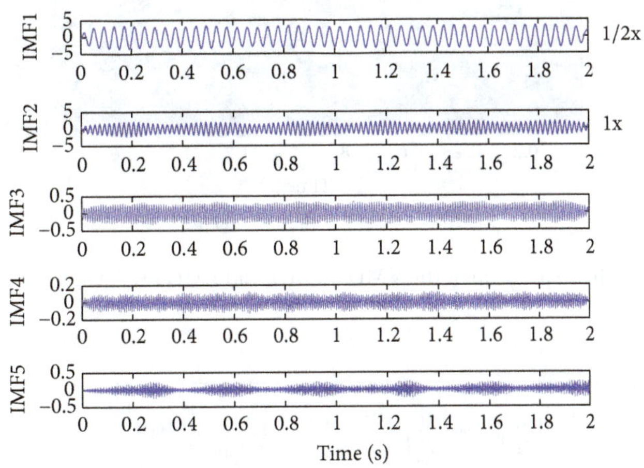

(a) VMD decomposed results

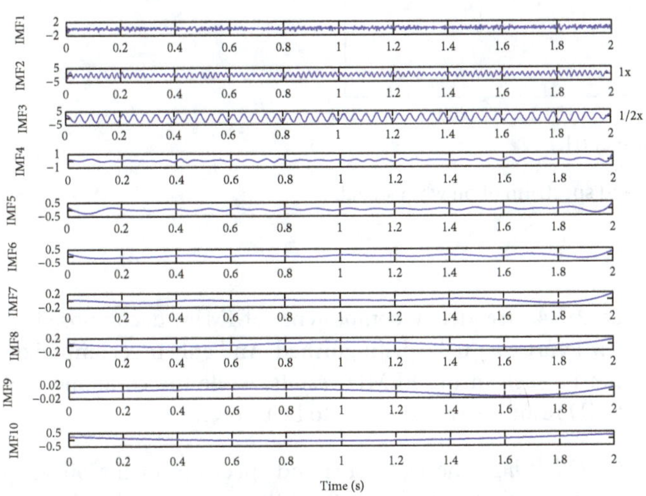

(b) EEMD decomposed results

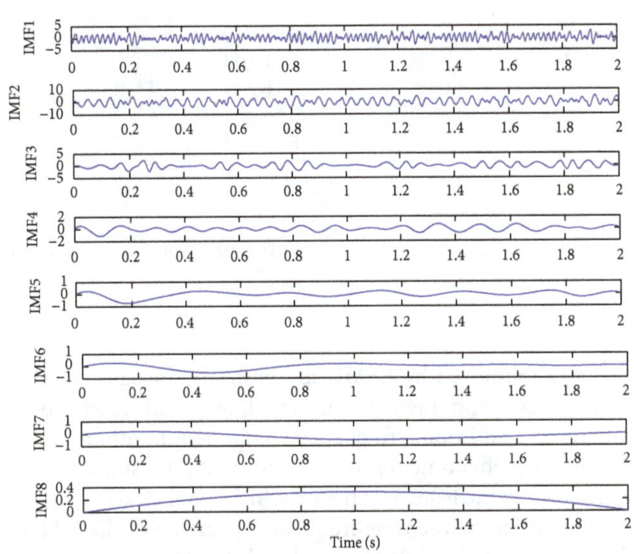

(c) EMD decomposed results

FIGURE 7: VMD, EEMD, and EMD decomposed results of oil whirl signal.

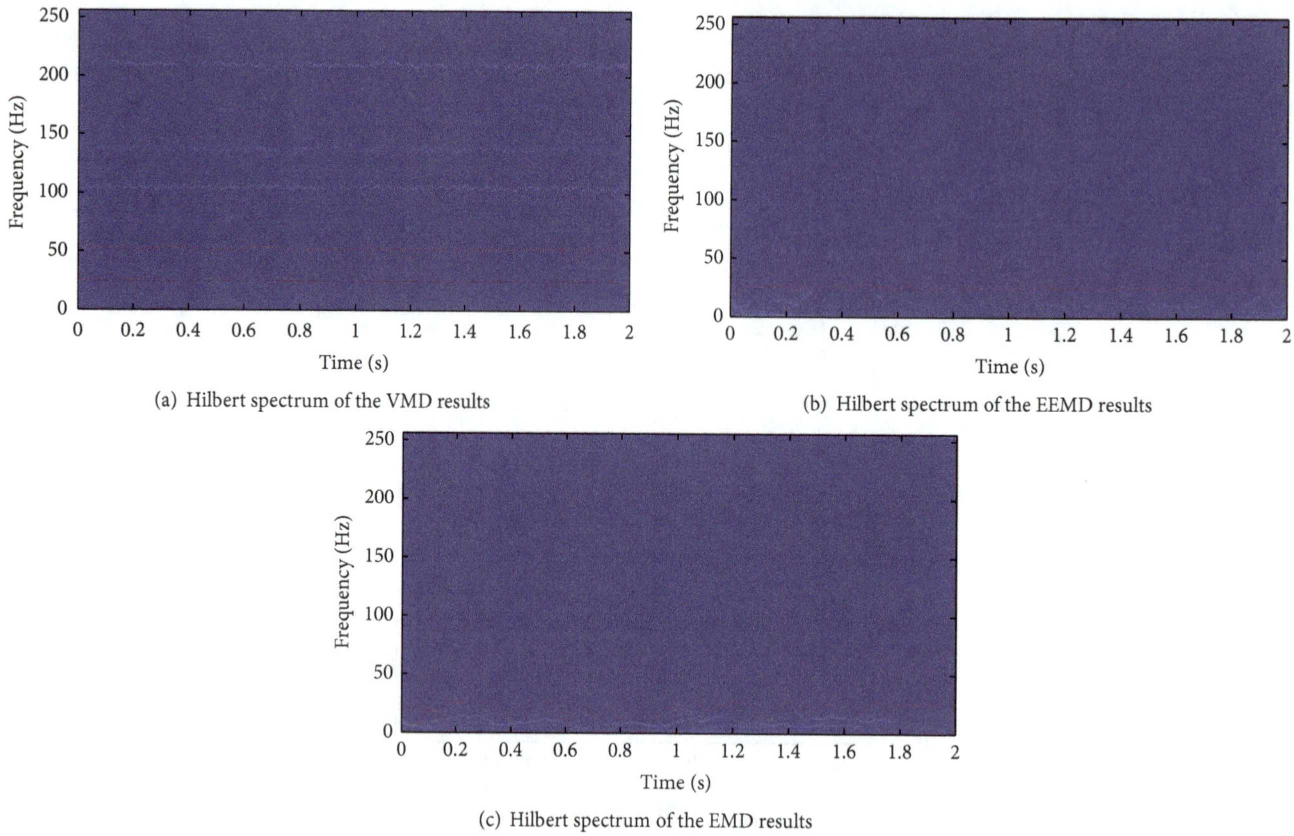

(a) Hilbert spectrum of the VMD results

(b) Hilbert spectrum of the EEMD results

(c) Hilbert spectrum of the EMD results

FIGURE 8: Hilbert spectrum of the VMD, EEMD, and EMD results of oil whirl signal.

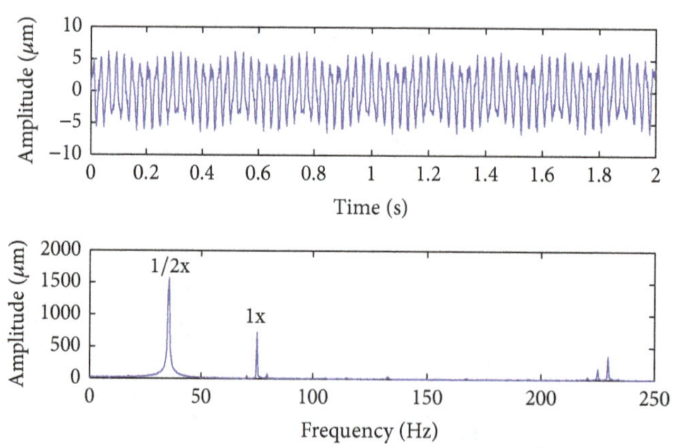

FIGURE 9: Time domain waveform and spectrum of oil whip signal.

IMF3 decomposed by EEMD are basically the same as the 1x and 1/2x components in the signal, there still exist mode mixing and additional frequency component in other IMFs. The mixing phenomenon is obvious in EMD results, which cannot show the feature of the oil whirl signal clearly.

Furthermore, the time-frequency spectrum of the VMD, EEMD, and EMD results was analyzed. The Hilbert spectrum of each IMF decomposed by the VMD, EEMD, and EMD is shown in Figure 8. It can be seen from Figure 8 that the 1/2x, 1x, 2x, 4x frequency components of IMFs, decomposed by VMD, are easy to be distinguished. And only 1/2x component can be seen in the EEMD results, while the component of EMD results is very difficult to be resolved.

(2) Oil Whip. The waveform and spectrum of the oil whip signal are shown in Figure 9. In the spectrum, the oil whirl frequency 1/2x (≈37 Hz) and rotating frequency 1x (≈75 Hz) can be seen. The value of the oil whirl frequency is about half

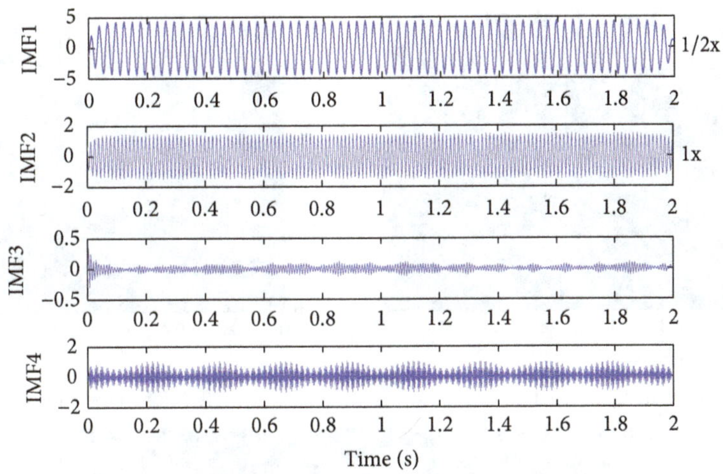

(a) VMD decomposed results

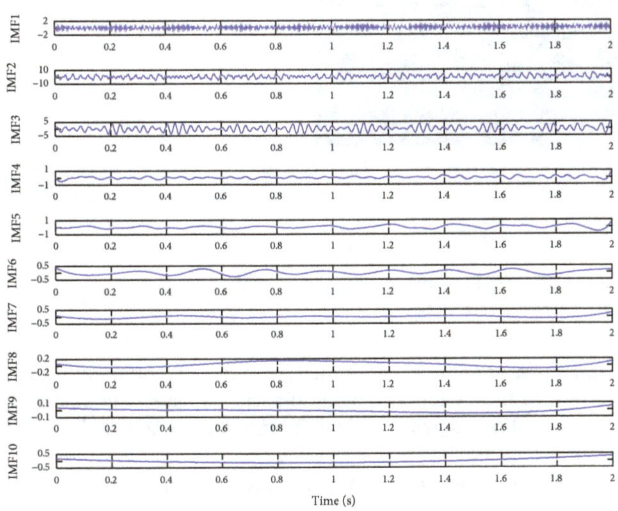

(b) EEMD decomposed results

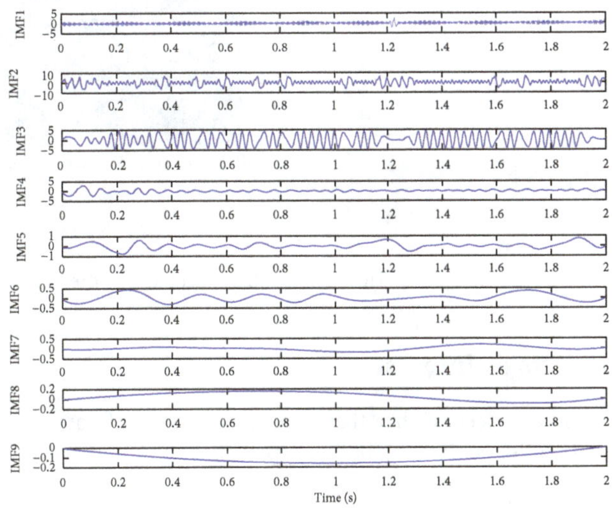

(c) EMD decomposed results

FIGURE 10: VMD, EEMD, and EMD decomposed results of oil whip signal.

value of rotating frequency and the amplitude is higher than rotating frequency.

Four IMFs obtained by VMD for oil whip signal are illustrated in Figure 10(a), and it is decomposed clearly. IMF1 and IMF2 are corresponding to 1/2x and 1x, respectively, and the 3x frequency components are also extracted. Figures 10(b) and 10(c) separately show the IMFs using EEMD and EMD, in which signatures are difficult to be identified.

The Hilbert spectrum of each IMF decomposed by VMD, EEMD, and EMD is shown in Figure 11. As can be seen from Figure 11, all time-frequency representations of each component decomposed by VMD are clearly segregated, while the component of EEMD and EMD results is very difficult to be resolved.

The analysis results in Figures 7, 8, 10, and 11 show better effectiveness of VMD than EEMD and EMD, demonstrating the validity of VMD in detecting rotor faults.

5. Conclusion

VMD is a newly developed technique for adaptive signal decomposition, which can decompose a multicomponent signal into a series of quasi-orthogonal intrinsic mode functions simultaneously. Unlike EMD and EEMD, VMD has theoretical support and can solve the problem of mode mixing and end effect more effectively. It also has the characteristic of adaptivity, reconfigurability, smoothing filtering, and orthogonality.

VMD is proposed to determine the fault of rotor systems, and an experimental analysis has been carried out. Comparative analysis results show that VMD method is more effective than EEMD and EMD in fault feature extraction of oil whirl and oil whip, so VMD method has important practical application value for rotor fault diagnosis and should be further considered in the future.

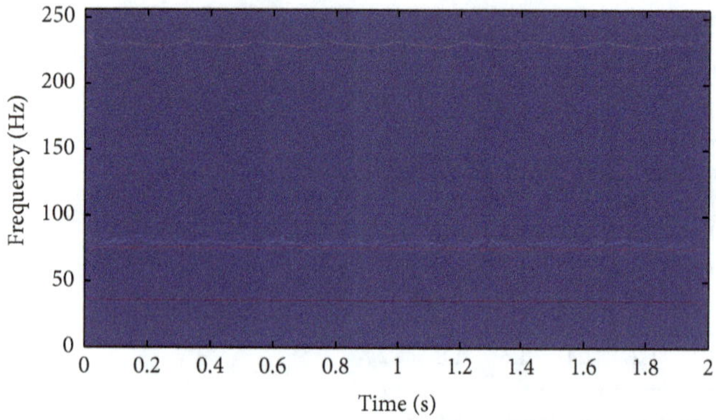

(a) Hilbert spectrum of the VMD results

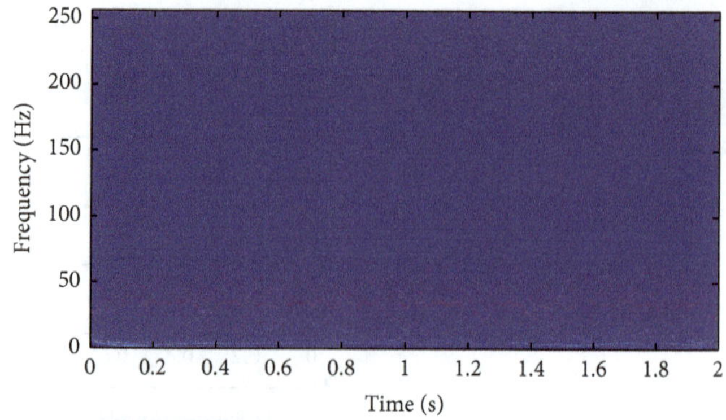

(b) Hilbert spectrum of the EEMD results

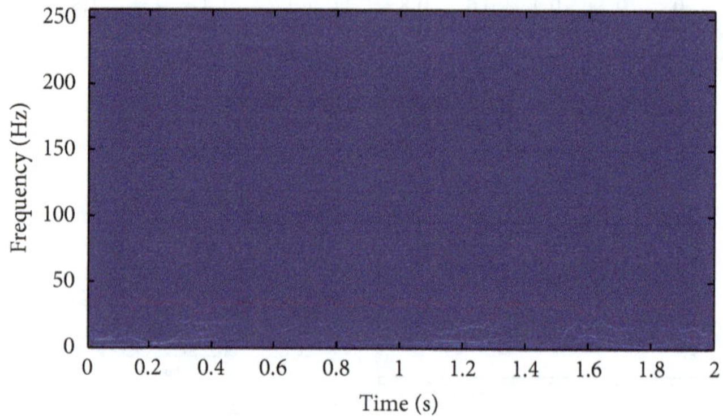

(c) Hilbert spectrum of the EMD results

FIGURE 11: Hilbert spectrum of the VMD, EEMD, and EMD results of oil whip signal.

Conflicts of Interest

The authors declare that there are no conflicts of interest regarding the publication of this paper.

Acknowledgments

This project is sponsored by the grants from the Ministry of Transport Applied Basic Research Project (no. 2013329811360), the National Natural Sciences Foundation of China (no. 51139005, no. 51509194), and the Fundamental Research Funds for the Central Universities (no. 2014-yb-018).

References

[1] D. Yongzuo and Q. Zhiying, "Holo-spectrum analysis of rotating machinery dynamic signals," *Journal of Vibration, Measurement & Diagnosis*, vol. 22, no. 2, pp. 81–88, 2002.

[2] Z. Li, Y. Jiang, C. Hu, and Z. Peng, "Recent progress on decoupling diagnosis of hybrid failures in gear transmission systems using vibration sensor signal: a review," *Measurement*, vol. 90, pp. 4–19, 2016.

[3] D. H. Gonsalves, R. D. Neilson, and A. D. S. Barr, "A study of the response of a discontinuously nonlinear rotor system," *Nonlinear Dynamics*, vol. 7, no. 4, pp. 451–470, 1995.

[4] N. Bachschmid, P. Pennacchi, and A. Vania, "Identification of multiple faults in rotor systems," *Journal of Sound and Vibration*, vol. 254, no. 2, pp. 327–366, 2003.

[5] T.-A. Ensslin, M. Frommert, and F.-S. Kitaura, "Information field theory for cosmological perturbation reconstruction and non-linear signal analysis," *Physical Review D Particles & Fields*, vol. 80, no. 10, pp. 281–287, 2009.

[6] G. Fele-Žorž, G. Kavšek, Ž. Novak-Antolič, and F. Jager, "A comparison of various linear and non-linear signal processing techniques to separate uterine EMG records of term and pre-term delivery groups," *Medical and Biological Engineering and Computing*, vol. 46, no. 9, pp. 911–922, 2008.

[7] Z. Li, X. Yan, Z. Tian, C. Yuan, Z. Peng, and L. Li, "Blind vibration component separation and nonlinear feature extraction applied to the nonstationary vibration signals for the gearbox multi-fault diagnosis," *Measurement: Journal of the International Measurement Confederation*, vol. 46, no. 1, pp. 259–271, 2013.

[8] J. Cheng, D. Yu, J. Tang, and Y. Yang, "Local rub-impact fault diagnosis of the rotor systems based on EMD," *Mechanism and Machine Theory*, vol. 44, no. 4, pp. 784–791, 2009.

[9] N. E. Huang, Z. Shen, S. R. Long et al., "The empirical mode decomposition and the Hilbert spectrum for nonlinear and non-stationary time series analysis," *The Royal Society of London. Proceedings. Series A. Mathematical, Physical and Engineering Sciences*, vol. 454, no. 1971, pp. 903–995, 1998.

[10] N. E. Huang, M. C. Wu, S. R. Long et al., "A confidence limit for the empirical mode decomposition and Hilbert spectral analysis," *Proceedings of the Royal Society A*, vol. 459, no. 2037, pp. 2317–2345, 2003.

[11] K. Dragomiretskiy and D. Zosso, "Variational mode decomposition," *IEEE Transactions on Signal Processing*, vol. 62, no. 3, pp. 531–544, 2014.

[12] Y. Wang and R. Markert, "Filter bank property of variational mode decomposition and its applications," *Signal Processing*, vol. 120, pp. 509–521, 2016.

[13] W. Yang, Z. Peng, K. Wei, P. Shi, and W. Tian, "Superiorities of variational mode decomposition over empirical mode decomposition particularly in time-frequency feature extraction and wind turbine condition monitoring," *IET Renewable Power Generation*, vol. 11, no. 4, pp. 443–452, 2017.

[14] S. Lahmiri, "Comparative study of ECG signal denoising by wavelet thresholding in empirical and variational mode decomposition domains," *Healthcare Technology Letters*, vol. 1, no. 3, pp. 104–109, 2014.

[15] N. Mohan, S. Sachin Kumar, P. Poornachandran, and K. P. Soman, "Modified variational mode decomposition for power line interference removal in ECG signals," *International Journal of Electrical and Computer Engineering*, vol. 6, no. 1, pp. 151–159, 2016.

[16] A. Mert, "ECG feature extraction based on the bandwidth properties of variational mode decomposition," *Physiological Measurement*, vol. 37, no. 4, pp. 530–543, 2016.

[17] P. Shi and W. Yang, "Precise feature extraction from wind turbine condition monitoring signals by using optimised variational mode decomposition," *IET Renewable Power Generation*, vol. 11, no. 3, pp. 245–252, 2017.

[18] V. Vishnu Pradeep, V. Sowmya, and K. P. Soman, "Variational mode decomposition based multispectral and panchromatic image fusion," *International Journal of Control Theory and Applications*, vol. 9, no. 16, pp. 8051–8059, 2016.

[19] G. Sun, T. Chen, Z. Wei, Y. Sun, H. Zang, and S. Chen, "A carbon price forecasting model based on variational mode decomposition and spiking neural networks," *Energies*, vol. 9, no. 1, article 54, 2016.

[20] S. Lahmiri, "Comparing variational and empirical mode decomposition in forecasting day-ahead energy prices," *IEEE Systems Journal*, no. 99, pp. 1–4, 2015.

[21] A. Upadhyay and R. B. Pachori, "Instantaneous voiced/non-voiced detection in speech signals based on variational mode decomposition," *Journal of the Franklin Institute*, vol. 352, no. 7, pp. 2679–2707, 2015.

[22] C. Yi, Y. Lv, and Z. Dang, "A fault diagnosis scheme for rolling bearing based on particle swarm optimization in variational mode decomposition," *Shock and Vibration*, vol. 2016, Article ID 9372691, 10 pages, 2016.

[23] Z. Lv, B. Tang, Y. Zhou, and C. Zhou, "A novel method for mechanical fault diagnosis based on variational mode decomposition and multikernel support vector machine," *Shock and Vibration*, vol. 2016, Article ID 3196465, 11 pages, 2016.

[24] Z. Li, Y. Jiang, Q. Guo, C. Hu, and Z. Peng, "Multi-dimensional variational mode decomposition for bearing-crack detection in wind turbines with large driving-speed variations," *Renewable Energy*, 2016.

[25] N. Huang, H. Chen, G. Cai, L. Fang, and Y. Wang, "Mechanical fault diagnosis of high voltage circuit breakers based on variational mode decomposition and multi-layer classifier," *Sensors*, vol. 16, no. 11, article 1887, 2016.

[26] Z. H. Wu and N. E. Huang, "Ensemble empirical mode decomposition: a noise-assisted data analysis method," *Advances in Adaptive Data Analysis*, vol. 1, no. 1, pp. 1–41, 2009.

17

Development and Application of Integrated Aerodynamic Protuberant Foil Journal and Thrust Bearing in Turboexpander

Tianwei Lai,[1,2] Yu Guo,[1] Wei Wang,[1] Yu Wang,[1] and Yu Hou[1]

[1] *State Key Laboratory of Multiphase Flow in Power Engineering, Xi'an Jiaotong University, Xi'an 710049, China*
[2] *Xi'an Jiaotong University Suzhou Institute, Suzhou 215123, China*

Correspondence should be addressed to Yu Hou; yuhou@mail.xjtu.edu.cn

Academic Editor: Lei Tan

Foil bearing provides compliant support and moderate Coulomb friction for rotor-bearing system, which is conducive to stability and reliability of high speed rotating machinery. In this paper, both hydrodynamic lubricated foil journal and foil thrust bearings are applied in a 150 $m^3 \cdot h^{-1}$ turboexpander for air separation. In the bearings, protuberant foil is chosen as the supporting subfoil due to its merits of easy fabrication and assembly. Static loading and deflection of the bearings are tested, respectively, before integration into the turboexpander. Afterwards, the loading and deflection curves of the journal and thrust bearings are polynomial fitted using least-square method. Then, performance tests are carried out on the rotor-bearing system, including transient speed-up, high speed, and speed-down processes. In the tests, the turboexpander supported by the hydrodynamic foil bearings operates smoothly with repeatability. Maximal rotor speed of the turboexpander reaches 52000 rpm with hydrodynamic lubricated protuberant foil bearings.

1. Introduction

Turboexpander is an indispensable component in energy conversion system. Higher speed is anticipated for the turboexpander in pursuit of high efficiency [1]. Its smooth operation depends, to a large extent, on the stability of the rotor-bearing system. In high speed or oil-free occasions, conventional oil lubrication is constrained due to its high viscosity and contamination issues [2]. Gas foil bearing, as an alternative, is a kind of gas bearing using pieces of metal foil as support. It is outperformed for high speed [3], higher reliability [4], soft failure [5], and extreme temperature adaptability [6, 7], as well as high levels of misalignment tolerance [8]. Additionally, aerodynamic lubrication using process gas in high speed operation is in favor of system conciseness and compactness.

For the foil bearing, multikinds of underlying supporting configurations have been developed since its first debut [9]. Among them, bump type foil bearing attracted considerable attention [10]. Besides, metal mesh [11] and viscoelastic [12, 13] configurations and so on have been presented. Foil bearing reduces the peak vibration amplitude notably when the rotor crosses the critical speed [14, 15]. Schiffmann and Spakovszky optimized the selective shimming in foil bearing and proposed a design guideline [4]. Pham and Ahn combined foil bearing and magnetic bearing to support a flexible rotor with vibration reduction of 26% [16].

In foil bearing, compliant surface and nonlinear friction are introduced due to the configuration of stacking foils. With the same load distribution, various foil deflections might be present that is ascribed to the hysteretic behavior of the friction force [17]. Subfoil stiffness of the elastic support affects the bearing load capacity [18]. The ultimate load of the bearing could not exceed that of underlying elastic structure [3].

For decades of development, foil bearing was applied in many high speed rotating machineries [19]. Lee et al. applied foil thrust bearing in automotive turbochargers with 24 mm diameter shaft at speed of 123000 rpm (round per minute) [20]. Dykas et al. adopted foil thrust bearing in micro turbomachinery with inner diameter of 50.8 mm and outer diameter of 101.6 mm with load capacity of 125 N around 23000 rpm [21]. Kim et al. evaluated the foil bearing performance in a 120 kW gas turbine generator with rated speed of 15800 rpm and generator weight of 84.2 N and turbine weight of 175.0 N [22]. Kim et al. applied foil thrust bearing and journal bearing in a 66 mm diameter 75 kW turbo blower with rotor speed of 30000 rpm, and subsynchronous rotor lateral motions appeared [23]. Kim et al. designed the mesoscale foil gas bearing for palm-sized turbomachinery with inner diameter of 5 mm and designed speed of 60000 rpm [24].

Comparing to existing hydrostatic bearing technology in cryogenic turboexpander, its configuration can be simplified to a large extent using hydrodynamic lubricated foil bearing. Two kinds of journal bearings, copper wire support and viscoelastic support, were compared for small cryogenic turboexpander of 12 mm diameter around 220000 rpm [6]. Recently, bump foil journal bearings was applied in the above turboexpander that could operate smoothly in a wide speed range [25]. Numerical results indicated that the journal bearing was stable in high speed range [26]. Subsynchronous vibrations were suppressed quite well [27, 28]. Static structure stiffness and load capacities of the thrust bearing were tested in transient start-up and speed-down processes [29, 30]. Dedicated tests on protuberant foil journal bearing or thrust bearing were conducted, respectively. However, hydrodynamic lubrication using both protuberant foil journal and thrust bearings (integrated aerodynamic foil bearing supporting technology) in the turboexpander has not been tested yet. In this paper, both protuberant foil journal bearing and foil thrust bearing are applied in a turboexpander with shaft diameter of 25 mm for air separation. In the bearings, protuberant foil is chosen as the supporting subfoils due to its merits of easy fabrication and assembly. Experimental tests are carried out to analyze the transient performance of the integrated aerodynamic foil bearing supporting technology. In addition, static loading and unloading of the foil journal bearing and thrust bearing are tested.

2. High Speed Turboexpander

The retrofitted turboexpander in this paper, as shown in Figure 1, is an existing commercial module with liquid nitrogen production of 30 L·h^{-1}. The turbine type expansion wheel converts high pressure gas into low pressure and low temperature states. Braking wheel is used to dissipate the power retrieved from the coaxial expansion wheel.

Rotor-bearing system of the turboexpander is shown in Figure 2. The rotor is horizontal layout and its weight is supported by two journal bearings. The net axial load from expansion wheel and braking wheel is balanced by two thrust bearings. Some geometrical parameters of the turboexpander are listed in Table 1. The former bearing supporting technology of the turboexpander is based on aerostatic journal and thrust bearing. The rated operating speed of the turboexpander is 100000 rpm.

3. Test Rig

In order to validate the applicability of integrated aerodynamic foil bearing supporting technology in this turboexpander, the test rig of high speed turboexpander is retrofitted based on the rig in reference [31]. The flow chart of gas driven test rig is shown in Figure 3. Compared to the aerostatic bearing supported turboexpander, there is no extra dedicated gas circuit for aerostatic bearings. Therefore, the system is simplified for lighter weight and compactness. High pressure air is supplied by a 75 kW screw compressor with maximal pressure of 1.0 MPa and maximal flow rate of 600 Nm^3·h^{-1}. Gas reservoir is implemented to stabilize the undulating pressurized gas from the compressor. Supply gas pressure and flow rate at the expansion wheel are regulated by proportional electric valve. The exhaust gas is vent to the ambient after expansion. Three eddy current displacement transducers (with linearity of <±2% and static resolution of 0.1 μm) in X, Y, and Z directions are installed to evaluate the vibrations of the high speed rotor. Sampling frequency of the displacement transducer is 100 kHz. The vibration data are collected by data acquisition system. Supply pressure is measured by pressure transducer (with resolution of 1000 Pa and FS ±1%).

The real picture of the test rig is shown in Figure 4 (expansion volute was removed from the turboexpander). The vibration transducers are installed as shown in the figure.

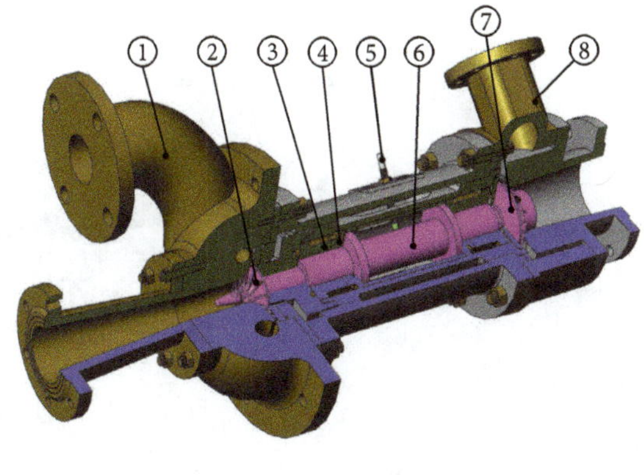

(1) Expansion volute
(2) Expansion wheel
(3) Journal bearing
(4) Thrust bearing
(5) Displacement transducer
(6) Shaft
(7) Braking wheel
(8) Braking volute

FIGURE 1: Assembly drawing of turboexpander for 150 m^3·h^{-1} air separation system.

TABLE 1: Geometrical parameters of the turbo-expander.

Parameter	Magnitude
Diameter of braking wheel (mm)	60
Diameter of expansion wheel (mm)	36.5
Outer diameter of thrust disc (mm)	44
Distance between journal bearings (mm)	127
Aspect ratio of journal bearing	≈1
Diameter of shaft (mm)	25
Length of rotor (mm)	250.5
Density of shaft (kg·m^{-3})	7830
Density of wheels (kg·m^{-3})	2730
Weight of rotor (g)	891

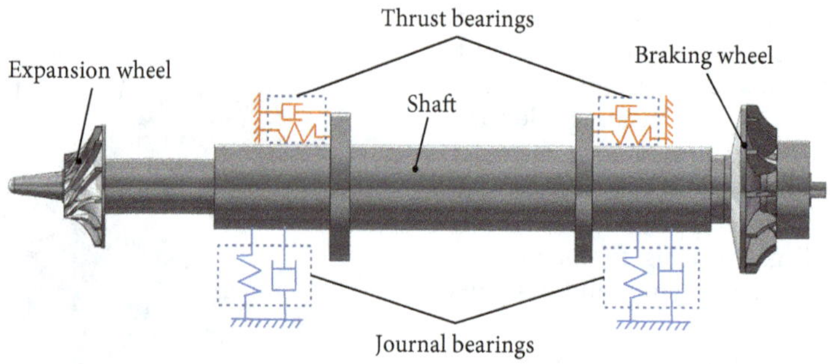

FIGURE 2: Rotor-bearing system of the turboexpander.

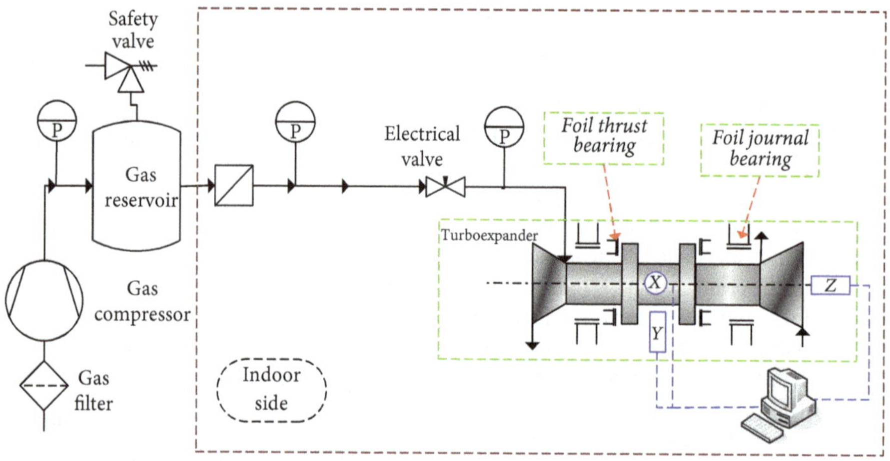

FIGURE 3: Flow chart of high pressure gas driven turboexpander test rig.

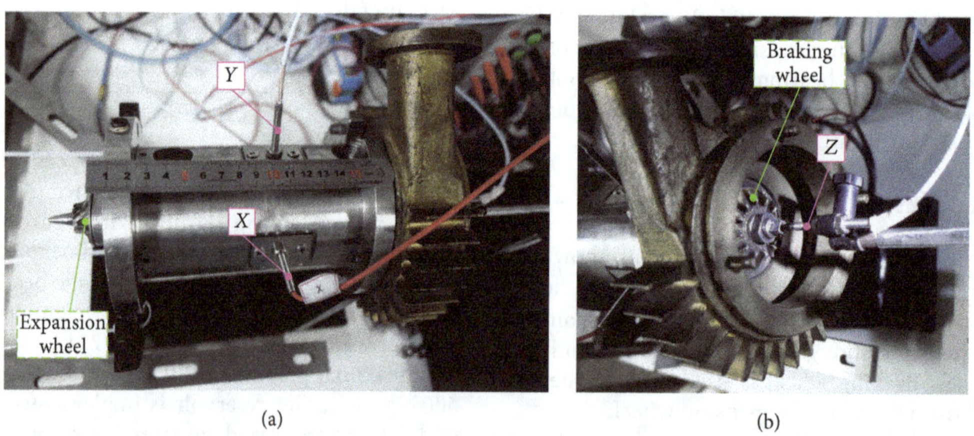

FIGURE 4: Real picture of high pressure gas driven turboexpander test rig: (a) upper view; (b) side view (with expansion volute removed).

4. Bearing Configurations

In pursuit of stability and compactness of turboexpander, aerodynamic foil bearing is a preferred choice. By virtue of the transient performances of the protuberant foil journal and thrust bearings [27–30], they are considered to be alternative candidates for the rotor-bearing system. In this paper, both the aerostatic journal and thrust bearings are replaced by protuberant foil bearings based on the previous configuration of the rotor-bearing system.

The schematic diagram and real picture of the protuberant foil journal bearing are shown in Figures 5(a) and 5(b), respectively. There are two protuberant foil layers and one top flat foil that stacked one layer over another in the bearing housing. The foils are fixed into the housing at one end by fixing pin. The detailed geometrical parameters and stacking layout of the foils could be found in Table 2.

Before application of the journal bearing and thrust bearing, deflection of the journal bearing and thrust bearing

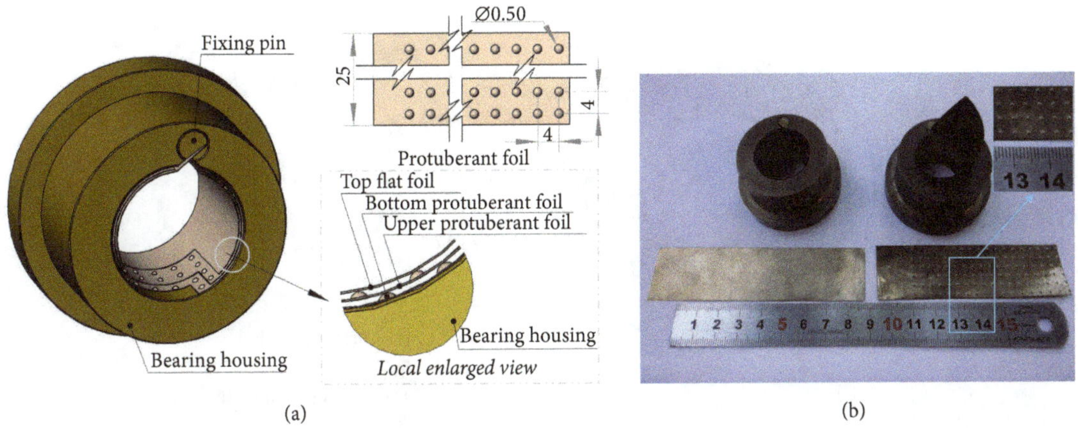

FIGURE 5: Schematic diagram and real picture of foil journal bearing in the turboexpander.

TABLE 2: Geometrical parameters of the protuberant foil journal bearing [31].

Parameter	Magnitude
Nominal G_m/g	830
Height of protuberant foil H_p/mm	0.2
Crown pitch in peripheral direction of upper layer S_{pu}/mm	4
Crown pitch in peripheral direction of lower layer S_{au}/mm	4
Protuberant foil thickness T_p/mm	0.05
Top foil thickness T_f/mm	0.07
Crown pitch in peripheral direction of upper layer S_{pb}/mm	4
Crown pitch in peripheral direction of lower layer S_{ab}/mm	4

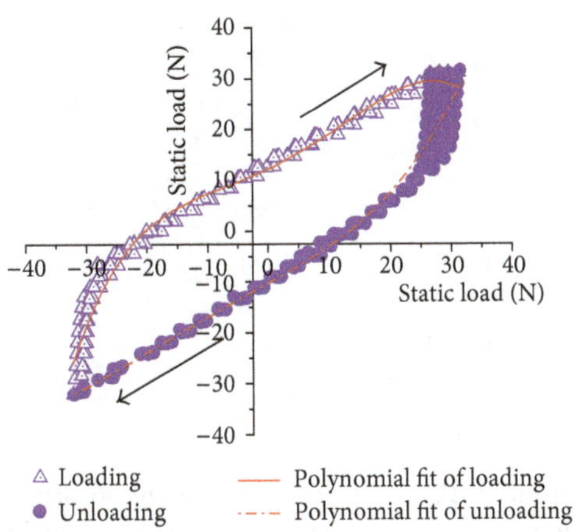

FIGURE 6: Deflection of protuberant foil journal bearing with static load.

under static load are performed on respective dedicated bearing test rigs.

Static load and deflection of the journal bearing is shown in Figure 6. In mild load range from −25 N to +25 N, the

TABLE 3: Standard error of the polynomials for load and deflection curves of journal bearing.

Coefficients of the polynomials	Loading	Unloading
Interception	0.233	0.1963
x	0.0172	0.0146
x^2	0.00138	0.00117
x^3	$2.61E-05$	$2.42E-05$
x^4	$1.53E-06$	$1.32E-06$

load and displacement exhibits linear relationship. Nonlinear characteristics appears in higher load range from +30 N to +25 N and from −30 N to −25 N. The displacement is about 30 μm at the maximal load of ±30 N. Under the same load, the absolute static deflection is higher in unloading process than that in loading process. For example, the displacement is about 10 μm when the static load is 0 N in the loading process. With the same load in unloading process, the displacement is about −20 μm, which is larger than that in loading process. This can be attributed to the hysteresis effect from Coulomb friction in loading and unloading processes. The area enclosed in hysteresis loop can be used to calculate the dissipated energy in one loading test cycle. Fourth-order polynomial was used to fit the loading curves based on weighted least-square method. The polynomial expressions for the loading and unloading processes for beryllium bronze bearing are shown in (1) and (2), respectively. The standard errors of the polynomials are listed in Table 3. The coefficients of displacement x are 0.67 N·μm^{-1} in the loading process, which is higher than that in the unloading process of 0.55 N·μm^{-1} from

$$y_{\text{loading}} = -10.2 + 0.67x + 0.0028x^2 + 0.00029x^3 + 0.000006x^4, \quad (1)$$

$$y_{\text{unloading}} = 12.2 + 0.55x + 0.0013x^2 + 0.00026x^3 + 0.000023x^4. \quad (2)$$

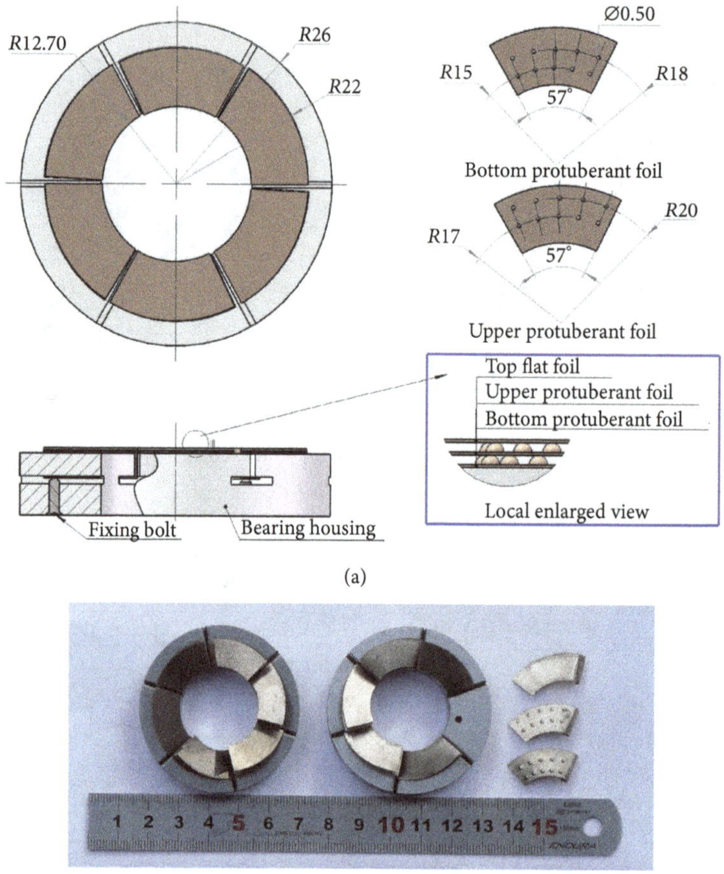

FIGURE 7: (a) Schematic diagram and (b) real picture of foil thrust bearing in the turboexpander.

TABLE 4: Geometrical parameters of the protuberant foil thrust bearing [32].

Parameter		Magnitude
Top foil thickness t_{tf}/mm		0.07
Protuberant foil thickness t_{pf}/mm		0.05
Number of supporting layers N_l		2
Height of protuberant convex h_{pf}/mm		0.3
Spherical radius of projections R_{sp}/mm		1.4
Radial location of inner protuberant projections R_{pi}/mm	Upper layer	10.4
	Bottom layer	8.2
Radial location of outer protuberant projections R_{po}/mm	Upper layer	14.8
	Bottom layer	12.6

According to the existing configuration of shaft and journal bearing, two thrust bearings are implemented to balance the axial forces from the wheels. In the foil thrust bearing, as is shown in Figure 7, six-pad configuration is adopted due to its preferable loading performance [32]. The configuration of the foil stacking is similar to the journal bearing above; detailed parameters of the thrust bearing can be found in Table 4. Protuberant projections of diameter of 0.5 mm on the foils are punched from flat foil. The stacking layout of upper protuberant foil and bottom protuberant foil is staggered in circumferential direction according to the locations of the supporting projections. The foils are fixed by screw bolt from the back of thrust bearing.

The deflection of the thrust bearing with static load is shown in Figure 8. Detailed testing method can refer to [33]. Unlike the protuberant foil journal bearing, the hysteresis loop of the deflection of the thrust bearing is smaller. Structural stiffness of six-pad multilayer protuberant foil thrust bearing is not constant and load dependent. The polynomial expressions for the loading and unloading

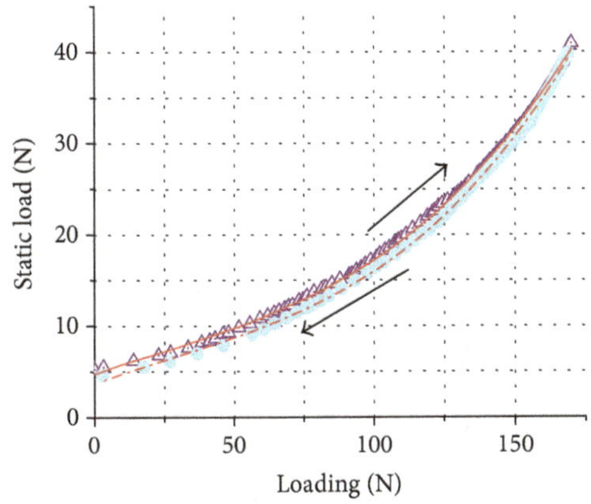

FIGURE 8: Deflection of protuberant foil thrust bearing with static load.

TABLE 5: Standard error of the polynomials for load and deflection curves of thrust bearing.

Coefficients of the polynomials	Loading	Unloading
Interception	0.259	0.178
x	0.0114	0.00736
x^2	$1.39E-04$	$9.05E-05$
x^3	$4.91E-07$	$3.26E-07$

processes for beryllium bronze bearing are shown in (3) and (4), respectively. The standard errors of the polynomials are listed in Table 5.

$$y_{\text{loading}} = -4.76 + 0.104x - 3.85x^2 + 0.00000587x^3, \quad (3)$$

$$y_{\text{unloading}} = 3.70 + 0.113x + 0.000598x^2 + 0.00000695x^3. \quad (4)$$

5. Testing Procedures

The testing process is shown in Figure 9. After checking the preparation for ready to start up, open up the electrical valve of supply gas at the expansion wheel gradually. The gas pressure increment should be small until the rotor started to rotate. Maintain the supply pressure to be constant when the rotor is in acceleration. Afterwards, increase the gas supply pressure further until the rotor speed is around 50000 rpm. After smooth operating for a period of time, close down the supply gas gradually to reduce the rotor speed. When the rotor stops operating, close the supplying gas valve. More than three times of start-up and speed-down processes are conducted to ensure the repeatability of collected data.

6. Test Results and Analysis

For this integrated bearing layout, there exist friction torques from the journal bearing and the thrust bearing, especially during the start-up process. When the preload on the thrust bearing is larger than a certain value, the torque at expansion wheel could not overcome the start-up friction torque [34]. Furthermore, during start-up and speed-down transient processes, friction torque of the foil thrust bearing increases linearly with rotor speed and bearing load. Therefore, the conventional start-up method of rotor by increasing the supply pressure does not fit for the integrated foil bearing supporting technology. In order to reduce the friction torque at the start-up stage, the nominal journal bearing clearance is increased to 100 μm.

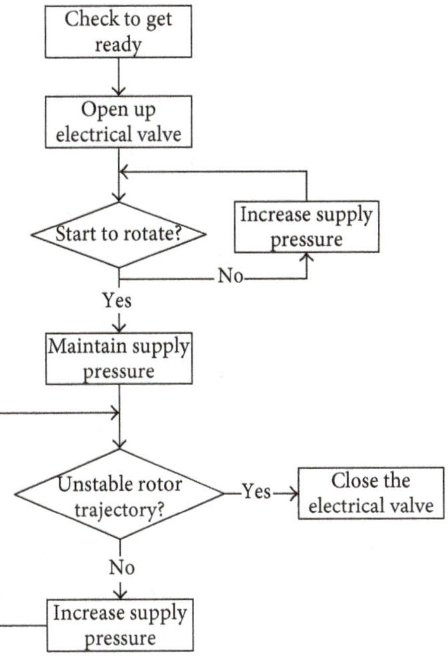

FIGURE 9: Testing process of the turboexpander with protuberant foil journal and thrust bearing.

Transient processes such as start-up, high speed, and speed-down of the rotor supported by integrated foil journal bearing and thrust bearing are tested after assembly of the turboexpander. In the transient processes, axial displacement is an important indicator of axial movement of the rotor due to net axial force from expansion wheel and braking wheel. Rotor speed and axial displacement of the rotor with supply pressure in the whole speed-up and speed-down processes are shown in Figure 10. At the beginning 8 s, there is no supply pressure. From 8 s to 18 s in region I, in which the supply pressure increases to around 0.13 MPa, there is no rotation yet due to the static friction between the rotor and foil surfaces. Because of the pressure difference between the front and back of expansion wheel, the rotor moves in the direction of expansion wheel with displacement around 180 μm before rotation. In region II, when the supply pressure exceeds 0.13 MPa, the rotor begins to rotate. The supply pressure is maintained at this value when the rotor is in accelerating process. In this region, the aerodynamic lubrication comes into effect due to the relative speed between the rotor and bearing surface. The hydrodynamic lubrication force pushes the rotor towards the braking wheel. Therefore, the thrust bearing clearance increases during this speed-up process. In region III, the

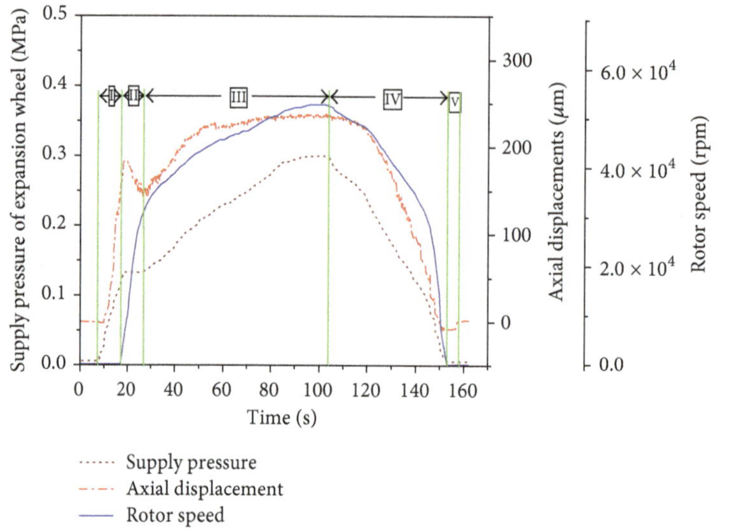

FIGURE 10: Axial displacement of the rotor in transient processes.

supply pressure increases further with expectation of higher rotor speed. The axial force from the wheels increases and the rotor moves axially in the direction of expansion wheel again. The axial displacement increases to 230 μm when higher supplying pressure is supplied continuously. The rotor speed reaches 52805 rpm with supply pressure of 0.3 MPa. Afterwards, the rotor speed is reduced gradually in region IV. In the speed-down process, the axial displacement of the rotor decreases correspondingly. It should be noted that there is minus axial displacement at the end of speed-down process. Furthermore, there are still minus axial displacements after the rotor stops running in region V. Recovery time of axial displacement is needed after one speed-up and coast-down process, which can be attributed to the hysteresis effect of the foils in the whole operating region. During the transient processes, the displacement is positive in most of the regions. That is to say, most of the axial force is balanced by the thrust bearing near expansion wheel.

During the transient operation, high speed rotor vibrates around its transient equilibrium position. The journal centerline during the above transient processes is shown in Figure 11. The rotor rotates in counterclockwise direction in the bearing. The circumferential angle θ initiates from upright direction of the bearing. Journal center location reflects its eccentricity and attitude angle (θ-270°). The journal center in speed-up process is marked by red hollow rectangles and speed-down process is marked by blue hollow circles. The initial static resting position of the rotor is assumed to be (100, 270°) in above polar coordinates. At lower speed before lift-off from the foil bearing, the rotor lifts up about 6 μm towards the bearing center and 9° in counterclockwise direction. At the speed around 9600 rpm, the journal center drifts about 5 μm towards the bearing center and 10° in counterclockwise direction that is marked as the lift-off speed of the bearing during the built-up of the aerodynamic film. The eccentricity and attitude angle of the rotor are 91 μm and 19°, respectively. The lift-off speed is around 9600 rpm. After lift-off from the

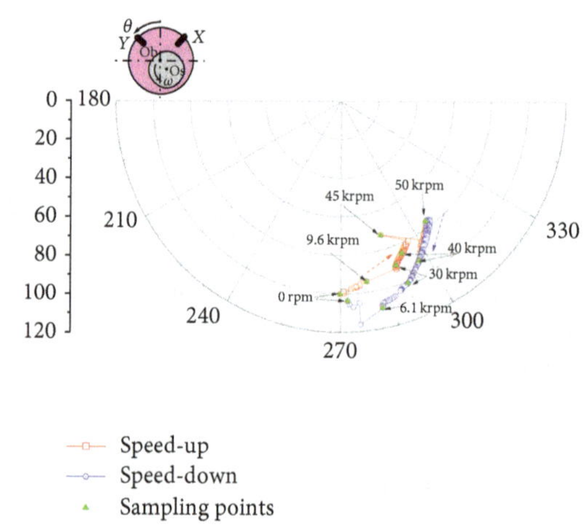

FIGURE 11: Polar plot of journal during the transient start-up and coast-down processes.

bearing, the rotor centerline moves smoothly from 9600 rpm to 30000 rpm. At around 45000 rpm, another unexpectedly rotor drift is observed. The eccentricity increases from 81 μm to around 89 μm and attitude angle undulates about 10°. This might be attributed to force and pressure undulation in the high speed rotating machinery [35–39]. Then the rotor drifts up 15 μm with increase of rotor speed. After the rotor reaches maximal speed of 52800 rpm, the speed-up process is terminated in avoidance of instability or bearing failure. Then, speed-down process commences. In the speed-down process, the rotor slides down in the bearing almost continuously from (77, 306°) to (110, 282°). Then the rotor touches down on the foil bearing at about 6100 rpm. Generally, the rotor location in the speed-down process is lower comparing to those in the speed-up process at the same rotor speed. For example, at the same rotor speed of 30000 rpm, the rotor centerline is at (90, 290°) in speed-up process and is at (100, 290°) in speed-down

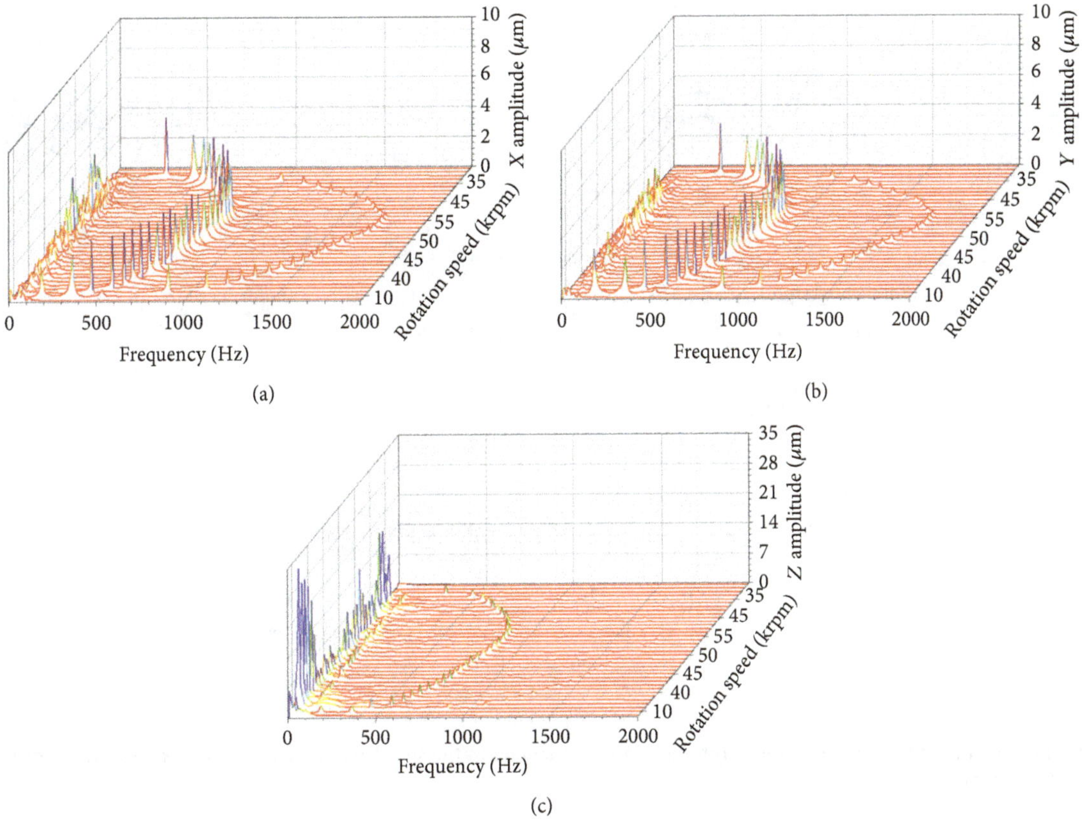

FIGURE 12: Spectrum cascade plot of the high speed rotor in the speed-up and coast-down process: (a) X direction; (b) Y direction; (c) Z direction.

process. When the rotor speed is smaller than 30000 rpm, it should also be noted that the eccentricity exceeds the nominal bearing clearance of 100 μm in the speed-down process. Indeed, the foil bearing could provide over-nominal bearing clearance with compliant surface. The landing speed of the rotor on the foil bearing is around 6000 rpm, which is smaller than that of lift-off speed of 9600 rpm. This can be attributed to the friction effect between the foils during the speed-up and speed-down processes [31]. The resting location of the rotor after coast-down is (103, 272°). Similar to the thrust bearing above, this might also be attributed to the static friction and hysteresis effect of the foils in the above speed-up and speed-down processes.

In this test, the rotor vibrations in X, Y, and Z directions are transformed into frequency domain by FFT. The FFT spectrum cascade plots of the vibrations are shown in Figures 12(a), 12(b), and 12(c), respectively. The main frequencies (1N) of the rotor are obvious comparing to other components. In all three directions, there are nonsynchronous vibrations around 100 Hz. In radial directions X and Y, the nonsynchronous vibration amplitudes are larger comparing to the cascade plot of the rotor supported with foil journal bearing and aerostatic thrust bearing in reference [31]. When the rotor speed is larger than 50000 rpm, the nonsynchronous vibration amplitude increases and exhibits speed dependency. These could be attributed to the introduction of six-pad protuberant foil thrust bearing and loose journal bearing preload. In higher speed range 4000 rpm~5000 rpm, the vibration amplitude in X direction is about 3 μm that is higher than that of <2 μm in Y direction, which can be attributed to higher attitude angle. In axial Z direction, the vibration amplitude of main frequency (1N) is around 2 μm. In smaller speed range, the vibration amplitude at low frequencies (20~30 Hz) is obvious larger. It reaches 30 μm around 30 Hz at the initial start-up stage. This might be attributed to configuration and assembly of the foil thrust bearing. With increase of supply gas pressure and the aerodynamic effect from braking wheel, the vibrations at low frequencies (20~30 Hz) are suppressed at higher rotor speed.

In order to get a clearer view of the transient processes, vibration amplitudes at main frequencies (1N) are extracted from the FFT cascade plot. The vibration amplitude at main frequencies in X, Y, and Z directions are shown in Figures 13(a), 13(b), and 13(c), respectively. When the rotor speed is lower than 10000 rpm before lift-off, there are critical speeds around 3700 rpm and 7200 rpm. For these lower order critical speeds, there is not much difference between the speed-up and speed-down processes. These two lower order critical speeds appear in all three directions. However, there are obvious differences in higher order critical speeds. For instance, the third critical speed is around 12000 rpm in the speed-up process and 17000 rpm in the speed-down process. The forth critical speed is around 21000 rpm in the speed-up process and 25000 rpm in the speed-down process. In

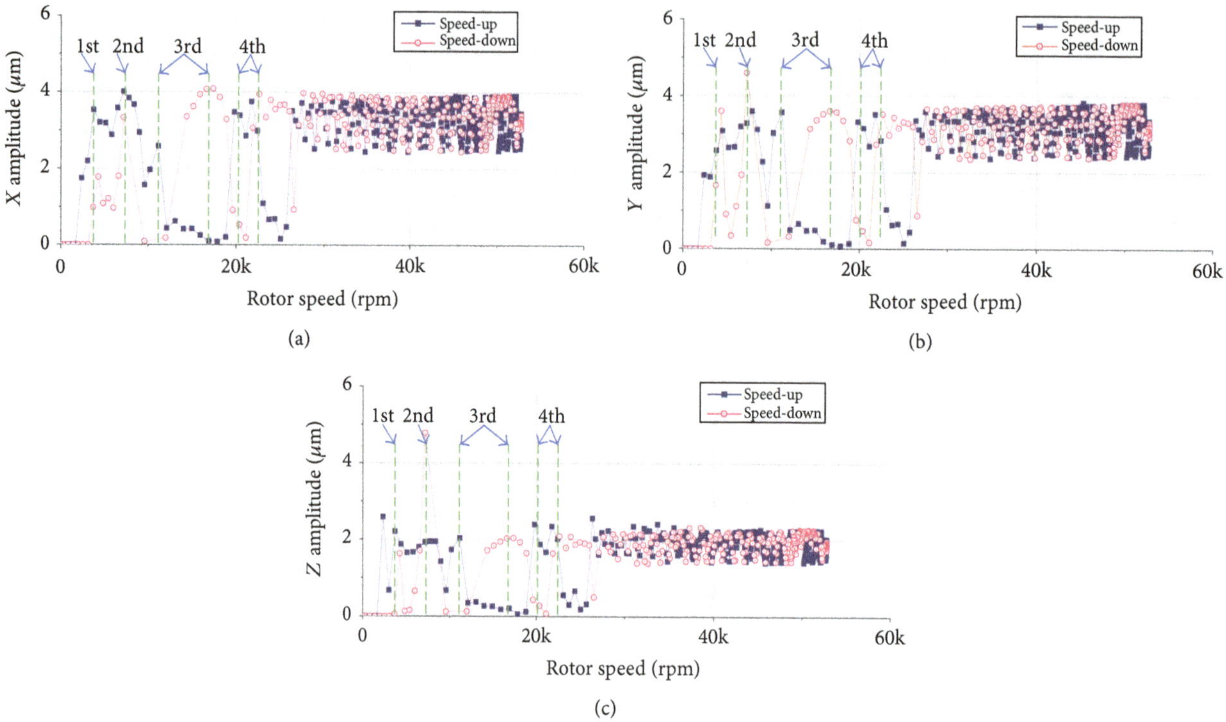

Figure 13: Vibration amplitude of the high speed rotor at main frequencies in the speed-up and coast-down process: (a) X direction; (b) Y direction; (c) Z direction.

operating speed range from 10000 rpm to 30000 rpm after lift-off, as shown in Figure 11, the eccentricity in speed-up process is smaller than nominal clearance of 100 μm and larger than the nominal clearance in speed-down process. The critical speeds of the rotor-bearing system are correlated to the supporting stiffness from bearing. With the rotor weight kept unchanged, the fluid film stiffness varies very little. Therefore, most of the supporting stiffness originates from foil supporting stiffness. In air lubrication regime, the stiffness of the foil bearing support in speed-down process is higher than that in speed-up process. These could also be attributed to the impeding effect of friction between the foils in speed-up process and speed-down process. Corresponding to the 1N main frequency vibration, there are 2N harmonic component vibrations with similar pattern. The vibration amplitude with 2 times the working frequency is relative smaller relative to the main frequency. There are many reasons for the 2N vibrations, such as misalignment of the rotor, asymmetric stiffness of the rotor, bearing nonlinearity, aerodynamic forces [40]. In this rotor-bearing system, misalignment and bearing nonlinearity in the journal and thrust bearings are inevitable.

The rotor locus in a time interval of 0.1 s around 30000 rpm, 40000 rpm, and 50000 rpm is shown in Figures 14(a), 14(b), and 14(c), respectively. The trajectory of the rotor is almost of circular shape and vibration amplitude is small being within 4 μm from the frequency and vibration amplitude relation. There are three vertices on the trajectory which might be attributed to the asymmetry of the foil journal bearing from leading edge to trailing edge. The repeatability of the rotor trajectory is quite well. With increase of rotor speed, the amplitude of nonsynchronous vibration around 100 Hz increases and cluttered trajectory appears at around 50000 rpm. With increase of rotor speed, its amplitude increases from less than 0.5 μm to around 1 μm. Similar to the tested results in [31], the rotor locus also becomes cluttered around 90 krpm when the bearing clearance is larger than 80 μm with protuberant foil thickness of 0.07 mm. In this paper, the nonsynchronous vibrations are more obvious, which could be attributed to the larger nominal bearing clearance. Preload of foil journal bearing suppresses the onset of nonsynchronous vibrations due to the direct stiffening and damping effects [38]. Therefore, higher damping foil bearing might be promising solutions to the nonsynchronous whirling motion. Thus, more sophisticated foil bearing designs are anticipated to reduce both vibration amplitudes comprehensively in future application.

In rotor-bearing system, higher vibration amplitudes (radial and axial directions) might lead to catastrophic collision in the rotor-bearing system. In this study, there is no bearing failure in the whole start-up and speed-down processes with maximal operating speed up to 52000 rpm. This indicates that the compliant supporting from the protuberant foil journal and thrust bearing performed quite well in this transient process, which can be attributed to the inherited characteristics of stiffness and damping of the foils. Foil bearing with lighter compliant surface is eminent in prevention of failure. In addition, the compliant structure could dissipate a certain amount of vibration through damping, which prevents the vibration energy accumulation

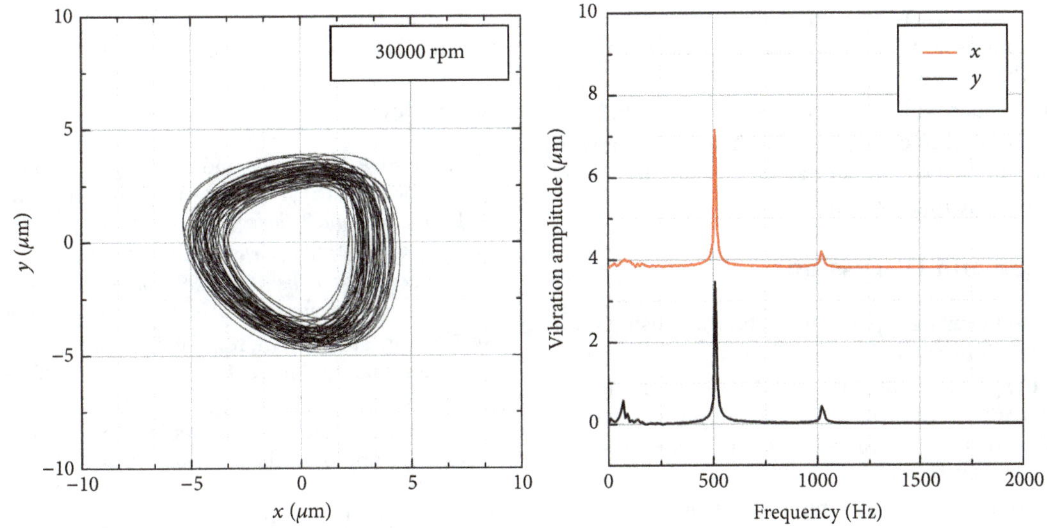

(a)

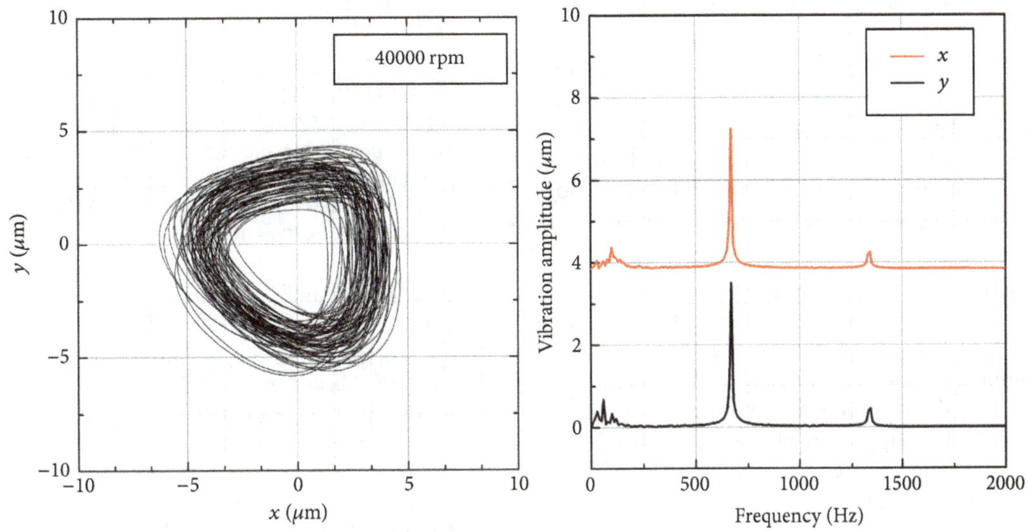

(b)

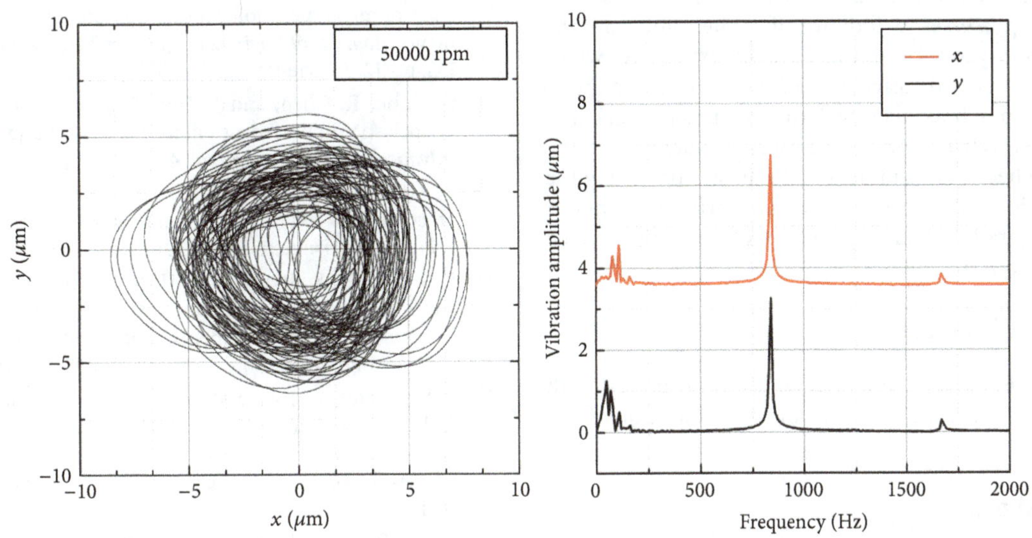

(c)

FIGURE 14: Rotor orbits and vibrations in X and Y directions: (a) 30000 rpm; (b) 40000 rpm; (c) 50000 rpm.

on the rotor. For the thrust bearing, larger subsynchronous rotor lateral motions could be suppressed by increasing the bearing load [23]. In this experiment, higher speed could not be reached under higher supply pressure and unbalance force on the rotor and so on. Optionally, thrust bearing with newer configurations is recommended for the next step rotor-bearing performance improvement.

7. Conclusion and Discussion

In this paper, foil bearing supporting technology using both protuberant foil journal and thrust bearings is developed for the turboexpander. For further simplification of the turboexpander, the dedicated aerostatic supply gas circuit is canceled out for process compactness and simplification. Experimental investigations are carried out on the rotor-bearing system and the following conclusions might be drawn.

(1) Nonlinear stiffness of the protuberant foil journal bearing and thrust bearing appears due to friction effect from the foils. Integrated protuberant foil journal bearing and thrust bearing is applied in turboexpander with maximal operating speed around 52000 rpm. Due to compliance of the foils in the bearing, the rotor supported by protuberant foil journal bearings could operate under over-nominal bearing clearances.

(2) Bearing stiffness can be reflected by the critical speeds of the rotor-bearing system. During the transient speed-up and speed-down processes, the stiffness of the compliant foil journal bearing is almost the same in lower speed range. However, in higher speed range after lift-off, the supporting stiffness is higher in the speed-down process due to the stiffening effect from the friction between the foils.

(3) Axial force on the rotor originates mainly from expansion wheel in the transient speed-up and speed-down processes. Before lift-off of the rotor from the bearing, pressure at the back of expansion wheel results in the displacement of the rotor in the direction from braking wheel to expansion wheel. For further retrofitting, aerodynamic matching of the two wheels is recommended to further balancing the axial load. Alternatively, thrust bearing with newer configurations is anticipated to withstand higher axial load from the rotor.

Conflicts of Interest

The authors declared no potential conflicts of interest with respect to the research, authorship, and/or publication of this article.

Acknowledgments

This project was supported by the National Natural Science Foundation of China (51406157), Natural Science Foundation of Jiangsu Province (BK20140408), the China Postdoctoral Science Foundation (2014M552438), and the Fundamental Research Funds for the Central Universities.

References

[1] S. Ghosh, P. Mukherjee, and S. Sarangi, "Development of bearings for a small high speed cryogenic turboexpander," *Industrial Lubrication and Tribology*, vol. 64, no. 1, pp. 3–10, 2012.

[2] G. L. Agrawal, "Foil air/gas bearing technology - An overview," in *Proceedings of the ASME 1997 International Gas Turbine and Aeroengine Congress and Exhibition, GT '997*, June 1997.

[3] T. H. Kim and L. S. Andrés, "Limits for high-speed operation of gas foil bearings," *Journal of Tribology*, vol. 128, no. 3, pp. 670–673, 2006.

[4] J. Schiffmann and Z. S. Spakovszky, "Foil bearing design guidelines for improved stability," *Journal of Tribology*, vol. 135, no. 1, Article ID 011103, 2012.

[5] K. Ryu and L. S. Andrés, "On the failure of a gas foil bearing: high temperature operation without cooling flow," *Journal of Engineering for Gas Turbines and Power*, vol. 135, no. 11, p. 112506, 2013.

[6] Y. Hou, Z. H. Zhu, and C. Z. Chen, "Comparative test on two kinds of new compliant foil bearing for small cryogenic turbo-expander," *Cryogenics*, vol. 44, no. 1, pp. 69–72, 2004.

[7] H. Heshmat, P. Hryniewicz, J. F. Walton II, J. P. Willis, S. Jahanmir, and C. DellaCorte, "Low-friction wear-resistant coatings for high-temperature foil bearings," *Tribology International*, vol. 38, no. 11-12, pp. 1059–1075, 2005.

[8] S. A. Howard, "Misalignment in gas foil journal bearings: an experimental study," *Journal of Engineering for Gas Turbines and Power*, vol. 131, no. 2, p. 022501, 2009.

[9] M. Branagan, D. Griffin, C. Goyne, and A. Untaroiu, "Compliant gas foil bearings and analysis tools," *Journal of Engineering for Gas Turbines and Power*, vol. 138, no. 5, Article ID 054001, 2015.

[10] D. Rubio and L. S. Andrés, "Bump-type foil bearing structural stiffness: experiments and predictions," *Journal of Engineering for Gas Turbines and Power*, vol. 128, no. 3, pp. 653–660, 2004.

[11] L. S. Andrés and T. A. Chirathadam, "Performance characteristics of metal mesh foil bearings: predictions versus measurements," *Journal of Engineering for Gas Turbines and Power*, vol. 135, no. 12, Article ID 122503, 2013.

[12] Q. Zhou, R. Chen, and Y. Hou, "Dynamic characteristics of viscoelastic supported air foil thrust bearing under axial whirl," *Advanced Science Letters*, vol. 4, no. 4-5, pp. 1552–1555, 2011.

[13] Q. Zhou, Y. Hou, and C. Chen, "Dynamic stability experiments of compliant foil thrust bearing with viscoelastic support," *Tribology International*, vol. 42, no. 5, pp. 662–665, 2009.

[14] L. S. Andrés, K. Ryu, and T. H. Kim, "Thermal management and rotordynamic performance of a hot rotor-gas foil bearings system—part I: measurements," *Journal of Engineering for Gas Turbines and Power*, vol. 133, no. 6, p. 062501, 2011.

[15] L. S. Andrés, K. Ryu, and T. H. Kim, "Thermal management and rotordynamic performance of a hot rotor-gas foil bearings system—Part II: predictions versus test data," *Journal of Engineering for Gas Turbines and Power*, vol. 133, no. 6, p. 062502, 2011.

[16] M. N. Pham and H.-J. Ahn, "Experimental optimization of a hybrid foil-magnetic bearing to support a flexible rotor," *Mechanical Systems and Signal Processing*, vol. 46, no. 2, pp. 361–372, 2014.

[17] D.-H. Lee, Y.-C. Kim, and K.-W. Kim, "The effect of Coulomb friction on the static performance of foil journal bearings," *Tribology International*, vol. 43, no. 5-6, pp. 1065–1072, 2010.

[18] M. Carpino and G. Talmage, "Subfoil stiffness effects in gas-lubricated foil journal bearings," *Tribology Transactions*, vol. 51, no. 5, pp. 602–608, 2008.

[19] H. Heshmat and P. Hermel, "Compliant foil bearings technology and their application to high speed turbomachinery," *Tribology and Interface Engineering Series*, vol. 25, no. C, pp. 559–575, 1993.

[20] Y.-B. Lee, S. B. Kwon, T. H. Kim, and K. Sim, "Feasibility study of an oil-free turbocharger supported on gas foil bearings via on-road tests of a two-liter class diesel vehicle," *Journal of Engineering for Gas Turbines and Power*, vol. 135, no. 5, Article ID 052701, 2013.

[21] B. Dykas, R. Bruckner, C. DellaCorte, B. Edmonds, and J. Prahl, "Design, fabrication, and performance of foil gas thrust bearings for microturbomachinery applications," *Journal of Engineering for Gas Turbines and Power*, vol. 131, no. 1, Article ID 012301, 2009.

[22] D. Kim, A. S. Lee, and B. S. Choi, "Evaluation of foil bearing performance and nonlinear rotordynamics of 120 kW oil-free gas turbine generator," *Journal of Engineering for Gas Turbines and Power*, vol. 136, no. 3, Article ID 032504, 2013.

[23] T. H. Kim, Y.-B. Lee, T. Y. Kim, and K. H. Jeong, "Rotordynamic performance of an oil-free turbo blower focusing on load capacity of gas foil thrust bearings," *Journal of Engineering for Gas Turbines and Power*, vol. 134, no. 2, Article ID 022501, 2011.

[24] D. Kim, A. Creary, S. S. Chang, and J. H. Kim, "Mesoscale foil gas bearings for palm-sized turbomachinery: design, manufacturing, and modeling," *Journal of Engineering for Gas Turbines and Power*, vol. 131, no. 4, Article ID 042502, 2009.

[25] Y. Hou, S. Yang, X. Chen, S. Chen, and T. Lai, "Study on the matching performance of a low temperature reverse Brayton air refrigerator," *Energy Conversion and Management*, vol. 89, pp. 339–348, 2015.

[26] Y. Hou, S. Chen, R. Chen, Q. Zhang, and H. Zhao, "Numerical study on foil journal bearings with protuberant foil structure," *Tribology International*, vol. 44, no. 9, pp. 1061–1070, 2011.

[27] Y. Hou, T. Lai, B. Ma, Y. Zheng, and S. Chen, "Experimental investigation on the multi-decked protuberant gas foil journal bearing," *Journal of Advanced Mechanical Design, Systems, and Manufacturing*, vol. 7, no. 4, pp. 791–799, 2013.

[28] S. Chen, Y. Hou, L. Niu, S. Yang, and T. Lai, "Study on double-layer protuberant gas foil journal bearings with different foil layers arrangement," *Journal of Advanced Mechanical Design, Systems, and Manufacturing*, vol. 9, no. 2, 2015.

[29] T. Lai, S. Chen, B. Ma, L. Liu, and Y. Hou, "Preliminary experimental study on static loading characteristics of multi-decked protuberant foil thrust bearing," *Journal of Advanced Mechanical Design, Systems, and Manufacturing*, vol. 10, no. 1, 2016.

[30] Y. Zheng, T. Lai, S. Chen, L. Chen, L. Liu, and Y. Hou, "Static characteristics of six pads multilayer protuberant foil thrust bearings," *Proceedings of the Institution of Mechanical Engineers, Part J: Journal of Engineering Tribology*, vol. 231, no. 2, pp. 158–164, 2016.

[31] T. Lai, S. Chen, B. Ma, Y. Zheng, and Y. Hou, "Effects of bearing clearance and supporting stiffness on performances of rotor-bearing system with multi-decked protuberant gas foil journal bearing," *Proceedings of the Institution of Mechanical Engineers, Part J: Journal of Engineering Tribology*, vol. 228, no. 7, pp. 780–788, 2014.

[32] T. Lai, Y. Guo, W. Wang, S. Chen, and Y. Hou, "Experimental study on multi-decked protuberant foil thrust bearing with different number of thrust pads," *Journal of Advanced Mechanical Design, Systems, and Manufacturing*, vol. 10, no. 9, 2016.

[33] Y. Zheng, T. Lai, S. Chen, L. Chen, L. Liu, and Y. Hou, "Static characteristics of six pads multilayer protuberant foil thrust bearings," *Proceedings of the Institution of Mechanical Engineers, Part J: Journal of Engineering Tribology*, vol. 231, no. 2, pp. 158–164, 2017.

[34] F. Balducchi, M. Arghir, R. Gauthier, and E. Renard, "Experimental analysis of the start-up torque of a mildly loaded foil thrust bearing," *Journal of Tribology*, vol. 135, no. 3, Article ID 031702, 2013.

[35] Y. Hao, L. Tan, Y. Liu, Y. Xu, J. Zhang, and B. Zhu, "Energy performance and radial force of a mixed-flow pump with symmetrical and unsymmetrical tip clearances," *Energies*, vol. 10, no. 1, article no. 57, 2017.

[36] Y. Liu, L. Tan, M. Liu, Y. Hao, and Y. Xu, "Influence of prewhirl angle and axial distance on energy performance and pressure fluctuation for a centrifugal pump with inlet guide vanes," *Energies*, vol. 10, no. 5, 2017.

[37] T. Lei, Z. B. Shan, C. S. Liang, W. Y. Chuan, and W. B. Bin, "Numerical simulation of unsteady cavitation flow in a centrifugal pump at off-design conditions," *Proceedings of the Institution of Mechanical Engineers, Part C: Journal of Mechanical Engineering Science*, vol. 228, no. 11, pp. 1994–2006, 2014.

[38] L. Tan, B. Zhu, Y. Wang, S. Cao, and S. Gui, "Numerical study on characteristics of unsteady flow in a centrifugal pump volute at partial load condition," *Engineering Computations (Swansea, Wales)*, vol. 32, no. 6, pp. 1549–1566, 2015.

[39] K. Sim, B. Koo, J. S. Lee, and T. H. Kim, "Effects of mechanical preloads on the rotordynamic performance of a rotor supported on three-pad gas foil journal bearings," *Journal of Engineering for Gas Turbines and Power*, vol. 136, no. 12, Article ID 122503, 2014.

[40] M. L. Adams, *Rotating Machinery Vibration: From Analysis to Trouble Shooting*, CRC Press Taylor and Francis Group, New York, NY, USA, 2nd edition, 2010.

18

Experimental and Numerical Investigations of Surge Extension on a Centrifugal Compressor with Vaned Diffuser using Steam Injection

Chuang Gao,[1] Weiguang Huang,[1] Tianhua Zheng,[2] Kang Yang,[2] Haosen Yang,[2] and Yun Cao[2]

[1]*Shanghai Advanced Research Institute, Chinese Academy of Sciences, Shanghai, China*
[2]*Division of Aerodynamics, Department of R&D, Helan Turbines, Co., Ltd., China*

Correspondence should be addressed to Chuang Gao; gaoc@sari.ac.cn

Academic Editor: Ryoichi Samuel Amano

This paper presents the first report on surge extension with steam injection through both experimental and numerical simulation. The experimental section covers the test facility, instrumentation, and prestall modes comparison with and without steam injection. It is found that surge extension is not in proportion to injected steam. There exists an upper bound above which deteriorates the margin. Injection of less than 1% of the designed mass flow can bring about over 10% margin improvement. Test results also indicated that steam injection not only damps out prestall waves, but also changes prestall modes and traveling direction. At 90% speed, injection changed the prestall mode from spike to modal, while at 80% speed line, it made the forward traveling wave become backward. Through numerical simulation, location and number of injectors, molecular weight, and temperature of injected gas are modified to explore their influences on surge margin. Similar to the test results, there exists an upper bound for the amount of steam injected. The flow field investigation indicates that this bound is caused by the early trigger of flow collapse due to the injected steam which is similar to the tip leakage flow spillage caused spike stall in axial compressors.

1. Introduction

Compressor flow instabilities may occur if the fluid flow decreases below a certain limit. Rotating stall and surge are such aerodynamic instabilities that occur in both axial and centrifugal compressors [1–24]. Surge, however, can result both in severe vibration and damage to compressor units and in reduced efficiency. Although many researches were published in the open literature on these two phenomena, the prediction and control of surge are still far away from the requirement of industrial application. In most cases, this can only be done case by case as it depends on the design.

For impellers with a vaned diffuser, most of the researches attribute rotating stall and surge to the increase of incidence angle at low flow rate; however, Spakovszky [2] pointed out that vaneless space just in front of the vaned diffuser is the root cause of rotating stall in the NASA CC3 compressor. Schleer et al. [3] measured the trajectories of tip clearance vortex at different flow rates and concluded that the tip clearance vortex caused the surge incipience. More recently, researchers from the von Karman Institute and University of Hannover [4] noticed the influences of volute on both performance and instability.

To control rotating stall and surge in a centrifugal compressor, various design features have been investigated: self-circulation device [13], variable inlet guide vanes (VIGV), adjustable vaned diffuser vanes, ported shroud [15], grooved diffuser [16], air injection, or bleeding [8].

The analyses for the prestall, stall inception, and the generation of fully developed stall were to explore the interrelationship among them and were then applied in the design of prewarning and stall control systems. Experimental exploration of compressor prestall or stall inception usually involves several transducers for unsteady pressure recording distributed circumferentially in one or more axial locations. The pressure transducer settings for axial compressors were

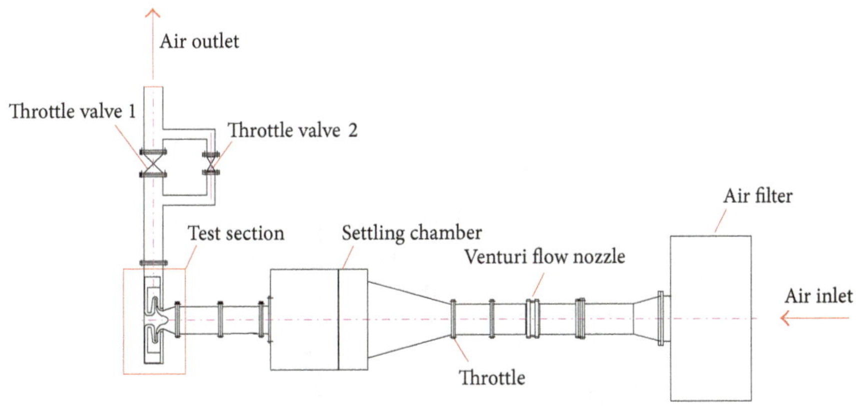

FIGURE 1: Schematic layout of the test facility.

often on the casing above the rotor [20] while the setup for centrifugal compressors sometimes may be considered on both the rotor and the diffuser to tell the stall inception location [24]. Pressure signals should be traced when compressors are throttled to stall. The definitions of prestall, stall inception, and fully developed stall (surge in some cases) were clearly described in [20] and can be understood according to the magnitude of pressure perturbation from relatively calm to large. Tryfonidis et al. [20] also indicated that the rotating stall influences the surge inception mechanism.

The stall wave was typically reported to travel in positive rotor direction as most of the axial compressor did, but in centrifugal compressors, negative traveling stall wave was also reported. Besides, two types of stall, modal stall and spike-like stall, were defined according to the stall length scale. With the theoretical background of this length scale study, various control methods were developed against different length scale stalls. Several methods have been conducted to analyze the circumferentially traveling pressure perturbation. Tryfonidis et al. [20] developed the spatial Fourier decomposition for stall inception area with the PSD analysis for traveling wave energy. In the past decades, the wavelet method was also quickly developed for further understanding of the physical mechanisms which lead to stall, but the result is sensitive to the parameter selection.

To sum up, researches on centrifugal compressor flow instabilities are still not enough for fully understanding and controlling rotating and surge.

Recently, with the stringent requirements on pollutant emissions, the Chinese government encourages distributed energy systems, especially combined cooling, heating, and power (CCHP) systems, in which the key component is a gas turbine. Considering the population of residential areas in most Chinese cities, it has been found that a gas turbine of 2 MW output level is best suited for CCHP systems. At this power level, the gas turbine efficiency is much lower than those used in power plants. Generally, small gas turbines are always used together with a HRSG (heat recovery steam generator) which could provide heating in winter and cooling in summer with absorption refrigerating machine.

In most parts of China, however, there are almost no requirements on both cooling and heating in transitional seasons, which means that the gas turbine has to be shut down due to its low simple cycle efficiency.

With such background, the authors intend to boost the power output and efficiency through reinjecting the steam from the HRSG. However, to fulfill this aim, the compressor surge line has to be moved to the left due to the increased back pressure caused by the steam injected. Then, a question arises: can we use part of the power boost steam to increase the surge margin without using an external air supply?

The current paper reports related experimental results about impeller and vaned diffuser. The structure of the paper could be summarized as follows. A brief introduction is provided in Section 1. Test facility, experimental procedure, and data acquisition are illustrated in Section 2. The influences of steam injection characteristics at different rotational speeds on stage performance and surge range are discussed in Section 3. Prestall modes are analyzed and compared between a solid casing and one with steam injection in Section 5. Numerical models, compressor performance, surge margin, and diffuser inlet flow angle variation with steam mass flow are investigated using the same configuration as test facility. Flow physics for surge enhancement are then discussed in Section 6. Section 7 covers influences of injected gas properties, injector configuration, and so forth in detail. Conclusion and acknowledgment are given in the last two sections.

2. Test Facilities and Instrumentation

2.1. Test Facilities. The experiments were conducted on the 500 kW DC motor driven test rig at the Product Laboratory in Shaanxi Blower Company. Figure 1 shows the schematic layout of the test facility.

Ambient air is drawn into the pipe through an air filter and through a flow nozzle to measure the flow rate. The settling chamber is used to lower the gas velocity and make the impeller incoming flow more uniform. After the test section, the compressed air is discharged into the atmosphere. The mass flow rate can be adjusted by two valves. The cross section of the test section is shown in Figure 2.

The test impeller is scaled down by a factor of 0.5 to reduce the power consumption so that it can be driven by the motor

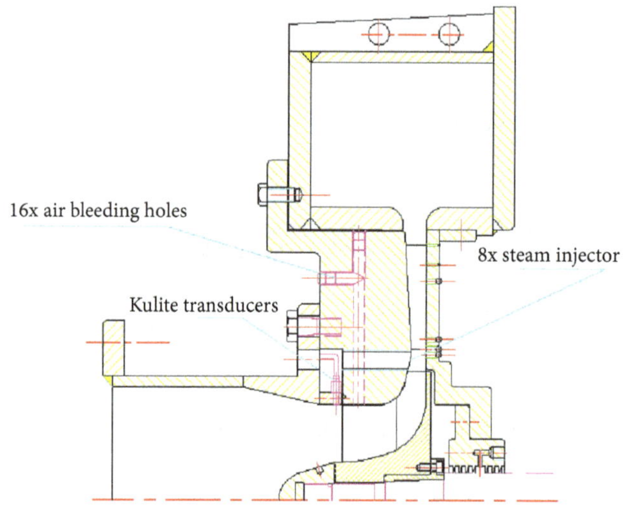

FIGURE 2: Cross-sectional view of the test section.

TABLE 1: Typical design parameters.

Total pressure ratio	—	3.55
Designed mass flow rate	kg/s	3.2
Impeller tip speed	m/s	433
Blade number	—	16 + 16
Impeller tip radius	m	0.115
Average blade exit angle	Deg	90
Impeller blade inlet angle	Deg	52
n_s (dimensionless)	—	0.83
Diffuser inlet angle	Deg	1.6
Diffuser vane number	—	23
Diffuser area ratio	—	2.1

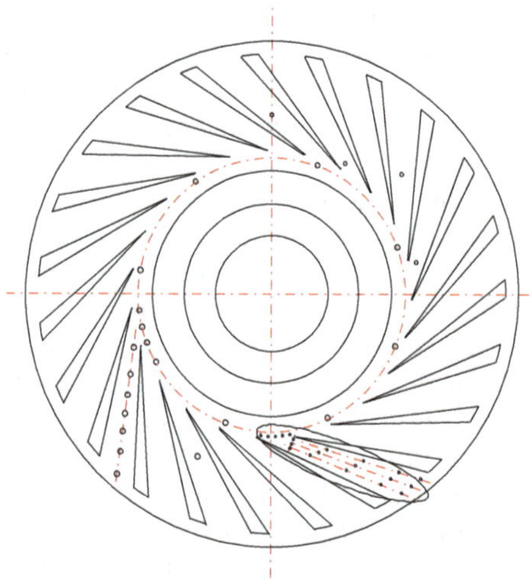

FIGURE 3: Steady and unsteady pressure taps in the diffuser.

on site. For the scaled version, the primary design parameters are given in Table 1.

In the present test, the compressor was operated at 60%, 70%, 80%, and 90% of the design speed to reduce the risk of potential surge damage at 100% speed. Eight steam injectors were installed midway between the impeller tip and the diffuser leading edge. Sixteen air bleeding holes formed by radial channel and axial channel were used for impeller throat air bleeding, the results of which are not covered in the current paper.

The collector after the diffuser has a constant cross-sectional area which is chosen to be 4 times the vaned diffuser outlet throat area to eliminate the asymmetric influences on upstream components.

2.2. Instrumentation. Compressor performance was evaluated by measuring the static pressure and temperature at the inlet of the impeller and the exit of the collector using a pressure transducer from Rosemount Engineering Co. A flow nozzle (shown in Figure 1) was used to measure the mass flow rate based on the pressure and temperature of the surroundings. A five-hole cobra probe with a thermocouple was used to measure the total pressure and temperature at the impeller outlet ($r/r_2 = 1.04$).

For the unsteady pressure measurements, 31 Kulite transducers were used. Six of them were uniformly installed 5 mm ahead of the impeller leading edge as shown in Figure 2. 23 of them were installed at the diffuser leading edge, in the diffuser throat, and inside the diffuser channel to record pressure data variation while throttling. The left two were installed in the exhaust collector. Besides the 23 unsteady transducers, 20 steady pressure taps circled by the ellipse shown in Figure 3 were in one channel to find pressure contour variation with reduction of flow rate.

The other two transducers are installed on the collector casing to study the flow behavior inside the collector. A photo of the test rig without shroud is shown in Figure 4 to show the location of the Kulite transducers in the diffuser.

During the test, two throttling valves were fully opened for steady operation at each speed line. Then, they are

FIGURE 4: A photo of the test impeller and transducers.

gradually throttled until the compressor surges. For each speed line, at least six operating points are measured. At each point, all the unsteady pressure data were acquired and stored for at least 20 seconds. After all six performance data measurements are taken, two valves are opened a little to make the compressor operate at the nearest stable point; then, the valves are throttled gradually to push the compressor into surge; after several surge cycles, the valves are quickly opened to 100% in order to pull the compressor out of surge. Unsteady pressure data were acquired all through this process in order to analyze stall/surge precursor and pathology. Sampling frequency of all the 31 transducers was chosen to be 50 kHz in order to avoid frequency alias and to provide enough time resolution.

3. Effects of Steam Injection on Performance and Surge Extension

Based on the analysis of component stability [24], accumulated shroud side separation due to centrifugal and Coriolis force, and positive results from Skoch [8], eight injectors are uniformly installed on the shroud side located in the middle of the impeller exit and diffuser leading edge to minimize impeller and diffuser mutual interaction. The position and tip segment of the injector are designed to take advantage of Coanda flow effects [18] as shown in Figure 5.

For the compressor test alone facility, the injected steam is generated from an electric boiler. The maximum steam mass flow rate is 200 kg/h. In the current experiment, the steam pressure is kept as 6 bar limited by the used electrical boiler and this number should be higher with HRSG used on site.

At constant impeller rotational speed, the injected steam mass flow is increased and kept at several different flow rates to investigate its influences on surge margin. At each steam flow rate, valves are gradually throttled to find the surge flow rate. In this paper, surge margin improvement (SMI) is defined as $(m_{so} - m_s)/m_{so} \times 100\%$, in which m_s and m_{so} are surge flow rates with and without steam injection, respectively.

Variations of SMI under 80% and 90% of the designed rotational speed for six different steam flow rates (I: 0.521 g/s, II: 1.042 g/s, III: 2.083 g/s, IV: 3.125 g/s, V: 4.167 g/s, and VI: 5.208 g/s) are shown in Figure 6.

In Figure 6, it is obviously observed that the trends for these two different rotational speeds are different especially at high steam flow rate. For the line with Mu = 0.96, there is an optimum steam flow for which the surge extension is maximized. This trend is kept the same for Mu = 1.08 except for the first point. For this point only, 0.15% designed mass flow rate can increase surge margin by over 11%. This effect is the same as the fifth point of which the mass flow rate is seven times as large as that of the first point. The strange phenomenon is that the surge margin extension decreased with increasing injected steam mass flow rate. This implies that the traditional explanation based on flow angle might not be correct for the surge inception of centrifugal compressor or at least for this compressor. Detailed explanation was given in Section 4 from the point of different stall mode about the different trends shown by 80% and 90% speed line.

After reviewing the test data, it is found that steam injection brings about loss to both pressure ratio and efficiency as shown in Figure 7. Stage efficiency is defined by the following equation:

$$\eta_{\text{stage}} = \frac{T_{t1}\left[\left(P_{3,\text{st}}/P_{t1}\right)^{(n-1)/n} - 1\right] m_1 + T_{t1,\text{inject}}\left[\left(P_{3,\text{st}}/P_{t1,\text{hole}}\right)^{(n-1)/n} - 1\right] m_{\text{inject}}}{m_1\left(T_{3,\text{st}} - T_{t1}\right) + m_{\text{inject}}\left(T_{3,\text{st}} - T_{t1,\text{hole}}\right)}. \tag{1}$$

Even with the same steam flow rate, the loss deviates greatly. For example, for the index equal to 5 (flow rate: 4.167 g/s), the loss at Mu = 0.96 is almost 8 times that of Mu = 1.08. Another interesting phenomenon is that good SMI points always coincide with good loss ones. The above phenomena shown by Figures 6 and 7 are repeatable at different ambient conditions from three test days.

4. Prestall Modes and Surge Characteristics

4.1. Unsteady Pressure Traces. Pressure traces at 80% and 90% with and without steam injection located at different streamwise locations (impeller inlet, semivaneless diffuser, diffuser throat, and diffuser midpassage) during stall ramp just before surge are presented in Figures 8(a)–8(d).

In Figure 8, it is observed that the prestall waves are pronounced at the diffuser throat which means that the dominant modes here are the diffuser modes. Thus, the following studies are focused on diffusers. The magnitude of pressure perturbations with injection is on the same level as the case without injection. Additionally, one interesting finding is that the prestall mode even changed from spike to be of modal type as shown in Figure 8(b) on 90% rotational speed with the existence of steam. It can be explained that

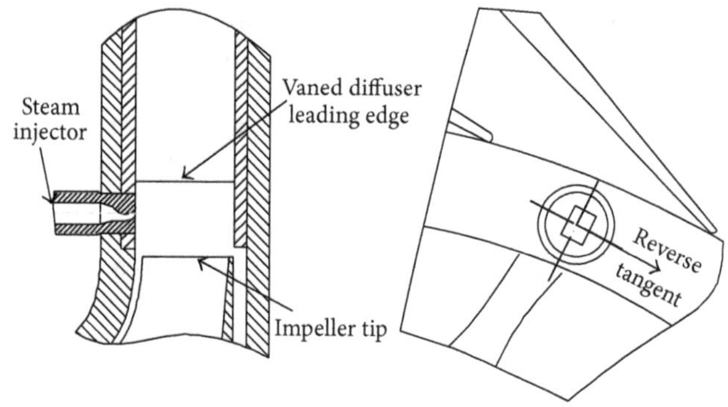

FIGURE 5: Steam injector.

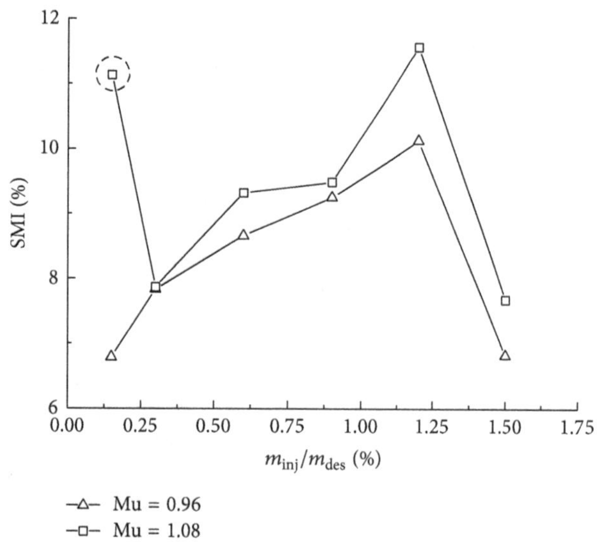

FIGURE 6: Steam injection percentages on surge extension.

the injected flow at shroud increases system damping which caused the prestall mode's variation from short wave (spike) to long wave (modal). Instead of flow injection in the current paper, endwall flow bleeding on the hub side can also generate similar effects as shown by Spakovszky and Roduner [5] in a turbocharger compressor.

With a closer view, it can also be seen that the steam injection damped out the pressure oscillation after surge (time = 0) occurred through comparison of the wave segment in the black circle with the same segment with injection. It is worth noting that, after surge, the pressure traces at the impeller inlet behave much less violently than those in other locations or in the same location with steam injection on. This also indicates that surge root comes from the diffuser area. Damping that pressure perturbation amplitude can reduce the surge intensity.

This can also explain the strange phenomenon shown by Figure 6 in which the largest surge margin improvement is at the smallest injection flow at 90% speed line while the surge margin improvement at 80% speed line is gradually increased with the amount of steam flow. This means that, at 90% speed line, the prestall wave is a spike which is nonlinear. With the increase of steam injection, the prestall wave changes from spike to be modal which is linear long wave. And at 80% speed line, the prestall wave is always linear modal wave.

4.2. Traveling Wave Energy Analysis of Prestall Behavior. Traveling wave energy method was used to analyze the prestall wave traveling frequency and direction. The x- and y-axis of the TWE spectra depicted in Figure 9 were normalized by rotor revolution and rotor frequency (N means one rotor 1x frequency in later description), respectively. To further analyze this prestall activity, the pressure signals are decomposed into spatial Fourier harmonics. For each harmonic, the evolution of power spectrum is then computed. The plotted spectrum is the difference between the positive and the negative frequency spectra of the complex spatial Fourier coefficient. Peaks and valleys therefore stand for the forward (with the rotor revolution direction) and backward (against the rotor revolution direction) traveling waves as described by Tryfonidis et al. [20]. In Figures 9 and 10, the time index equal to zero is assigned for fully developed stall (surge).

Figure 9 shows the 90% TWE results without steam injection. The impeller rotating frequency which is 1 N was clearly shown. The 0.3 N frequency was the first stall frequency emerging at relatively high mass flow rate compared with that of surge point, and 0.13 N that appeared in the 3rd harmonic just before surge is the second stall frequency. The 0.3 N frequency in the 1st harmonic is clearly the dominant frequency and mode before surge, and 0.13 N frequency is the dominated stall precursor in the 3rd mode. Both stall precursors traveled in the forward direction.

However, with 120 kg/h steam injection, the 0.3 N frequency was suppressed and almost invisible. The 0.13 N frequency dominated all the other harmonics as shown in Figures 10 and 11. In Figure 11, due to the in-phase or unchanged phase features of the 4 pressure traces at different circumferential angles, the authors claim that these circumferential areas are covered by one stall cell. This means that, after steam injection, there exists one big stall cell that moves in the impeller rotation direction in the speed ratio of 13%.

For the 80% TWE results with/without steam injection, the results are similar. The 0.22 N frequency was the first stall wave emerging at relatively high mass flow rate and the 0.13 N appeared in the 2nd harmonic just before surge. The only difference and strange thing is that the prestall modal wave traveling direction of 0.13 N shifts from forward (rotation direction) to be backward (against rotation direction). This phenomenon can be observed by listing all circumferential unsteady pressure traces in circumferential direction as shown in Figure 12. This further complicates surge scenario and was also never been reported before in public references. Similarly, as shown in Figure 11, Figure 12 also indicates a stall cell moving slowly against impeller rotation direction with injected steam but with larger size than the case of 90%.

Detailed influences of the amount of steam injection (0 kg/h, 30 kg/h, 60 kg/h, and 120 kg/h) on prestall mode activity at two different rotational speeds (80% and 90%) in a semivaneless space are summarized in Figure 13. At 80% speed, the 0.22 N content only existed at 0 kg/h and was suppressed in other cases. With the increase of injection flow,

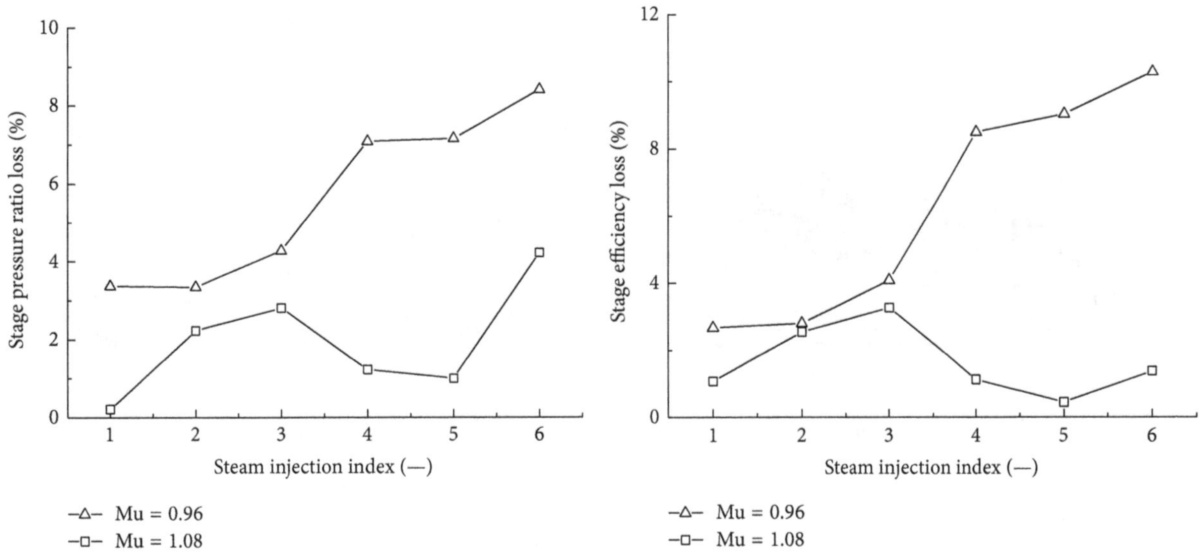

(a) Stage pressure loss versus steam injection index

(b) Stage efficiency loss versus steam injection index

FIGURE 7

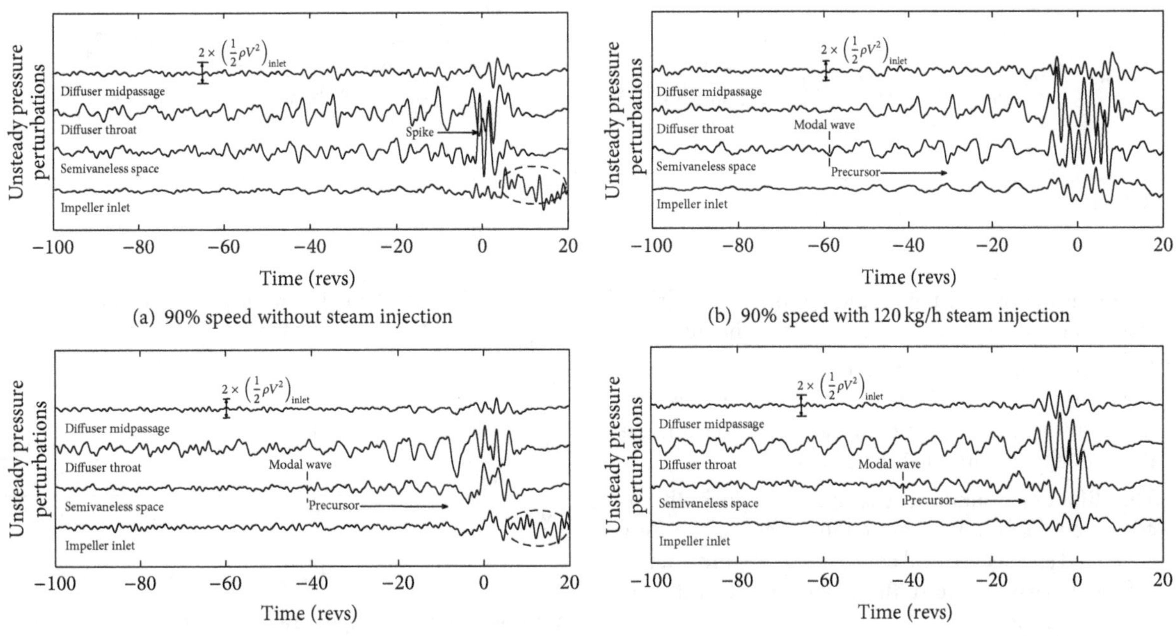

(a) 90% speed without steam injection

(b) 90% speed with 120 kg/h steam injection

(c) 80% speed without steam injection

(d) 80% speed with 60 kg/h steam injection

FIGURE 8: Pressure perturbation analysis for streamwise locations.

the amplitude of prestall waves was gradually attenuated. At 90% speed, the frequency of the original 0.3 N prestall wave decreased from 0.3 N to be 0.23 N with 120 kg/h. With 150 kg/h injection flow (not shown in the figure), this content is completely suppressed. From the figure, nonlinear coupling between different modes can be easily found which means one mode always has contributions from other harmonics by seeing that the same frequency exists in different modes. It can be observed that 30 kg/h injection damps out almost all the prestall modes compared to other amounts of injections. If, further, more steam is injected, some higher modes can be damped further, but not that obviously. These observations agree well with test results shown in Figure 6.

4.3. Discussion on Experimental Results. Depending upon steam quality (superheated/saturated), nozzle parameter (position/number/angle), and other factors, there are several things that happen with steam injection into the flow path. These things are as follows:

(a) The injection of steam into the flow path has an effect on the properties of the flow, especially in changes

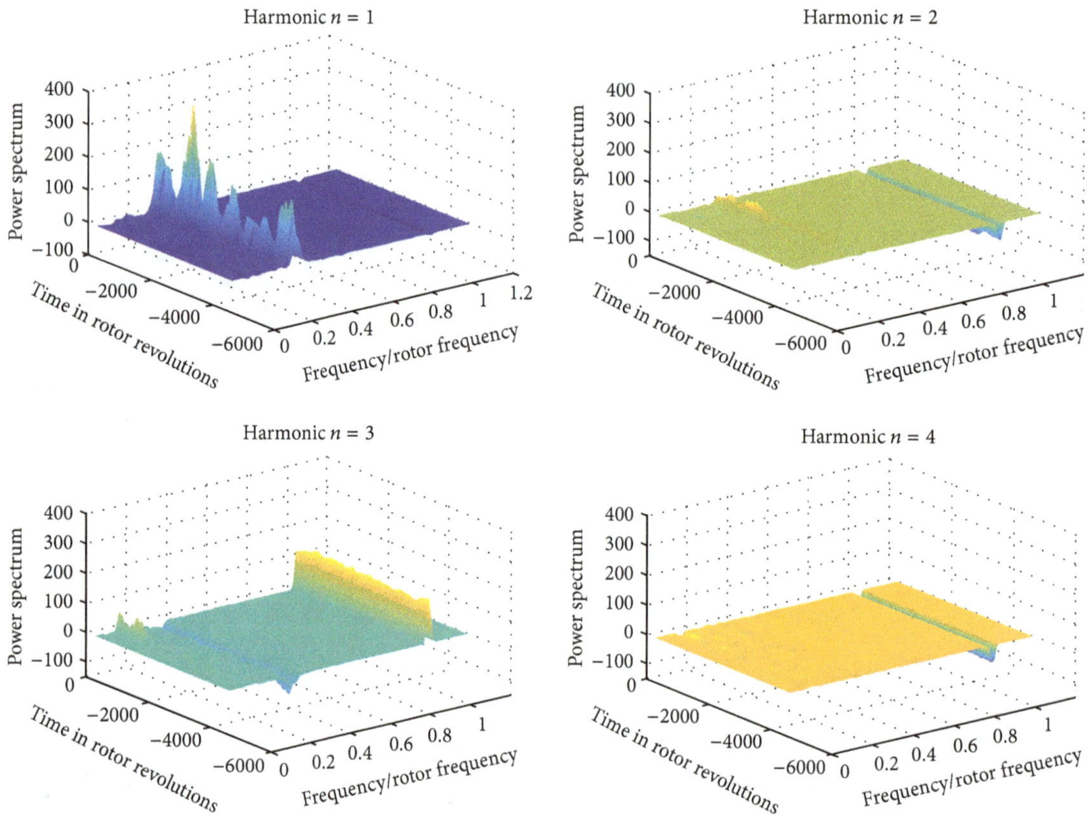

FIGURE 9: Spectrograms of the 1st, 2nd, 3rd, and 4th spatial harmonic pressure perturbations during stall ramp at 90% design speed without steam injection.

to the gas constant, theta, and gamma. This finally changes the operating characteristics of the diffuser.

(b) Depending upon temperature difference between steam and air, the heat transfer effect between the two fluids influences the incidence angle of the diffuser through changes in velocity vectors.

(c) The different nonuniform temperature in both circumferential and axial directions changes the local corrected speed and local corrected flow which makes different parts of the diffuser work under different inlet flow conditions.

(d) The steam injection can enhance the flow momentum of the flow near the shroud where wake and separation always exist and delay the flow separation which improves diffuser stability.

(e) At subsonic diffuser inlet conditions, steam injection can also have certain influences on the impeller performance which might also be beneficial for the stability.

To answer the above questions, numerical simulations and detailed flow field analysis were performed.

5. Numerical Procedures

Full annulus numerical grids are meshed with impeller, injector, vaned diffuser, and volute. The grids are blade centered multiblocks with hexahedral structure. The stator and rotor grids had ATM optimized topology which was a proprietary blocking strategy for obtaining good quality grids. The grids were made finer at the regions where high gradients were expected in the flow passage. The 3D computational grids used in the analysis are shown in Figure 14. The mesh in the boundary layers was refined to ensure Y+ ranged from 30 to 40, the total grids number was about 5.2 million, and mesh densities after grid independency check for each component are given in Table 2.

TABLE 2: Mesh densities of CFD domains.

Domain	Numbers of grids
Rotor	2328384
Diffuser	2435424
Injector	180168
Collector	289856
Total	5233832

The k-ϵ turbulence model with a scalable wall function was chosen. A total pressure and total temperature inlet boundary condition were used at the rotor inlet, a mass flow outlet boundary condition was set at the diffuser outlet, and a second mass flow inlet was defined at the injectors inlet. The rotor domain was solved in rotational coordinate frame with rotational speed of 30780 rpm. The rotor domain and

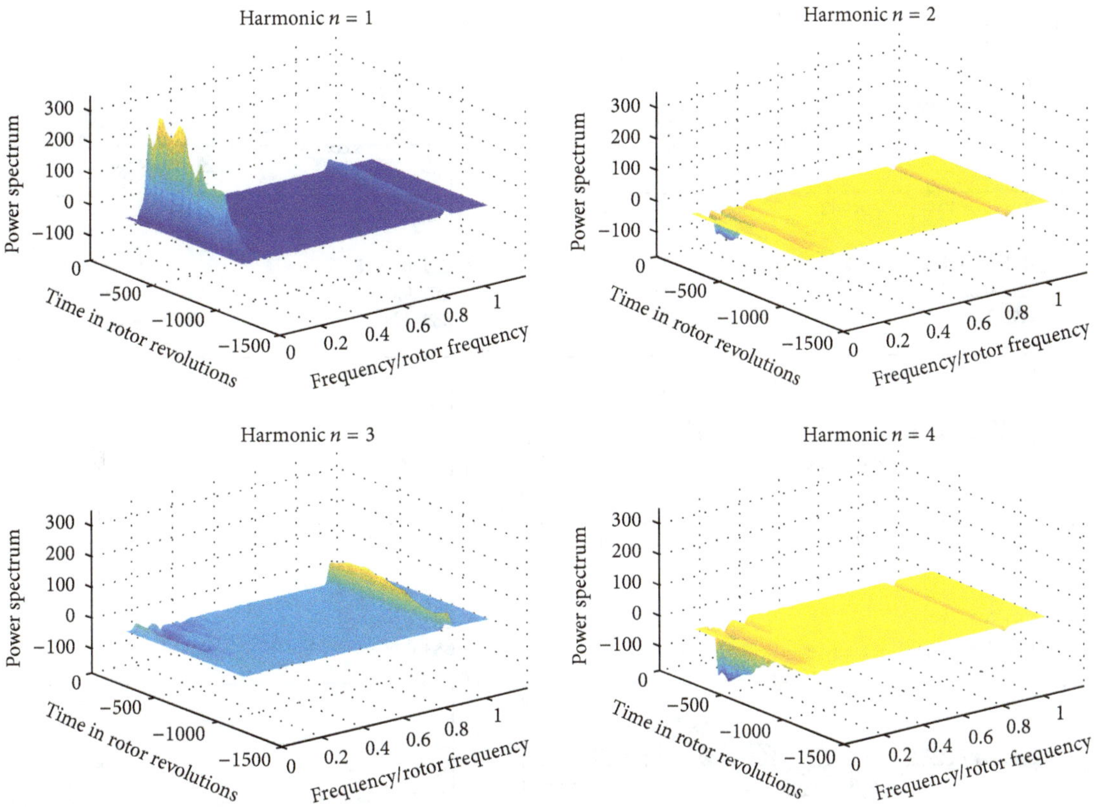

FIGURE 10: Spectrograms of the 1st, 2nd, 3rd, and 4th spatial harmonic pressure perturbations during stall ramp at 90% design speed with 120 kg/h steam injection.

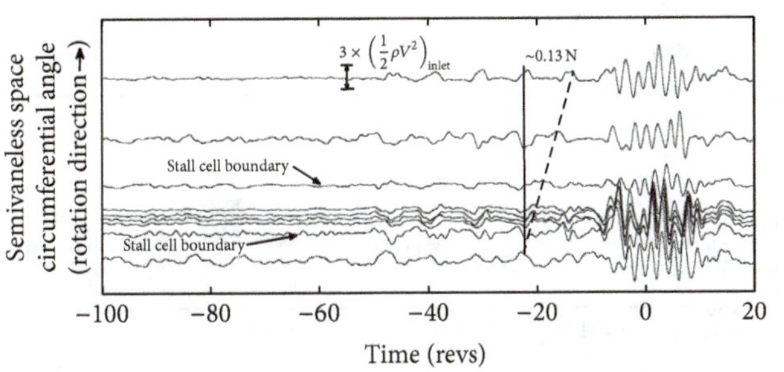

FIGURE 11: Unsteady pressure traces at 90% speed with 120 kg/h steam injection (Kulite transducers all located in a semivaneless space).

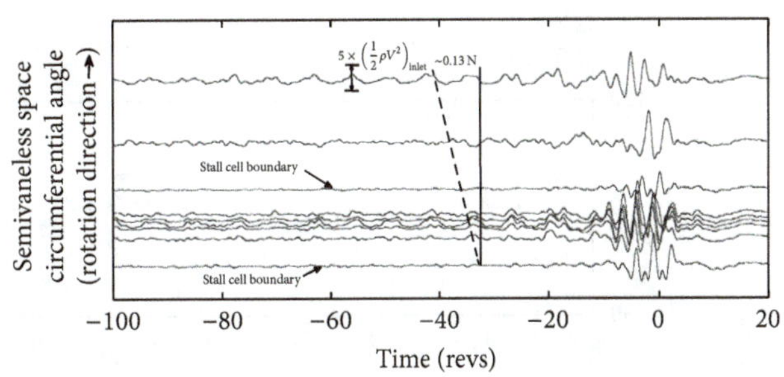

FIGURE 12: Unsteady pressure traces at 80% speed with 60 kg/h steam injection (Kulite transducers all located in a semivaneless space).

injection domain exchanged boundary conditions across the mixing plane interface. The downstream side was treated as an inlet face with the boundary conditions defined by averaged flow properties at the impeller exit. The upstream impeller side could be seen as an outlet face with the exit flow condition being provided by the diffuser inlet flow. The case was considered to be converged when the RMS was below $1e - 5$ and the variation of the TTE (total-to-total efficiency) was within 3%.

Compressor performance was evaluated by measuring the static pressure and temperature at the inlet of the impeller and exit of the gas collector using a pressure transducer from Rosemount Engineering Co., Ltd. A flow nozzle was used to measure the mass flow rate based on the pressure and temperature of the surroundings. A five-hole cobra probe with a thermoresistor was used to measure the total pressure and temperature just at the impeller outlet.

Figure 15 presents the comparisons between computed and measured compressor characteristics with 120 kg/h steam injection and without steam injection at 90% speed, which plotted both the TTPR and the TTE as a function of diffuser exit mass flow corrected to sea level total pressure and temperature; the choke limit and the surge limit frame the operating range on both ends of the abscissa. Both the CFD predictions and the experimental datum with steam injection extended the stable flow range, while around 5% drops in

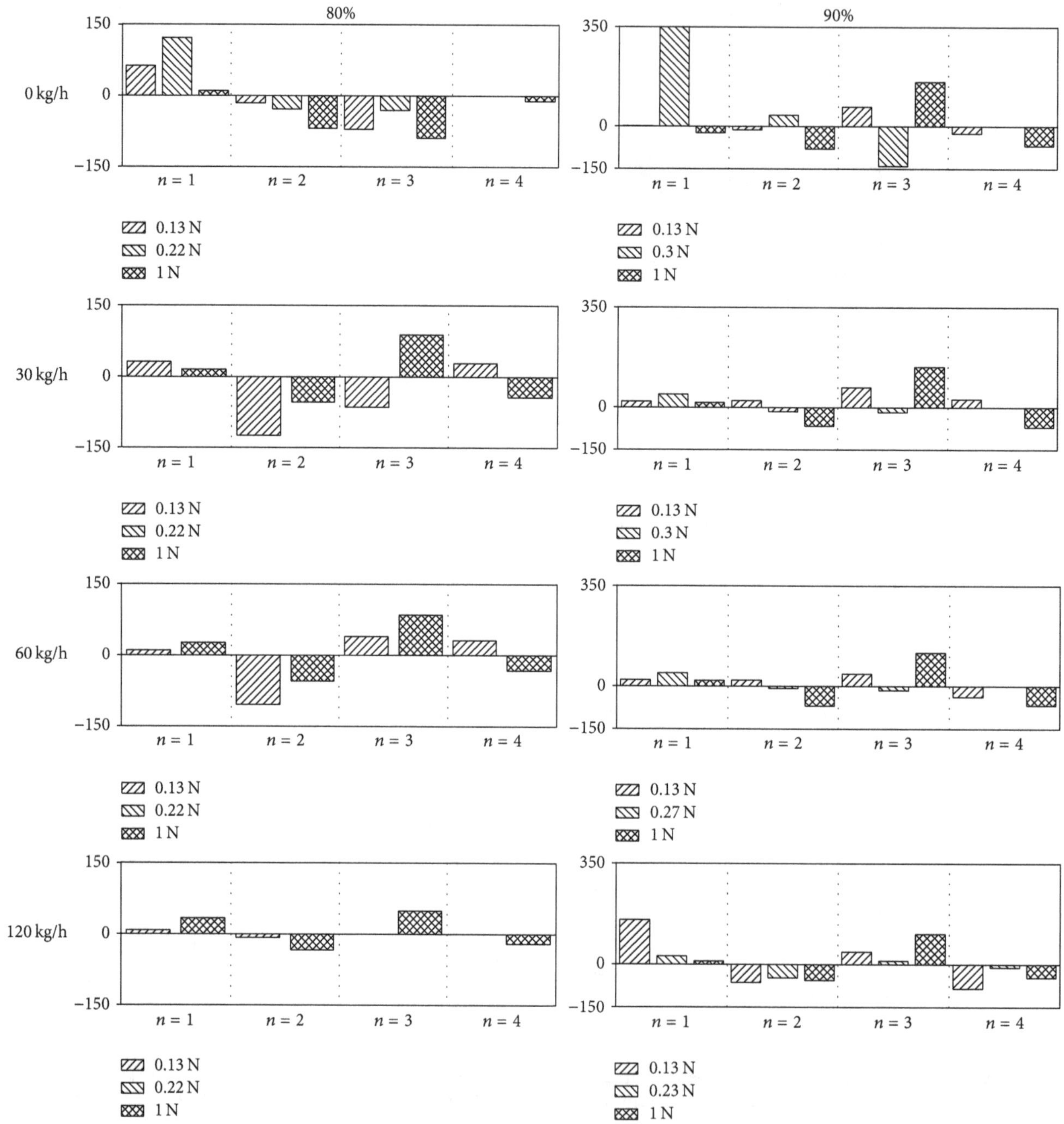

FIGURE 13: Harmonic content of traveling energy of the first four prestall modes at the diffuser inlet.

pressure ratio and efficiency were observed in steam injected technology compressors.

The overall agreement between experiments and computational data was reasonable. The simulation characteristics indicated a higher pressure rise and higher efficiency than that observed in experiment. The difference in pressure ratio and efficiency between the measurements and CFD calculations could possibly be attributed to the turbulence model. As the full flow passage model cases were simulated, κ-ϵ turbulence model was selected in order to procure convergence in the calculations. However, some loss may be ignored and the compressor performance was improved.

6. Mechanisms of Surge Enhancement with Steam Injection

In order to clearly identify the effect of steam injection on impeller and vaned diffuser, the individual components of the stage were studied separately. Impeller total efficiency and pressure recovery coefficient (Cp) of the vaned diffuser based on the CFD results are used to depict the performance of the impeller and the diffuser. Calculated results with different steam flow rates at the same operating point are shown in Figure 16. Each curve is composed of 20 different span stations from hub to shroud, that is, from 0.0 to 1.0. Thus, the

(a) Rotor mesh details

(b) Stator mesh details

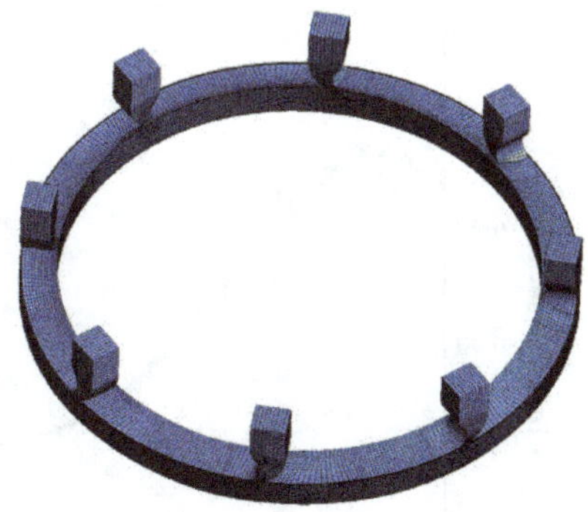

(c) Injection mesh details

(d) Volute mesh details

FIGURE 14: CFD computational grids.

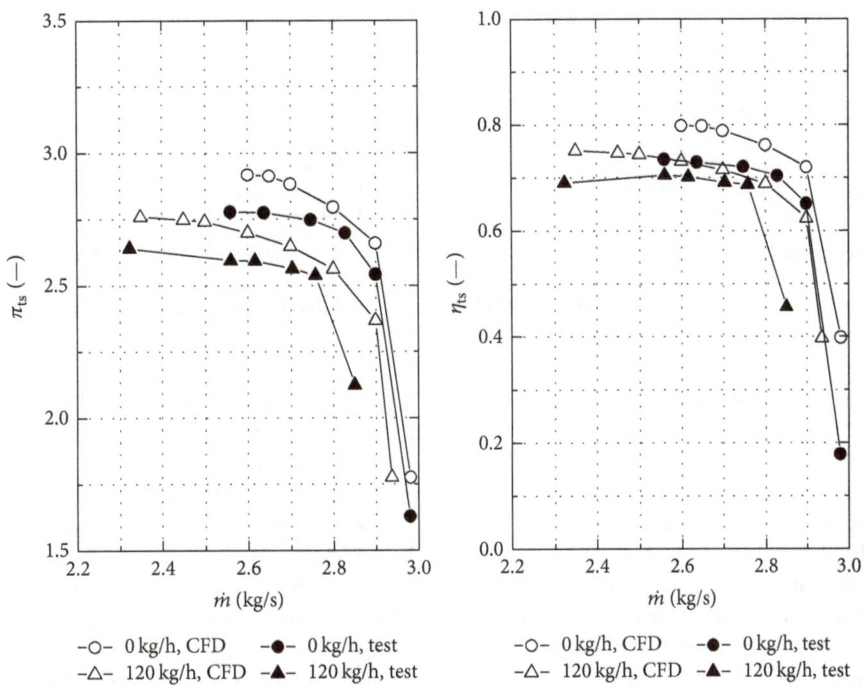

FIGURE 15: Performance comparison between numerical simulation and test results.

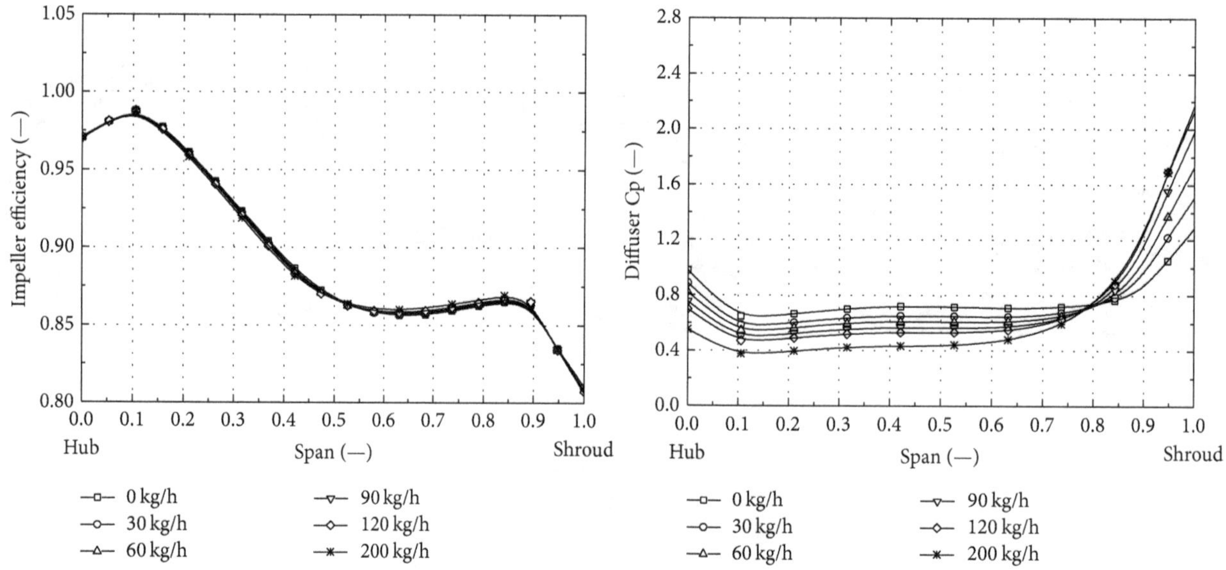

FIGURE 16: Characteristics of the impeller and the diffuser.

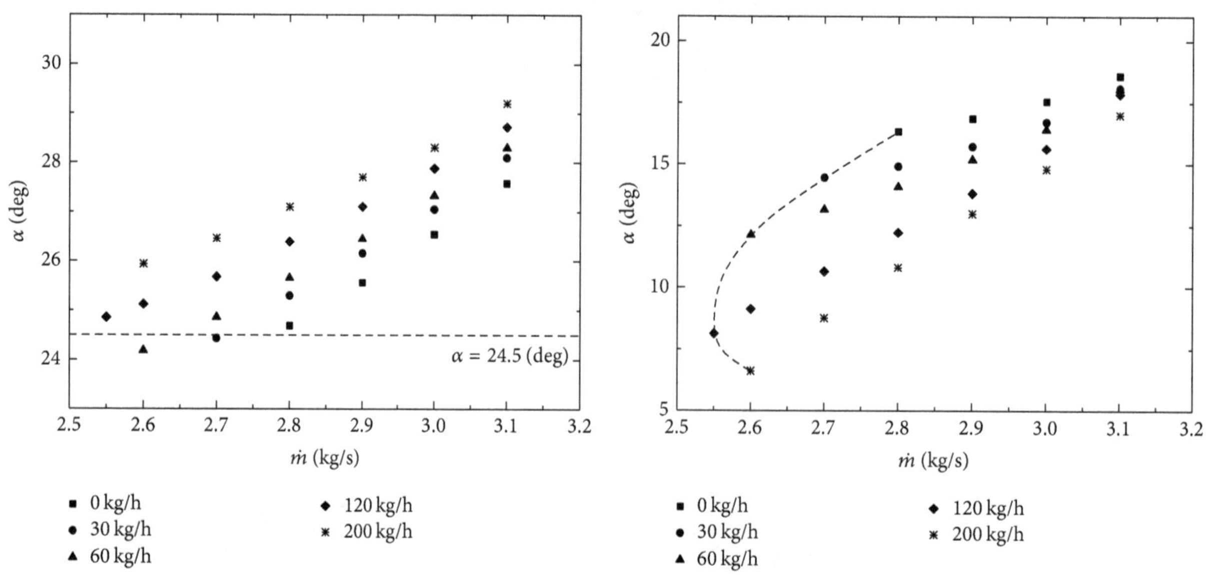

(a) 0 (hub)–0.8 span averaged flow angle

(b) 0.8–1 (shroud) span averaged flow angle

FIGURE 17: Flow angle at diffuser inlet of different span ranges.

diffuser can be decomposed to 20 small diffuser segments and each of them operates at its own working condition. Results show that the impeller efficiency stays constant at different steam injected flow rates, which means that the injected steam hardly affects the performance of the impeller.

However, the diffuser Cp near the shroud side shows a large increase with injected steam, while the change of Cp from hub to 80% blade height is much smaller. The diffuser Cp value varies obviously as the steam flow rate increases from 30 kg/h to 120 kg/h.

From the above observed trends, a conjecture can be reached that the vaned diffuser is the dominant component that affects the stage stability.

The trends of diffuser inlet absolute velocity flow angle at different operating conditions and different flow rate steam injection are plotted in Figure 17. In the 0–0.8 span analysis, as the mass flow rate through the compressor decreases from choke to surge, the average absolute flow angle at the diffuser inlet reduces. At the same operating condition, the flow angle increases after steam is injected. It could also increase velocity flow angle at the same operating condition and improve surge margin. Compressor surge occurred when the flow angle reached about 24.5 under all different steam injection flow rate conditions except 200 kg/h, at which the compressor surge occurred at larger flow rate. Therefore, the surge margin improvement was not in proportion to the steam flow rate.

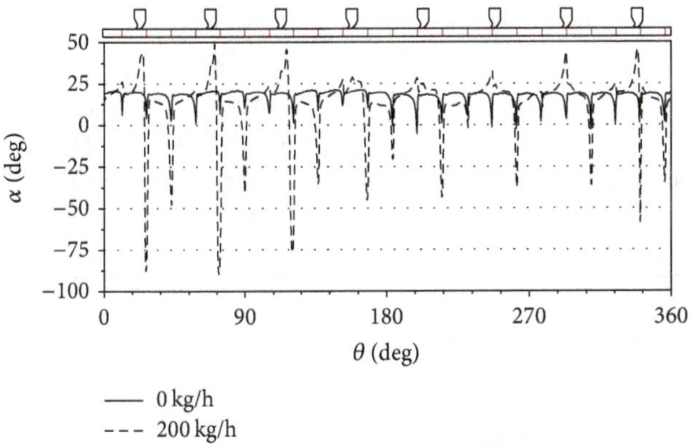

FIGURE 18: Absolute velocity flow angle at 0.9 span at the diffuser inlet.

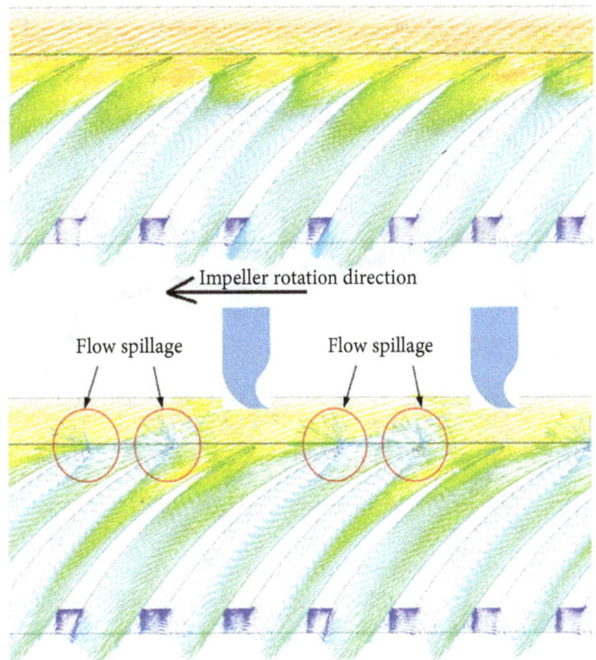

FIGURE 19: Surface streamline at 0.9 span around the diffuser inlet.

In certain cases, the large steam injection rate has a smaller surge margin improvement than small steam injection. This phenomenon could be explained in the 0.8–1 span analysis. Steam injection could decrease the velocity flow angle at the same operating conditions and deteriorate the flow structure in this area. The velocity flow angle reached the critical value under 200 kg/h steam injection condition. Therefore, surge occurred earlier when steam flow rate was 200 kg/h. This could be attributed to the excessively small velocity flow angle in 0.8–1 span and will be explained in the following section.

The absolute velocity flow angle at 0.9 span at the diffuser inlet with different steam injection rates was plotted relative to the peripheral direction in Figure 18. Eight steam injectors were uniformly installed upstream of 23 diffuser blades. Their relative positions in the peripheral direction are also shown in the figure. In the current case, the steam is injected in the reverse direction of impeller rotation. Therefore, the direct impact of steam injection was to decrease the circumferential velocity at high span regions and increase the flow angle at the 8 injectors' locations. However, Figure 18 also indicated that flow angle was decreased suddenly as fluid passed through an injector adjacent downstream blade.

If the steam injector is located just between two adjacent diffuser vanes in peripheral direction, flow angle suddenly increased just after the steam injector and flow angle decreased excessively at the neighboring diffuser vane behind this injector. And if the steam injector is located close to a diffuser vane or coincided with a blade in theta direction due to odd number of diffuser vanes, flow angle increased unintelligibly at the steam injection locations, while flow angle decreased slightly at the neighboring diffuser vane behind this injector. In general, as the steam injection flow rate increased from 0 kg/h to 200 kg/h, the flow angle at the steam injector location increased greatly while the flow angle at the neighboring diffuser vane behind this injector decreased excessively.

Detailed flow structures and flow mechanism were studied to explain the feature of the flow angle discussed above. Figure 19 presents the velocity vector field comparison at 90% span around the diffuser inlet area including two certain steam injectors and their downstream neighboring diffuser vanes with 0 kg/h at numerical surge point and 200 kg/h steam injection. As indicated in red circles, with injected steam against the direction of impeller rotation, the circumferential velocity decreased, and thus the diffuser inlet flow angle at high span regions increased. Steam injection causes streamline to turn toward the blade passage, and the flow will spill over adjacent blade leading edge, thus triggering system instability. However, the streamline flowed into the next downstream blade passage instead of the current one affected by the neighboring diffuser vane. This was the main reason that flow angle decreased excessively at the neighboring diffuser vane behind an injector, and separation or blockage in injector positioned diffuser passage may be responsible for the observed behavior. This is quite similar to those found in axial compressor.

To figure out how the steam injection affected the flow angle, the circumferential velocity and the radial velocity versus the normalized span location at the diffuser inlet with a main flow rate of 2.8 kg/s are shown in Figure 20, respectively. In general, large injection mass flow generated a stronger influence on the velocity. But the influence was not linear with respect to the span location. The variation of V_θ was limited to the high span space due to the Coanda effect, and a uniform V_θ is obtained for different injection flow rate below 0.5 span, while V_r decreased near the shroud side and increased near the hub side. This was because the main flow rate, which is directly associated with V_r, was set to be a constant value. If the mass through the high span space was reduced, the mass through the low span space must increase to maintain a constant mass flow. The coupled influence by V_θ and V_r increased the flow angle in the low span and decreased it in the high span.

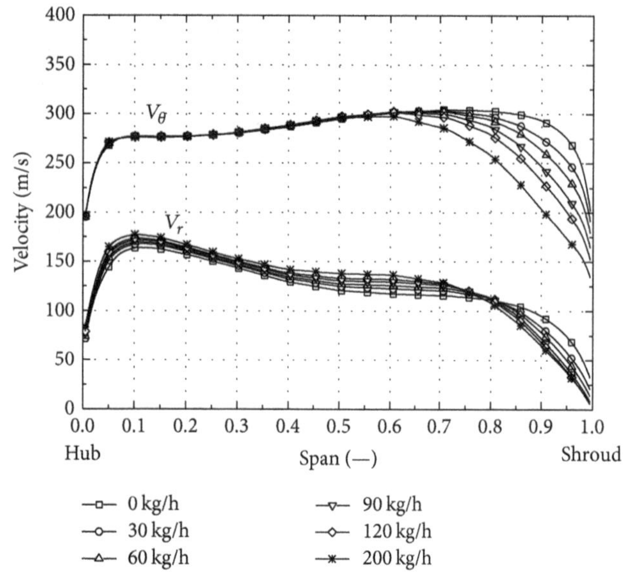

FIGURE 20: V_θ and V_r with different steam injection at the diffuser inlet.

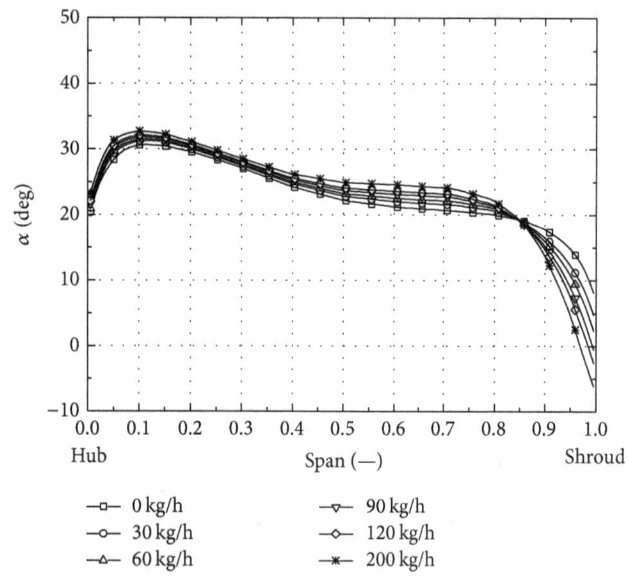

FIGURE 21: Flow angle with different steam injection at the diffuser inlet.

However, this could not explain why there was a cross point of the flow angles versus the normalized span location at the diffuser inlet at different steam injection rate conditions, as illustrated in Figure 21. The authors believe that it should be caused by the Coanda effect which means the radial attached span will remain unchanged regardless of the change in the injected mass flow.

7. Influences of Thermal Physics Properties

7.1. Molecular Weight. To study the material influence, three ideal gases were chosen: steam ($M < 29$), N_2 ($M \approx 29$), and CO_2 ($M > 29$). To eliminate the influences of other factors involved, the authors keep the momentum of injected gas constant. The related injected gas properties are given in Table 3.

TABLE 3: Boundary conditions of different materials.

Gas	Compressor mass flow (kg/s)	Injected gas static temperature (K)	Injected gas mass flow (kg/s)
Steam ideal gas	2.80	438	0.0333
N_2 ideal gas	2.80	438	0.0416
CO_2 ideal gas	2.80	438	0.0521

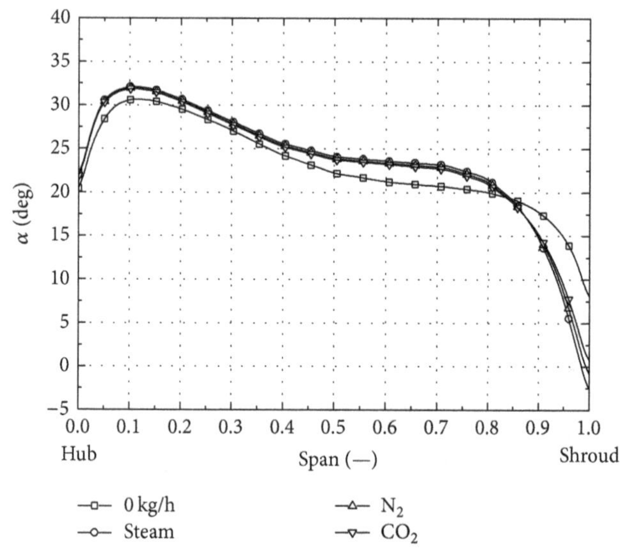

FIGURE 22: Flow angle in span direction.

Figure 22 compares the flow angle from hub to shroud. Slight discrepancies can be observed. The results show that the diffuser inlet flow angle decreases from hub to shroud. The light gas can reduce 0.87–1 span flow angle and enhance it at 0–0.87 span and increase span averaged flow angle.

Figure 23 illustrates the volume fraction from hub to shroud obtained from the meridional surface at the same radius of the injectors. Mass flow in Table 4 and volume fraction in Figure 23 indicate that the larger the mass flow is, the larger the volume fraction injected is.

TABLE 4: Boundary conditions of different temperatures.

Injected gas static temperature (K)	Compressor mass flow (kg/s)	Gas injected	Injected gas mass flow (kg/s)
573	2.80	N_2	0.0364
438	2.80	N_2	0.0416
293	2.80	N_2	0.0508
223	2.80	N_2	0.0583
173	2.80	N_2	0.0662

7.2. Temperature and Heat Transfer. Figure 24 compares the flow angle from hub to shroud. The results show that the

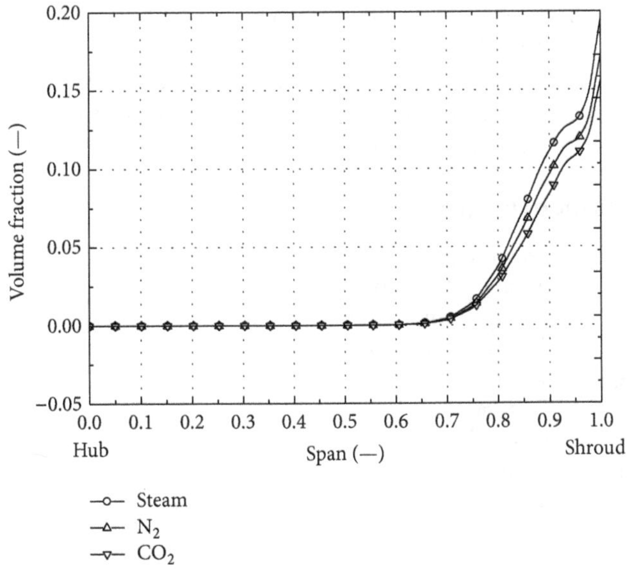

FIGURE 23: Volume fraction obtained from the meridional surface at the same radius of the injectors.

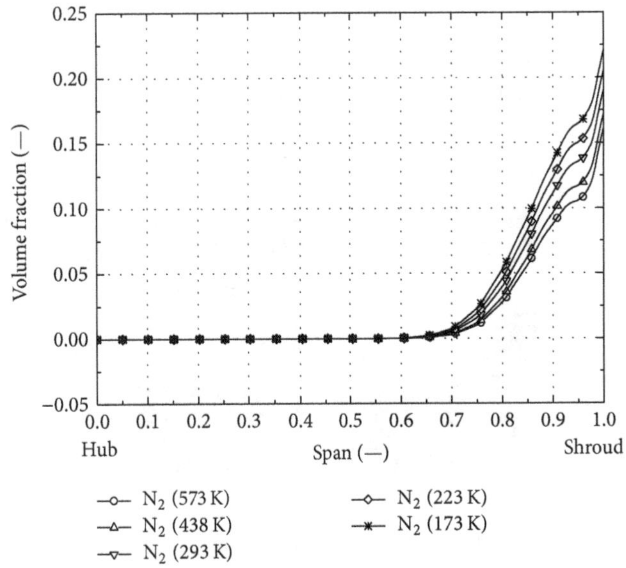

FIGURE 25: Volume fraction obtained from the meridional surface of different temperatures.

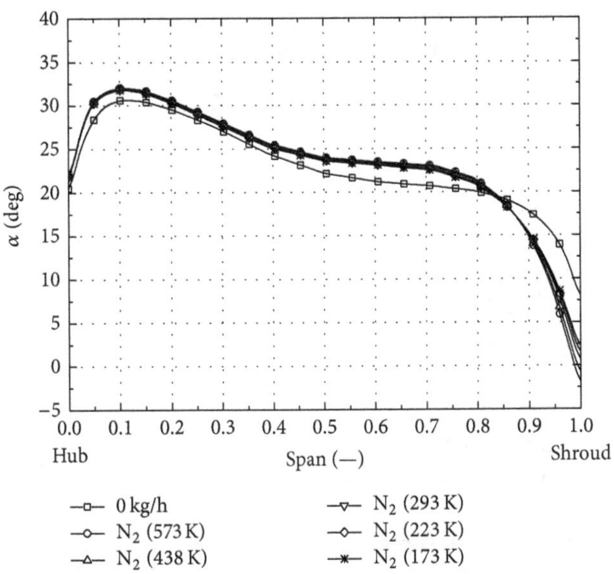

FIGURE 24: Flow angle in span direction of different temperatures.

diffuser inlet flow angle decreases from hub to shroud. The higher gas temperature can reduce 0.87–1 span flow angle and enhance it at 0–0.87 span and thus increase span averaged flow angle.

Figure 25 depicts the volume fraction from hub to shroud obtained from the meridional surface at the same radius of the injectors. It shows that the larger the mass flow is, the smaller the volume fraction is.

The volume flow as the dominant factor that changes the flow angle can be concluded by the comparisons of mass and volume flow influence above. Therefore, with the same momentum injection, higher temperature or smaller molecular weight gas can both increase the diffuser inlet flow angle and extend the compressor stable operation range. For example, if the injected gas temperature is higher than the air temperature, then it will heat up the air, cause it to expand, and thus increase radial velocity near the shroud side.

7.3. Asymmetric and Symmetric Injector, Air Bleeding, and Air Recirculation. As discussed above, the injector position would affect the flow pattern downstream of that injector significantly. In order to figure out the influence on the enhancement, cases of 8 asymmetrically distributed injectors and 23 uniformly distributed injectors were simulated here. $\eta\Delta$ and $\alpha\Delta$ represent the efficiency variation rate under the surge limit and the change rate of the mean flow angle at the inlet of the diffuser vane space compared to those of cases without injection, respectively. The results demonstrate that the change of the injector position does not improve the enhancement compared with the former cases, and a distinct decrease of efficiency was captured for the case of 23 injectors.

At last, the air bleeding and air recirculation were tested, as shown in Figure 26. Air bleeding out from the pipes installed at each vane passage throat was discharged to the environment or reinjected to the diffuser vaneless space. The total air bleeding mass flow was set to be 30 kg/h. The air recirculation flow was driven by the differential pressure between the vane throat and vaneless space, measured as about 135 kg/h. But SMI is only about 1/3 of the case of steam injection. It shows that the air bleeding configuration enhances the stability with an increase of efficiency, while the injection configurations decrease it. But it is confused that the air recirculation, which can be regarded as a combination of bleeding out and injection, reduces the surge margin distinctly. A positive influence on the flow angle, which the air recirculation had, always indicates an increase of the surge margin; however, it failed in this case. The reason may be the complex flow in the bleeding tubes that affects the convergence of the simulations.

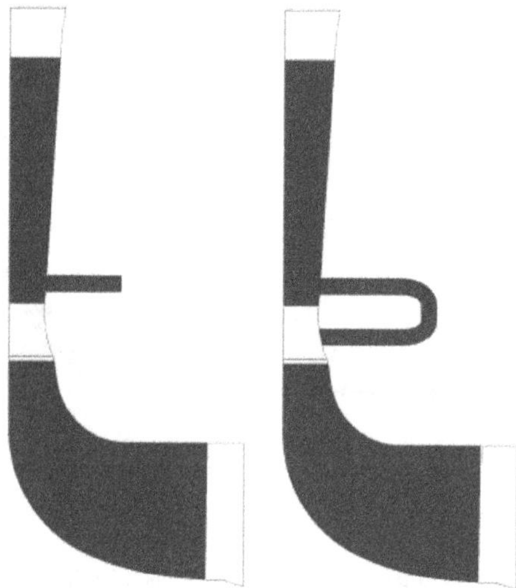

FIGURE 26: Air bleeding and air recirculation configuration.

8. Conclusions

A centrifugal compressor with vaned diffuser was analyzed in detail through both experimental and numerical methods. After comparing the steady CFD results and unsteady test data, some conclusions could be made as follows:

(a) Steam injection is shown to be an effective method to improve stage stability. 0.3% of compressor designed mass flow rate steam can increase the surge margin by over 9.0% at Mu = 1.08. At each speed line, there is an optimum steam flow rate.

(b) Steam injection changes the type of stall precursor from spike to modal waves at 90% speed line. In addition, it changes the stall cell traveling direction from forward to backward at 80% speed.

(c) Both efficiency and pressure ration drop with steam injection.

(d) The root cause of the compressor surge is the flow angle at the diffuser inlet, and the surge margin could be improved by decreasing the incidence angle at the diffuser inlet.

(e) At the same momentum injection, higher temperature or lighter gas can increase the diffuser inlet flow angle and extend the compressor stable operation range.

(f) The 8 asymmetrical injectors and the 23 circumferentially symmetrical injectors performed less and equal enhancement compared with the normal configuration, respectively. A distinct decrease of efficiency occurred for the 23-injector case, which demonstrated a more pressure loss. The air bleeding increased the stability and the efficiency. But the air reinjection had a negative effect on the surge margin.

(g) Traditional flow instability theory based on incidence caused flow separation can still be applied for vaned diffuser by viewing its spanwise distribution.

(h) Symmetric injector is better than asymmetric case from the point of SMI.

(i) Steam injection is more effective than simple bleeding at throat area or recirculation in SMI but will result in a drop of efficiency.

Nomenclature

p: Pressure (kPa)
b: Axial width (mm)
r: Radius (mm)
f_s: Sampling frequency (kHz)
Q: Volumetric flow rate (m^3/s)
T: Temperature (K)
V: Absolute velocity (m/s)
W: Relative velocity (m/s)
U: Impeller circumferential velocity (m/s)
n: Impeller rotational speed (rpm)
N: Number of blades (—)
TWE: Traveling wave energy.

Greek Symbols

α: Absolute flow angle from tangential direction (°)
β: Blade angle from tangential direction (°)
ϕ: Flow coefficient defined as $Q/\pi r_2^2 u_2$
ψ: Pressure rise coefficient defined as $\Delta P/(1/2)\rho_\infty {u_2}^2$
ρ: Density.

Subscripts

0: Stage inlet
1: Impeller inlet
2: Impeller outlet/vaned diffuser inlet
3: Vaned diffuser outlet
4: Collector outlet
r: Radius direction
θ: Tangential direction
∞: Atmospheric condition.

Conflicts of Interest

The authors declare that they have no conflicts of interest.

Acknowledgments

This project is supported by the National Natural Science Foundation of China under Grants nos. 51306199 and 51405130. The authors would like to acknowledge the kind comments and discussions on the design of the test facility by Colin Rodgers.

References

[1] E. Greitzer, "The stability of pumping systems," *ASME Journal of Fluids Engineering*, vol. 103, no. 11, pp. 193–242, 1981.

[2] Z. S. Spakovszky, "Backward traveling rotating stall waves in centrifugal compressors," *Journal of Turbomachinery*, vol. 126, no. 1, pp. 1–12, 2004.

[3] M. Schleer, S. J. Song, and R. S. Abhari, "Clearance effects on the onset of instability in a centrifugal compressor," *Journal of Turbomachinery*, vol. 130, no. 3, Article ID 031002, 11 pages, 2008.

[4] D. Hagelstein, K. Hillewaert, R. A. Van Den Braembussche, A. Engeda, R. Keiper, and M. Rautenberg, "Experimental and numerical investigation of the flow in a centrifugal compressor volute," *Journal of Turbomachinery*, vol. 122, no. 1, pp. 22–31, 2000.

[5] Z. S. Spakovszky and C. H. Roduner, "Spike and modal stall inception in an advanced turbocharger centrifugal compressor," *Journal of Turbomachinery*, vol. 131, no. 3, Article ID 031012, 2009.

[6] H. Tsurusaki and T. Kinoshita, "Flow control of rotating stall in a radial vaneless diffuser," *Journal of Fluids Engineering, Transactions of the ASME*, vol. 123, no. 2, pp. 281–286, 2001.

[7] W. Huang, S. Geng, J. Zhu, and H. Zhang, "Numerical simulation of rotating stall in a centrifugal compressor with vaned diffuser," *Journal of Thermal Science*, vol. 16, no. 2, pp. 115–120, 2007.

[8] G. J. Skoch, "Experimental investigation of diffuser hub injection to improve centrifugal compressor stability," *Journal of Turbomachinery*, vol. 127, no. 1, pp. 107–117, 2005.

[9] T. Hirano, T. Uchida, and H. Tsujita, "Control of surge in centrifugal compressor by using a nozzle injection system: universality in optimal position of injection nozzle," *International Journal of Rotating Machinery*, vol. 2012, Article ID 259293, 8 pages, 2012.

[10] C. Gao, C. Gu, T. Wang, and B. Yang, "Analysis of geometries' effects on rotating stall in vaneless diffuser with wavelet neural networks," *International Journal of Rotating Machinery*, vol. 2007, Article ID 76476, 2007.

[11] K. B. Abidogun, "Effects of vaneless diffuser geometries on rotating stall," *Journal of Propulsion and Power*, vol. 22, no. 3, pp. 542–549, 2006.

[12] X. Sun, D. Sun, and W. Yu, "A model to predict stall inception of transonic axial flow fan/compressors," *Chinese Journal of Aeronautics*, vol. 24, no. 6, pp. 687–700, 2011.

[13] F. B. Fisher, "Application of map width enhancement devices to turbocharger compressor stages," SAE255 Paper No. 8809794. Power boost: light, medium and heavy duty engines. SP-780, 1989.

[14] M. Werner, R. Baar, P. Haluska et al., "Bidirectional flow measurement based on the differential pressure method for surge analysis on a small centrifugal compressor", proceedings of the institute of mechanical engineers, part A," *Journal of Power and Energy, Turbocharger and Turbocharging Special Issue*, 2016.

[15] A. Stein, S. Niazi, and L. N. Sankar, "Computational analysis of stall and separation control in centrifugal compressors," *Journal of Propulsion and Power*, vol. 16, no. 1, pp. 65–71, 2000.

[16] C. Gao, C. Gu, T. Wang, and B. Yang, "Passive control of rotating stall in vaneless diffuser with radial grooves: detailed numerical study," in *Proceedings iof the 2009 ASME Turbo Expo*, pp. 1225–1232, Orlando, Fla, USA, June 2009.

[17] R. Taghavi-Zenouz, E. Solki, and H. Afshari, "Stall margin improvement of a centrifugal compressor utilizing various stepped tip gap configurations," *Proceedings of the Institution of Mechanical Engineers, Part A: Journal of Power and Energy*, vol. 228, no. 7, pp. 772–781, 2014.

[18] G. W. Felsing and Moller P. S., "Coanda flow over circular cylinder with injection normal to surface," *AIAA Journal*, vol. 7, no. 5, pp. 842–846, 1969.

[19] R. L. Davis and J. Yao, "Computational approach for predicting stall inception in multistage axial compressors," *Journal of Propulsion and Power*, vol. 23, no. 2, pp. 257–265, 2007.

[20] M. Tryfonidis, O. Etchevers, J. D. Paduano, A. H. Epstein, and G. J. Hendricks, "Pre-stall behavior of several high-speed compressors," *Journal of Turbomachinery*, vol. 117, no. 1, pp. 62–80, 1995.

[21] C. S. Tan, I. Day, S. Morris, and A. Wadia, "Spike-type compressor stall inception, detection, and control," *Annual Review of Fluid Mechanics*, vol. 42, pp. 275–300, 2010.

[22] Y.-S. Yoon and S. J. Song, "Analysis and measurement of the impact of diffuser width on rotating stall in centrifugal compressors," *Journal of Mechanical Science and Technology*, vol. 28, no. 3, pp. 895–905, 2014.

[23] Y. Bousquet, X. Carbonneau, G. Dufour, N. Binder, and I. Trebinjac, "Analysis of the unsteady flow field in a centrifugal compressor from peak efficiency to near stall with full-annulus simulations," *International Journal of Rotating Machinery*, vol. 2014, Article ID 729629, 11 pages, 2014.

[24] C. Gao, W. G. Huang, H. W. Zhang, Y. L. Yao, and J. C. Duan, "Surge characteristics and range extension with steam injection in a centrifugal compressor with vaned diffuser," in *Proceedings of the ASME, 2013, International Mechanical Engineering Congress & Exposition (IMECE '13)*, San Diego, Calif, USA, November 2013.

19

Influence of Varnish on Bearing Performance and Vibration of Rotating Machinery

Chu Zhang,[1] Jian-Gang Yang,[2] Shi Liu,[1] Qing-Shui Gao,[1] and Yi Yang[1]

[1]*Electric Power Research Institute of Guangdong Power Grid Co. Ltd., Guangzhou 510080, China*
[2]*National Engineering Research Center of Turbo-Generator Vibration, Southeast University, Nanjing 210096, China*

Correspondence should be addressed to Chu Zhang; zhangchu718@163.com

Academic Editor: Ryoichi Samuel Amano

Varnish is a kind of fouling adhered on bearing inner face near the load zone. The model to analyze its influence on bearing performance and rotor vibration was set up. Pressure and temperature distribution were solved on the basis of the Reynolds equation by the iteration method simultaneously. It is found that bearing performance is changed by the step flow effect at the leading edge of the varnish. Its influence is more obvious when the rotating speed is low. Bearing varnish results in a large decrease of minimum clearance and an increase in the oil film temperature and pressure. Loading capacity of bearing is also decreased. It is harmful to the bearing. With the varnish, the minimum oil film thickness decreases. This leads to an increase of bearing stiffness and damping coefficients. Its influence is larger in the horizontal direction than that in the vertical direction. The trend of unbalance response to rotating speed is similar. There is little influence of the varnish on the modal frequency. However, system stability is improved due to the increased bearing damping coefficients.

1. Introduction

Journal bearings have a great influence on the safe operation of rotating machinery. Machine works reliably under the squeeze flow effect formed in the convergent gap between bearing inner face and the journal.

With the development of industry, journal surface speed becomes higher and bearing load becomes more and more heavy. Varnish problem has occurred on journal bearings [1]. Varnish is a kind of fouling adhered on journal bearing inner face nearby the load zone, resulting in poor bearing lubrication. It could lead to unacceptable rises in temperature and vibration [1]. Now, the research on the formation mechanism, detection, prevention, and removal of the varnish is being carried out [2]. The formation of varnish is related to the quality of lubricating oil and the condition of journal bearing. The high temperature generated by the variable load operation of bearing will accelerate the formation of the varnish [3].

The clearance of bearing with varnish is irregular. Research on the performance of bearing with irregular clearance is mainly focused on groove or microtexture [4–8]. Bearing performance can be improved through an appropriate arrangement of the textured area on the contact surface [9, 10]. The texture arrangement, mainly the location domain on the journal bearing, is the main criterion for journal bearing performance enhancement. It would be interesting to put dimples in the bearing first angular part (between 0 and 180) which is the maximum pressure region [11]. Appropriately choosing groove length, width, or interval can reduce the maximum radius, collapse pressure, and collapse temperature of the bubble. There exists a critical groove depth minimizing the bubble pressure and temperature [12]. The influence of surface variations such as texturing or grooving on the bearing performances was also compared [13]. The microgrooving reduces the friction coefficient and average temperature is maximum in comparison with spherical texture. Influence of texture distributions on vibration was investigated [14]. Three kinds of textures configurations are designed and applied on the lower pad surface. There is a significant decrease in acceleration amplitude of the textured bearings compared to nontextured bearings.

From inspection results of many bearings, varnish area is larger than expected. Unlike the groove or the microtexture,

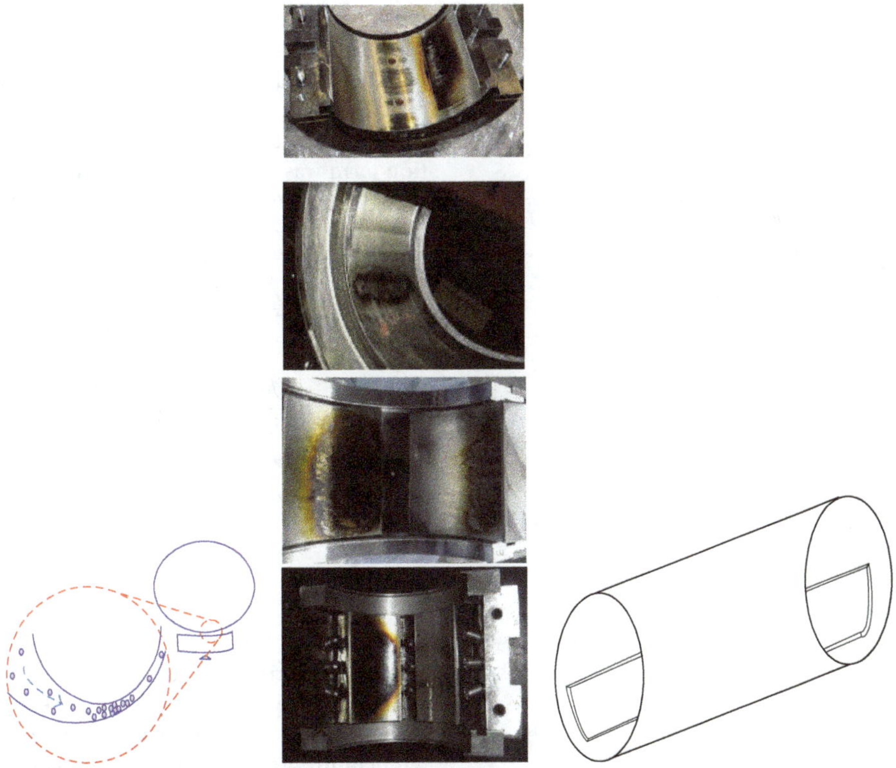

(a) Varnish location in the bearing (b) Varnish photographs (c) The simplified varnish model

FIGURE 1: Bearing varnish model.

varnish is a kind of protrusion on bearing inner face. It is mainly located in bearing load area, which is near the minimum gap and is the most dangerous zone of the bearing. Its influence on bearing performance may be large. Research in this field should be concerned.

In this paper, influences of varnish on bearing properties and rotor vibration are studied. Loading capacity, temperature rise, oil film thickness, and dynamic coefficients were compared. Conclusions can be applied to bearing condition monitoring.

2. Model of Bearing with Varnish

2.1. Varnish Model. Varnish is a kind of fouling adhered on journal bearing inner face nearby the load zone, as shown in Figure 1(a). Figure 1(b) shows 4 photographs of bearing with varnish.

For ease of analysis, varnish shape is simplified. A trapezoid distribution in the boundary and a uniform distribution in the inner face were assumed, as shown in Figure 1(c). This assumption guarantees continuity of the varnish at the boundary approximately.

Clearance of bearing with varnish can be expressed as [15]

$$H = 1 + \varepsilon \cos\varphi - H_1. \tag{1}$$

The partial derivative of clearance to circumferential angle is given by

$$\frac{\partial H}{\partial \varphi} = -\varepsilon \sin\varphi - \frac{\partial H_1}{\partial \varphi}. \tag{2}$$

2.2. The Lubrication Equation. A two-dimensional (2D) Reynolds equations were used. Taking into account the axial symmetry; the solution region was selected as the rectangular region ABCD, as shown in Figure 2. The varnish area is represented by a grid of lines in the figure.

The dimensionless form of the Reynolds equation is [15]

$$\frac{\partial}{\partial \varphi}\left(H^3 \frac{\partial P}{\partial \varphi}\right) + \left(\frac{d}{l}\right)^2 \frac{\partial}{\partial \lambda}\left(H^3 \frac{\partial P}{\partial \lambda}\right) = 3\frac{\partial H}{\partial \varphi}. \tag{3}$$

The Reynolds equation with the perturbation pressure $P_\varepsilon, P_\theta, P_{\varepsilon'}, P_{\theta'}$ as the variables is given as [16]

$$\frac{\partial}{\partial \varphi}\left(H^3 \frac{\partial P_\varepsilon}{\partial \varphi}\right) + \left(\frac{d}{l}\right)^2 \frac{\partial}{\partial \lambda}\left(H^3 \frac{\partial P_\varepsilon}{\partial \lambda}\right) = -3\sin\varphi - \frac{9}{H}$$

$$\cdot \cos\varphi \frac{\partial H}{\partial \varphi} + 3H\left[\left(\cos\varphi \frac{\partial H}{\partial \varphi} + \sin\varphi H\right)\frac{\partial P}{\partial \varphi}\right.$$

$$\left. + \left(\frac{d}{l}\right)^2 \cos\varphi \frac{\partial H}{\partial \lambda}\frac{\partial P}{\partial \lambda}\right] \frac{\partial}{\partial \varphi}\left(H^3 \frac{\partial P_\theta}{\partial \varphi}\right)$$

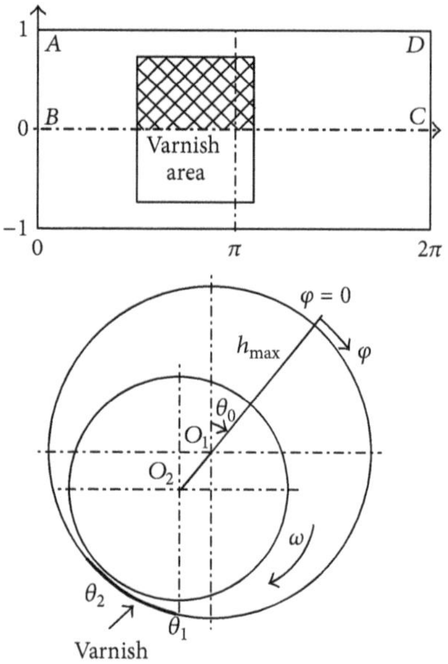

FIGURE 2: Lubrication equation solution model.

$$+\left(\frac{d}{l}\right)^2\frac{\partial}{\partial\lambda}\left(H^3\frac{\partial P_\theta}{\partial\lambda}\right)=3\cos\varphi$$

$$-\frac{9}{H}\sin\varphi\frac{\partial H}{\partial\varphi}+3H\left[\left(\sin\varphi\frac{\partial H}{\partial\varphi}-\cos\varphi H\right)\frac{\partial P}{\partial\varphi}\right.$$

$$\left.+\left(\frac{d}{l}\right)^2\sin\varphi\frac{\partial H}{\partial\lambda}\frac{\partial P}{\partial\lambda}\right]$$

$$\frac{\partial}{\partial\varphi}\left(H^3\frac{\partial P_{\varepsilon'}}{\partial\varphi}\right)+\left(\frac{d}{l}\right)^2\frac{\partial}{\partial\lambda}\left(H^3\frac{\partial P_{\varepsilon'}}{\partial\lambda}\right)=6\cos\varphi$$

$$\frac{\partial}{\partial\varphi}\left(H^3\frac{\partial P_{\theta'}}{\partial\varphi}\right)+\left(\frac{d}{l}\right)^2\frac{\partial}{\partial\lambda}\left(H^3\frac{\partial P_{\theta'}}{\partial\lambda}\right)=6\sin\varphi$$

$$P_\varepsilon=\frac{\partial P}{\partial\varepsilon},$$

$$P_\theta=\frac{\partial P}{\varepsilon\partial\theta},$$

$$P_{\varepsilon'}=\frac{\partial P}{\partial\varepsilon'},$$

$$P_{\theta'}=\frac{\partial P}{\varepsilon\partial\theta'}. \quad (4)$$

The boundary conditions of (3) and (4) are

$$\begin{aligned}
&P=P_\varepsilon=P_\theta=P_{\varepsilon'}=P_{\theta'}=0,\ \text{Line AB}\\
&P=P_\varepsilon=P_\theta=P_{\varepsilon'}=P_{\theta'}=0,\ \text{Line CD}\\
&P=P_\varepsilon=P_\theta=P_{\varepsilon'}=P_{\theta'}=0,\ \text{Line AD}\\
&\frac{\partial P}{\partial\lambda}=\frac{\partial P_\varepsilon}{\partial\lambda}=\frac{\partial P_\theta}{\partial\lambda}=\frac{\partial P_{\varepsilon'}}{\partial\lambda}=\frac{\partial P_{\theta'}}{\partial\lambda}=0,\ \text{Line BC.}
\end{aligned} \quad (5)$$

During the iteration process, pressure is set to 0 if the negative pressure appears.

The varnish thickness is much smaller than the thickness of the pad. It has little influence on the heat conduction along the thickness direction of the pad. The heat conduction term was thus omitted. The simplistic two-dimensional energy equation with the adiabatic boundary condition is [13]

$$\begin{aligned}
&\rho\left[\left(\frac{Uh}{2}-\frac{h^3}{12\mu}\frac{\partial p}{\partial x}\right)\frac{\partial(c_vT)}{\partial x}-\frac{h^3}{12\mu}\frac{\partial p}{\partial z}\frac{\partial(c_vT)}{\partial z}\right]\\
&\quad=\mu\frac{U^2}{h}+\frac{h^3}{12\mu}\left[\left(\frac{\partial p}{\partial x}\right)^2+\left(\frac{\partial p}{\partial z}\right)^2\right].
\end{aligned} \quad (6)$$

The symmetry boundary condition is adopted in the axis direction:

$$\frac{\partial T}{\partial\lambda}=0,\quad \lambda=0. \quad (7)$$

For the oil mixing condition at groove, the oil temperature is assumed as the mixing temperature between the recirculating oil and inlet oil. The average value of the recirculating temperature and the inlet oil temperature is taken as the mixing temperature. It is expressed as

$$T_{1,j}^{k+1}=\frac{T_{1,j}^k+T_{n,j}^k}{2}. \quad (8)$$

Lubricating oil viscosity variation with temperature is taken care of by the following expression [13]:

$$\mu=\mu_0e^{-\alpha(T-T_0)}. \quad (9)$$

Equations (3), (6), and (9) were solved by iterative method. Assuming the oil temperature distribution in the bearing, oil viscosity is calculated with (9) and oil film pressure is solved with (3). Substituting the calculated pressure into (6), the new temperature distribution is obtained. For the kth iteration process, initial temperature $T_{1,j}^k$ was known. Temperature $T_{i,j}^k$, $i=2,\ldots,n$ was calculated with (6) from the start angle to the end angle step by step. Then it was averaged using (8) to obtain the new temperature $T_{1,j}^{k+1}$. The iteration process stops when the solved temperature field is fully converged. The total iteration process stops when the solved pressure and temperature field are fully converged.

Bearing load is obtained by integral in axial and circumferential directions [15]:

$$\begin{aligned}
F_x&=-\frac{\mu\omega Rl}{(c/R)^2}\int_{\varphi_1}^{\varphi_2}\int_{-1}^{1}(pd\lambda)\sin\varphi\,d\varphi\,d\lambda\\
F_y&=-\frac{\mu\omega Rl}{(c/R)^2}\int_{\Phi_1}^{\Phi_2}\int_{-1}^{1}(pd\lambda)\cos\varphi\,d\varphi\,d\lambda.
\end{aligned} \quad (10)$$

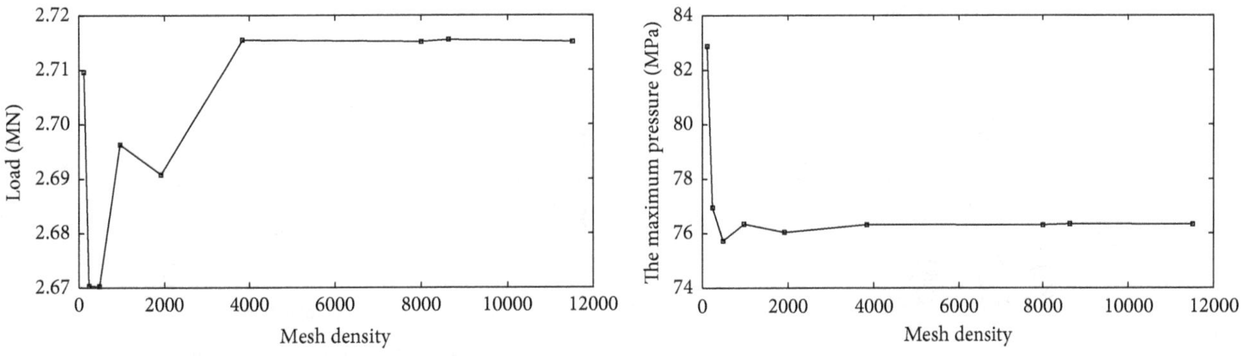

FIGURE 3: Variation of the load and the maximum pressure with mesh density.

The 8 dimensionless dynamic coefficients K_{ij}, C_{ij}, $i, j = x, y$ can be calculated by [16]

$$\begin{bmatrix} K_{xx} & K_{xy} \\ K_{yx} & K_{yy} \end{bmatrix} = \begin{bmatrix} \sin\theta_0 & \cos\theta_0 \\ \cos\theta_0 & -\sin\theta_0 \end{bmatrix} \cdot \begin{bmatrix} K_{ee} & K_{\theta e} \\ K_{e\theta} & K_{\theta\theta} \end{bmatrix} \cdot \begin{bmatrix} \sin\theta_0 & \cos\theta_0 \\ \cos\theta_0 & -\sin\theta_0 \end{bmatrix}$$

$$\begin{bmatrix} C_{xx} & C_{xy} \\ C_{yx} & C_{yy} \end{bmatrix} = \begin{bmatrix} \sin\theta_0 & \cos\theta_0 \\ \cos\theta_0 & -\sin\theta_0 \end{bmatrix} \cdot \begin{bmatrix} C_{ee} & C_{\theta e} \\ C_{e\theta} & C_{\theta\theta} \end{bmatrix} \cdot \begin{bmatrix} \sin\theta_0 & \cos\theta_0 \\ \cos\theta_0 & -\sin\theta_0 \end{bmatrix}$$

$$\left.\begin{matrix} K_{ee} \\ K_{\theta e} \end{matrix}\right\} = -\int_{-1}^{1}\int_{\varphi_1}^{\varphi_2} P_{\varepsilon} \begin{Bmatrix} \cos\varphi \\ \sin\varphi \end{Bmatrix} d\varphi\, d\lambda$$

$$\left.\begin{matrix} K_{e\theta} \\ K_{\theta\theta} \end{matrix}\right\} = -\int_{-1}^{1}\int_{\varphi_1}^{\varphi_2} P_{\theta} \begin{Bmatrix} \cos\varphi \\ \sin\varphi \end{Bmatrix} d\varphi\, d\lambda$$

$$\left.\begin{matrix} C_{ee} \\ C_{\theta e} \end{matrix}\right\} = -\int_{-1}^{1}\int_{\varphi_1}^{\varphi_2} P_{\varepsilon'} \begin{Bmatrix} \cos\varphi \\ \sin\varphi \end{Bmatrix} d\varphi\, d\lambda$$

$$\left.\begin{matrix} C_{e\theta} \\ C_{\theta\theta} \end{matrix}\right\} = -\int_{-1}^{1}\int_{\varphi_1}^{\varphi_2} P_{\theta'} \begin{Bmatrix} \cos\varphi \\ \sin\varphi \end{Bmatrix} d\varphi\, d\lambda. \quad (11)$$

2.3. Grid Independence Test. The computational accuracy depends on mesh density. The load and the maximum pressure were used to test grid independence. With the increase of journal eccentricity ratio in the bearing, mesh density should be increased. Thus, the grid independence test was done at large eccentricity ratio $\varepsilon = 0.90$.

Table 1 gives bearing data. Figure 3 shows variation of the load and the maximum pressure with the mesh density. When the grid density is greater than 4000, the load and the maximum pressure change a little. To ensure the accuracy, mesh with 60 in the axial direction by 180 in the circumferential direction was used at last.

TABLE 1: Bearing data.

Parameter	Value
Radius R	300 mm
Radial clearance c	0.45 mm
Bearing length l	300 mm
Bearing load F_y	20.0 t
Oil density ρ	890 kg/m^3
Oil viscosity μ	0.025 Pa s
Rotating speed ω	314 rad/s
Specific heat capacity c_v	1.944 kJ/kg K

2.4. Journal Position in the Bearing. Oil film force at different eccentricity ratio and attitude angle was calculated in advance. The back propagation neural network is well known for its strong nonlinear mapping ability. A three-layer backpropagation neural network with 8 nodes in the hidden layer, as shown in Figure 4, was trained to map the functional relationship between the force and the journal position in the bearing. The input of the neural network is bearing load in the horizontal and the vertical directions. The output of the neural network is journal eccentricity ratio and the attitude angle. For improving mapping accuracy, neural network for small bearing load case and for large bearing load case was trained, respectively. Once the bearing load is given, the position of the journal in the bearing can be calculated easily.

3. Influence of Varnish on Bearing Performance

Bearing data is provided in Table 1. In the analysis, the dimensionless thickness of the varnish is changed from 0.0 to 0.3. Varnish is located in the load zone. The leading edge angle and the trailing edge angle are 165° and 192° and the arc length of varnish is about 27°. Influence of varnish on bearing performance was carried out assuming that bearing load is the same.

3.1. Oil Film Thickness. Figure 5 compares oil film thickness distribution in the central section along the circumferential direction for bearing with and without varnish. Figure 6

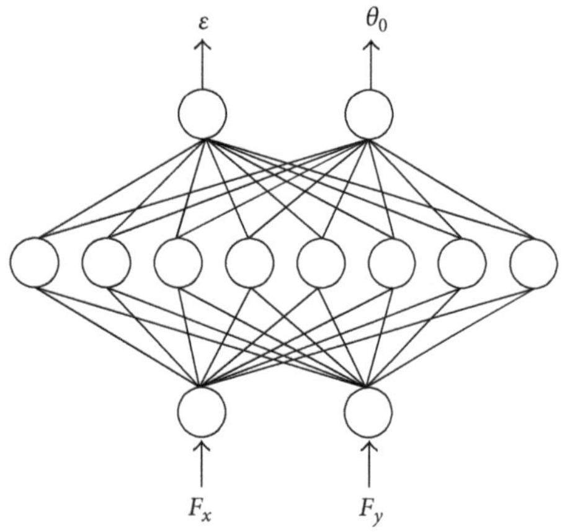

FIGURE 4: Neural network model mapping the function between the force and journal position.

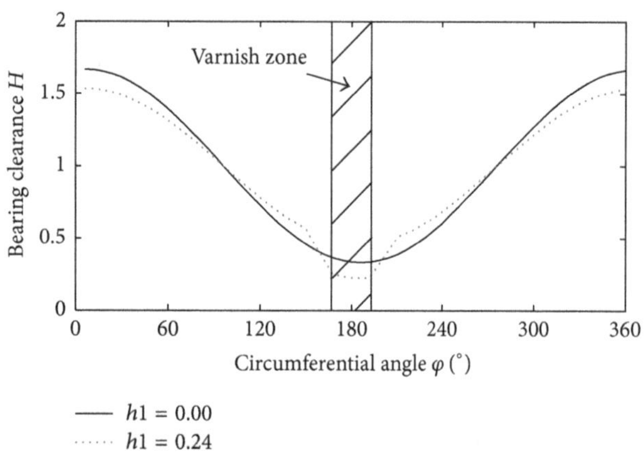

FIGURE 5: Oil film thickness distribution.

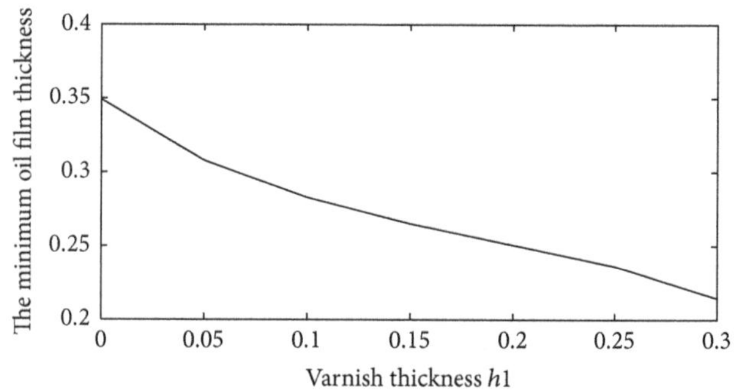

FIGURE 6: Variation of the minimum oil film thickness with varnish thickness.

shows variation of the minimum oil film thickness with varnish thickness. The oil film thickness is smaller in the varnish zone. The minimum oil film thickness decreases with the increase of varnish thickness.

3.2. Bearing Pressure Distribution. Figure 7 compares oil film pressure distribution for bearings. Beyond the varnish zone, oil pressure increases gradually under the effect of squeeze film formed in the converged region of the bearing. At the leading edge of the varnish, the step flow effect is formed for that bearing clearance becomes smaller suddenly. The pressure increases quickly and the maximum pressure appears. The performance of bearing with vanish depends on the joint action of the squeeze film effect and the step flow effect.

3.3. Journal Position. Figure 8 shows variation of journal eccentricity ratio with bearing load. Journal eccentricity ratio of bearing without varnish is larger because that bearing load is supported by the squeeze film effect only. Figure 9 shows variation of journal eccentricity ratio with varnish thickness. Figure 10 shows variation of the eccentricity ratio and the attitude angle with rotational speed. For bearing with varnish, bearing load is partially compensated by the step flow effect caused by the varnish. So, the thicker the varnish, the smaller the journal eccentricity ratio and the larger the attitude angle.

3.4. Bearing Temperature. Figure 11 compares temperature distribution of bearing with different varnish thickness. Oil film temperature increases quickly at the leading edge of the varnish and reaches the maximum value at the trailing edge. Figure 12 shows variation of the maximum oil temperature with varnish thickness. The maximum oil temperature increases with the increase of varnish thickness. In the early stage of the varnish, temperature growth trend is slow. However, the trend of temperature increasing is more and more obvious with the increase of thickness.

Figure 13 shows temperature trend in a year for the third bearing of Figure 1(b). Although the bearing temperature is not high, the trend of growth is obvious. And the temperature is getting higher and higher. Bearing was overhauled for inspection. Vanish is found on the inner surface of the bearing, as shown in Figure 1(b). The temperature increase phenomenon found in the bearing is almost similar to the simulated result of Figure 12.

3.5. Loading Capacity. Figure 14 shows the variation of the minimum clearance with bearing load. With the increase of varnish thickness, the load corresponding to the same minimum clearance decreases. It means that loading capacity of bearing with varnish is smaller than that of bearing without varnish.

Varnish in the bearing results in a large decrease of minimum clearance and an increase in the oil film temperature and the pressure. Loading capacity of bearing is decreased. It is harmful to the safe operation of bearing.

3.6. Bearing Dynamic Coefficients. Figure 15 shows the variation of bearing stiffness and damping coefficients with frequency. The stiffness and the damping coefficients are larger for bearing with varnish. Its influence is larger in the horizontal direction than that in the vertical direction. The dynamic pressure effect caused by the step flow is relatively large at low speed. This leads to the greater impact on bearing dynamics at low speed.

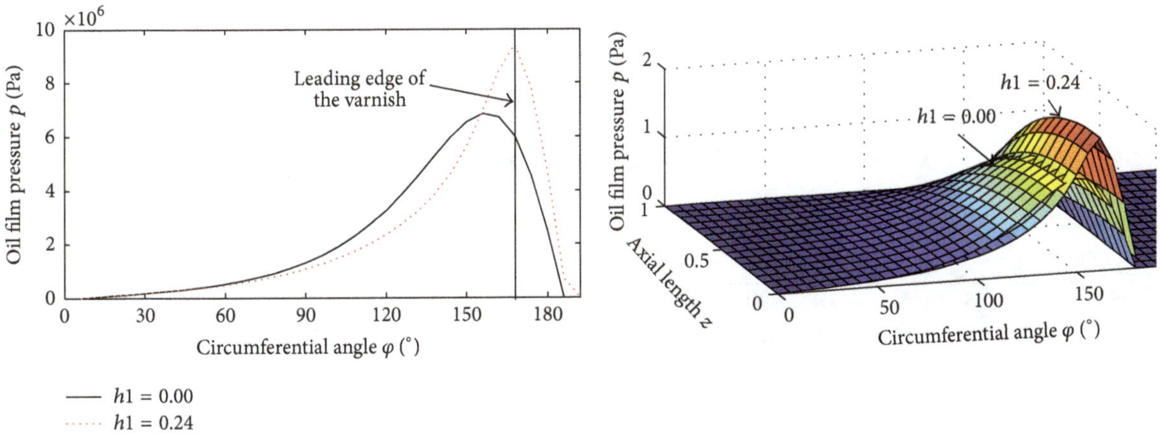

(a) Pressure distribution in the central section

(b) The three-dimensional pressure distribution

FIGURE 7: Oil film pressure distribution.

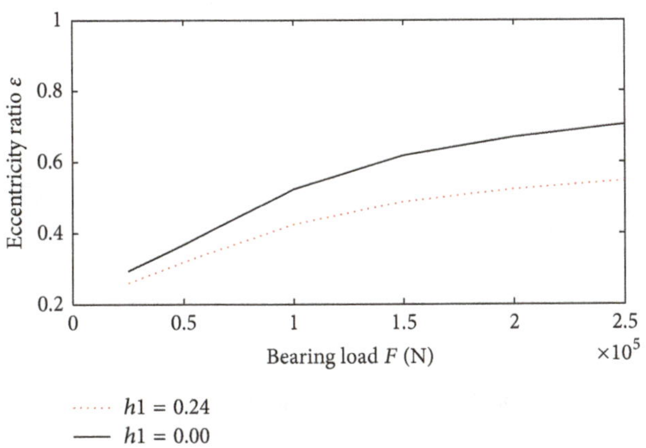

FIGURE 8: Variation of journal eccentricity ratio with bearing load.

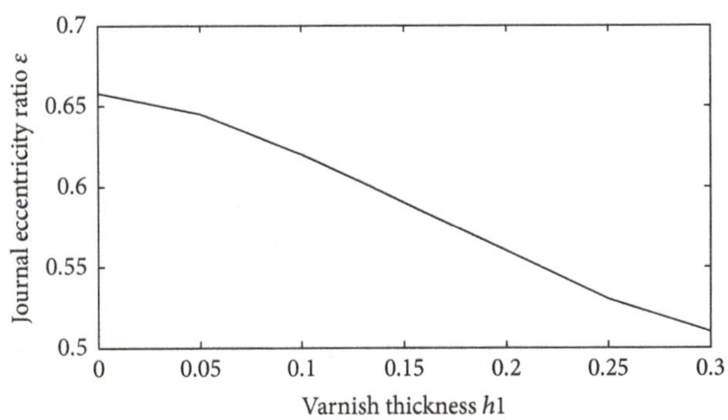

FIGURE 9: Variation of journal eccentricity ratio with varnish thickness.

For stability study, bearing equivalent stiffness and critical whirl ratio can be presented using bearing coefficients. They are given as [17]

$$K_{eq} = \frac{K_{xx}C_{yy} + K_{yy}C_{xx} - K_{xy}C_{yx} - K_{yx}C_{xy}}{C_{xx} + C_{yy}}$$
$$\gamma_{st}^2 = \frac{\left(K_{xx} - K_{eq}\right)\left(K_{yy} - K_{eq}\right) - K_{xy}K_{yx}}{C_{xx}C_{yy} - C_{xy}C_{yx}}. \quad (12)$$

The calculation results are given in Table 2. For bearings studied in the case, $K_{eq} > 0$, $\gamma_{st}^2 < 0$. It means that bearings are stable. However, for bearings with varnish, its equivalent stiffness is higher and its critical whirl ratio is lower. It means that varnish has certain positive effect on bearing stability.

4. Influence of Varnish on Rotor Unbalance Response and System Stability

4.1. Dynamic Equation. The dynamic equation of rotor bearing system is given as [18]

$$[\mathbf{M}]\left\{\ddot{\mathbf{U}}(t)\right\} + [\mathbf{D}]\left\{\dot{\mathbf{U}}(t)\right\} + [\mathbf{K}]\{\mathbf{U}(t)\} = \{\mathbf{F}(t)\}. \quad (13)$$

TABLE 2: Bearing equivalent stiffness and critical whirl ratio.

Parameter	With varnish	Without varnish
Equivalent stiffness	1.8314	1.7542
Critical whirl ratio	−0.2956	−0.2146

Assume that

$$\mathbf{U}(t) = \mathbf{U}_0(t)e^{i\omega t}$$
$$\mathbf{F}(t) = \mathbf{F}_0(t)e^{i\omega t}. \quad (14)$$

Substituting (14) into (13), we can get

$$\left[\mathbf{K} + i\omega\mathbf{D} - \omega^2\mathbf{M}\right]\{\mathbf{U}_0\} = \{\mathbf{F}_0\}. \quad (15)$$

The unbalance response can be calculated with the equation.

The homogeneous equation corresponding to (13) can be written as

$$[\mathbf{M}]\left\{\ddot{\mathbf{U}}(t)\right\} + [\mathbf{D}]\left\{\dot{\mathbf{U}}(t)\right\} + [\mathbf{K}]\{\mathbf{U}(t)\} = \{\mathbf{0}\}. \quad (16)$$

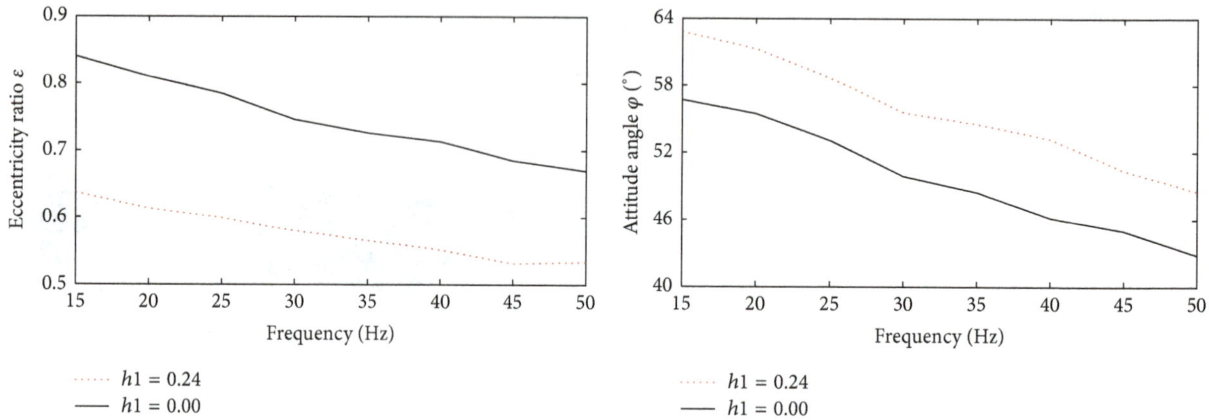

(a) The eccentricity ratio

(b) The attitude angle

FIGURE 10: Comparison of journal eccentricity ratio and attitude angle.

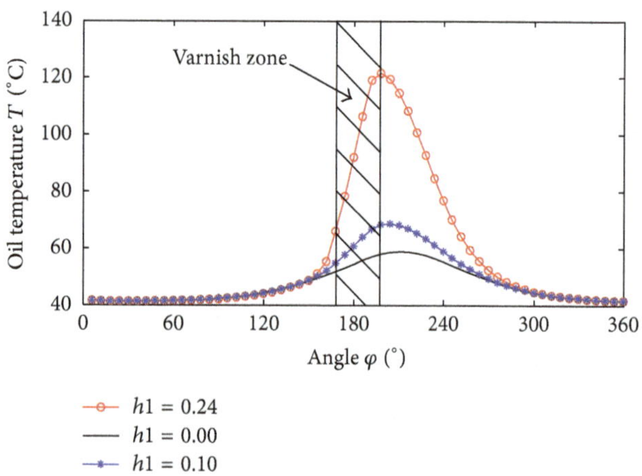

FIGURE 11: Comparison of temperature distribution.

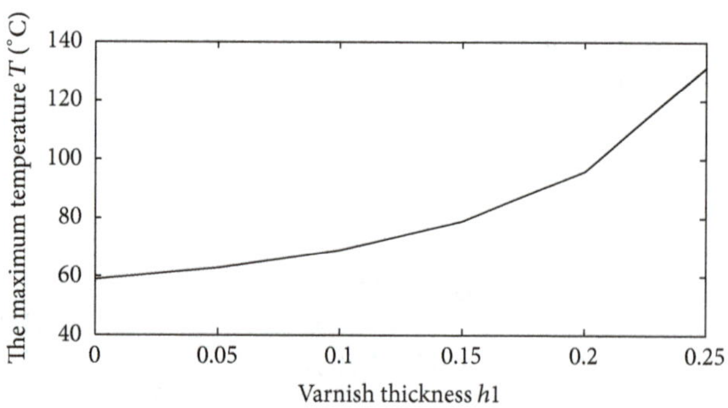

FIGURE 12: Variation of the maximum oil temperature with varnish thickness.

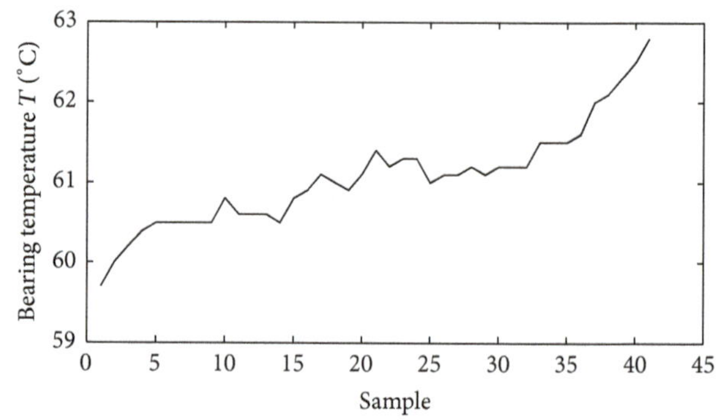

FIGURE 13: Bearing temperature trend in a year.

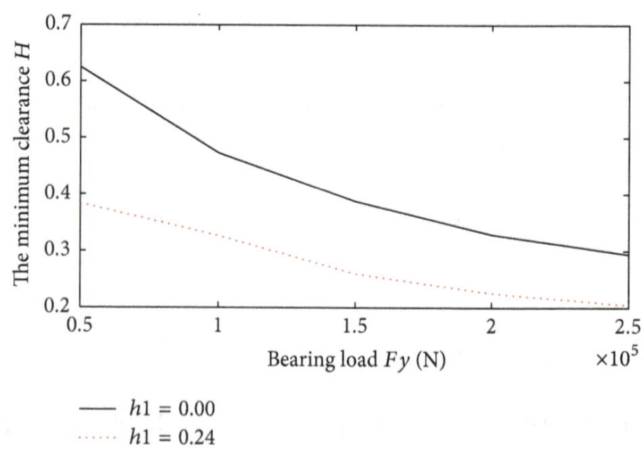

FIGURE 14: Variation of the minimum oil film thickness with bearing load.

The first-order differential equations of (16) are

$$
\begin{gathered}
[\mathbf{A}]\{\dot{\mathbf{V}}\} + [\mathbf{B}]\{\mathbf{V}\} = \{\mathbf{0}\} \\
\mathbf{A} = \begin{bmatrix} \mathbf{M} & \mathbf{0} \\ \mathbf{0} & \mathbf{K} \end{bmatrix}, \\
\mathbf{B} = \begin{bmatrix} \mathbf{D} & \mathbf{K} \\ -\mathbf{K} & \mathbf{0} \end{bmatrix}, \\
\mathbf{V} - \begin{Bmatrix} \dot{\mathbf{U}} \\ \mathbf{U} \end{Bmatrix}.
\end{gathered} \tag{17}
$$

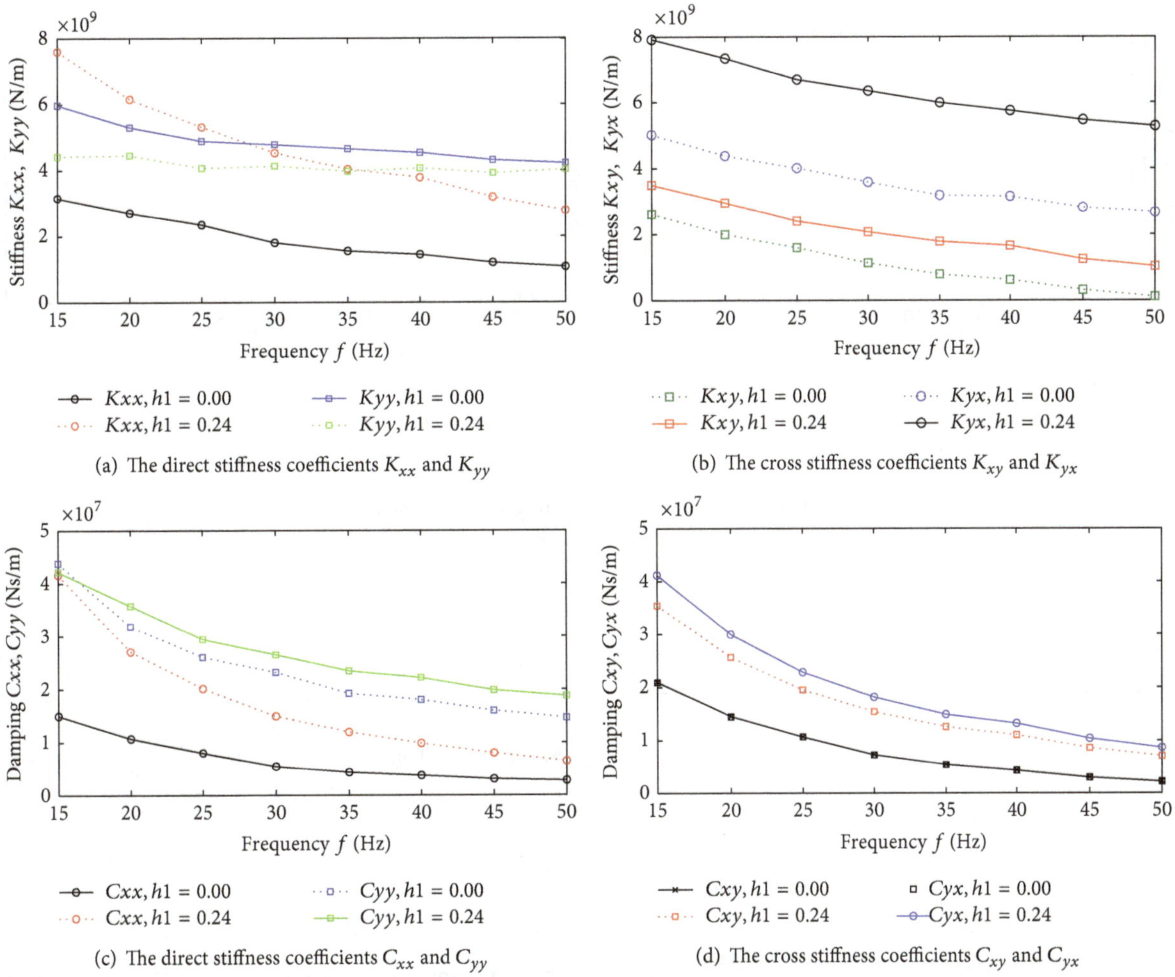

(a) The direct stiffness coefficients K_{xx} and K_{yy}

(b) The cross stiffness coefficients K_{xy} and K_{yx}

(c) The direct stiffness coefficients C_{xx} and C_{yy}

(d) The cross stiffness coefficients C_{xy} and C_{yx}

FIGURE 15: Comparison of bearing dynamic coefficients.

The eigenvalues and eigenvectors are

$$(\mathbf{E} - \nu \mathbf{I}) \mathbf{V}_0 = \mathbf{0}$$

$$\mathbf{E} = \begin{bmatrix} -\mathbf{M}^{-1}\mathbf{D} & -\mathbf{M}^{-1}\mathbf{K} \\ \mathbf{I} & \mathbf{0} \end{bmatrix}. \tag{18}$$

The stability of the system is represented by the logarithmic decrement rate δ. It can be calculated from system eigenvalue ν:

$$\delta = -2\pi \frac{\Re(\nu)}{\Im(\nu)}. \tag{19}$$

The greater the value δ is, the more stable the system is. In order to ensure the stability of the system, $\delta \geq 0.2$.

4.2. Model of a Low Pressure Rotor. A low pressure rotor of a large steam turbine was used. Figure 16 shows the model. Bearing data is provided in Table 1. Shaft data is given in Table 3. Considering the axial symmetry, only half of the shaft data is given.

4.3. Unbalance Response Analysis. Three unbalance masses as shown in Table 4 were exerted on the rotor. The first and the third set of masses represent the coupled unbalance and the second set of mass represents the static unbalance in the middle plane.

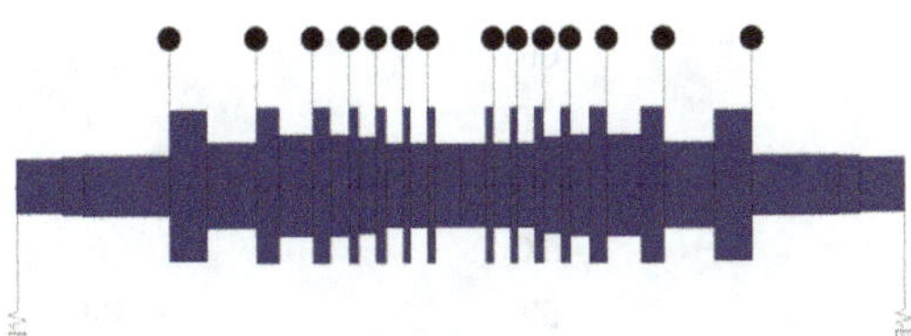

FIGURE 16: Shaft model.

The influence of varnish on the vertical and the horizontal vibration during the run up process is shown in Figure 17. For bearing with varnish or without varnish, the trend of the response to rotating speed is similar. Influence of varnish on the horizontal vibration is larger than that on the vertical vibration, especially in the low speed zone. This is consistent with the influence of the varnish on bearing dynamic characteristics. The damping coefficient in the horizontal direction of the bearing with varnish is large. Its effect on vibration control near the first critical speed zone is obvious.

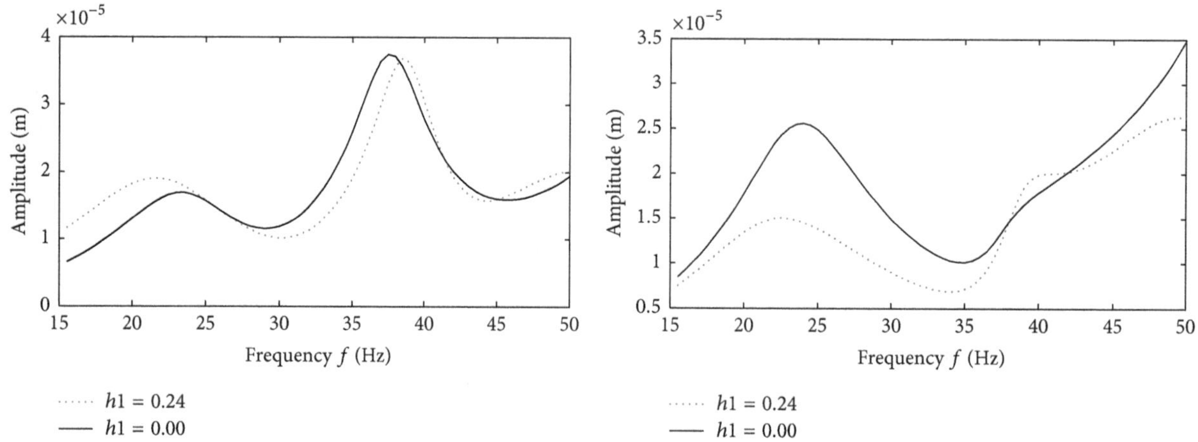

(a) Vertical vibration

(b) Horizontal vibration

FIGURE 17: Variation of vibration with rotating speed during run-up process.

TABLE 3: Shaft model.

Diameter D (mm)	Length L (mm)	Additional mass M (kg)
600	300	0
630	142	0
660	594	0
1668	256	1850
951	340	0
1720	156	900
1116	236	0
1718	120	300
1114	130	0
1718	65	280
1068	120	0
1718	66	180
934	120	0
1718	50	110
934	120	0
1718	50	100
915	175	0

TABLE 4: Unbalance mass.

Node	Product of mass and radius (kg m)	Angle (°)
4	0.5	45
17	1	135
32	0.5	225

4.4. System Stability. The lowest order modes are prone to be unstable and are of great concern. In this paper, stability of the first and the second mode is analyzed.

Table 5 shows the influence of the varnish on system stability. Influence of the varnish on modal frequency is small. However, system stability is improved due to the increased bearing damping coefficients. The conclusion is the same as that in Section 3.6.

TABLE 5: Comparison of system stability.

Mode	Parameter	With varnish	Without varnish
Mode 1	Frequency/Hz	23.48	23.31
	Logarithmic decrement rate	0.5029	0.4645
Mode 2	Frequency/Hz	34.45	34.41
	Logarithmic decrement rate	0.5327	0.4614

5. Conclusions

In this paper, the influence of the varnish on bearing performance and rotor vibration is studied. The results show the following.

(1) Performance of bearing with varnish is changed by the step flow effect at the leading edge of the varnish. Compared with the squeeze flow effect in the convergent clearance zone of bearing, influence of varnish is more obvious at the low speed case.

(2) Varnish in the bearing results in a large decrease of minimum clearance and an increase in the oil film temperature and the pressure. Loading capacity of bearing is also decreased. It is harmful to the safe and reliable operation of bearing.

(3) For bearing with varnish, the minimum oil film thickness decreases. This leads to an increase of bearing stiffness and damping coefficients. Its influence is larger in the horizontal direction than that in the vertical direction.

(4) For bearing with varnish or without varnish, the trend of the response to rotating speed is similar. Its Influence on the horizontal vibration is larger than that on the vertical vibration, especially in the low speed zone. This is consistent with the influence of the varnish on bearing dynamic

characteristics. There is little influence of the varnish on the modal frequency. However, system stability is improved due to the increased bearing damping coefficients.

Nomenclature

Symbols

H:	Dimensionless bearing clearance
h:	Dimensional bearing clearance
H_1:	Dimensionless varnish thickness
φ:	Circumferential angle
ε:	Journal eccentricity ratio
e:	Journal eccentricity
θ_0:	Journal attitude angle
c:	Bearing radial gap
d:	Bearing diameter
R:	Bearing radius
l:	Bearing length
λ:	Bearing axial dimensionless coordinate
P:	Dimensionless oil pressure
p:	Dimensional oil pressure
$P_\varepsilon, P_\theta, P_{\varepsilon'}, P_{\theta'}$:	Perturbation oil pressure
T:	Oil temperature
ρ:	Oil density
μ:	Oil viscosity
α:	Oil viscosity coefficient
c_v:	Specific heat capacity
U:	Journal velocity
F_x, F_y:	Bearing loads in horizontal and vertical direction
K_{ij}:	Bearing stiffness coefficients
C_{ij}:	Bearing damping coefficients
K_{eq}:	Bearing equivalent stiffness
γ_{st}:	Critical whirling ratio
ω:	Rotating speed
δ:	Logarithmic decrement rate
M, K, D:	System mass, stiffness, and gyroscopic/damping matrices
U:	Degree of freedom vector
F:	External force vector
$\boldsymbol{\nu}$:	System eigenvalues
$\mathbf{V}_0$:	System eigenvectors
$\mathbf{U}_0$:	Amplitude of displacement vector
$\mathbf{F}_0$:	Amplitude of the forces vector
$\Re(\cdot)$:	Real part of variable
$\Im(\cdot)$:	Imaginary part of variable.

Conflicts of Interest

The authors declare that there are no conflicts of interest regarding the publication of this paper.

References

[1] F. Yokoyama, Y. Iwama, M. Maruyama, and M. Sano, "Deposits on bearing pad caused by particulate contamination in turbine oil," *Tribology Online*, vol. 10, no. 2, pp. 162–171, 2015.

[2] A. Yano, S. Watanabe, Y. Miyazaki, M. Tsuchiya, and Y. Yamamoto, "Study on sludge formation during the oxidation process of turbine oils," *Tribology Transactions*, vol. 47, no. 1, pp. 111–122, 2004.

[3] A. Sasaki, S. Uchiyama, and M. Kawasaki, "Varnish formation in the gas turbine oil systems," *Journal of ASTM International*, vol. 5, no. 2, 2008.

[4] T. V. V. L. N. Rao, A. M. A. Rani, T. Nagarajan, and F. M. Hashim, "Analysis of slider and journal bearing using partially textured slip surface," *Tribology International*, vol. 56, pp. 121–128, 2012.

[5] N. Tala-Ighil and M. Fillon, "A numerical investigation of both thermal and texturing surface effects on the journal bearings static characteristics," *Tribology International*, vol. 90, pp. 228–239, 2015.

[6] C. Sinanoğlu, F. Nair, and M. B. Karamış, "Effects of shaft surface texture on journal bearing pressure distribution," *Journal of Materials Processing Technology*, vol. 168, no. 2, pp. 344–353, 2005.

[7] F. M. Meng and T. Yang, "Preliminary study on mechanism of cavitation in lubricant of textured sliding bearing," *Proceedings of the Institution of Mechanical Engineers, Part J: Journal of Engineering Tribology*, vol. 227, no. 7, pp. 695–708, 2013.

[8] M. Qiu, B. R. Minson, and B. Raeymaekers, "The effect of texture shape on the friction coefficient and stiffness of gas-lubricated parallel slider bearings," *Tribology International*, vol. 67, pp. 278–288, 2013.

[9] A. De Kraker, R. A. J. Van Ostayen, and D. J. Rixen, "Development of a texture averaged Reynolds equation," *Tribology International*, vol. 43, no. 11, pp. 2100–2109, 2010.

[10] C. B. Khatri and S. C. Sharma, "Influence of textured surface on the performance of non-recessed hybrid journal bearing operating with non-Newtonian lubricant," *Tribology International*, vol. 95, pp. 221–235, 2016.

[11] N. Tala-Ighil, M. Fillon, and P. Maspeyrot, "Effect of textured area on the performances of a hydrodynamic journal bearing," *Tribology International*, vol. 44, no. 3, pp. 211–219, 2011.

[12] F. M. Meng, L. Zhang, and T. Long, "Effect of Groove Textures on the Performances of Gaseous Bubble in the Lubricant of Journal Bearing," *Journal of Tribology*, vol. 139, no. 3, Article ID 031701, 2017.

[13] S. Kango, R. K. Sharma, and R. K. Pandey, "Comparative analysis of textured and grooved hydrodynamic journal bearing," *Proceedings of the Institution of Mechanical Engineers, Part J: Journal of Engineering Tribology*, vol. 228, no. 1, pp. 82–95, 2014.

[14] J. Dong, X. Wang, and J. Zhang, "An experimental research on the vibration of surface-textured journal bearings," *Shock and Vibration*, vol. 2017, Article ID 1261826, 9 pages, 2017.

[15] V. Brizmer and Y. Kligerman, "A laser surface textured journal bearing," *Journal of Tribology*, vol. 134, no. 3, Article ID 031702, 2012.

[16] O. Ebrat, Z. P. Mourelatos, N. Vlahopoulos, and K. Vaidyanathan, "Calculation of journal bearing dynamic characteristics including journal misalignment and bearing structural deformation," *Tribology Transactions*, vol. 47, no. 1, pp. 94–102, 2004.

[17] C. Y. Chen, R. H. Yen, and C. S. Liu, "Characteristics of dynamic coefficients on stability for herringbone-grooved journal bearings," *Applied Mathematics & Information Sciences*, vol. 7, no. 3, pp. 1215–1223, 2013.

[18] K. L. Cavalca, P. F. Cavalcante, and E. P. Okabe, "An investigation on the influence of the supporting structure on the dynamics of the rotor system," *Mechanical Systems and Signal Processing*, vol. 19, no. 1, pp. 157–174, 2005.

20

Application of Artificial Neural Networks to Calculation of Oil Film Reaction Forces and Dynamics of Rotors on Journal Bearings

Alexey V. Kornaev,[1] Nickolay V. Kornaev,[2] Elena P. Kornaeva,[3] and Leonid A. Savin[2]

[1] *Modeling of Hydromechanical Systems Research Laboratory, The Orel State University Named after I.S. Turgenev, Orel 302026, Russia*
[2] *Department of Mechatronics and International Engineering, The Orel State University Named after I.S. Turgenev, Orel 302026, Russia*
[3] *Department of Information Systems, The Orel State University Named after I.S. Turgenev, Orel 302026, Russia*

Correspondence should be addressed to Alexey V. Kornaev; rusakor@inbox.ru

Academic Editor: Fadi Dohnal

Increase of energy efficiency and level of information system development of rotor machines in general requires improvement of theoretical approaches to research. In the present paper the problem of high-precision and high-performance computing programs development has been considered to simulate rotor vibrations. Based on two-layer feed-forward neural networks, numerical models have been developed to calculate oil film reaction forces to solve the rotor dynamics problems. Comparison has been done of linear and nonlinear approaches to solution of rotor dynamics problems, and a qualitative evaluation has been presented of accuracy and performance of a neural network approach compared to conventional approaches to rotor dynamics.

1. Introduction

Dynamic behavior of a rotating shaft on journal bearings is defined by the resonance vibration at critical speeds, self-excited vibrations due to the internal damping of the shaft's material, self-excited vibrations due to the oil film of journal bearings, oil whip, and flow-induced vibrations [1–4]. Since the authors are closely connected with hydrodynamic lubrication, this paper is dedicated to oil whirl and oil whip problems. Oil film forces are obtained by means of solving the hydrodynamics equations and depend on many design and operational parameters of a bearing as well as on various specific physical effects that may occur, such as vibration, cavitation, and turbulence [2, 4, 5]. The number of input parameters can reach several dozen [6]. At the same time, development of technology in the direction of informatization requires simultaneous increase of accuracy and performance of the numerical models. One of the possible ways to provide both accuracy and high performance is using deterministic and stochastic approaches to modeling. Nowadays the methods and resources of implementation of neural network algorithms are developing quite intensively, which, based on the known data about the subject of study, allows building multifactor models of input and output parameters relations. Three basic levels of developed models could be distinguished.

Models of the first level have no direct interaction with the rotating machine; their training is based on the given numerical or experimental data. In [7] the results of study of an elastic rotor's dynamics on rigid bearings have been presented. Neural networks that have been applied are feed-forward with two hidden layers. High accuracy has been demonstrated for models of rotor systems with roller-element bearings based on neural networks: accuracy from 82% up to 100% has been presented. Dynamics has been studied in a certain range of speed and for every case of speed separately, so the neural network had to be retrained. Some works are known, where neural network models are used to calculate the distributed characteristics, for example, pressure distribution in the oil film of the journal bearings [8, 9]. It should be noted that application of stochastic methods of rotor dynamics modeling allows skipping the stage of distributed values calculation. Instead, one could instantly determine the components of a resulting oil film force. In [10] dynamics of an imbalanced rigid rotor on fluid-film bearings has been studied. Authors determined the relation between three kinematic characteristics and two components of the resulting oil film force using a model of a neural network. The numerical tests were carried out for various angular speed and allowed determination of rotor trajectories. The synchronous, periodic, quasiperiodic, and chaotic motions for rotation speed within the range from 1000 to 10000 rpm were presented as the results of calculation. The authors suggest that there is a limitation to using linearized models in solving the problems of dynamics; however, they do not present a quantitative comparison of the numerical results obtained using linear and nonlinear models. Moreover, the reduction of the number of input variables has been done according to Poincare transformation of Reynolds equation, but possible outcomes of such substitution were not discussed in the paper. It is also not quite clear how one introduced kinematic variable allowed accounting for three given variables (two projections of rotor's lateral vibration speed and its rotating speed). It could be that the neural network had to be retrained for every particular case of the rotating speed. So, the results of this indeed interesting and important work remain in need of additional verification. A number of quite perspective papers should also be highlighted where neural network algorithms are used to account for complex specific phenomena [5, 11] and to optimize the operational conditions of fluid-film bearings [12]. Models of higher level can learn in real time using the processed measurement data. One of the most popular applications of such models is using neural networks to flaw detection of bearings in rotor machines [13–16]. In [17] authors investigate the inverse model of the squeeze-film damper bearing (force input/displacement output) from empirical data and its application to a nonlinear inverse rotor-bearing problem. Comparison of experimental and numerical data showed high accuracy of the model based on recurrent neural networks. Models of highest level of complexity are able to generate some control impact [18, 19]. Today such approach is used to solve the problems of control in magnetic bearings [20–22]. It should be, however, noted that the values of clearance in such bearings are an order of magnitude more than the same value in fluid-film bearings.

The present paper considers development of neural network algorithms of first level, that is, models with no direct contact to the rotor machine. Matters of quantitative evaluation of accuracy of application of neural network in comparison to known methods of oil film forces calculation and applicability of neural networks during simulation of nonlinear self-excited oscillations—oil whip—and during simulation of transient regimes of rotor's operation have been considered.

2. Mathematical Model

A rigid symmetrical imbalanced rotor is considered. Both ends of the rotor are supported by journal bearings of the same specification (see Figure 1). Given the chosen direction of coordinate axes, the pressure distribution and the resulting

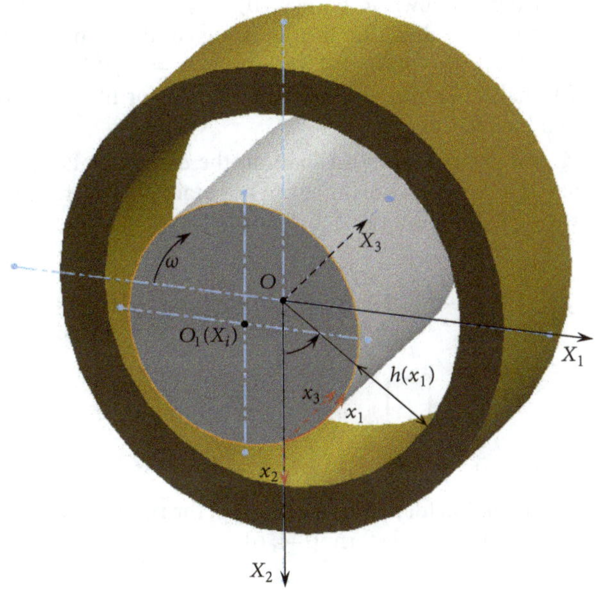

FIGURE 1: Calculation diagram of a journal bearing.

oil film force can be determined by means of numerically solving the Reynolds equation for a case of an isothermal flow of a viscous incompressible fluid [6]:

$$\frac{\partial}{\partial x_1}\left[\frac{h^3}{\mu}\frac{\partial p}{\partial x_1}\right] + \frac{\partial}{\partial x_3}\left[\frac{h^3}{\mu}\frac{\partial p}{\partial x_3}\right] = 6\frac{\partial}{\partial x_1}(u_1 h) + 12u_2, \quad (1)$$

where $h = h(x_1)$ is the oil film thickness (radial gap) function, μ is the dynamic viscosity coefficient (assumed to be constant), $p = p(x_1, x_3)$ is the pressure function, and $u_i = u_i(x_1)$ is accordingly the tangential and normal components of the journal's surface velocity, $i = 1, 2$.

Radial gap function $h(x_1)$ could be determined geometrically (see Figure 1) (coordinates and velocities of the center of the rotor are indicated with the capital letters, while the coordinates and the velocities of the fluid medium are indicated with the lower case letters):

$$h(x_1) = h_0 - X_1 \sin(\alpha) - X_2 \cos(\alpha), \quad (2)$$

where h_0 is the mean oil film thickness, (X_1, X_2) are coordinates of the center of the journal, $\alpha = 2x_1/D$ is the angular coordinate, and D is the bearing's diameter.

The values of u_i over the surface of the journal could be determined using the following kinematic relations:

$$\begin{aligned} u_1 &= \frac{\omega D}{2} + V_1 \cos(\alpha) - V_2 \sin(\alpha), \\ u_2 &= V_1 \sin(\alpha) + V_2 \cos(\alpha), \end{aligned} \quad (3)$$

where ω is angular speed of the rotor and (V_1, V_2) are projections of the velocity vector of the center of the rotor.

Calculation of pressure distribution $p(x_1, x_3)$ at some specific point in time is a boundary problem of solution of Reynolds equation (1) considering (2) and (3) with the following boundary conditions: the pressure on each side of the bearing is known to be $p(x_1,0) = p_0$ and $p(x_1,L_3) = p_1$; along the x_1 direction the following condition has to be met: $p(0,x_3) = p(\pi D,x_3)$; $\partial p/\partial x_1|_{x_1=0} = \partial p/\partial x_1|_{x_1=\pi D}$. Additionally, Gumbel's boundary condition for film rupture is used [6, 23].

Given the pressure distribution, the components of the resulting oil film reaction force can be obtained as projections on the X_i (see Figure 1):

$$R_1 = -\int_0^{L_3}\int_0^{\pi D} p(x_1,x_3)\cos(\alpha)\,dx_1 dx_3,$$
$$R_2 = -\int_0^{L_3}\int_0^{\pi D} p(x_1,x_3)\sin(\alpha)\,dx_1 dx_3, \qquad (4)$$

where L_3 is the length of the bearing.

The equation for lateral vibrations of the rotor's center can be presented in simple form [1–4, 6]:

$$m\frac{d\vec{V}}{dt} = \vec{R} + \vec{F}, \qquad (5)$$

where m is the mass of the rotor per one bearing, t is time, $\vec{F} = m\Delta\omega^2\left[\begin{smallmatrix}\cos\omega t\\ \sin\omega t\end{smallmatrix}\right] + m\left[\begin{smallmatrix}0\\ -g\end{smallmatrix}\right]$ is the bearing load (centrifugal and inertia forces), $m\Delta$ is the rotor's imbalance, and g is free fall acceleration.

Oscillations' velocity $\vec{V}$ of the rotor is significantly less than tangential velocity of journal's surface $\omega D/2$, so the solution of (5) is linked to solution of the steady-state fluid's flow problem solution. So, time as variable could be neglected during calculation of oil film force (4). This significantly simplifies the procedure of numerical solution of rotor dynamics problems [1, 6]. Three basic approaches have been considered and are characterized by different methods of oil film forces determination.

The first approach, called trajectory method [1–4, 6], is based on solution of Reynolds equation (1) followed by integration of pressure distribution and determination of R_i (4). Such approach is considered referential due to high precision and however has low performance in computational terms, since the Reynolds equation has to be solved at each time step.

The second approach, called the method of linearization, of the oil film force determination [1–4, 6] is based on the decomposition of reaction force components R_i into the Taylor series in the proximity of the equilibrium position X_i^E. The matrices of stiffness K and damping B, calculated in the equilibrium position, shall be introduced as follows:

$$K_{ij} = -\frac{\partial R_i}{\partial X_j},$$
$$B_{ij} = -\frac{\partial R_i}{\partial V_j}. \qquad (6)$$

Then the motion equation of the rotor (5) with given operational conditions of the rotor system and given the linearization of the film's reaction forces will be as follows [1–4, 6, 24]:

$$m\begin{bmatrix}\dfrac{dV_1}{dt}\\[2mm] \dfrac{dV_2}{dt}\end{bmatrix} = \begin{bmatrix}R_1^E\\ R_2^E\end{bmatrix} - \begin{bmatrix}K_{11} & K_{12}\\ K_{21} & K_{22}\end{bmatrix}\cdot\begin{bmatrix}X_1 - X_1^E\\ X_2 - X_2^E\end{bmatrix} - \begin{bmatrix}B_{11} & B_{12}\\ B_{21} & B_{22}\end{bmatrix}\cdot\begin{bmatrix}V_1\\ V_2\end{bmatrix} + m\Delta\omega^2\begin{bmatrix}\cos\omega t\\ \sin\omega t\end{bmatrix} + m\begin{bmatrix}0\\ -g\end{bmatrix}. \qquad (7)$$

The equilibrium point X_i^E can be determined using the trajectory approach. In many practical applications the equilibrium locus curve is calculated to solve problems of rotor dynamics [1–4, 6]. The points on this curve correspond to the equilibrium positions at various rotational speeds. Necessity to determine the equilibrium position or the equilibrium locus curve complicates the method of linearization.

The third approach is based on a neural network program module to approximate the oil film force's components R_i and to solve (5). A two-layer feed-forward neural network with inputs U_k ($k = 1,\ldots,N_{\text{inp}}$) and outputs Y_i ($i = 1,\ldots,N_{\text{out}}$) is used (see Figure 2). The functionality of the network is determined by specifying the strengths of the connections, called weights w, and the neuron transfer function f [25]. The mathematical model of the used neural network is

$$Y_i = f^{\text{out}}\left(\sum_{j=1}^{N_{\text{hid}}} w_{i,j}^{\text{out}} z_j + b^{\text{out}}\right),$$
$$z_j = f^{\text{hid}}\left(\sum_{k=1}^{N_{\text{inp}}} w_{j,k}^{\text{hid}} U_k + b^{\text{hid}}\right), \qquad (8)$$

where $f^{\text{out}}(x) = x$ is a linear transfer function and $f^{\text{hid}}(x) = -1 + 2/(1+\exp(-2x))$ is a sigmoid symmetric transfer function, $w_{i,j}^{\text{out}}$ and $w_{j,k}^{\text{hid}}$ are weights, b^{out} and b^{hid} are thresholds of output and hidden layers, respectively, and z_j are the intermediate outputs of hidden neurons ($j = 1,\ldots,N_{\text{hid}}$).

In the present paper the components of the oil film force $Y_i = R_i$ are considered as outputs, and four or five kinematic values, depending on the problem, are considered inputs: $\vec{U} = [X_1, X_2, V_1, V_2]$ or $\vec{U} = [n, X_1, X_2, V_1, V_2]$, where n is the rotor's speed.

Neural network learning is essentially the solution of optimization problem of a minimum of the mean squared error function search:

$$\text{MSE} = \frac{1}{Q}\sum_{i=1}^{N_{\text{out}}}\left(\sum_{q=1}^{Q}\left(Y_i^q - \tilde{Y}_i^q\right)^2\right) \Longrightarrow \min, \qquad (9)$$

where Q is a number of samples of inputs and outputs and $\tilde{Y}_i^q$ is the target value of an ith output in the qth sample.

The mean squared error function (9) is the function of a number of variables $w_{i,j}^{\text{out}}$ and $w_{j,k}^{\text{hid}}$. The search for an

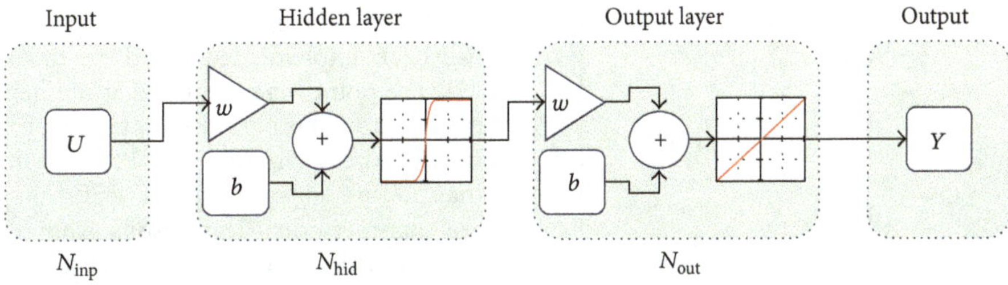

FIGURE 2: Feed-forward neural network with N_{inp} number of inputs, with N_{hid} neurons in hidden layer and with N_{out} outputs.

extremum of such function is implemented numerically. One of the criteria of neural network's quality is the correlation coefficient between outputs and targets $r_{Y\widetilde{Y}}$ [26]. Target data $\widetilde{Y}_i$ in the present paper is based on the numerical results, obtained using the trajectory approach.

Stability analysis of the obtained results could be implemented based on the rotor's locus analysis using Poincare maps [10, 27–29]. Moreover, the Routh–Hurwitz criterion could be applied to the linearization approach [1, 10].

3. Simulation Results

Three basic approaches have been considered and characterized by different methods of oil film forces determination: the trajectory method, the linearization method, and the neural network method. Numerical solution of Reynolds equation (1) was implemented using the finite difference method. The obtained set of linear equations was solved using the Gauss method [30]. The Levenberg-Marquardt backpropagation algorithm was used in training neural networks [25, 31, 32], which, in its turn, was based on the numerical results, obtained using the trajectory method. The rotor dynamics calculation program was based on the numerical solution of (5) or (7) using the explicit Runge-Kutta method of 4th and 5th order [30]. During the development of the programs the GNU-Octave software and Octave's Neural Network Package [31] have been used. The GNU-Octave products are free software and use the Matlab compatible language. During simulations a PC with following characteristics has been used: Intel® Core™ i5-4690K CPU 3.5 GHz, RAM 12 GB, x64 OS.

The results presented below correspond to various types of problems solved.

3.1. Rotor Dynamics with Fixed Speed and Low Vibrations. A rotor system with the following parameters has been subjected to the study: rotor with 8.21 kg mass rotates at a constant speed of 4000 rpm under the load mg and centrifugal force $1.8mg$ and is supported by two plain journal fluid-film bearings with 40 mm diameter and 20 mm length with a mean radial gap of 100 μm; lubricant is mineral oil with dynamic viscosity of 13 mPa·s; the pressure on each side of the bearing is 0.1 MPa.

First, trajectory approach has been applied. A preliminary numerical test neglecting the centrifugal force successfully

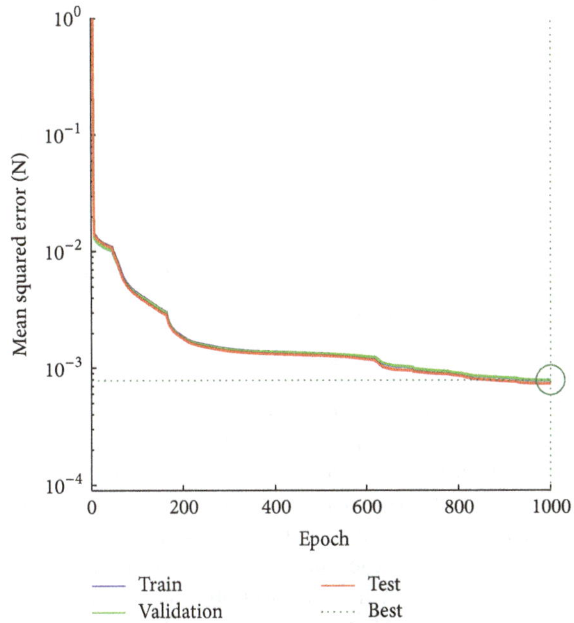

FIGURE 3: Mean squared error.

converged to the equilibrium point $\vec{X}^E = [\,-69.4 \quad 31.8\,]\ \mu$m. Next, linearization approach was used.

Then, a neural network has been developed for training with a number of neurons in the hidden layer $N_{hid} = 15$. As the input data a matrix of 4×10^5 elements was used, which consisted of numerical values of the coordinates X_i and the velocities V_i, with $t_0 = 1$ s calculated time of rotor's motion. The target data matrix of 2×10^5 elements consisted of the numerical values of oil film forces R_i. The matrices of input and output data were divided into three groups for training, validation and testing in fractions 0.6, 0.3, and 0.1, respectively. The procedure of adjustment has been implemented using application software [31]. Training was done in 100 iterations, so-called epochs [25], in approximately 10 minutes. The convergence graph of a mean squared error of the network during training is presented in Figure 3. Training quality of a neural network is characterized by the following values of the mean squared error's value and correlation coefficients: MSE $= 7.5 \cdot 10^{-4}$ N, $r_{Y\widetilde{Y}} = 9.9999 \cdot 10^{-1}$.

The numerical tests for comparative calculation of the rotor's trajectories have been carried out for various values of

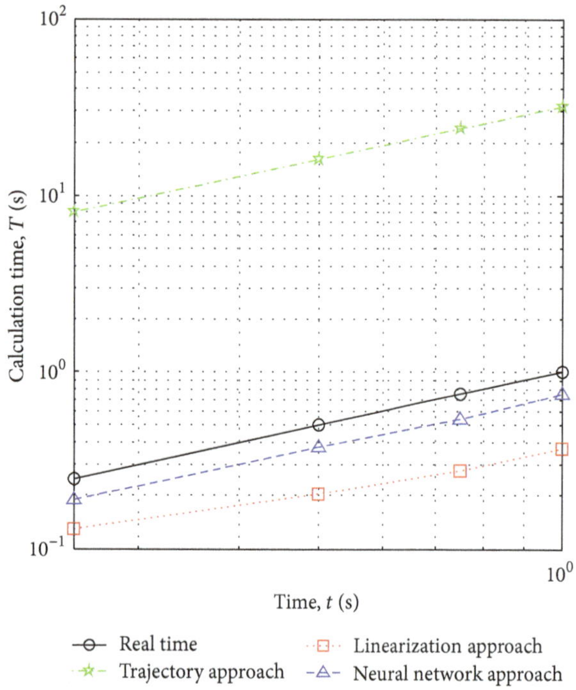

FIGURE 4: Calculation time.

the relative centrifugal force $\widehat{F}_c = m\Delta\omega^2/mg = \Delta\omega^2/g$. Other input parameters remained as during the training procedure.

As a result of numerical tests a calculation time of 1 sec of rotor motion was determined given the action of dimensionless centrifugal force $\Delta\omega^2/g = 1.6$ (see Figure 4). It could be seen that the calculation time of rotor dynamics calculation using the linearization approach is the least in the whole range. Application of the neural network module increased the machine time up to 1.5–2 times under given conditions. Application of trajectory approach increased the calculation time up to 60–90 times compared to the linearization approach. It is indicative that calculation time of rotor dynamics using linearization approach and neural network approach is faster than the flow of real time in the whole range. This fact means that the application of linearization or neural network approaches in a real rotational machine control system would make it possible to obtain results of prognostic modeling [20, 33, 34].

As a result of another series of numerical tests, the trajectories have been calculated for a dimensionless centrifugal force $\Delta\omega^2/g = 0.8, \ldots, 2.5$ (see Figure 5). Calculated rotor motion's time was set to 1 sec. The simulation started at the equilibrium point X_i^E, the trajectory approach taken as referential. The relative error of calculation has been chosen as quantitative criteria of accuracy evaluation:

$$\delta X_i = \frac{X_i^{\text{Traj}} - X_i^{\text{Lin(NN)}}}{h_0} \cdot 100, \qquad (10)$$

where X_i^{Traj} are coordinates, obtained using the trajectory approach, and $X_i^{\text{Lin(NN)}}$ are coordinates, obtained using the linearization or neural network approach, $i = 1, 2$.

The quality analysis of stability of calculated trajectories has been implemented based on the Poincare maps [27–29]. The point were displayed at moments of time that were multiples of the period of rotation of the rotor $T = 1/n$. It could be seen from the Poincare maps (see Figure 5) that in the given range of loads the rotor's trajectories are quasiperiodic. This results from the fact that after a short transient process the trajectory's mapping becomes a dense pointed line, which closes if the calculation time increases by 100 times or more. This shows the presence of incommensurable frequencies in the oscillations, while the motion is not chaotic [27]. It could also be seen from the Figure 5 that all three approaches to calculation of reaction forces gave the same result. The relative error of the center of mass location calculation with $\Delta\omega^2/g = 0.8$ during steady-state regime was 0.4% and 0.3% for the linearization approach and the neural network approach, respectively. With $\Delta\omega^2/g = 1.6$ the relative error remains small but accuracy differs by 10 times: 0.3% and 3% for the linearization approach and the neural network approach, respectively. Preservation of accuracy of the neural network approach could be explained by the fact that the model of training was carried out with value of the dimensionless centrifugal force $\Delta\omega^2/g = 1.8$ close to 1.6. Finally, with $\Delta\omega^2/g = 2.5$ the error was no more than 2% for the neural network approach and no more than 6% for the linearization approach. So, the neural network approach compared to the linearization approach given that the amplitude of oscillations is more than $0.1h_0$ has been proven to be significantly more accurate.

Further increase of the imbalance was not studied as it usually results in unstable trajectories. Increase of imbalance leads to increase of vibration's amplitude. That is why the linearization method could give inadequate results due to actual nonlinearity of the reaction forces at large eccentricities. The neural network approach gives good interpolation results, that is, within the learning sample. During extrapolation, deviation from the learning conditions could lead to unpredictable results and rapid increase of the error.

3.2. Rotor Dynamics for Various Speed Regimes and Transition from Low to High Vibrations. The same rotor system as in Section 3.1 of the present paper has been studied. Imbalance was constant $m\Delta = 10^{-4}$ kg·m, and tests were run for various rotation speeds: 1000 rpm, 3000 rpm, 5000 rpm, and 7000 rpm.

The first series of simulations of rotor's behavior at various rotation speeds was implemented using the trajectory approach. Calculated motion time was set to 1 sec. Apart from trajectories, the location of equilibrium points has been determined, marked "+" in Figure 6(a). The second series of simulations has been implemented using the linearization approach. For that, matrices of stiffness and damping have been determined at four points on the equilibrium locus curve.

Then, four neural networks have been developed for training with $N_{\text{hid}} = 15$ neurons in the hidden layer. For every network, matrices $\vec{U} = [X_1, X_2, V_1, V_2]$ and $\vec{R} = [R_1, R_2]$ of 4×10^4 and 2×10^4 elements were used as input and output

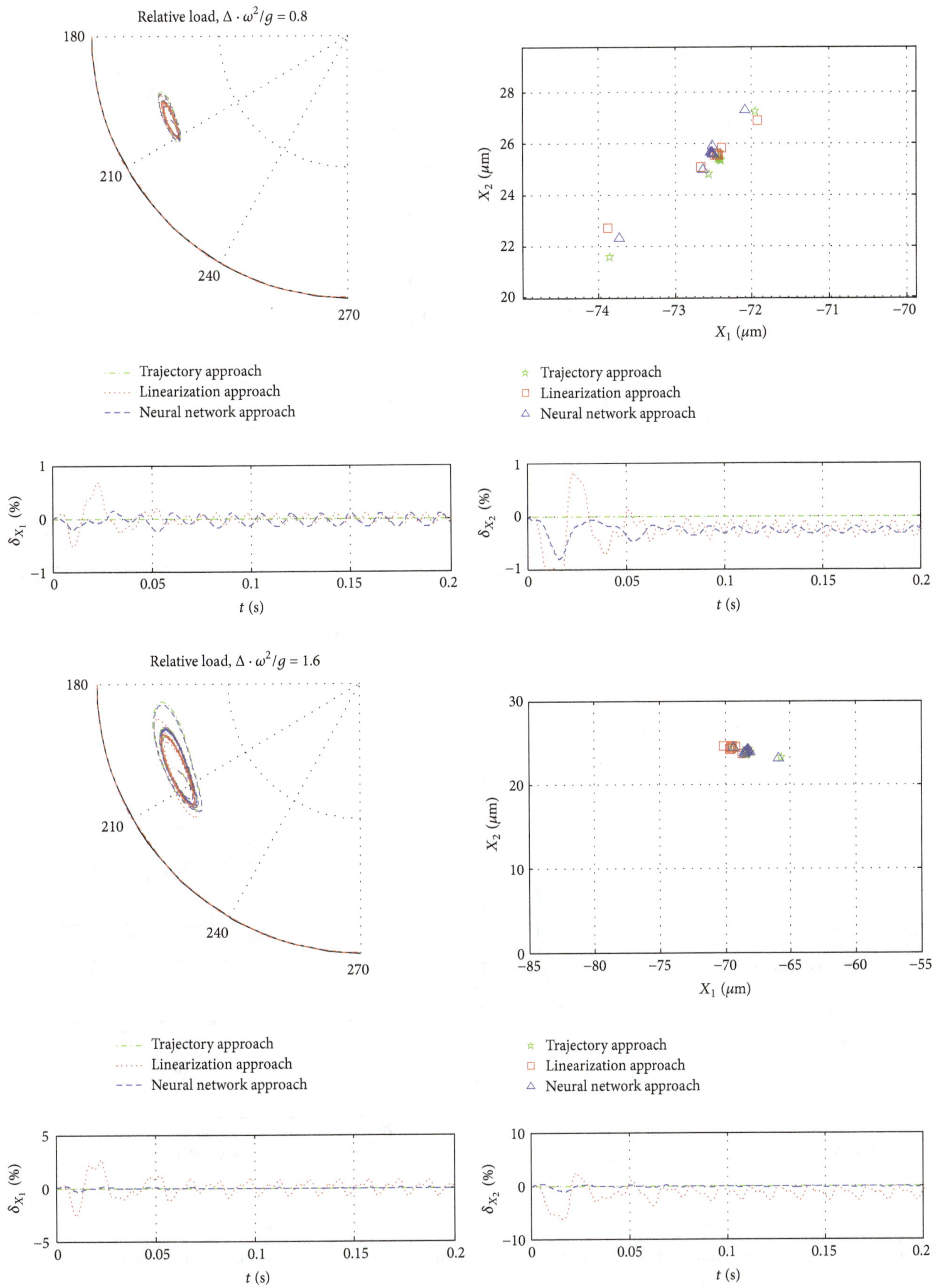

FIGURE 5: Continued.

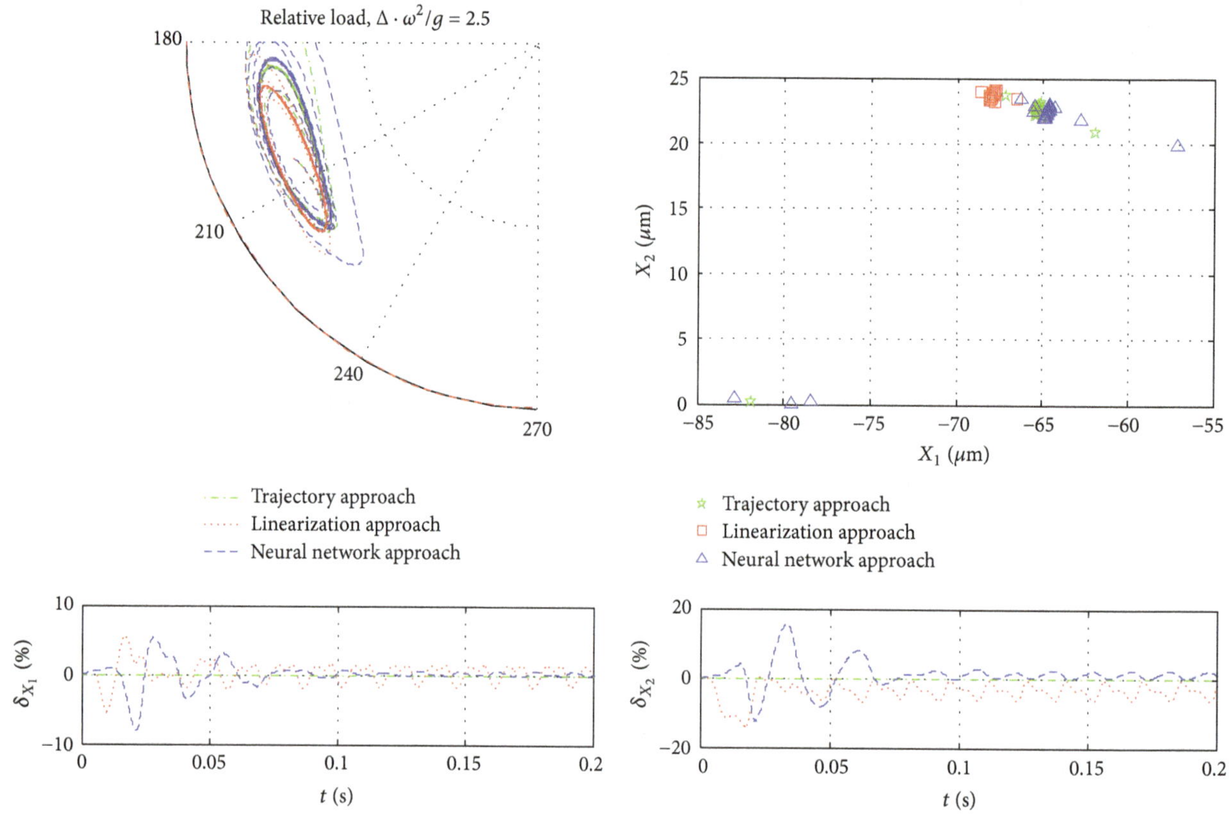

FIGURE 5: Comparative calculation results in the form of trajectories, Poincare maps, and error plots with different values of dimensionless load.

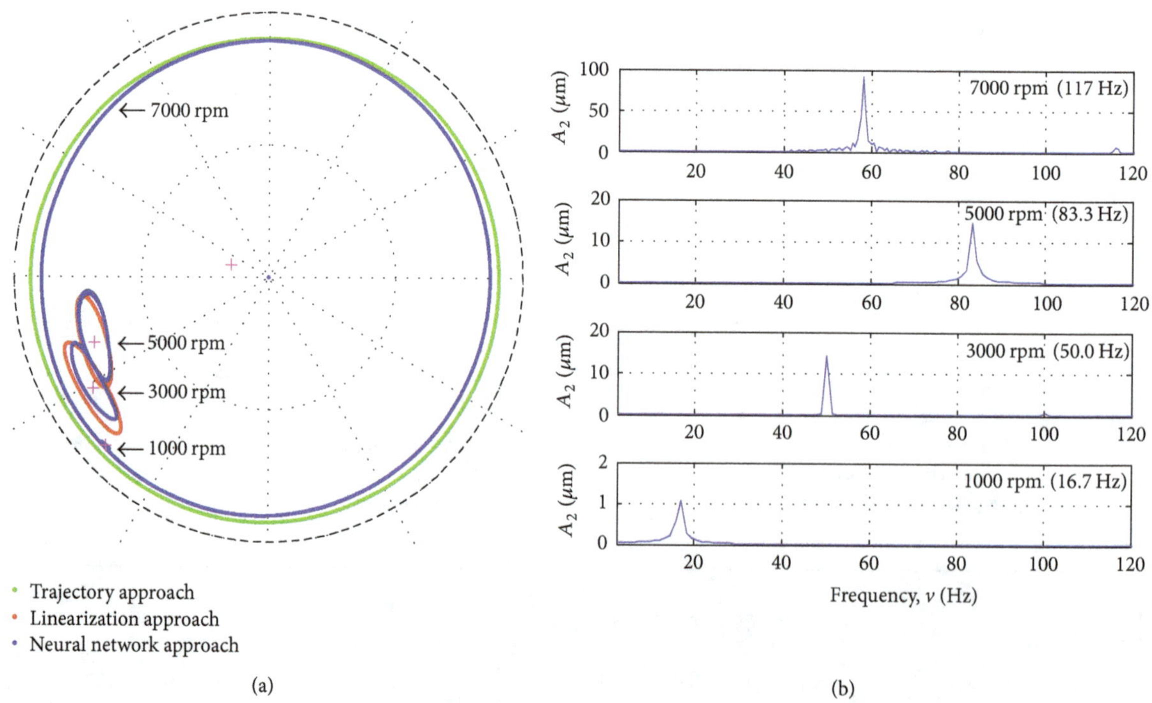

FIGURE 6: Comparative results of rotor's vibrations calculation (a) and frequency responses at various rotation speed in a $1000, \ldots, 7000$ rpm range.

TABLE 1: Comparative results of rotor's trajectory calculation at various rotation speed in 1000, ..., 7000 rpm range.

n, rpm	max δX_i, %	
	Linearization approach	Neural network approach
1000	0.5	0.04
3000	9	0.6
5000	6	0.1
7000	inf.	4

data, respectively. The matrices of input and output data were divided into three groups for training, validation and testing in fractions 0.6, 0.3, and 0.1, respectively. Training of each network was done in 1000 epochs in approximately 1–3 min. Training quality is characterized by the following values of mean squared errors and correlation coefficients: MSE $= [2.4\ \ 3.1\ \ 2.7\ \ 14] \cdot 10^{-4}$ N, $r_{Y\tilde{Y}} = [9.99\ \ 9.99\ \ 9.99\ \ 9.99] \cdot 10^{-1}$.

The calculation results showed that in the range of speeds from 1000 to 5000 rpm the vibrations of the rotor have the frequency equal to the rotation frequency (see Figure 6(b)). In this range of speed the oil whirl does not appear under given conditions, and the main reason of the occurrence of vibrations is the imbalance of the rotor. Additional calculations have shown that the critical speed of the rotor is 3300 rpm. In the range of speed from 6000 to 7000 rpm some increased vibrations occur that correspond to approximately twice critical speed (see Figures 6(a) and 6(b)). The vibrations of the rotor have the frequency close to the critical frequency (see Figure 6(b)). This is typical of oil whip [6]. Here, linearization approach becomes unstable, and the neural network approach, on the other hand, remains stable with sufficient accuracy (see Table 1).

So, the neural network approach allows modeling of significantly nonlinear vibrations, such as oil whip. The necessity of development of a separate neural network for every rotation speed; however, it is a drawback of the applied approach. Consequently, in the next paragraph a neural network is considered with an additional input value, rotation speed of the rotor.

3.3. Rotor Dynamics during Transient Regimes. Modeling of rotor dynamics during transient start/stop regime is more complex and costly in terms of calculation resources. The main advantages of the linearization approach that are essentially simple implementation and good computational performance manifest themselves weakly in this case. This is due to necessity to implement a number of intermediate calculations of equilibrium position, of stiffness and damping matrices, and due to interpolation procedures. In case of the neural network approach, such additional interpolation procedures are not needed. There is also a possibility of training a neural network based on one experiment and not on a series of them. This is why application of the linearization approach has not been considered in this section.

The transient regime has been studied using the test rig, shown in the Figure 7. The main parameters of the test rig

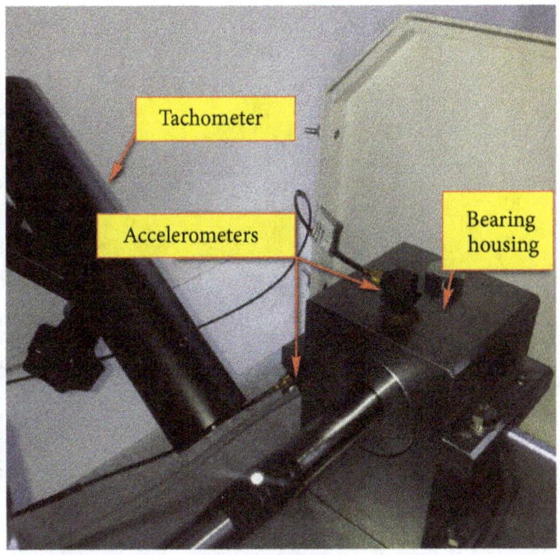

FIGURE 7: Part of the test rig with accelerometers DeltaTron 4507 and the contactless tachometer M0024 [24].

are the same as in the Section 3.2. During the experiment the rotor was steadily accelerated up to 4000 rpm and then the electromotor was shut down and disconnected. During the deceleration the data from tachometer and acceleration sensors was acquired. Deceleration of the rotor took about 1 min. Based on the results of rotation speed measurements, rotor's speed was approximated as a linear function of time, which describes the deceleration of the rotor during hydrodynamic lubrication regime (down to 500 rpm) accurately enough. Additional experiments at constant rotation speed in the 1000–4000 rpm range showed that in this range the rotor vibrates at frequencies equal to the rotation frequency. This result matches the results in the Section 3.2.

In this case, one neural network has been developed for training with $N_{\text{hid}} = 35$ neurons in the hidden layer. Additionally, the rotation speed was taken as an input, so the $5 \times 4 \cdot 10^4$ matrix of inputs was taken as $\vec{U} = [n, X_1, X_2, V_1, V_2]$. The combined sample of trajectories of the rotor at eight different rotation speeds $n = 1000, \ldots, 5000$ rpm was used for training. In other words, the training function had a step-like form. The matrices of input and output data were divided into three groups for training, validation, and testing in fractions 0.7, 0.15, and 0.15, respectively. Training took no more than 10 minutes. Training quality is characterized by the following values of mean squared errors and correlation coefficients: MSE $= 4.0 \cdot 10^{-4}$ N; $r_{Y\tilde{Y}} = 9.998 \cdot 10^{-1}$.

It has to be noted that one of the alternative ways to train a neural network during monotonous smooth change of speed was not successful. The results obtained based on such neural networks were unstable. This could be explained by fact that every value of speed occurs in a monotonous function only once. In a step-like function, many intermediate values are absent, but the neural network manages to successfully interpolate them.

In Figures 8(a) and 8(b) the results of deceleration of the rotor simulation are shown for trajectory and neural networks approaches. In Figure 8(c) the comparison of

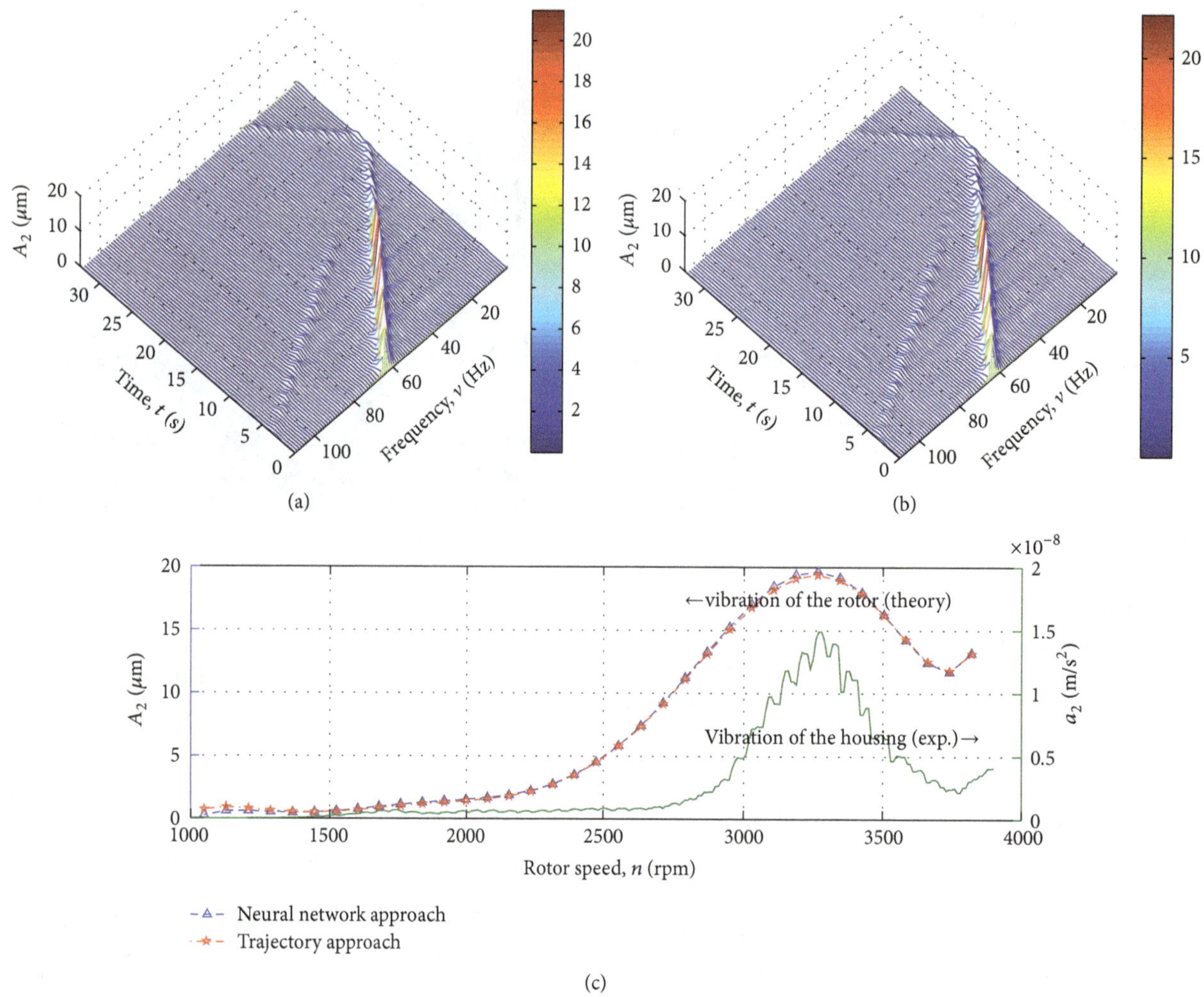

FIGURE 8: Cascade frequency response during deceleration of the rotor from 3900 rpm down to 1000 rpm during 32 sec., calculated using the trajectory approach (a), neural network approach (b), and comparative results of the housing's vibration measurement and the rotor's vibration calculation (c).

numerical and experimental results is shown. It could be seen from the graphs that numerical results match well quantitatively (see Figures 8(a) and 8(b)) and qualitatively the experimental results (see Figure 8(c)). The calculated value of the critical speed matches the experimentally obtained value quite well (see Figure 8(c)).

So, the solution of the rotor dynamics problems during transition regimes could be successfully implemented using neural networks.

4. Conclusion

An approach to neural network program module formation has been presented aimed at calculation of the reaction force's components of the fluid film depending on the position and oscillation velocities of the center of a rotor. Three approaches used in solving the rotor dynamics problem have been compared: the trajectory approach, the linearization approach, and the neural network approach to approximation of the oil film reaction forces. As a result of quantitative analysis, it has been determined that the use of linear and neural approximation allows decreasing the calculation time up to two orders of magnitude. It has also been determined that if the amplitude of the rotor's oscillation exceeds 10% of the mean radial gap, the neural network approach is up to 3 times more accurate compared to the linearization approach. Also, neural networks were used to simulate significantly nonlinear regimes of rotor's vibration with high vibrodisplacement and demonstrated good results in simulation of transient regimes. The considered models of a rotor and the fluid film were relatively simple, which significantly improved performance using conventional methods and aided sampling for neural network's training. Further complication of the models, as well as training under condition of interaction with a real rotor machine, shall not significantly influence the main characteristics of artificial neural networks.

Conflicts of Interest

The authors declare that they have no conflicts of interest.

Authors' Contributions

Alexey Kornaev developed the fluid flow and rotor's dynamic mathematical model, scripted the simulation program, and wrote the manuscript. Nickolay Kornaev developed the artificial neural network program module and performed the numerical experiment. Elena Kornaeva and Leonid Savin analyzed the experiment's results. Leonid Savin was in charge of supervising this work.

Acknowledgments

This work was supported by the Russian Science Foundation under Project no. 16-19-00186 (formulation and solution of the problem in all sections of the present paper) and by the Ministry of Education and Science of the Russian Federation under Project no. 9.2952.2017/4.6 (development and modernization of the test rig in Section 3.3). The authors gratefully acknowledge this support. The authors also express gratitude to Alexander Fetisov for carrying out the experiment and to Alexander Babin for translation.

References

[1] D. Childs, *Turbomachinery Rotordynamics. Phenomena, Modeling, and Analysis*, John Wiley and Sons, New York, NY, USA, 1993.

[2] M. G. Vance, *Rotordynamics of Turbomachinery*, John Wiley and Sons, New York, NY, USA, 1988.

[3] T. Yamamoto, *Linear and Nonlinear Rotordynamics. A Modern Treatment with Applications*, John Wiley and Sons, New York, NY, USA, 2001.

[4] L. Savin, O. Solomin, and D. Ustinov, "Rotor dynamics on friction bearings with cryogenic lubrication," in *Proceedings of the Tenth World Congress on the Machines and Mechanisms*, vol. 4, pp. 1716–1721, June 1999.

[5] S. P. Asok, K. Sankaranarayanasamy, T. Sundararajan, K. Rajesh, and G. Sankar Ganeshan, "Neural network and CFD-based optimisation of square cavity and curved cavity static labyrinth seals," *Tribology International*, vol. 40, no. 7, pp. 1204–1216, 2007.

[6] Y. Hori, *Hydrodynamic lubrication*, Yokendo Ltd., Tokyo, 2006.

[7] M. Kalkat, S. Yildirim, and I. Uzmay, "Rotor dynamics analysis of rotating machine system using artificial neural networks," *International Journal of Rotating Machinery*, p. 9, 2003.

[8] M. Karkoub and A. Elkamel, "Modelling pressure distribution in a rectangular gas bearing using neural networks," *Tribology International*, vol. 30, no. 2, pp. 139–150, 1997.

[9] A. D. Dongare and A. D. Kachare, "Predictive tool: an artificial neural network," *International Journal of Engineering and Innovative Technology (IJEIT)*, vol. 2, pp. 209–2014, 2012.

[10] P. Qin, Y. Shen, J. Zhu, and H. Xu, "Dynamic analysis of hydrodynamic bearing-rotor system based on neural network," *International Journal of Engineering Science*, vol. 43, no. 5-6, pp. 520–531, 2005.

[11] A. O. Kurban and Ş. Yildirim, "Analysis of a hydrodynamic thrust bearing with elastic deformation using a recurrent neural network," *Tribology International*, vol. 36, no. 12, pp. 943–948, 2003.

[12] J. Ghorbanian, M. Ahmadi, and R. Soltani, "Design predictive tool and optimization of journal bearing using neural network model and multi-objective genetic algorithm," *Scientia Iranica*, vol. 18, no. 5, pp. 1095–1105, 2011.

[13] M. Simsir, R. Bayir, and Y. Uyaroglu, "Real-time monitoring and fault diagnosis of a low power hub motor using feedforward neural network," *Computational Intelligence and Neuroscience*, vol. 2016, Article ID 7129376, 2016.

[14] B. Samanta, K. R. Al-Balushi, and S. A. Al-Araimi, "Bearing fault detection using artificial neural networks and genetic algorithm," *Eurasip Journal on Applied Signal Processing*, vol. 2004, no. 3, pp. 366–377, 2004.

[15] M. Subrahmanyam and C. Sujatha, "Using neural networks for the diagnosis of localized defects in ball bearings," *Tribology International*, vol. 30, no. 10, pp. 739–752, 1997.

[16] K. M. Saridakis, P. G. Nikolakopoulos, C. A. Papadopoulos, and A. J. Dentsoras, "Identification of wear and misalignment on journal bearings using artificial neural networks," *Proceedings of the Institution of Mechanical Engineers, Part J: Journal of Engineering Tribology*, vol. 226, no. 1, pp. 46–56, 2012.

[17] S. G. Tores, C. Bonello, and P. Bonello, "Empirical identification of the inverse model of a squeeze-film damper bearing using neural networks and its application to a nonlinear inverse problem," *Journal of Vibration and Control*, p. 22, 2016.

[18] S. Morishita and T. Ura, "ER fluid applications to vibration control devices and an adaptive neural-net controller," *Journal of Intelligent Material Systems and Structures*, vol. 4, no. 3, pp. 366–372, 1993.

[19] R. Jha and C. h. He, "Adaptive neurocontrollers for vibration suppression of nonlinear and time varying structures," *Journal of Intelligent Material Systems and Structures*, vol. 15, p. 11, 2004.

[20] S. Y. Chen, Y. C. Hung, Y. H. Hung, and C. H. Wu, "Application of a recurrent wavelet fuzzy-neural network in the control of a magnetic-bearing mechanism," *Computers Electrical Engineering*, vol. 54, pp. 147–158, 2016.

[21] P.-Y. Couzon and J. der Hagopian, "Neuro-fuzzy active control of rotor suspended on active magnetic bearing," *Journal of Vibration and Control*, vol. 13, no. 4, pp. 365–384, 2007.

[22] C. R. Burrows, P. S. Keogh, and M. N. Sahinkaya, "Progress towards smart rotating machinery through the use of active bearings," *Proceedings of the Institution of Mechanical Engineers, Part C: Journal of Mechanical Engineering Science*, vol. 223, p. 11, 2009.

[23] Q. J. Wang and Y. Chung, *Encyclopedia of Tribology*, Springer, Boston, Mass, USA, 2013.

[24] A. Kornaev, L. Savin, E. Kornaeva, and A. Fetisov, "Influence of the ultrafine oil additives on friction and vibration in journal bearings," *Tribology International*, vol. 101, pp. 131–140, 2016.

[25] S. Haykin, *Neural Networks. A Comprehensive Foundation*, Prentice Hall, NJ, USA, 2nd edition, 1999.

[26] G. A. Korn and T. M. Korn, *Mathematical Handbook for Scientists and Engineers*, Dover Publications, New York, NY, USA, 2000.

[27] F. C. Moon, *Chaotic Vibrations. An Introduction for Applied Scientists and Engineers*, John Wiley and Sons, New York, NY, USA, 1987.

[28] L. Cui and J. Zheng, "Nonlinear vibration and stability analysis of a flexible rotor supported on angular contact ball bearings," *JVC/Journal of Vibration and Control*, vol. 20, no. 12, pp. 1767–1782, 2014.

[29] Y. Wang and X. Wang, "Nonlinear vibration analysis for a Jeffcott rotor with seal and air-film bearing excitations," *Mathematical Problems in Engineering*, vol. 2010, Article ID 657361, 15 pages, 2010.

[30] M. G. Salvadori and M. L. Baron, *Numerical Methods in Engineering*, Englewood Cliffs, NJ, USA, 2nd edition, 1961.

[31] http://www.gnu.org/software/octave.

[32] http://www.mathworks.com/help/nnet/ref/feedforwardnet.html.

[33] A. Harnoy, *Bearing Design in Machinery: Engineering Tribology and Lubrication*, Marcel Dekker Inc, New York, NY, USA, 2003.

[34] A. Chasalevris, "An investigation on the dynamics of high-speed systems using nonlinear analytical floating ring bearing models," *International Journal of Rotating Machinery*, vol. 2016, Article ID 7817134, 22 pages, 2016.

Permissions

All chapters in this book were first published in IJRM, by Hindawi Publishing Corporation; hereby published with permission under the Creative Commons Attribution License or equivalent. Every chapter published in this book has been scrutinized by our experts. Their significance has been extensively debated. The topics covered herein carry significant findings which will fuel the growth of the discipline. They may even be implemented as practical applications or may be referred to as a beginning point for another development.

The contributors of this book come from diverse backgrounds, making this book a truly international effort. This book will bring forth new frontiers with its revolutionizing research information and detailed analysis of the nascent developments around the world.

We would like to thank all the contributing authors for lending their expertise to make the book truly unique. They have played a crucial role in the development of this book. Without their invaluable contributions this book wouldn't have been possible. They have made vital efforts to compile up to date information on the varied aspects of this subject to make this book a valuable addition to the collection of many professionals and students.

This book was conceptualized with the vision of imparting up-to-date information and advanced data in this field. To ensure the same, a matchless editorial board was set up. Every individual on the board went through rigorous rounds of assessment to prove their worth. After which they invested a large part of their time researching and compiling the most relevant data for our readers.

The editorial board has been involved in producing this book since its inception. They have spent rigorous hours researching and exploring the diverse topics which have resulted in the successful publishing of this book. They have passed on their knowledge of decades through this book. To expedite this challenging task, the publisher supported the team at every step. A small team of assistant editors was also appointed to further simplify the editing procedure and attain best results for the readers.

Apart from the editorial board, the designing team has also invested a significant amount of their time in understanding the subject and creating the most relevant covers. They scrutinized every image to scout for the most suitable representation of the subject and create an appropriate cover for the book.

The publishing team has been an ardent support to the editorial, designing and production team. Their endless efforts to recruit the best for this project, has resulted in the accomplishment of this book. They are a veteran in the field of academics and their pool of knowledge is as vast as their experience in printing. Their expertise and guidance has proved useful at every step. Their uncompromising quality standards have made this book an exceptional effort. Their encouragement from time to time has been an inspiration for everyone.

The publisher and the editorial board hope that this book will prove to be a valuable piece of knowledge for researchers, students, practitioners and scholars across the globe.

List of Contributors

Djamal Hissein Didane, Nurhayati Rosly, Mohd Fadhli Zulkafli and Syariful Syafiq Shamsudin
Department of Aeronautical Engineering, Faculty of Mechanical and Manufacturing Engineering, Universiti Tun Hussein Onn Malaysia (UTHM), Parit Raja, 86400 Batu Pahat, Johor, Malaysia

Nobumichi Fujisawa and Yutaka Ohta
Department of Applied Mechanics and Aerospace Engineering, Waseda University, 3-4-1 Okubo, Shinjuku-ku, Tokyo 169-8555, Japan

Si Huang, Jing He, Xueqian Wang and Guangqi Qiu
School of Mechanical and Automotive Engineering, South China University of Technology, Guangzhou 510640, China

Xiaojiao Gu
School of Mechanical Engineering, Shenyang University of Technology, Shenyang 110870, China

Changzheng Chen
School of Mechanical Engineering, Shenyang University of Technology, Shenyang 110870, China
Liaoning Engineering Center for Vibration and Noise Control, Shenyang 110870, China

Tzong-Hann Shieh
Department of Aerospace and Systems Engineering, Feng Chia University, No. 100,Wenhwa Rd., Seatwen, Taichung 40724, Taiwan

Wei Cai, Yuan Li and Chunbao Liu
School of Mechanical Science and Engineering, Jilin University, Changchun 130022, China

Xingzhong Li
State Key Laboratory of Automotive Simulation and Control, Jilin University, Changchun, China

Mahdi Majidniya and Kobra Gharali
Department of Mechanical Engineering, University of Tehran, Tehran, Iran

Kaamran Raahemifar
Department of Electrical & Computer Engineering, Ryerson University, Toronto, ON, Canada

Aurélien Marsan and Stéphane Moreau
Department of Mechanical Engineering, Université de Sherbrooke, Sherbrooke, QC, Canada J1K 2R1

Isabelle Trébinjac
LMFA, Ecole Centrale de Lyon, 36 Avenue Guy de Collongue, 69134 Ecully, France

Sylvain Coste
Safran Helicopter Engines, Avenue Joseph Szydlowski, 64510 Bordes, France

Jose M. Garcia-Bravo
Purdue University, West Lafayette, IN, USA

Ivo N. Ayala-Garcia
Universidad Tecnológica de Querétaro, Santiago de Querétaro, QRO, Mexico

Juan L. Cepeda-Aguilar
Universidad de los Andes, Bogotá, Colombia

Matthias Rolfes, Martin Lange and Ronald Mailach
Technische Universität Dresden, Institute of Fluid Mechanics, Chair of Turbomachinery and Flight Propulsion, 01062 Dresden, Germany

Adam R. Hickman and Scott C. Morris
University of Notre Dame, Notre Dame, IN 46556, USA

Thomas Hagemann, Christopher Zeh, Maximilian Prölß and Hubert Schwarze
Institute of Tribology and Energy Conversion Machinery, Clausthal University of Technology, Leibnizstraße 32, 38678 Clausthal-Zellerfeld, Germany

Johannes Schreiber, Benoit Paoletti and Xavier Ottavy
Laboratoire de Mécanique des Fluides et d'Acoustique, UMR CNRS 5509, Université de Lyon, Ecole Centrale de Lyon,UCB Lyon I, INSA, 36 Av. Guy de Collongue, 69134 ´ Ecully Cedex, France

Zhiheng Feng
Guangxi Liugong Machinery Co., Ltd, Liuzhou, Guangxi 545007, China

Chaosheng Song
The State Key Laboratory of Mechanical Transmissions, Chongqing University, Chongqing 400030, China

Li Cui, Wenxia Wang and Zhang Yanlei
College of Engineering, Shanghai Polytechnic University, Shanghai, China

Kai Chen, Xin-Cong Zhou and Jun-Qiang Fang
Key Laboratory of High Performance Ship Technology (Wuhan University of Technology), Ministry of Education, Wuhan 430063, China
Reliability Engineering Institute, School of Energy and Power Engineering, Wuhan University of Technology, Wuhan 430063, China

Li Qin
School of Machinery and Automation, Wuhan University of Science and Technology, Wuhan 430081, China

Yu Guo, Wei Wang, Yu Wang and Yu Hou
State Key Laboratory of Multiphase Flow in Power Engineering, Xi'an Jiaotong University, Xi'an 710049, China

Tianwei Lai
State Key Laboratory of Multiphase Flow in Power Engineering, Xi'an Jiaotong University, Xi'an 710049, China
Xi'an Jiaotong University Suzhou Institute, Suzhou 215123, China

Chuang Gao and Weiguang Huang
Shanghai Advanced Research Institute, Chinese Academy of Sciences, Shanghai, China

Tianhua Zheng, Kang Yang, Haosen Yang and Yun Cao
Division of Aerodynamics, Department of R&D, Helan Turbines, Co., Ltd., China

Chu Zhang, Shi Liu, Qing-Shui Gao and Yi Yang
Electric Power Research Institute of Guangdong Power Grid Co. Ltd., Guangzhou 510080, China

Jian-Gang Yang
National Engineering Research Center of Turbo-Generator Vibration, Southeast University, Nanjing 210096, China

Alexey V. Kornaev
Modeling of Hydromechanical Systems Research Laboratory, The Orel State University Named after I.S. Turgenev, Orel 302026, Russia

Nickolay V. Kornaev and Leonid A. Savin
Department of Mechatronics and International Engineering, The Orel State University Named after I.S. Turgenev, Orel 302026, Russia

Elena P. Kornaeva
Department of Information Systems, The Orel State University Named after I.S. Turgenev, Orel 302026, Russia

Index

www.ingramcontent.com/pod-product-compliance
Lightning Source LLC
LaVergne TN
LVHW060536210926
848485LV00005B/1207